Carbon Nanotubes and Graphene

Carbon Nanotubes and Graphene

Edition 2

Edited by

K. Tanaka
Department of Molecular Engineering,
Kyoto University, Nishikyo-ku,
Kyoto 615-8510, Japan

S. Iijima
Department of Materials Science and Technology,
Meijo University, Tempaku-ku, Nagoya 468-8502, Japan
and
Nanotube Research Center,
National Institute of Advanced Industrial Science
and Technology (AIST), Tsukuba 305-8565, Japan

AMSTERDAM • BOSTON • HEIDELBERG • LONDON
NEW YORK • OXFORD • PARIS • SAN DIEGO
SAN FRANCISCO • SINGAPORE • SYDNEY • TOKYO

ELSEVIER

Elsevier
Radarweg 29, PO Box 211, 1000 AE Amsterdam, Netherlands
The Boulevard, Langford Lane, Kidlington, Oxford OX5 1GB, UK
225 Wyman Street, Waltham, MA 02451, USA

ISBN: 978-0-08-098232-8

British Library Cataloguing in Publication Data
A catalogue record for this book is available from the British Library

Library of Congress Cataloging-in-Publication Data
A catalog record for this book is available from the Library of Congress

For information on all Elsevier publications
visit our web site at store.elsevier.com

This book has been manufactured using Print On Demand technology. Each copy is produced to order and is limited to black ink. The online version of this book will show color figures where appropriate.

Contents

List of Contributors

Elena Bekyarova
Departments of Chemistry and Chemical & Environmental Engineering, Center for Nanoscale Science and Engineering, University of California, Riverside, CA 92521-0403, USA

Alexey Cherevan
Institut für Physikalische Chemie, Westfälische Wilhelms-Universität Münster, Corrensstrasse 28-30, 48149 Münster, Germany

Daniela Dragoman
Physics Department, University of Bucharest, P.O. Box MG-11, 077125 Bucharest, Romania

Mircea Dragoman
National Institute for Research and Development in Microtechnology (IMT), P.O. Box 38-160, 023573 Bucharest, Romania

Mildred S. Dresselhaus
Department of Physics; Department of Electrical Engineering and Computer Science, Massachusetts Institute of Technology, 77 Massachusetts Avenue, Cambridge, MA 02139, USA

Dominik Eder
Institut für Physikalische Chemie, Westfälische Wilhelms-Universität Münster, Corrensstrasse 28-30, 48149 Münster, Germany

Caitlin Fisher
CSIRO Materials Science and Engineering, Bradfield Road, Lindfield, New South Wales 2070; School of Physics, The University of Sydney, Sydney, New South Wales 2006, Australia

Robert C. Haddon
Departments of Chemistry and Chemical & Environmental Engineering, Center for Nanoscale Science and Engineering, University of California, Riverside, CA 92521-0403, USA

Zhao Jun Han
CSIRO Materials Science and Engineering, Bradfield Road, Lindfield, New South Wales 2070, Australia

Zhong Hu
Department of Mechanical Engineering, South Dakota State University, Brookings, South Dakota 57007, USA

Jiaxing Huang
Department of Materials Science and Engineering, Northwestern University, 2220 Campus Drive, Evanston, IL 60208, USA

Akihiro Ito
Department of Molecular Engineering, Kyoto University, Nishikyo-ku, Kyoto 615-8510, Japan

Jing Kong
Department of Electrical Engineering and Computer Science, Massachusetts Institute of Technology, 77 Massachusetts Avenue, Cambridge, MA 02139, USA

Shailesh Kumar
CSIRO Materials Science and Engineering, Bradfield Road, Lindfield, New South Wales 2070, Australia

Hans Kuzmany
Universität Wien, Fakultät für Physik, Strudlhofgasse 4, A-1090 Wien, Austria

Igor Levchenko
CSIRO Materials Science and Engineering, Bradfield Road, Lindfield, New South Wales 2070; School of Physics, The University of Sydney, Sydney, New South Wales 2006, Australia

Xiaoxing Lu
Department of Materials Science and Engineering, University of California, Los Angeles, California 90095, USA

Brendan Meany
Department of Chemistry and Biochemistry, University of Maryland, College Park, MD 20742, USA

Toshiya Okazaki
Nanotube Research Center, National Institute of Advanced Industrial Science and Technology (AIST), Tsukuba, 305-8565, Japan

Kostya (Ken) Ostrikov
CSIRO Materials Science and Engineering, Bradfield Road, Lindfield, New South Wales 2070; School of Physics, The University of Sydney, Sydney, New South Wales 2006; Australian Institute for Innovative Materials, The University of Wollongong, New South Wales 2522; Faculty of Science, University of Technology, Sydney, New South Wales 2007; School of Chemistry, Physics, and Mechanical Engineering, Queensland University of Technology, Brisbane, QLD 4000, Australia

Amanda E. Rider
CSIRO Materials Science and Engineering, Bradfield Road, Lindfield, New South Wales 2070, Australia

Kalyan Raidongia
Department of Materials Science and Engineering, Northwestern University, 2220 Campus Drive, Evanston, IL 60208, USA

Riichiro Saito
Department of Physics, Tohoku University, Sendai 980-8578, Japan

Takeshi Saito
Nanotube Research Center, National Institute of Advanced Industrial Science and Technology (AIST), Tsukuba, Ibaraki 305-8565, Japan

Santanu Sarkar
Departments of Chemistry and Chemical & Environmental Engineering, Center for Nanoscale Science and Engineering, University of California, Riverside, CA 92521-0403, USA

Cameron J. Shearer
Institut für Physikalische Chemie, Westfälische Wilhelms-Universität Münster, Corrensstrasse 28-30, 48149 Münster, Germany

Yong Cheol Shin
Department of Materials Science and Engineering, Massachusetts Institute of Technology, 77 Massachusetts Avenue, Cambridge, MA 02139, USA

Chuan-Fu Sun
Department of Chemistry and Biochemistry, University of Maryland, College Park, MD 20742, USA

Alvin T.L. Tan
Department of Materials Science and Engineering, Northwestern University, 2220 Campus Drive, Evanston, IL 60208, USA

Kazuyoshi Tanaka
Department of Molecular Engineering, Kyoto University, Nishikyo-ku, Kyoto 615-8510, Japan

Timothy van der Laan
CSIRO Materials Science and Engineering, Bradfield Road, Lindfield, New South Wales 2070; School of Physics, The University of Sydney, Sydney, New South Wales 2006, Australia

YuHuang Wang
Department of Chemistry and Biochemistry, University of Maryland, College Park, MD 20742, USA

Kazuhiro Yanagi
Department of Physics, Tokyo Metropolitan University, Hachioji, Tokyo, 192-0397, Japan

Preface

Novel carbon allotropes often mentioned as nanocarbon materials due to their size dimension have been giving a ceaseless impact to carbon science and technology. At the same time, these materials, without doubt, construct main core of nanoscience and nanotechnology as well.

Carbon nanotube (CNT) is one of the main members of nanocarbon family, and the initial stage of the development for CNT was mostly established in the 1990s. In 1999 we launched a monograph 'The Science and Technology of Carbon Nanotubes' from Elsevier (herein called the 1st edition), which covered most of the themes researched on CNT by that time.

Researches on CNT have further been carried out in many ways thereafter from the fundamental aspects to the applied field including the improvement of synthesis, modification of its structure, detailed characterization of properties and making miscellaneous devices. In particular, more advanced developments have been made including a large scale of its production, effective fractioning of CNT with different chiralities and various aspects such as its incorporation into organic reactions, biological interaction with protein or DNA, utilization of the functionalized hybrids towards miscellaneous catalysis and so on. Based on these we have come to an idea to review the 'second-phase' developments of CNT researches after 2000. In this sense, we rather suppress the same old subjects that have already been described in the 1st edition.

Meanwhile, in 2004, graphene was explicitly introduced to nanocarbon material group. This material is very thin by itself with atomic-size thickness and, hence, has purely two-dimensional (2D) structure. In this sense, specific 2D electronic properties of graphene were explicitly checked in its early stage. It is of interest to note that the electronic property such as the band structure of graphene had been quite familiar to the researchers since the 1940s, for graphene was taken as the simplest model for graphite when neglecting the interlayer interaction therein.

After 2004, researches on applications of graphene and its fragments (nanographene and graphene nanoribbon) have been conducted. Moreover, large-area production of graphene using the chemical vapour deposition (CVD) technique is being intensely performed towards its actual application, for instance, touch panel of the electronic devices and so on.

The situations mentioned above make us consider that it is time to unify the new aspects and developments of these two materials together so as to offer a concrete viewpoint to overlook these in a parallel sense. The editors hope this unification results in preferable interaction and exchange of knowledge and information between the researchers dealing with CNT and/or graphene.

On this occasion we would like to thank Sharmila Vadivelan, Christine McElvenny, Jessica Vaughan, Louisa Hutchins, Graham Nisbet and Adrian Shell in the Editorial Office of Elsevier for their management and incessant encouragement.

Spring, 2014
The Editors

Chapter 1

Classification of Carbon

Kazuyoshi Tanaka
Department of Molecular Engineering, Kyoto University, Nishikyo-ku, Kyoto 615-8510, Japan

One of the reasons of fertile chemical nature of carbon materials comes from its hybridization state, which also basically determines the features of the variety of organic compounds. In association with this fact, classification of carbon materials would be of importance, which can be simply seen in terms of a certain domain map as shown in Figure 1.1. In this map there are two axes: one is the degree of sp^n hybridization with taking n as variable and the other [H]/[C] atomic ratio of the material. Typical carbon materials and the relatives occupying specific points on this domain map are shown in Figure 1.2.

Since graphene has complete sp^2 hybridization at every carbon atom and hydrogen atoms are ideally not involved in the skeleton, it occupies the point (2, 0) on the map, being the same for graphite when neglecting the small interlayer interaction. On the other hand, the point (2, 1) for polyacetylene and benzene has complete sp^2 hybridization with the [H]/[C] ratio of unity. On the line between the points (2, 1) and (2, 0) exists the graphene-fragments group including graphene nanoribbon and nanographene. Hence, the members belonging to this group also consist of sp^2 hybridization but their [H]/[C] ratios are between 1 and 0 (mostly between 0.5 and 0) along with the development of carbon skeleton.

Fullerenes and carbon nanotubes (CNTs), moreover, belong to an interesting family without any hydrogen atoms but not have genuine sp^2 hybridization due to the curved surfaces. In other words, they have the hybridization of $sp^{2+\delta}$, a bit closer to sp^3 (between $n = 2$ and 3) and, hence, they are on the broken line on the horizontal axis at the right to the point for graphene and graphite. It is, however, generally considered that these two materials are of π conjugation system in actuality. In this sense, π conjugation on the carbon atoms is rather robust even on the curved surface. This feature has been mentioned as σ-bond hybridization due to more mixing of 2s atomic orbital of carbon atom into the original $2p\pi$ orbital [1].

Around the line of the graphene fragments is an ambiguous area surrounded by broken line called amorphous carbon (a-C) including coal, charcoal, coke, soot, carbon black and so on. These members are sometimes of importance in the industrial field and often consist of not only carbon and hydrogen atoms but

Carbon Nanotubes and Graphene. DOI: 10.1016/B978-0-08-098232-8.00001-2

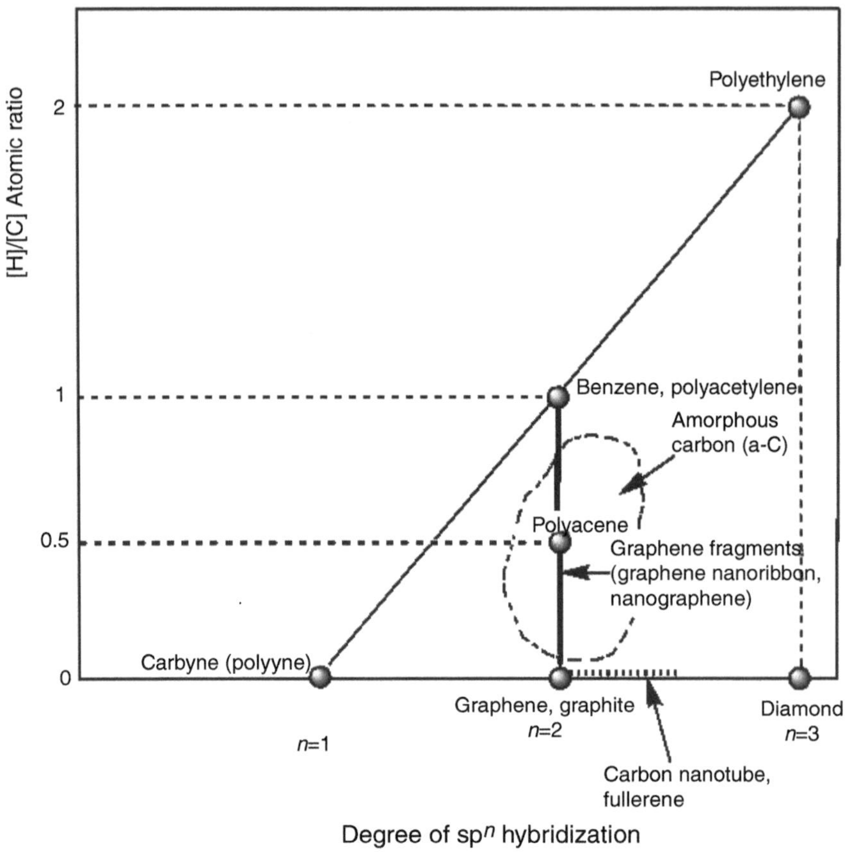

FIGURE 1.1 Domain map of carbon-material family.

also heteroatoms, such as oxygen, nitrogen or sulfur. Carbon fibres also belong to this group although they are rather near to graphite from the viewpoints of atomic arrangement. It is of interest to note that the precursor of vapour-grown carbon fibre (VGCF) quite looks like CNT [2].

On the other hand, the sp^3 hybridization members are paraffinic including polyethylene and diamond with the [H]/[C] ratio values of two and zero, respectively. The materials of this group have no π conjugation but only chemical bonds of σ type. Note that diamond is less energetically stable than graphite by 1.895 kJ mol^{-1} (the standard heat of formation) at 298.15 K. In this sense, the phase diagram of carbon material in Figure 1.3 shows that graphite is the most thermodynamically stable morph at ambient condition. On the contrary, for both silicon and germanium the most energetically stable morph is the diamond structure as is well known.

An sp hybridization should be seen in carbyne (or polyyne) at the point (1, 0), which has a typical linear structure with two orthogonal π conjugations.

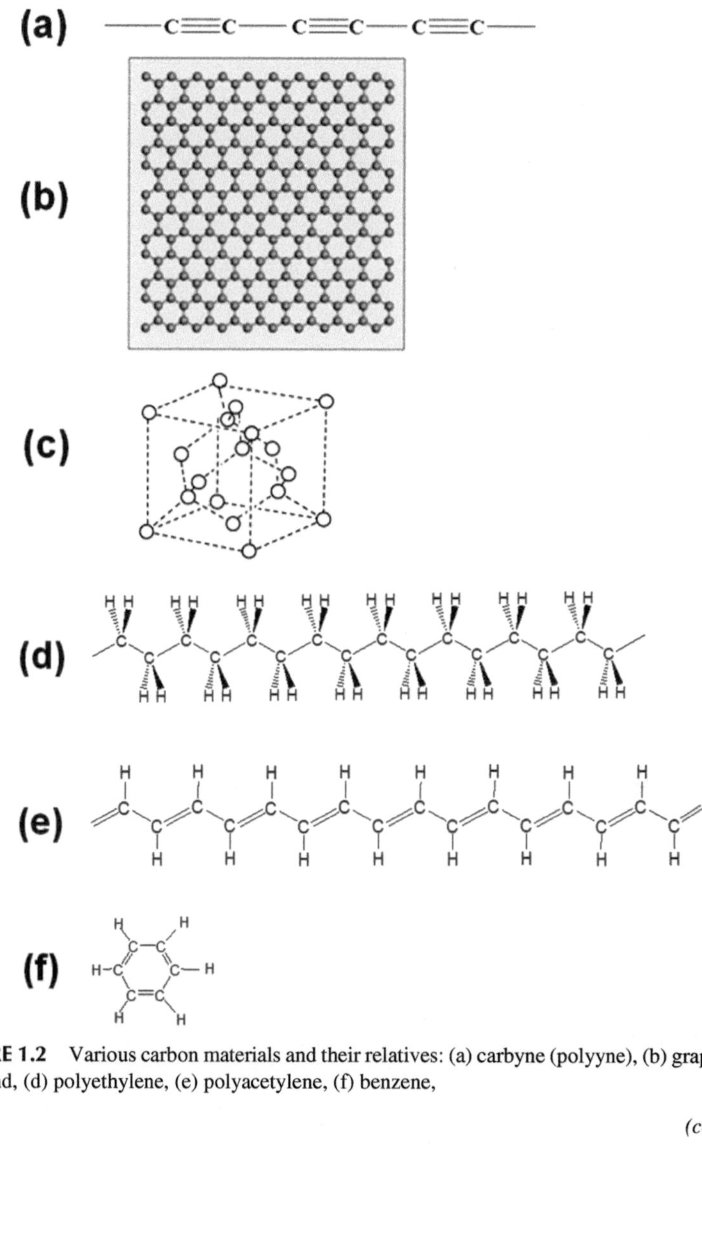

FIGURE 1.2 Various carbon materials and their relatives: (a) carbyne (polyyne), (b) graphene, (c) diamond, (d) polyethylene, (e) polyacetylene, (f) benzene,

(continued)

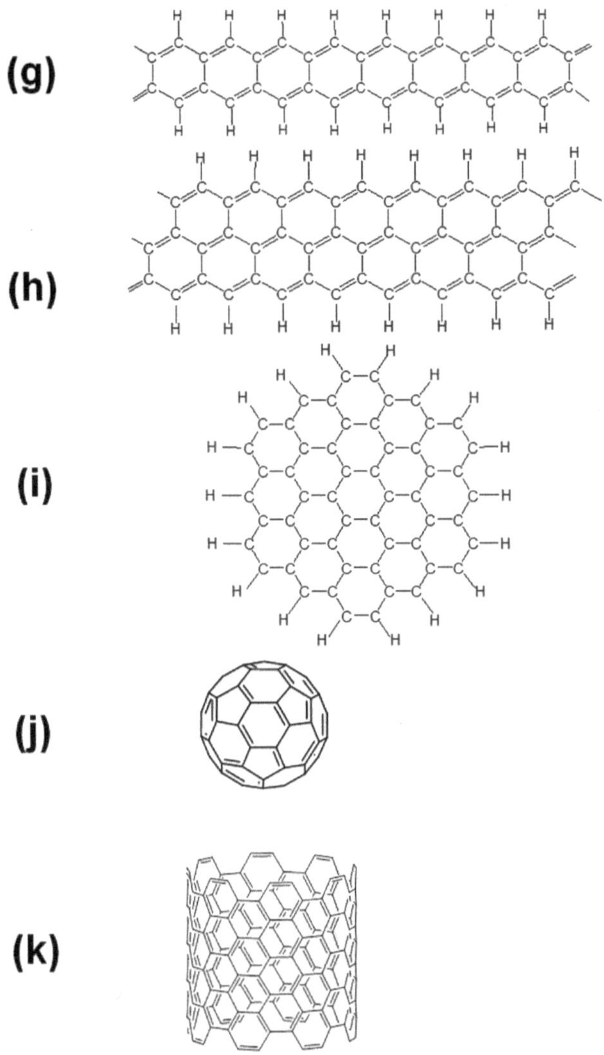

FIGURE 1.2 ***(cont.)*** (g) polyacene, (h) graphene nanoribbon, (i) nanographene, (j) fullerene and (k) CNT.

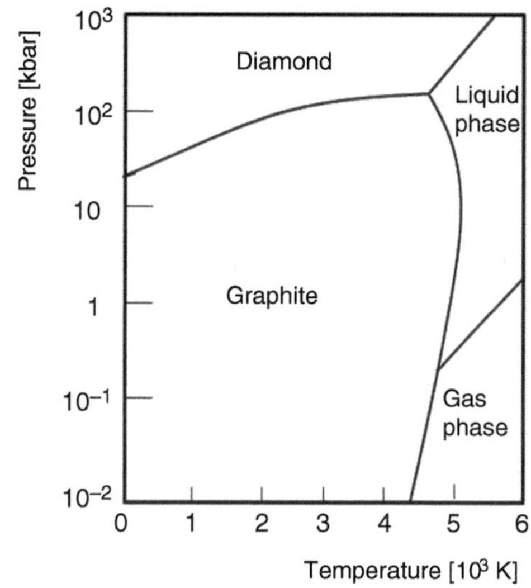

FIGURE 1.3 Phase diagram of carbon.

This material, however, has not been found in nature or yet chemically synthesized [3,4]. In recent years, syntheses of oligoynes including 10, 12 or 16 triple bonds have been reported [5].

CNT and graphene described above are central themes of this book and will be intensively described in the following chapters.

REFERENCES

[1] Haddon RC. Electronic structure, conductivity and superconductivity in alkali metal doped C_{60}. Acc Chem Res 1992;25:127.

[2] Endo M, Fujiwara H, Fukunaga E. Abst. 2nd Fullerene Symp. 1992;101 [in Japanese].

[3] Akagi K, Kinoshita I, Shirakawa H. One-dimensional conjugated carbyne – synthesis and properties. Synth Met 1987;17:557.

[4] Kijima M, Kinoshita I, Shirakawa H. Syntheses and properties of conjugated polymers with cumulenic bond in the main chain. Synth Met 1999;101:145.

[5] Tykwinski RR, Chalifoux W, Eisler S, Lucotti A, Tommasini M, Fazzi D, et al. Toward carbine: synthesis and stability of really long polyynes. Pure Appl Chem. and references therein 2010;82:891.

Chapter 2

Multidimensional Aspects of Single-Wall Carbon Nanotube Synthesis

Takeshi Saito

Nanotube Research Center, National Institute of Advanced Industrial Science and Technology (AIST), Tsukuba, Ibaraki 305-8565, Japan

2.1 VARIOUS SYNTHESIS METHODS FOR SINGLE-WALL CARBON NANOTUBES

In 1993, the first single-wall carbon nanotubes (SWCNTs) were discovered [1] in the soot produced by the arc discharge using graphite-rod electrodes. However, it is well-known that the arc-discharge method in those days could produce only an extremely low yield of SWCNTs with lots of carbonaceous impurities. After the discovery, vigorous researches on a variety of production processes, such as laser vapourization [2] and chemical vapour deposition (CVD) [3–8], have been carried out, which have resulted in the significant increase in the SWCNT productivity. Progresses in the SWCNT production have promoted various researches in the broad field of study on the physical and chemical properties of SWCNTs. In this sense, the development of production processes for high-quality SWCNTs is the crucial key technology for SWCNT researches.

In general, the SWCNT production processes, such as arc-discharge and laser-vapourization methods, generate the high-crystallinity of graphene-like network in SWCNTs because of the high growth temperature. Accordingly, SWCNTs produced by these methods have afforded to show excellent potential in the application research [9]. For example, Figure 2.1, showing the recently reported performances of transparent conducting SWCNT films, supports the superiority of SWCNTs produced using high reaction temperature [9,10].

On the other hand, the quality of SWCNTs produced by CVD has also been becoming better due to the technology maturity. As shown in Figure 2.1, the transparent conductivity performance of SWCNTs produced by CVD is steadily

Carbon Nanotubes and Graphene. DOI: 10.1016/B978-0-08-098232-8.00002-4

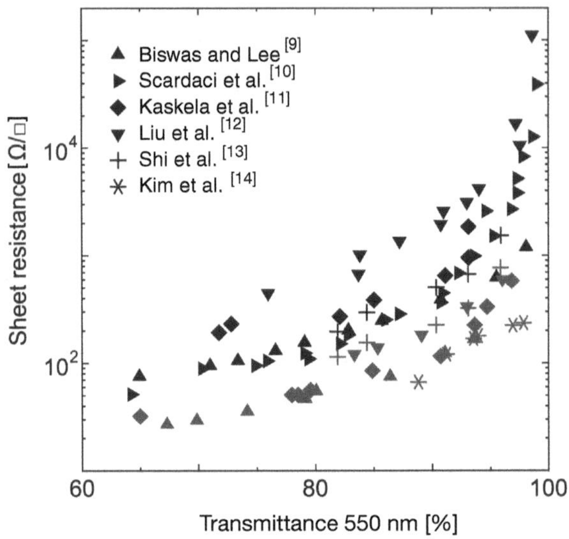

FIGURE 2.1 Recently reported performances of transparent conducting films (TCFs) of SWCNTs produced by arc-discharge methods (▲ and ►) and CVD methods (◆, ▼, + and *). The grey plots are the doped TCFs potentials.

advancing, and to date it reaches to the level of the high-temperature production processes [11–14].

Arc-discharge and laser vapourization methods need lots of energy consumption for high reaction temperature and large-equipment configurations [15]. Furthermore, in these methods, the development of mass-production technology from lab to commercial scale is difficult in principle. On the other hand, the CVD method is generally known to be easy and well suited for the mass production. CVD method affords to produce SWCNTs from raw materials containing carbon-source molecules and catalysts, i.e. SWCNTs grow by supplying carbon generated by the chemical decomposition of carbon-source molecules to the catalyst nanoparticles. For the efficient carbon supply on the catalyst nanoparticles in the CVD growth of SWCNTs, various techniques decomposing carbon-source molecules, such as plasma and hot filament, have been utilized. Therefore, we frequently address separately these different CVD processes, such as thermal-, plasma- and hot-filament CVDs.

2.2 CATALYSTS FOR THE SWCNT GROWTH

Regardless of production processes, nanoparticles as the catalyst are indispensable for the growth of SWCNTs as a general rule, and the nanoparticles of iron-group transition metal, such as iron, cobalt and nickel, are frequently used [4]. It is well known that these iron-group transition metals possess the catalytic activity for the decomposing hydrocarbons and forming the graphite on the surface of their bulk crystals [16]. The carbon-source molecules in the gas-phase decompose on the surface of the catalyst nanoparticles and generate carbon atoms that would dissolve in the catalyst metal. As a result, according to the supersaturation of carbon in the metal, the depositing carbons nucleate on the surface of catalyst nanoparticles to form SWCNTs. It is expected that the catalyst deactivation in the above-mentioned growth mechanism can be caused by the carbonization and/or oxidization of nanoparticles due to the excess supply of the carbon source. Because SWCNTs in the solid phase are produced from carbon-source molecules in the gas phase and the catalyst nanoparticles in the liquid phase due to the effects of their size and eutectic reaction, this growth mechanism is called as vapour–liquid–solid (VLS) model [17]. However, it is still controversial whether the catalysts nanoparticles are in the liquid or solid phases, especially in CVD growth of SWCNTs because the reaction temperature is generally lower than that in the arc-discharge and laser vapourization methods [18].

Beside the iron-group metal, it is known that nanoparticles of aluminium [19] and noble metals, such as palladium, platinum, gold, silver and copper [20], can work as the catalyst for CVD growth of SWCNTs. Furthermore, studies on SWCNT growth by using non-metallic nanoparticles, such as silicon, germanium, silicon carbide [21], alumina [22] and diamond [23], have been also reported. Because the melting points of these non-metallic catalysts are relatively high in their bulk states, the growth mechanism of SWCNTs on these catalysts is expected to be different from the above-mentioned VLS model, although the detail is still an open question.

2.3 THE WAY TO INTRODUCE THE CATALYST ON THE CVD GROWTH OF SWCNTS

In terms of the way to introduce the catalysts in the reaction system for SWCNT growth, we can broadly classify lots of CVD processes into two general categories, i.e. supported- [3–5,7] and floating-catalyst [6,8,24–29] methods. In the former, catalyst nanoparticles are supported upon inert substrates or porous materials, while in the latter, they exist as aerosol within a reactive environment.

Because the supported-catalyst method has its advantage in the inexpensive equipment and wide process-condition range, a lot of commercially available SWCNTs are produced by this method. However, the following problems are generally known in the supported-catalyst method: firstly, the Ostwald ripening of catalyst nanoparticles due to the high temperature in the reactor expands their size distribution, which causes not only the deactivation of catalysts but also the large variability of the quality in diameter, length and crystallinity [30]. Secondly, the supported-catalyst method is basically batch process and it is difficult to develop the process for continuous production. Furthermore, the supported-catalyst method generally needs larger amounts of substrates or porous materials than the amount of produced SWCNTs, which can result in a significant increase in the cost. Although the supported-catalyst method is becoming improved by vigorous researches on the technological development, most of these problems are inherently difficult to be solved.

On the other hand, the floating-catalyst method has an advantage in its continuity of synthesizing products. That is, the floating-catalyst CVD synthesis, which is represented by HiPco [8], is believed to be an ideal continuous process for the production of SWCNTs at commercial scale. SWCNTs grown from the aerosol of catalyst nanoparticles floating in the furnace within the reactive environment flowed out and collected continuously. Continuous supply of fresh catalysts for the growth of SWCNTs and their quick outflow from the reactor prevent broadening the size distribution of the catalyst nanoparticles, which can lead to the production of SWCNTs with homogeneous quality.

As much as the author knows, the first study on the SWCNT growth by floating-catalyst method was reported by Cheng et al. in 1998 [6]. Similar to that reported for the VGCF production by Endo and Shikata [31], they produced high-quality SWCNTs by using the floating-catalyst CVD reactor claiming the relatively large-scale and low-cost production of SWCNTs. Thereafter, a variety of studies on the floating-catalyst method, including growth processes with different catalysts and carbon sources, have been reported [8,24–29], and some of them have demonstrated the specific controllability of structural features as described in subsequent sections.

2.4 CARBON SOURCES LEADING TO THE EFFICIENT CVD GROWTH OF SWCNTS

Although CVD production from hydrocarbons involves a wide variety of chemical reactions, the role of such chemical reactions has not received sufficient attention. In the limited reports focusing on the chemical reactions [32–37], the CVD growth of SWCNTs and/or multi-walled CNTs (MWCNTs) from hydrocarbons have been suggested to be initiated via the decomposition of hydrocarbons into reactive species, free radicals and intermediates. This is strongly supported by the presence of many hydrocarbons as by-products in the exhaust gas [32,38–41]. These studies clearly indicate the involvement of gas-phase chemical reactions in the CVD growth of CNTs. However, the importance of gas-phase chemical reactions in the carbon nanotube (CNT) growth has not been fully explored yet. Due to the reaction complexities in the CVD process, it is not so easy to clarify the gas-phase reactions directly. Therefore, the analysis of thermal decomposition phenomena of used hydrocarbons, in comparison to the experimental observation of SWCNT growth, should be taken into account for extracting some useful information about the key carbon sources and gas-phase intermediates initiating the real growth process.

Very recently the molecular structure of the hydrocarbons used as a carbon source has been found to greatly affect the efficiency of SWCNT growth by producing specific intermediates with varying concentrations from the gas-phase decompositions [42]. Other studies have similar opinions about the impact of the gas-phase composition of the reaction mixture on growth efficiency [43,44]. These studies indicate that growth efficiency strongly depends on the reactivity and concentration of gas-phase intermediates produced as a result of complex gas-phase reactions. On this basis, it can be expected that the most-efficient intermediates, which have the potential of chemisorption or physisorption on the catalyst surface to initiate CNT growth, should be produced in the gas phase. The overall kinetics of the growth process depends on the interaction/competition of gas-phase and surface reactions.

By some recent studies [45–49], the gas-phase carbon precursor enabling CNT growth has been speculated on the basis of the presence/absence of particular species in the mass spectrometric analyses of gas-phase composition of the exhaust/reaction mixture. For example, Franklin and Dai [45] and Tian et al. [46] proposed C_6H_6 while Kim et al. [47] proposed radical species: C_5H_9, C_6H_{13} and C_6H_9. However, other studies [48,49] reported that C_1 or C_2 radical/neutral species should be the key species because of their greater activity caused by fewer hydrogens and smaller size [33], facilitating higher mobility on the catalyst surface compared to larger species. Since many groups have realized that the significant enhancement in the growth of CNTs upon direct supply of C_2 species as a primary carbon source [47,50] or by the addition of an enhancing agent [43,45,51,52], the smaller radical/intermediate species seems to be the more probable key species.

Of course, the key intermediate species for SWCNT growth should vary on the reaction dynamics/environment, such as the gas composition and the reaction temperature. For example, although a lot of studies on the supported-catalyst CVD for CNT growth suggest the efficiency of C_2H_2 as the carbon source [53–56], it has also been reported that the reaction kinetics in the CNT growth significantly depends on the composition of environment gas that can affect the rate of decomposition reaction of C_2H_2 [57]. Furthermore, in the floating-catalyst CVD in which the reaction temperature is much higher than that of supported-catalyst CVD, hydrocarbons capable of producing sp C_2 species (C_2H/C_2H_2) are known to be less efficient [42], probably because of the passivation of the catalyst caused by the excessive deposition of carbon on the catalyst surface from the overactive sp C_2 species [57]. Instead, sp^2 C_2 species (C_2H_3/C_2H_4) has been concluded as a final and efficient species capable of initiating the SWCNT growth in the floating-catalyst CVD, deductively based on systematic experimental and theoretical studies [42,58].

2.5 STRUCTURAL CONTROLLABILITY IN THE CVD SYNTHESIS OF SWCNTS

In general, the properties of carbon nanotubes strongly depend on their structural characteristics [59], i.e. wall number, diameter, length, number of defects and so forth. Hence, there is a need for methods to control their structure for a wide variety of potential technological applications. The selective production of SWCNTs is as sort of the CNT growth with the structural controllability in terms of the wall number [60]. CVD methods for the CNT production can satisfy the structure-control requirements and have therefore been the subject of considerable research efforts. To date it has been reported that the distribution of the SWCNT length [61], diameter [52,62–64], chirality [65–67] and metallic/semiconducting types [68–71] can be controlled by modifying the reaction system and conditions, such as catalysts, substrates, reaction temperature, additives, carbon sources and/or precursors, adopted in CVD syntheses.

Above all the structural characteristics in SWCNTs to be controlled for the realization of their many potential applications, the diameter control is fundamentally yet indispensable because the tube diameter deeply correlates with the bandgap that is the determining factor in the SWCNT application using their excellent semiconducting property [72]. In addition, the precise control of the tube diameter is also important for the applications using the molecular encapsulation into SWCNTs as the containers, such as peapods (the SWCNTs encapsulating fullerenes [73]) and drug delivery systems (DDS). The size matching between encapsulated molecules and the diameter of SWCNTs can contribute to the efficiency of encapsulation [74]. Therefore, an excellent method for tuning the diameter is still eagerly anticipated.

Recently, an efficient floating-catalyst CVD method called enhanced direct injection pyrolytic synthesis (eDIPS) has been developed [52], in which at least two kinds of hydrocarbons with different decomposition properties are used as the carbon sources, as shown in Figure 2.2. Because of their high throughput and productivity [28], eDIPS-CVD is one of the most suitable CVD processes for the industrial production of CNTs. It has been reported that the eDIPS-CVD provides easy synthesis of diameter-controlled SWCNTs by conditioning multiple carbon sources [52]. As demonstrated in Figure 2.2, eDIPS-CVD using ethylene (C_2H_4) as the secondary carbon source allows the SWCNT diameters to be precisely controlled.

Hitherto, the mechanism for diameter-controlled growth during eDIPS-CVD is proposed as follows [52]: from a static view point, the iron-nanoparticle catalysts are distributed according to particle size within the reactor, with smaller particles upstream and the larger particles downstream. Therefore, the position in the reactor where the catalysts start to grow SWCNTs can be tuned by controlling the reaction gas atmosphere by means of manipulating the flow rates of two kinds of carbon sources with different pyrolytic properties.

Optical analyses in Figure 2.3 support the controllable range of the diameter from ~1 to 2 nm [64,67]. Absorption spectra (Figure 2.3(a)) clearly show that the SWCNTs with the four types of diameters possessing the distinctly different electronic structures respectively that reflect the importance of precise diameter control of SWCNTs for their electronics applications [72]. In addition, Raman spectra (Figure 2.3(b)) suggest that the CNTs produced by eDIPS-CVD are well graphitized and have qualities comparable with those produced by the

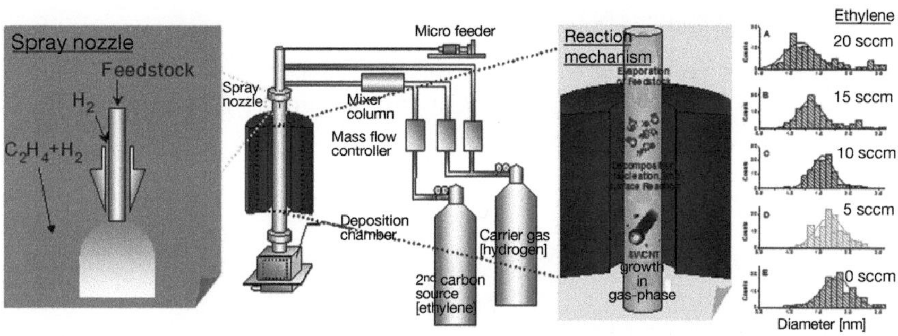

FIGURE 2.2 Setup of the (eDIPS) CVD [52] and the diameter distribution histograms of SWCNTs synthesized under different conditions of ethylene flow rate [64].

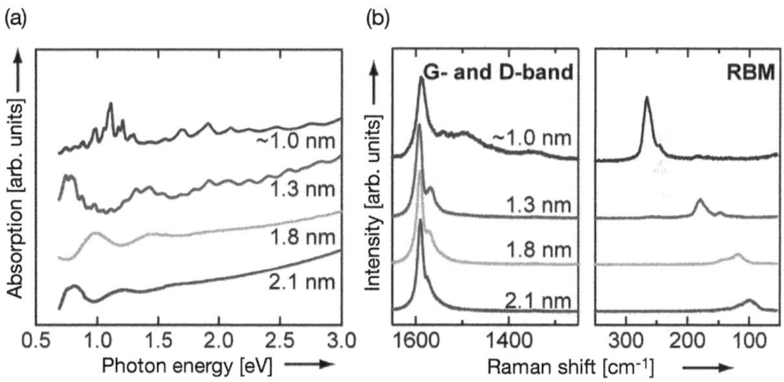

FIGURE 2.3 Optical analyses (absorption spectra (a) and resonance Raman spectra (b) measured using excitation wavelength of 514.5 nm) of SWCNTs synthesized by eDIPS-CVD under different conditions of ethylene flow rate.

arc-discharge and laser vapourization methods in terms of the high G/D bands' intensity ratio.

2.6 SUMMARY AND OUTLOOK

Various scientific and technological aspects of the SWCNT production have been overviewed in the context of reaction systems, catalysts, carbon sources, growth mechanisms and mass productivity. Furthermore, in the CVD synthesis of SWCNTs, recent progresses of the structural controllability have been reviewed. Still the development of new technology for the SWCNT production continues to be a major challenge in this research field. Toward the promising applications of SWCNTs, hereafter advances of the preferential growth of SWCNTs with the specific structure will become vastly more important. "Special SWCNTs" with favourable properties brought by the preferential growth can significantly boost the R&D of their promising applications in the next decade.

ACKNOWLEDGEMENTS

The author wishes to acknowledge the financial supports from the New Energy and Industrial Technology Development Organization (NEDO) and the Japan Science and Technology Agency (JST-PRESTO).

REFERENCES

[1] Iijima S, Ichihashi T. Single-shell carbon nanotubes of 1-nm diameter. Nature 1993;363:603–5.
[2] Thess A, Lee R, Nikolaev P, Dai H, Petit P, Robert J, et al. Crystalline ropes of metallic carbon nanotubes. Science 1996;273:483–7.
[3] Dai H, Rinzler AG, Nikolaev P, Thess A, Colbert DT, Smalley RE. Single-wall nanotubes produced by metal-catalyzed disproportionation of carbon monoxide. Chem Phys Lett 1996;260: 471–5.
[4] Kong J, Cassell AM, Dai H. Chemical vapor deposition of methane for single-walled carbon nanotubes. Chem Phys Lett 1998;292:567–74.
[5] Hafner JH, Bronikowski MJ, Azamian BR, Nikolaev P, Rinzler AG, Colbert DT, et al. Catalytic growth of single-wall carbon nanotubes from metal particles. Chem Phys Lett 1998;296:195–202.
[6] Cheng HM, Li F, Su G, Pan HY, He LL, Sun X, et al. Large-scale and low-cost synthesis of single-walled carbon nanotubes by the catalytic pyrolysis of hydrocarbons. Appl Phys Lett 1998;72:3282–4.
[7] Colomer J-F, Bister G, Willems I, Kónya Z, Fonseca A, Nagy JB, et al. Synthesis of single-wall carbon nanotubes by catalytic decomposition of hydrocarbons. Chem Commun 1999; 1343–4.
[8] Nikolaev P, Bronikowski MJ, Bradley RK, Rohmund F, Colbert DT, Smith KA, et al. Gas-phase catalytic growth of single-walled carbon nanotubes from carbon monoxide. Chem Phys Lett 1999;313:91–7.
[9] Biswas C, Lee YH. Graphene versus carbon nanotubes in electronic devices. Adv Funct Mater 2011;21:3806–26.
[10] Scardaci V, Coull R, Coleman JN. Very thin transparent, conductive carbon nanotube films on flexible substrates. Appl Phys Lett 2010;97:023114.
[11] Kaskela A, Nasibulin AG, Timmermans MY, Aitchison B, Papadimitratos A, Tian Y, et al. Aerosol-synthesized SWCNT networks with tunable conductivity and transparency by a dry transfer technique. Nano Lett 2010;10:4349–55.
[12] Liu Q, Fujigaya T, Cheng H-M, Nakashima N. Free-standing highly conductive transparent ultrathin single-walled carbon nanotube films. J Am Chem Soc 2010;132:16581–6.
[13] Shi Z, Chen X, Wang X, Zhang T, Jin J. Fabrication of superstrong ultrathin free-standing single-walled carbon nanotube films via a wet process. Adv Funct Mater 2011;21:4358–63.
[14] Kim Y, Chikamatsu M, Azumi R, Saito T, Minami N. Industrially feasible approach to transparent, flexible, and conductive carbon nanotube films: cellulose-assisted film deposition followed by solution and photonic processing. Appl Phys Express 2013;6:025101.
[15] Jorio A, Dresselhaus G, Dresselhaus MS, editors. Carbon nanotubes, advanced topics in the synthesis, structure, properties and applications. Berlin: Springer; 2008. ISBN:978-3-540-72864-1.
[16] Hamilton JC, Blakely JM. Carbon segregation to single crystal surfaces of Pt, Pd, and Co. Surf Sci 1980;91:199–217.
[17] Gavillet J, Loiseau A, Journet C, Willaime F, Ducastelle F, Charlier J-C. Root-growth mechanism for single-wall carbon nanotubes. Phys Rev Lett 2001;87:275504.

[18] Harutunyan AR. The catalyst for growing single-walled carbon nanotubes by catalytic chemical vapor deposition method. J Nanosci Nanotechnol 2009;9:2480–95.
[19] Yuan D, Ding L, Chu H, Feng Y, McNicholas TP, Liu J. Horizontally aligned single-walled carbon nanotube on quartz from a large variety of metal catalysts. Nano Lett 2008;8:2576–9.
[20] Takagi D, Homma Y, Hibino H, Suzuki S, Kobayashi Y. Single-walled carbon nanotube growth from highly activated metal nanoparticles. Nano Lett 2006;6:2642–5.
[21] Takagi D, Hibino H, Suzuki S, Kobayashi Y, Homma Y. Carbon nanotube growth from semiconductor nanoparticles. Nano Lett 2007;7:2272–5.
[22] Liu H, Takagi D, Ohno H, Chiashi S, Chokan T, Homma Y. Growth of single-walled carbon nanotubes from ceramic particles by alcohol chemical vapor deposition. Appl Phys Express 2008;1:014001.
[23] Takagi D, Kobayashi Y, Homma Y. Carbon nanotube growth from diamond. J Am Chem Soc 2009;131:6922–3.
[24] Bladh K, Falk LKL, Rohmund F. On the iron-catalysed growth of single-walled carbon nanotubes and encapsulated metal particles in the gas phase. Appl Phys A 2000;70:317–22.
[25] Ago H, Ohshima S, Uchida K, Yumura M. Gas-phase synthesis of single-wall carbon nanotubes from colloidal solution of metal nanoparticles. J Phys Chem B 2001;105:10453–6.
[26] Zhu HW, Xu CL, Wu DH, Wei BQ, Vajtai R, Ajayan PM. Direct synthesis of long single-walled carbon nanotube strands. Science 2002;296:884–6.
[27] Nasibulin AG, Moisala A, Brown DP, Jiang H, Kauppinen EI. A novel aerosol method for single walled carbon nanotube synthesis. Chem Phys Lett 2005;402:227–32.
[28] Saito T, Xu WC, Ohshima S, Ago H, Yumura M, Iijima S. Supramolecular catalysts for the gas-phase synthesis of single-walled carbon nanotubes. J Phys Chem B 2006;110:5849–53.
[29] Li Y-L, Kinloch IA, Windle AH. Direct spinning of carbon nanotube fibers from chemical vapor deposition synthesis. Science 2004;304:276–8.
[30] Amama PB, Pint CL, McJilton L, Kim SM, Stach EA, Murray PT, et al. Role of water in super growth of single-walled carbon nanotube carpets. Nano Lett 2009;9:44–9.
[31] Endo M, Shikata M. Growth of vapor-grown carbon fibers using fluid ultra-fine particles of metals. Ouyoubuturi 1985;54:507–10.
[32] Endo H, Kuwana K, Saito K, Qian D, Andrews R, Grulke EA. CFD prediction of carbon nanotube production rate in a CVD reactor. Chem Phys Lett 2004;387:307–11.
[33] Grujicic M, Cao G, Gersten B. Optimization of the chemical vapor deposition process for carbon nanotubes fabrication. Appl Surf Sci 2002;191:223–39.
[34] Becker A, Huttinger KJ. Chemistry and kinetics of chemical vapor deposition of pyrocarbon—II pyrocarbon deposition from ethylene, acetylene and 1,3-butadiene in the low temperature regime. Carbon 1998;36:177–99.
[35] Descamps C, Vignoles GL, Feron O, Langlais F, Lavenac J. Correlation between homogeneous propane pyrolysis and pyrocarbon deposition. J Electrochem Soc 2001;148:C695–708.
[36] Somenath M, Brukh R. Mechanism of carbon nanotube growth by CVD. Chem Phys Lett 2006;424:126–32.
[37] Becker A, Hu Z, Huttinger KJ. A hydrogen inhibition model of carbon deposition from light hydrocarbons. Fuel 2000;79:1573–80.
[38] Tian Y, Hu Z, Yang Y, Chen X, Ji W, Chen Y. Thermal analysis–mass spectroscopy coupling as a powerful technique to study the growth of carbon nanotubes from benzene. Chem Phys Lett 2004;388:259–62.
[39] Kuwana K, Endo H, Saito K, Qian D, Andrews R, Grulke EA. Catalyst deactivation in CVD synthesis of carbon nanotubes. Carbon 2005;43:253–60.

[40] Norinaga K, Deutschmann O, Huttinger KJ. Analysis of gas phase compounds in chemical vapor deposition of carbon from light hydrocarbons. Carbon 2006;44:1790–800.

[41] Plata DL, Hart AJ, Reddy CM, Gschwend PM. Early evaluation of potential environmental impacts of carbon nanotube synthesis by chemical vapor deposition. Environ Sci Technol 2009;43: 8367–73.

[42] Shukla B, Saito T, Yumura M, Iijima S. An efficient carbon precursor for gas phase growth of SWCNTs. Chem Commun 2009;3422–34.

[43] Qian WZ, Tian T, Guo CY, Wen Q, Li KJ, Zhang HB, et al. Enhanced activation and decomposition of CH_4 by the addition of C_2H_4 or C_2H_2 for hydrogen and carbon nanotube production. J Phys Chem C 2008;112:7588–93.

[44] Xiang R, Einarsson E, Okawa J, Miyauchi Y, Maruyama S. Acetylene-accelerated alcohol catalytic chemical vapor deposition growth of vertically aligned single-walled carbon nanotubes. J Phys Chem C 2009;113:7511–5.

[45] Franklin NR, Dai H. An enhanced CVD approach to extensive nanotube networks with directionality. Adv Mater 2000;12:890–4.

[46] Tian Y, Hu Z, Yang Y, Wang X, Chen X, Xu H, et al. In situ TA-MS study of the six-membered-ring-based growth of carbon nanotubes with benzene precursor. J Am Chem Soc 2004;126: 1180–3.

[47] Kim S-M, Zhang Y, Teo KBK, Bell MS, Gangloff L, Wang X, et al. SWCNT growth on Al/Fe/Mo investigated by in situ mass spectroscopy. Nanotechnology 2007;18:185709.

[48] Vinciguerra V, Buonocore F, Panzera G, Occhipinti L. Growth mechanisms in chemical vapour deposited carbon nanotubes. Nanotechnology 2003;14:655–60.

[49] Lee TY, Han JH, Choi SH, Yoo JB, Park CY, Jung T, et al. Effects of source gases on the growth of carbon nanotubes. Diamond Relat Mater 2003;12:851–5.

[50] Hata K, Futaba DN, Mizuno K, Namai T, Yumura M, Iijima S. Water-assisted highly efficient synthesis of impurity-free single-walled carbon nanotubes. Science 2004;306:1362–4.

[51] Qi H, Yuan DN, Liu J. Two-stage growth of single-walled carbon nanotubes. J Phys Chem C 2007;111:6158–60.

[52] Saito T, Ohshima S, Okazaki T, Ohmori S, Yumura M, Iijima S. Selective diameter control of single walled carbon nanotubes in the gas phase synthesis. J Nanosci Nanotechnol 2008;8: 6153–7.

[53] Eres G, Kinkhabwala AA, Cui H, Geohegan DB, Puretzky AA, Lowndes DH. Molecular beam-controlled nucleation and growth of vertically aligned single-wall carbon nanotube arrays. J Phys Chem B 2005;109:16684–94.

[54] Zhong G, Hofmann S, Yan F, Telg H, Warner JH, Eder D, et al. Acetylene: a key growth precursor for single-walled carbon nanotube forests. J Phys Chem C 2009;113:17321–5.

[55] Xiang R, Einarsson E, Okawa J, Miyauchi Y, Maruyama S. Acetylene-accelerated alcohol catalytic chemical vapor deposition growth of vertically aligned single-walled carbon nanotubes. J Phys Chem C 2009;113:7511–5.

[56] Sugime H, Noda S. Millimeter-tall single-walled carbon nanotube forests grown from ethanol. Carbon 2010;48:2203–11.

[57] Jung M, Eun KY, Lee JK, Baik YJ, Lee KR, Park JW. Growth of carbon nanotubes by chemical vapor deposition. Diamond Relat Mater 2001;10:1235–40.

[58] Shukla B, Saito T, Ohmori S, Koshi M, Yumura M, Iijima S. Interdependency of gas phase intermediates and chemical vapor deposition growth of single wall carbon nanotubes. Chem Mater 2010;22:6035–43.

[59] Saito R, Fujita M, Dresselhaus G, Dresselhaus MS. Electronic structure of chiral graphene tubules. Appl Phys Lett 1992;60:2204–6.

[60] Kobayashi K, Shukla B, Ohmori S, Kiyomiya M, Hirai T, Kuwahara Y, et al. Wall-number selectivity in single/double-wall carbon nanotube production by enhanced direct injection pyrolytic synthesis. Jpn J Appl Phys 2013;52:105102.

[61] Zhong G, Iwasaki T, Robertson J, Kawarada H. Growth kinetics of 0.5 cm vertically aligned single-walled carbon nanotubes. J Phys Chem B 2007;111:1907–10.

[62] Cheung CL, Kurtz A, Park H, Lieber CM. Diameter-controlled synthesis of carbon nanotubes. J Phys Chem B 2002;106:2429–33.

[63] Liu Q, Ren W, Chen Z-G, Wang D-W, Liu B, Yu B, et al. Diameter-selective growth of single-walled carbon nanotubes with high quality by floating catalyst method. ACS Nano 2008;2:1722–8.

[64] Saito T, Ohmori S, Shukla B, Yumura M, Iijima S. A novel method for characterizing the diameter of single-wall carbon nanotubes by optical absorption spectra. Appl Phys Express 2009;2:095006.

[65] Bachilo SM, Balzano L, Herrera JE, Pompeo F, Resasco DE, Weisman RB. Narrow (n,m)-distribution of single-walled carbon nanotubes grown using a solid supported catalyst. J Am Chem Soc 2003;125:11186–7.

[66] Miyauchi Y, Chiashi S, Murakami Y, Hayashida Y, Maruyama S. Fluorescence spectroscopy of single-walled carbon nanotubes synthesized from alcohol. Chem Phys Lett 2004;387:198–203.

[67] Ohmori S, Saito T, Tange M, Shukla B, Okazaki T, Yumura M, et al. Fundamental importance of background analysis in precise characterization of single-walled carbon nanotubes by optical absorption spectroscopy. J Phys Chem C 2010;114:10077–81.

[68] Li Y, Mann D, Rolandi M, Kim W, Ural A, Hung S, et al. Preferential growth of semiconducting single-walled carbon nanotubes by a plasma enhanced CVD method. Nano Lett 2004;4:317–21.

[69] Qu L, Du F, Dai L. Preferential synthesis of semiconducting vertically aligned single-walled carbon nanotubes for direct use in FETs. Nano Lett 2008;8:2682–7.

[70] Harutyunyan AR, Chen G, Paronyan TM, Pigos EM, Kuznetsov OA, Hewaparakrama K, et al. Preferential growth of single-walled carbon nanotubes with metallic conductivity. Science 2009;326:116–20.

[71] Sundaram RM, Koziol KKK, Windle AH. Continuous direct spinning of fibers of single-walled carbon nanotubes with metallic chirality. Adv Mater 2011;23:5064–8.

[72] Asada Y, Ohmori S, Nihey F, Shinohara H, Saito T. Diameter-dependent performance of single-walled carbon nanotube thin-film transistors. Adv Mater 2011;23:4631–5.

[73] Bandow S, Takizawa M, Kato H, Okazaki T, Shinohara H, Iijima S. Smallest limit of tube diameters for encasing of particular fullerenes determined by radial breathing mode Raman scattering. Chem Phys Lett 2001;347:23–8.

[74] Okada S, Saito S, Oshiyama A. Energetics and electronic structures of encapsulated C_{60} in a carbon nanotube. Phys Rev Lett 2006;86:3835–8.

Chapter 3

Differentiation of Carbon Nanotubes with Different Chirality

Kazuhiro Yanagi
Department of Physics, Tokyo Metropolitan University, Hachioji, Tokyo, 192-0397, Japan

3.1 INTRODUCTION AND BRIEF HISTORY OF DIFFERENTIATION OF SINGLE-WALLED CARBON NANOTUBES (SWCNTS) WITH DIFFERENT ELECTRONIC TYPES

Single-walled carbon nanotubes (SWCNTs) are cylindrical graphitic tubules with diameters of approximately 1.0 nm, and they have a variety of electronic properties that depend on how the graphene sheet is rolled, which is referred to as chirality [1]. The chirality is given by a single vector called the chiral vector (n, m) [2]. The (n, m) vector is very helpful to specify the structures of SWCNTs and also to predict their electronic properties. When $n - m$ is a multiple of 3, the SWCNTs are metallic; in other cases, the SWCNTs are semiconducting. During the production of SWCNTs, both metallic and semiconducting SWCNTs are inevitably grown at the same time. Therefore, in the as-grown samples, various SWCNTs with different chiralities are included. The mixed states of different (n, m) SWCNTs have created a bottleneck that has impeded the correct understanding of the basic physical and chemical properties of SWCNTs as well as their device applications. For example, in the case of semiconducting SWCNTs used in electric field effect transistor channels, the presence of metallic SWCNTs significantly decreases the on/off ratio and degrades device performances [3–5]; in contrast, in the case of metallic SWCNTs used for conductive films, the presence of semiconducting SWCNTs degrades the resistivity [6,7] and the stability [8] of the films. The resistivity of the network of semiconducting SWCNTs is 1–10 Ωcm at room temperature, which is a thousand times greater than that of metallic SWCNTs (10^{-3}–10^{-2} Ωcm) [7]. Therefore, the ability to differentiate between SWCNTs with different chiralities and to obtain

Carbon Nanotubes and Graphene. DOI: 10.1016/B978-0-08-098232-8.00003-6

samples in metallic, semiconducting or single-chiral states is important with respect to basic research as well as with respect to the applications of SWCNTs.

Numerous purification techniques have been suggested as methods to overcome this problem since the discovery [9] of SWCNTs in 1993. Here, I provide a brief description of the recent advancements in purification techniques (I also refer interested readers to a good review [10] by Prof. Hersam). At the time SWCNTs were developed, "purification" referred to any technique used to remove amorphous carbon or residual catalyst metals attached to SWCNTs and typically involved several chemical reagents such as an acid, and annealing procedures [11,12]. The first account of SWCNTs being sorted according to their length and diameter was reported in 1998 and involved the use of size-exclusion column chromatography [13]. In 2002, a technique to obtain monodispersed SWCNTs through the use of surfactants was reported [14]. A monodispersed solution was obtained by ultracentrifugation, which is based on the difference between the densities of monodispersed and multi-bundled SWCNTs. Soon after that, in 2003, remarkable purification methods for SWCNTs based on their electronic properties were reported [15,16]. One of these methods is metallic/semiconducting (MS) sorting via DNA-wrapping techniques [15]; the other is MS sorting via AC electric field applications [16]. In the DNA-wrapping techniques, the SWCNTs are dispersed by DNA and their electronic types are sorted by ion exchange column chromatography. Metallic and semiconducting types were sorted on the basis of differences in the DNA-binding properties of SWCNTs with different electronic properties. These techniques have now been advanced to single-chiral extractions [17,18]. In the case of AC electric field sorting, metallic and semiconducting SWCNTs were sorted on the basis of the difference in the relative dielectric constants between metallic and semiconducting SWCNTs. Metallic SWCNTs are attracted to an electrode array, where semiconducting tubes remain in the solvent. The techniques that involve AC electric field sorting have been extended to continuous extraction using microfluidic channels [19].

In 2003, semiconducting and metallic SWCNTs were reported to exhibit different chemical reactivities toward amines [20], and this difference in chemical reactivity was used to separate semiconducting SWCNTs from metallic SWCNTs [20]. Similar differences in the chemical reactivity of SWCNTs with different electronic properties were clearly observed in 2005, and several purification techniques that involve amines and nitronium ions have been reported for the extraction of metallic or semiconducting types of SWCNTs [21–23].

In 2006, Arnold et al. reported remarkable breakthrough in the purification of SWCNTs [4]. They reported that metallic and semiconducting SWCNTs can be sorted by differences in their densities. In this technique, they dispersed SWCNTs using sodium dodecyl sulfate (SDS) and sodium cholate (SC) as co-surfactants, the surfactants induced slight differences in the solvated densities of metallic and semiconducting SWCNTs, and strong centrifugation force under density-gradient ultracentrifugation (DGU) resulted in sorting of the metallic and the semiconducting types. The authors suggested two important concepts: (1) SDS and SC are important for differentiating densities of SWCNTs with different electronic types and (2) a co-surfactant system is a good approach for tuning the sorting capability. Numerous studies related to the differentiation of densities have been reported since the initial report of this method, including studies on length sorting [24], metallic extraction using sodium deoxycholate (DOC) for dispersion [25] and specific chirality extractions using a co-surfactant approach [26]. In 2007, single-chirality extraction of (7, 5) SWCNT was achieved using poly(9,9-dioctylfluorenyl-2,7-diyl) polymers (PFO polymers) [27]. In 2008, single-chirality extraction of (8, 6) SWCNT was achieved with a flavin monomer [28]. In 2009, the field of chromatographic sorting was significantly advanced when Moshammer et al. reported the sorting of metallic and semiconducting SWCNTs using gel-chromatography Sephacryl gel [29]; soon afterwards, Tanaka et al. reported a sorting method based on the use of agarose gel [30]. Moshammer et al. [29] dispersed SWCNTs with SDS and used Sephacryl gel (covalently cross-linking allyl dextran with *N,N'*-methylene bisacrylamide, GE Health Care) as a filtration medium. Remarkably, they found that SDS is a key surfactant for the separation of metallic and semiconducting SWCNTs. Notably, SDS has been used for the suspension/differentiation of SWCNTs via various purification techniques including AC electric field separations [16], DGU separations [31] and gel separations [29]. As will be discussed later, SDS is a key surfactant for such separation. In 2009, single-chirality extractions for various kinds of SWCNTs – (9, 1), (8, 3), (6, 5), (7, 5), (10, 2), (8, 4), (9, 4), (7, 6), (8, 6), (9, 5), (10, 5) and (8, 7) – were achieved by DNA sorting [18]. In 2010, single-chirality extractions for various chiral SWCNTs – (6, 4), (7, 3), (6, 5), (9, 1), (8, 3), (9, 2), (7, 5), (8, 4), (10, 2), (7, 6) – and enantiomers separations were achieved with the DGU technique [32]. Enantiomer sorting via DGU was also reported by Hersam's group [33]. The purification of optically active SWCNTs through the use of diporphyrin molecules was first reported by Peng et al. in 2007 [34]; however, the method developed by Peng et al. requires the preparation of specific molecules for the extraction of enantiomers, whereas the extraction of enantiomers using surfactants described in the previously discussed studies involved more accessible and general approaches. In 2011, large-scale single-chirality separation of 13 different (n, m) SWCNTs was achieved by the gel-chromatography method using Sephacryl gel and SDS solutions [35], and single-chirality extraction of SWCNTs with a diameter of 1.4 nm, (11, 10), was recently achieved through the use of DGU [36].

As previously described, remarkable advancements in differentiation techniques have been reported almost every year. Here, I focus primarily on sorting using DGU and gel chromatography/filtration; however, various other sorting techniques, such as differentiation using specific self-assembled monolayers [37], selective joule heating [38,39] and radical/ion reaction processes [40,41], have also been reported. The characteristics of differentiation techniques and the reported single chirality of (n, m) sample are summarized in Tables 3.1 and 3.2,

TABLE 3.1 Methods for differentiating SWCNTs with different electronic properties

	Differentiation method	Characteristics
Surfactant differentiation	Density-gradient sorting	MS sorting, single-chirality extraction, enantiomer extraction, length sorting
	Gel chromatography/ filtration	MS sorting, single-chirality extraction, enantiomer extraction, length sorting
DNA wrapping		MS sorting, single-chirality extraction
Polymer wrapping		Single-chirality extraction; (7, 5)
Monomer extraction		Single-chirality extraction; (8, 6)
Chemical reactivity		MS sorting

TABLE 3.2 Reported extracted single chiralities

Diameter (nm)	~0.6	0.7–0.8	0.8–0.9	0.9–1.0	1.0–1.1	1.4
(*n*, *m*)	(6, 4) [35]	(7, 3)[35], (6, 5) [18,35], (9, 1) [18], (8, 3) [18,35]	(9, 2) [32], (7, 5) [18,35], (8, 4) [18,35], (10, 2) [18,35], (7, 6) [18,35]	(9, 4) [18,35], (10, 3) [35], (8, 6) [18,28,35], (9, 5) [18], (12, 1) [35]	(10, 5) [18,46], (8, 7) [18,35]	(11, 10) [36]

and Figure 3.1. As shown in Figure 3.1, many (*n*, *m*) SWCNTs can be obtained as single-chiral state samples. However, the degree of single chirality is evaluated on the basis of SWCNTs' photoluminescence characteristics; thus, to the base of my knowledge, single-chiral purification of metallic SWCNTs that do not exhibit photoluminescence has not been reported (although some reports [42,43] have indicated that metallic SWCNTs obtained by DGU tend to be near-arm-chair types). In addition, single chiralities of semiconducting zig-zag SWCNTs have not yet been obtained (one of the probable reasons for this fact might be easy bundling of zig-zag SWCNTs [44]). These problems remain to be solved. As a result of chirality purification, SWCNT solutions with various colours can be obtained, where the colour of the solution depends on the chirality [25,45]. As an example, the colours of metallic and semiconducting SWCNTs with different diameters are shown in Figure 3.2. Currently, two main techniques have been developed for the preparation of electrically selected SWCNTs: DGU and gel-chromatography/filtration using surfactants. Here, I focus on the details of these two techniques.

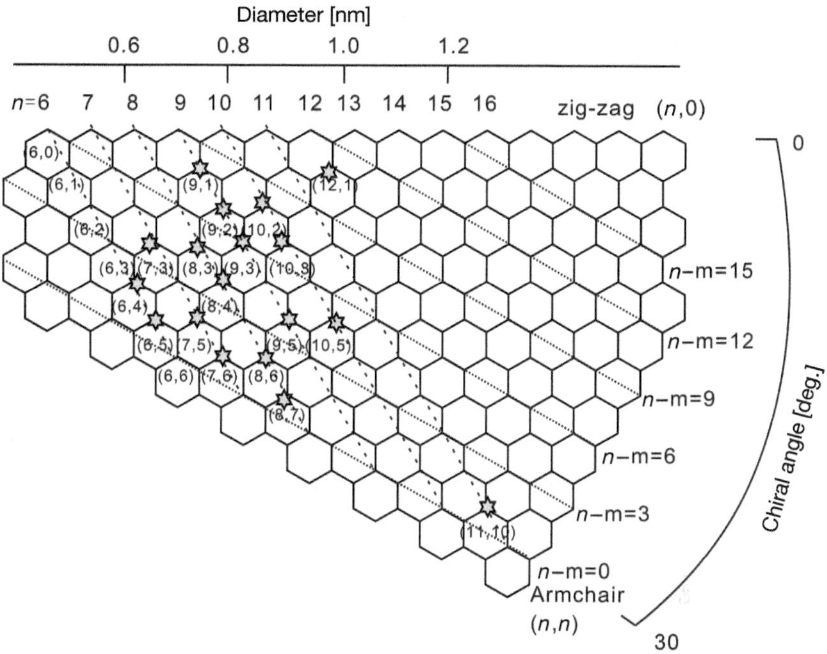

FIGURE 3.1 Distribution of extracted single-chiral (n, m) SWCNTs. Stars indicate the extracted (n, m).

FIGURE 3.2 Colours of metallic and semiconducting SWCNTs with different diameters (unfortunately, this figure is printed in monochrome. They exhibit systematically different colours (see Ref. [25] as an example)).

3.2 DIFFERENTIATION OF DENSITIES OF SWCNTS WITH DIFFERENT CHIRALITIES

The density-gradient centrifugation technique is a commonly used technique for the purification of biological materials, such as proteins, DNA and RNA [47]. The first successful application of this technique to the differentiation of SWCNTs was reported by Arnold et al. in 2005 [31]. In their first report, they succeeded in the enrichment of several SWCNTs with specific diameters using DNA-wrapped SWCNTs; however, 1 year later, they reported a substantial breakthrough in purification. They succeeded in clearly sorting metallic and semiconducting types of SWCNTs wrapped with typical surfactants, such as SDS and SC [4]. After their discovery, numerous purification studies that involve DGU have been reported [7,24–26,32,36,45,48–53]. For example, DOC has been used to disperse and extract metallic SWCNTs [25], and SWCNTs have also been sorted by length [24]. In 2009, several (*n*, *m*) chiral SWCNTs were obtained by DGU; moreover, left-handed and right-handed types of (6, 5) were also obtained (i.e. enantiomer sorting) [32,33]. Currently, the DGU technique is commonly used in the purification of SWCNTs, double-walled SWCNTs [50,52] and SWCNTs encapsulated molecules [36,48,54,55].

3.2.1 Typical procedure to sort metallic and semiconducting types of SWCNTs by DGU

First, I describe a typical procedure to sort metallic and semiconducting types of SWCNTs using DGU techniques [7]. A schematic procedure is shown in Figure 3.3. The purification process for SWCNTs with a diameter of 1.4 nm produced by arc-discharge method (Meijyo Nanocarbon Co., ArcSO) is described.

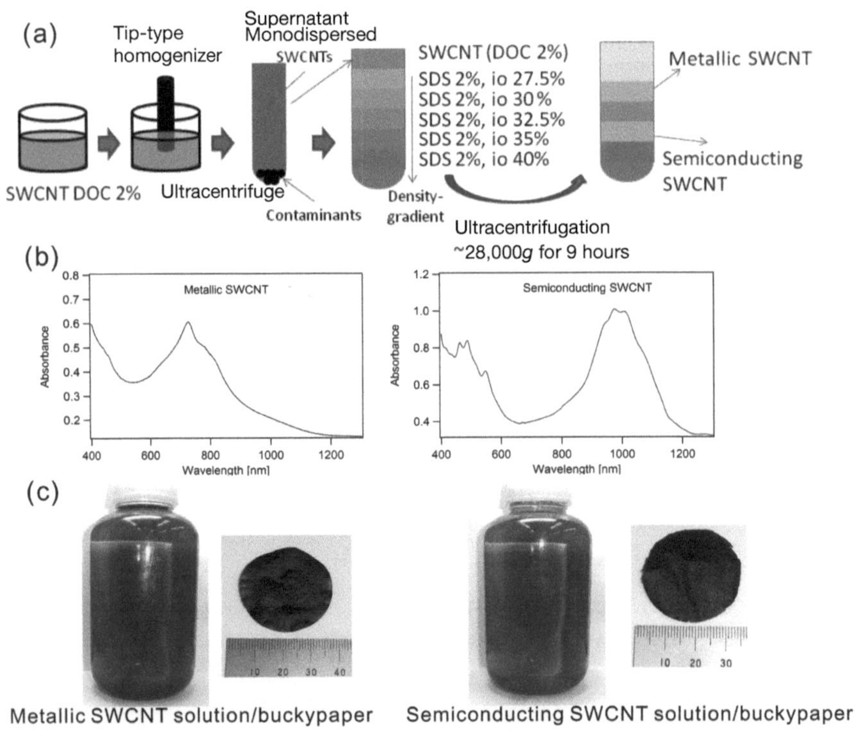

FIGURE 3.3 (a) Schematic of the procedure to sort metallic and semiconducting types of SWCNTs by DGU. (b) Optical absorption spectra of the metallic and semiconducting SWCNTs. (c) Pictures of metallic and semiconducting SWCNT solutions and buckypaper. (Adapted with permission from [7]. Copyright 2010 American Chemical Society).

First, SWCNTs are dispersed in a DOC 2% (w/w) solution using a bath-type ultrasonic generator. Typically, 1 mg of SWCNTs is dispersed into 1 ml of solution. Then, the solution is dispersed by a tip-type ultrasonic homogenizer (17 W, Sonifier, model 250D, Branson Co.). The dispersed solution is centrifuged for 2 hours, and the supernatant is obtained. In this centrifugation process, metal catalysts, which are used in the growth of SWCNTs, and metallic particles, which are contaminants generated during the ultrasonic procedures, are well removed due to the difference in their densities: the densities of SWCNTs are 1.0–1.2 g cm^{-3}, whereas those of metallic contaminants are 2–3 g cm^{-3}. Usually, the duration of centrifugation should be longer than 1 hour to achieve good sorting results in the next step. However, when the centrifugation is performed for more than 20 hours, most of the nanotubes are concentrated at the bottom of the centrifuge tubes because the density of SWCNTs wrapped by DOC is greater than that of water (1.0 g cm^{-3}).

A density-gradient solution is then prepared in centrifuge tubes. In density-gradient sorting, the samples are sorted according to the difference in their densities. Gradient media are used to adjust the densities of the solvents. Typically, the densities of the gradient solutions for the sorting of SWCNTs are adjusted by the addition of iodixanol (Optiprep, Axis-Shield PoC). The following gradient solutions are prepared: SDS 2%/iodixanol 40%, SDS 2%/iodixanol 35%, SDS 2%/iodixanol 32.5%, SDS 2%/iodixanol 30% and SDS 2%/iodixanol 27.5%. The densities of these solutions range from 1.22 to 1.15 g cm^{-3}. These solutions are placed in centrifuge tubes, as shown in Figure 3.3. The supernatant solution is added to the top of the gradient solution. The centrifuge tubes are then centrifuged for 9 hours at 243,000*g* (rotor P50VT2, Hitachi Koki Co.). After the centrifugation, a blue-coloured region is observed in the top parts of the tubes, whereas a red-coloured region is observed in the bottom parts. When small liquid fractions of the solutions are extracted from the tops of the tubes, metallic SWCNTs are obtained from the blue-coloured region, and semiconducting SWCNTs are obtained from the red-coloured regions. Using this sorting procedure, a large amount of purified SWCNTs is obtained, and, as a result, preparation of a buckypaper of metallic and semiconducting SWCNTs is possible (Figure 3.3).

3.2.2 Sorting mechanisms in DGU

Here, the sorting mechanism of DGU is briefly discussed (more details are provided in Ref. 55). In general, during DGU, particles are separated due to the difference in their movements in the presence of a strong centrifugal force. The separations of particles can thus be performed in two ways: through differences in velocity and through differences in equilibrium positions (Figure 3.4). In velocity separations, particles move in the direction of the centrifugal force and separate according to their size and density, whereas in equilibrium separations, particles move to a position in the gradient that reflects their solvated density.

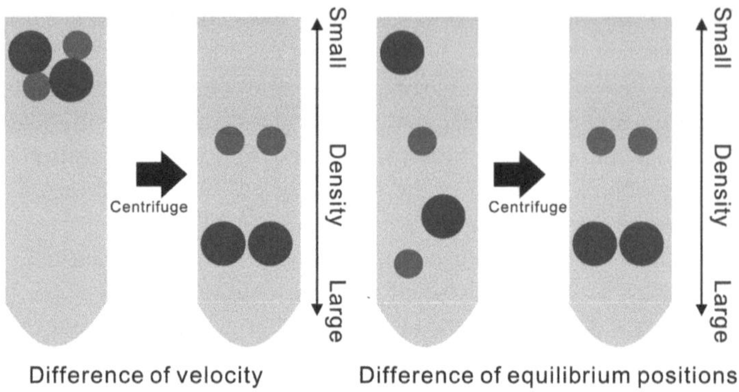

FIGURE 3.4 Two different sorting mechanisms in DGU.

The basic equation of movements of particles during centrifugation is described as follows:

$$M(1-\bar{v}\rho_m)\omega^2 r = Nf(dr/dt) \tag{3.1}$$

where M is the weight of the particle, $\bar{v}$ the solvated volume of a unit mass of particle, ρ_m the density of the solvent, N Avogadro's number, f the frictional coefficient of a single particle described as $f = 6\pi\eta a$ for a spherical particle with radius a in a solution of viscosity η and (dr/dt) the velocity of the particle. As shown in Equation (3.1), differences in the sizes and densities of particles cause a difference in their velocities, and if the densities of the solvent are varied in the centrifugation tubes, the equilibrium positions of particles will be different, as shown in Figure 3.4.

Therefore, the differentiation of the solvated densities (sometimes referred to as the buoyant densities) of SWCNTs with different electronic types is the key to sorting SWCNTs by DGU. However, identification of the solvated densities of SWCNTs is rather difficult. A schematic model is described in Figure 3.5(a). SWCNTs were wrapped by surfactants; thus, with respect to the solvated density, the contribution of surfactants must be included. As shown in Figure 3.5, the adsorption of these surfactants, i.e. the type and the number of surfactants adsorbed on the surface of the SWCNTs, significantly influences the solvated density. Therefore, several parameters, such as the temperature, the concentration of surfactants and the concentrations of other chemical reagents, significantly influence the solvated densities. Several groups have attempted to analyse the details of the solvated densities of SWCNTs; however, the arrangements of the surfactants on the surface of SWCNTs are still under discussion [49,57–63]. For example, in the case of SDS adsorption, three models: the cylindrical micelle model, the hemimicellar adsorption model and the random adsorption model [57,60–62], have been suggested, and no consensus has been reached concerning the arrangements of SDS on the

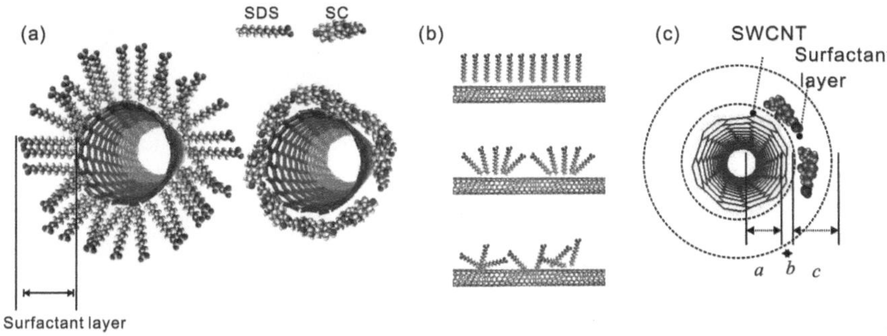

FIGURE 3.5 (a) Schematic pictures of solvated densities of SWCNTs. The thickness of the surfactant layer, which depends on various parameters such as the type and concentration of surfactants, the temperature, the type of salt, the type of alcohol and so on, influences the density. (b) Three different models for SDS arrangements on the surface of SWCNTs are shown: (top) a cylindrical micelle, (middle) hemimicellar adsorption and (bottom) random adsorption. (c) A simple model for evaluation of the solvated densities of SWCNTs. (Adapted with permission from [36]. Copyright 2012 American Chemical Society).

surface of SWCNTs (Figure 3.5(b)). The arrangements of surfactants, such as the number of adsorbed surfactants per unit length, clearly influence the solvated densities of SWCNTs. Therefore, the correct theoretical prediction of the solvated densities of SWCNTs is difficult.

However, a very simple model has been proposed to roughly evaluate characteristics of the solvated densities of SWCNTs. If we assume a simple model as shown in Figure 3.5(c), the solvated (buoyant) density of the surfactant–SWCNT complex ρ_{complex} is written as,

$$\rho_{\text{complex}} = \rho_{\text{in}}\left(\frac{a}{a+b+c}\right)^2 + \rho_C \frac{2ab+b^2}{(a+b+c)^2} + \rho_S \frac{2c(a+b)+c^2}{(a+b+c)^2} \tag{3.2}$$

Here, ρ_{in}, ρ_C, ρ_S, a, b and c are the density of the inner space of the SWCNT, the density of the graphene sheet of the SWCNT, the density of the surfactant layer, the radius of the SWCNT, the thickness of the graphene sheet and the thickness of the surfactant layer, respectively. According to this equation, if ρ_{in}, ρ_C and ρ_S do not depend on the SWCNT chiralities, then the packing structures of the surfactants on the SWCNTs do not change, and the solvated density decreases as the SWCNT diameter ($2a$) increases. As an example, a typical sorting result in a monosurfactant system is shown in Figure 3.6. In this case, larger-diameter SWCNTs tend to be located in a smaller-density region. Therefore, if the diameters of SWCNTs are similar, then the densities of the SWCNTs should also be the same. To differentiate the densities of SWCNTs with different electronic properties and with similar diameters, the adsorption arrangements of the surfactants on the surfaces of SWCNTs must be differentiated. Several analytical studies on this topic have been reported [58,59]; however, additional studies are required to achieve control of the arrangements.

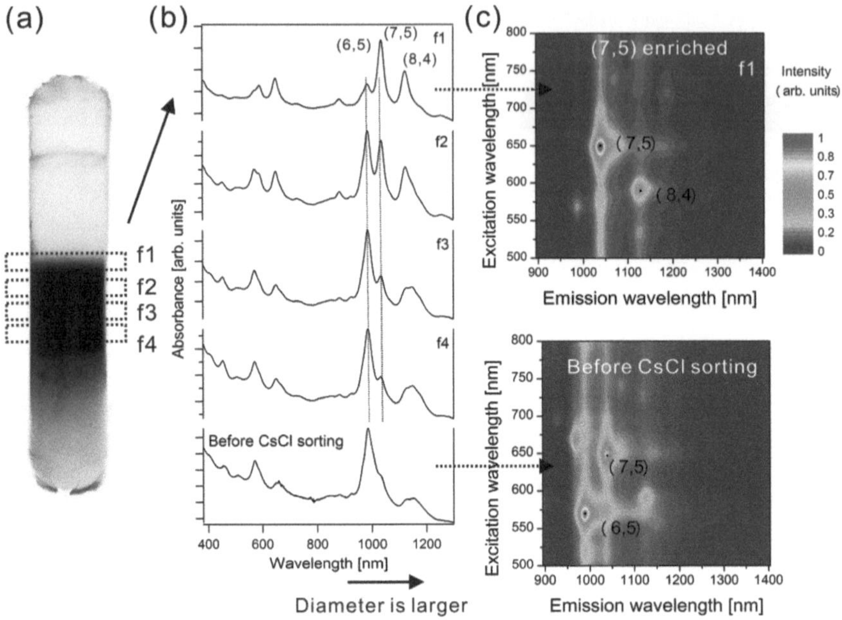

FIGURE 3.6 Typical density-gradient sorting with a monosurfactant system in which CsCl is used as the gradient medium. The results of sorting of CoMoCAT (Aldrich) SWCNTs are shown. (a) Picture of a centrifugation tube after ultracentrifugation; (b) the optical absorption spectrum of each fraction; (c) the photoluminescence spectra before and after sorting. Large-diameter SWCNTs, such as (7, 5), were located in the f1 fraction, which was the lower-density fraction. (Adapted with permission from [36]. Copyright 2012 American Chemical Society).

In the case of SDS, under the assumption of a cylindrical micelle model, the density of a single rope of SWCNT is approximately 1.0 g cm^{-3}; however, that of a bundled of SWCNTs is approximately 1.2 g cm^{-3} [14]. Therefore, O'Connell et al. [14] separated a single rope of SWCNT from bundled SWCNTs through centrifugation in a D_2O SDS 1% solution with medium density (1.1 g cm^{-3}). Differentiation of the solvated densities of metallic and semiconducting SWCNTs is possible using SDS-co-surfactant systems; however, the densities strongly depend on the concentration ratio of SDS to the other surfactant [25,31]. Therefore, in the differentiation of SWCNTs by DGU, two important parameters significantly influence the sorting results: (1) the types of surfactants used for dispersion and separation and (2) the type of the gradient medium and the temperature. Here, I discuss the details of the influences of these parameters.

3.2.2.1 *The types of surfactants used for dispersion and separation*

For the sorting of SWCNTs by their electronic properties, the SWCNTs must first be isolated from each other. If the SWCNTs are bundled, then the sorting of SWCNTs by their electronic properties cannot be achieved. Therefore, the efficient procurement of monodispersed SWCNTs is a critical step for efficient sorting. Wenseleers et al. have reported that bile salts, such as SC and DOC, are the most efficient surfactants for the dispersion of SWCNTs [64]. However, the sorting of MS SWCNTs is difficult (i.e. not efficient) when only DOC or SC is used. When a bile salt is used for density-gradient sorting, SWCNTs can be sorted according to their diameters. Large-diameter SWCNTs are located in a larger-density region than smaller SWCNTs, as previously discussed [4]. However, when SDS is added as a co-surfactant, the results drastically change. In their initial paper, Arnold et al. [4], dispersed SWCNTs in a 2% w/v SC surfactant solution and then diluted this solution into 2% w/v co-surfactant solutions, such as 1:4 SDS/SC and 3:2 SDS/SC. As a result, a clear difference between metallic and semiconducting SWCNTs was observed. Therefore, SDS is a key surfactant to differentiate the densities of SWCNTs with different electronic properties. For example, in the previously discussed typical sorting procedure, DOC is used to efficiently obtain a monodispersed SWCNT solution, and SDS is used to differentiate the densities of SWCNTs with different electronic types. During centrifugation, the SWCNTs move to the region of SDS gradient solution, and then differentiation of densities occurs; as a result, the densities of metallic and semiconducting SWCNTs become different.

3.2.2.2 The type of gradient medium and the temperature

Usually, iodixanol is used as gradient medium for the separation of SWCNTs. However, other gradient media, such as sucrose [51] and cesium chloride [36], are also available for sorting. Sucrose is the most popular and inexpensive gradient medium for DGU in biological sciences. In the case of iodixanol, the difference in densities is the primary mechanism; however, in the case of a sucrose gradient, the difference in the diffusion velocities of particles becomes the main factor for sorting. Thus, the sorting mechanisms are different; however, if the ratio of SDS and SC and the temperature are well adjusted, then sorting of metallic and semiconducting types of SWCNTs is also achieved in the sucrose gradient [51]. Temperature was observed to significantly influence the sorting capability.

Cesium chloride can also be used as a density medium. Inorganic salts are popular density-gradient media. However, the density of solvated SWCNTs in cesium chloride solutions was found to be approximately 1.4 g cm^{-3} [36]. Therefore, the concentration of CsCl should be adjusted to be approximately 42.5% (w/w) [36]. However, at such high salt concentration, SDS cannot be used because SDS cannot be dissolved in such solutions. Typically, in CsCl sorting, the larger-diameter SWCNTs are located in a region of greater density than are smaller-diameter SWCNT. However, the single-chiral state of (11, 10) SWCNT was unexpectedly obtained as shown in Figure 3.7 [36]. The physical mechanism is still under discussion. However, CsCl can clearly also be used to sort SWCNTs.

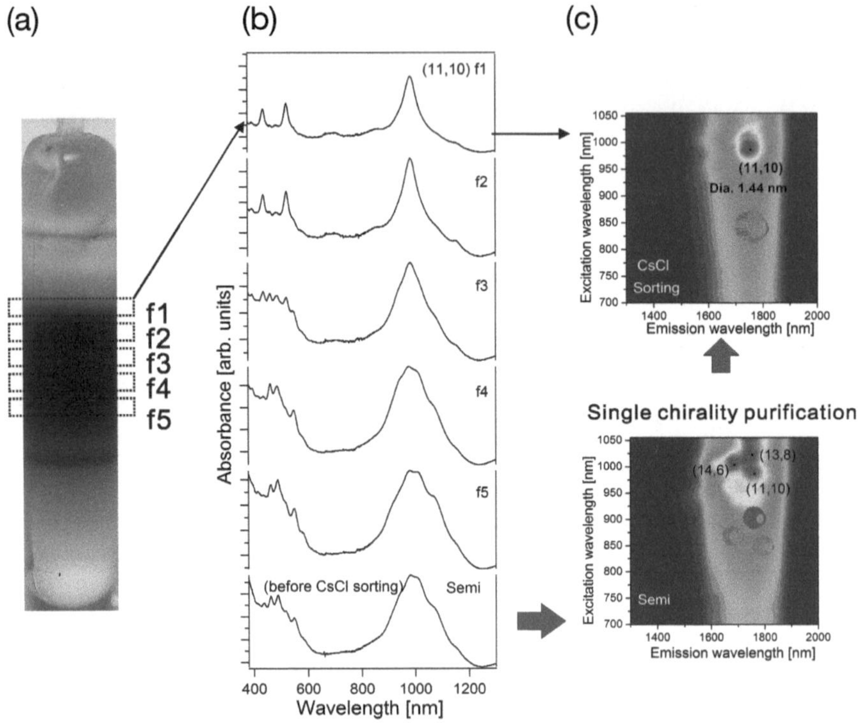

FIGURE 3.7 (11, 10) Single-chirality sorting using CsCl as a gradient medium. Picture of a centrifugation tube after ultracentrifugation (a); the optical absorption spectrum of each fraction (b) and photoluminescence spectra before and after sorting (c). In this case, (11, 10) is the middle diameter, 1.44 nm, compared to the other SWCNTs ((14, 6), 1.41 nm and (13, 8), 1.46 nm); however, (11, 10) was extracted at the top fraction, which means it exhibited the lowest density. (Adapted with permission from [36]. Copyright 2012 American Chemical Society).

3.3 DIFFERENTIATION OF SWCNTS WITH DIFFERENT CHIRALITIES THROUGH SIZE EXCLUSION CHROMATOGRAPHY OR GEL FILTRATION

Chromatography and gel filtration are very common techniques for the purification of nanomaterials. Soon after the discovery of SWCNTs [65], size separation of SWCNTs was performed through size exclusion chromatography using controlled-pore glass (with an average pore size of 300 nm, CPG 3000A, Fluka) in SDS 1 wt% solutions in 1998 by Duesberg et al. [13]. Since then, the size separation of SWCNTs using size exclusion chromatography has been reported by several groups [66]. In 2004, Heller et al. successfully sorted SWCNTs by length and diameter using surfactants [67]. They used Sephacryl S-500 gel (GE Healthcare) as a filler and dispersed SWCNTs into SC solutions [67]. Notably, in their paper, the enrichment of metallic and semiconducting types was observable in their optical absorption spectra, although they did not discuss this enrichment

[67]. In their paper, they also presented results for the electrophoresis of SC dispersed SWCNTs; however, they could not identify the MS sorting phenomenon. With respect to the sorting of SWCNTs by electrophoresis, Wakizaka et al. first reported the success of MS sorting of SWCNTs through electrophoresis using SDS as surfactants in 2006 [68], Tanaka et al. later reported MS sorting through the use of agarose gel and SDS [69], and more recently, Ihara et al. reported successful MS sorting using polyoxyethylene stearyl ether (Brij 700, Aldrich) [70]. Notably, Heller et al. also used electrophoresis with agarose gel in 2004 [67]; however, they did not achieve MS separation because they used SC instead of SDS as a surfactant. This fact indicates that SDS is a key surfactant for differentiating SWCNTs with different (*n*, *m*). In 2009, a significant advancement was reported by Moshammer et al. [29], who clearly succeeded in sorting metallic and semiconducting SWCNTs using Sephacryl S-200 gel filtration medium (GE Healthcare) with SDS. Soon after that, MS sorting using agarose gel was also reported [30,71]. In the study of Moshammer et al., SWCNTs were dispersed in an SDS 1 wt% solution. Metallic SWCNTs were eluted with an SDS solution, and then an SC 1 wt% solution was used to extract semiconducting SWCNTs, which were more strongly adsorbed to the gel. The authors suggested that metallic tubes have a greater affinity towards SDS than do semiconducting SWCNTs, and as a result, metallic SWCNTs have weak adsorption to the gel. In 2011, Liu et al. successfully extracted various (*n*, *m*) single-chiral states of SWCNTs using a similar Sephacryl S-200 gel and SDS solutions [35]. In their study, SWCNTs were dispersed into a 2 wt% SDS solution, the dispersed solution was subjected to gel filtration, and strongly adsorbed SWCNTs on the gel were collected with a 5 wt% SDS solution. Through repetitive adsorption and dissipation processes 13 different (*n*, *m*) species were separated in high yield [35].

The previous discussion provides a brief history of the separation of SWCNTs by gel chromatography/filtration in dispersion surfactants, such as SDS, SC and DOC. However, in the case of DNA dispersion, the differentiation of SWCNTs with different electronic properties was achieved more rapidly than in the case of surfactant dispersion. In 2003, single-stranded DNA (ssDNA) was used to disperse SWCNTs, and the SWCNTs wrapped by ssDNA were sorted into fractions with different electronic properties by anion-exchange column chromatography [15]. Thus, through the use of ssDNA, metallic and semiconducting types of SWCNTs were first sorted using a chromatography method. The underlying mechanism in this sorting process is the differential binding of DNA to SWCNTs with different diameters and chiralities. As this technique has advanced, single-chirality SWCNTs have been sorted through adjustments to the appropriate DNA sequence used for the separation [17,18]. Twelve different (*n*, *m*) chiral species have been obtained [18,46]. Therefore, DNA is a very important dispersant for the differentiation of SWCNTs with different electronic types; however, surfactant dispersion is a more accessible, simple and low-cost approach for the purification of SWCNTs. Thus, here I discuss how to differentiate SWCNTs through chromatography/gel filtration processes with surfactant dispersion.

3.3.1 Typical procedure to extract single-chiral state (6, 5) SWCNTs by the gel filtration method

The typical process to obtain single-chiral state SWCNTs can be described as follows (Figure 3.8). First, SWCNTs (CoMoCAT, Aldrich Co.) are dispersed in SDS 2 wt% solution, and then the SWCNTs are dispersed using a bath-type ultrasonic generator and tip-type homogenizer (17 W, 1 hour, Branson Co.) for approximately 1 hour. The duration of ultrasonication strongly influences the sorting results. The SWCNT dispersed solution is then centrifuged for 2 hours (284,000*g*, P40ST rotor, Hitachi Koki Co.), and the supernatant is collected. A gel (Sephacryl S-200) is packed in a column tube (to a height of approximately 3 cm). The gel solution is first exchanged into SDS 2% because the gel is stored in ethanol 20% solutions. Then, the supernatant solution is applied to the top of the column. Afterwards, a solution of 2 wt% SDS solution is poured onto the column with sufficient pressure applied with compressed air to increase the flow. When the SDS 2 wt% solution is applied, most of the unbounded SWCNTs are collected. The SDS 2.5 wt% solution is then poured onto the column, followed by the 5 wt% solution, which is used to collect the most strongly adsorbed SWCNTs. As a result of this process, a (6, 5) highly enriched

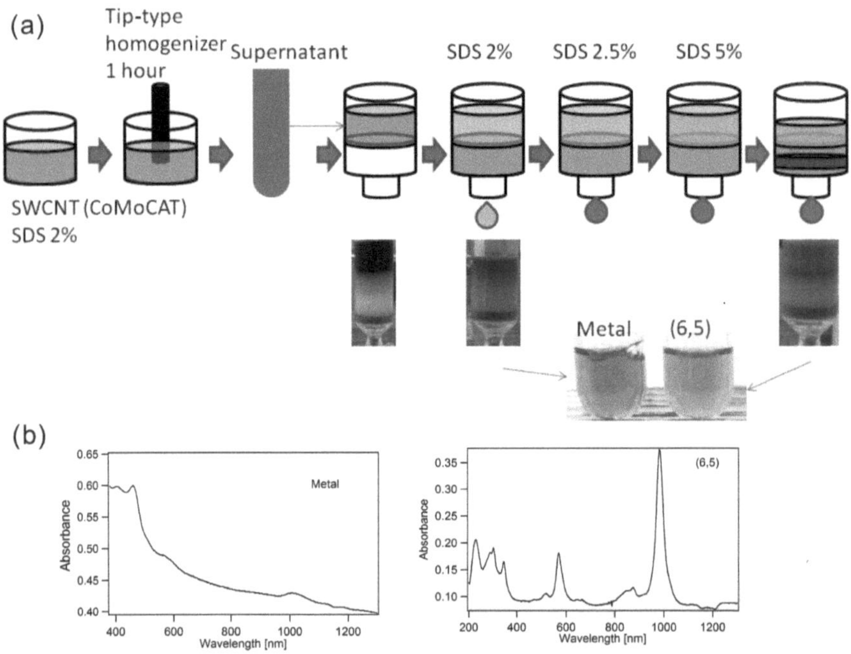

FIGURE 3.8 (a) Schematic procedure of (6, 5) single-chirality sorting using the gel filtration method. (b) Optical absorption spectra of metallic (left) and (6, 5) SWCNTs (right), which were obtained through gel filtration processes. In the case of metallic SWCNTs, DGU was also used to further improve the purity.

SWCNT solution is collected (Figure 3.8). By repeating this procedure, high-purity (6, 5) SWCNTs can be obtained.

3.3.2 Sorting mechanisms in gel chromatography/filtration

A typical gel chromatography process sorts particles according to differences in their size (size exclusion chromatography/filtration). However, the process of sorting SWCNTs with different electronic properties through a gel is based on differences in the adsorption characteristics of the different SWCNTs on the gel [29,35,72]. SWCNTs are wrapped by surfactants; thus, similar to the case of DGU, the arrangements of the surfactants on the surfaces of SWCNTs and the adsorption characteristics of the surfactant–SWCNT complexes on the gel significantly influence the sorting capability. Here, I discuss the two important parameters that affect the sorting results: (1) the type and concentration of surfactant and (2) the type and temperature of fillers.

3.3.2.1 Type and concentration of surfactant

As the previously discussed history clearly indicates, SDS plays a key role in differentiating SWCNTs with different electronic properties on a gel. When SWCNTs were dispersed in SC solutions, and the solutions were applied to Sephacryl gel, no remarkable selectivity between metallic and semiconducting SWCNTs was observed [67]. In this case, the SWCNTs were sorted based solely on differences in their length, which indicates a size exclusion process. However, when the SWCNTs were dispersed in SDS solutions and the solutions were subsequently passed through Sephacryl gel, clear separation of metallic and semiconducting SWCNTs was observed [29]. Therefore, SDS plays a key role in sorting SWCNTs according to their electronic properties via gel filtration. Typically, metallic-type SWCNTs that are wrapped by SDS exhibit weaker interactions with the gel than do semiconducting-type SWCNTs. As a result, SDS-wrapped metallic SWCNTs are eluted first with a low-concentration SDS solution (approximately 1–2 wt%). In contrast, semiconducting SWCNTs tend to bind more strongly to the gel; as a result, the semiconducting SWCNTs tend to remain in the gel. To collect the bound semiconducting SWCNTs, SC, DOC and high-concentration SDS solutions are used. Liu et al. reported that the concentration of the SDS solution used for dispersion strongly influences the binding characteristics between SDS-wrapped SWCNTs and the gel. SWCNTs with different electronic properties exhibit slight differences in their binding properties [35]. Therefore, by adjusting the concentration of SDS, Liu et al. obtained various kinds of SWCNTs with different electronic properties [35].

3.3.2.2 Type of filler and temperature

Single-chiral state SWCNTs were obtained through the use of Sephacryl gel, but metallic and semiconducting SWCNTs can also be sorted by other gels, such as agarose [69]. However, Sephacryl gel allows high-purity SWCNTs to be obtained efficiently. The differentiation of the binding properties between surfactant-wrapped SWCNTs and the filler is key to the sorting of SWCNTs using chromatography and gels. The adsorption characteristics were quantitatively evaluated on the basis of the amount of SWCNTs adsorbed [73], and slight differences between the adsorption free energies of the metallic and semiconducting types of SWCNTs (approximately 4 kJ mol^{-1} for agarose and 9 kJ mol^{-1} for Sephacryl gel) were reported [73]. These results also suggest that Sephacryl gel is better for sorting than agarose gel. Glass with a controlled pore diameter [13] or other fillers [66] can also be used to sort SWCNTs according to their length.

Temperature significantly influences the morphology of surfactant-wrapped SWCNTs, and, as a result, sometimes plays a key role in their sorting. In the case of sucrose DGU, at room temperature, no separation of metallic and semiconducting types of SWCNTs was observed. However, when the temperature was decreased, metallic and semiconducting types of SWCNTs were well sorted [51]. In the case of gel filtration, when the temperature is varied, the sorting results drastically change. As the temperature was increased, most of the SWCNTs were adsorbed on the gel (Sephacryl S-200), and single-chiral SWCNTs – (6, 4), (6, 5), (7, 5), (8, 3), (8, 4), (7, 6) and (8, 6) – were easily extracted [72]. Therefore, temperature is a very important parameter for successful sorting.

3.4 SUMMARY

As discussed above, the key for sorting in the DGU and gel chromatography/filtration processes is the differentiation of the morphology of surfactant-wrapped SWCNTs with different (n, m) values. In the case of DGU, such differences in morphology influence the solvated densities of SWCNTs; as a result, SWCNTs with different (n, m) values exhibit different densities, and they are sorted under strong centrifugation forces. In the case of chromatography/gel filtration processes, such differences in morphology differentiate the adsorption characteristics between SWCNTs with different (n, m) values and the fillers; as a result, differentiation according to (n, m) values becomes possible when the solutions are passed through adsorptive fillers. Various experimental results suggest that SDS is the most promising surfactant for achieving these differentiations. The use of SDS or an SDS co-surfactant system, such as SC or DOC, enables the modification of the adsorption characteristics of surfactants on SWCNTs; as a result, the sorting processes of DGU and gel chromatography/filtration can be tuned. The temperature and the concentration ratio of co-surfactants are important parameters for achieving good purification results. Moreover, the density-gradient media and fillers for chromatography/filtration must be carefully selected. Substantial experimental knowledge and experiences have been accumulated; however, we still lack precise knowledge concerning the arrangements of surfactants on the surface of SWCNTs. A theoretical understanding of the differentiation of arrangements of surfactants on the surface of SWCNTs is insufficient. In addition, the preparation conditions and properties of commercially available as-grown samples, such as the production techniques (arc discharge, laser vapourization or CVD), the amount of defects and the pre-purification procedures, often influence the purification results. Therefore, researchers rely on trial and error procedures for the purification of SWCNT samples based on their accumulated experimental knowledge. However, as detailed in this chapter, various techniques for the preparation of high-purity (*n*, *m*) SWCNTs are now available, and advances in purification methods continue to be made. I hope that this chapter contributes to the understanding of purification mechanisms and to the future investigations of readers.

REFERENCES

[1] Saito R, Dresselhaus G, Dresselhaus MS. Physical properties of carbon nanotubes. London: Imperial College Press; 1998.

[2] Saito R, Fujita M, Dresselhaus G, Dresselhaus MS. Electronic structure of chiral graphene tubules. Appl Phys Lett 1992;60:2204–6.

[3] Izard N, Kazaoui S, Hata K, Okazaki T, Saito T, Iijima S, et al. Semiconductor-enriched single wall carbon nanotube networks applied to field effect transistors. Appl Phys Lett 2008;92(243112):1–3.

[4] Arnold M, Green A, Hulvat J, Stupp S, Hersam M. Sorting carbon nanotubes by electronic structure using density differentiation. Nat Nanotechnol 2006;1:60–5.

[5] Fuji S, Tanaka T, Miyata Y, Suga H, Naitoh Y, Minari T, et al. Performance enhancement of thin-film transistors by using high-purity semiconducting single-wall carbon nanotubes. Appl Phys Express 2009;2(071601):1–3.

[6] Blackburn J, Barnes T, Beard M, Kim Y, Tenent R, McDonald T, et al. Transparent conductive single-walled carbon nanotube networks with precisely tunable ratios of semiconducting and metallic nanotubes. ACS Nano 2008;2:1266–74.

[7] Yanagi K, Udoguchi H, Sagitani S, Oshima Y, Takenobu T, Kataura H, et al. Transport mechanisms in metallic and semiconducting single-wall carbon nanotube networks. ACS Nano 2010;4:4027–32.

[8] Miyata Y, Yanagi K, Maniwa Y, Kataura H. Highly stabilized conductivity of metallic single-wall carbon nanotube thin films. J Phys Chem C 2008;112:3591–6.

[9] Iijima S, Ichihashi T. Single-shell carbon nanotubes of 1-nm diameter. Nature 2003;363:603–5.

[10] Hersam M. Progress towards monodisperse single-walled carbon nanotubes. Nat Nanotechnol 2008;3:387–94.

[11] Wang Y, Shan H, Hauge R, Pasquali M, Smalley R. A highly selective, one-pot purification method for single-walled carbon nanotubes. J Phys Chem B 2007;111:1249–52.

[12] Yang C, Kaneko K, Yudasaka M, Iijima S. Effect of purification on pore structure of HiPco single-walled carbon nanotube aggregates. Nano Lett 2002;2:385–8.

[13] Duesberg GS, Muster J, Krstic V, Burghard M, Roth S. Chromatographic size separation of single wall carbon nanotubes. Appl Phys A 1998;67:117–9.

[14] O'Connell MJ, Bachilo S, Huffman C, Moore V, Strano M, Haroz E, et al. Band gap fluorescence from individual single-walled carbon nanotubes. Science 2002;297:593–6.

[15] Zheng M, Jagota A, Semke E, Diner B, Mclean R, Lustig S, et al. DNA assisted dispersion and separation of carbon nanotubes. Nat Mater 2003;2:338–42.

[16] Krupke R, Hennrich F, Lohneysen H, Kappes M. Separation of metallic from semiconducting single-walled carbon nanotubes. Science 2003;301:344–7.

[17] Zheng M, Semke E. Enrichment of single chirality carbon nanotubes. J Am Chem Soc 2007;129:6084–5.

[18] Tu X, Manohar S, Jagota A, Zheng M. DNA sequence motifs for structure-specific recognition and separation of carbon nanotubes. Nature 2009;460:250–3.

[19] Shin D, Kim J, Shim H, Song J, Yoon J, Kim J, et al. Continuous extraction of highly pure metallic single-walled carbon nanotubes in a microfluidic channel. Nano Lett 2008;8:4380–5.

[20] Chattopadhyay D, Galeska I, Papadimitrakopoulos F. A route for bulk separation of semiconducting from metallic single-wall carbon nanotubes. J Am Chem Soc 2003;125:3370–5.

[21] Maeda Y, Kimura S, Kanda M, Hirashima Y, Hasegawa T, Wakahara T, et al. Large-scale separation of metallic and semiconducting single-walled carbon nanotubes. J Am Chem Soc 2005;127:10287–90.

[22] An K, Park J, Yang C, Jeong S, Lim S, Kang C, et al. A diameter-selective attack of metallic carbon nanotubes by nitronium ions. J Am Chem Soc 2005;127:5196–203.

[23] Maeda Y, Kanda M, Hashimoto M, Hasegawa T, Kimura S, Lian Y, et al. Dispersion and separation of small-diameter single-walled carbon nanotubes. J Am Chem Soc 2006;128: 12239–42.

[24] Fagan JA, Becker ML, Chun J, Hobbie E. Length fractionation of carbon nanotubes using centrifugation. Adv Mater 2007;20:1609–13.

[25] Yanagi K, Miyata Y, Kataura H. Optical and conductive characteristics of metallic single-wall carbon nanotubes with three basic colors; cyan, magenta, and yellow. Appl Phys Express 2008;1(034003):1–3.

[26] Wei L, Wang B, Goh T, Li L, Yang Y, Chan-Park M, et al. Selective enrichment of (6,5) and (8,3) single-walled carbon nanotubes via cosurfactant extraction from narrow (n, m) distribution samples. J Phys Chem B 2008;112:2771–4.

[27] Nish A, Hwang J, Doig J, Nicholas R. Highly selective dispersion of single-walled carbon nanotubes using aromatic polymers. Nat Nanotechnol 2007;2:641–6.

[28] Ju S, Doll J, Sharma I, Papadimitrakopoulos F. Selection of carbon nanotube with specific chiralities using helical assemblies of flavin mononucleotide. Nat Nanotechnol 2008;3:356–62.

[29] Moshammer K, Hennrich F, Kappes M. Selective suspension in aqueous sodium dodecyl sulfate according to electronic structure type allows simple separation of metallic from semiconducting single-wall carbon nanotubes. Nano Res 2009;2:599–606.

[30] Tanaka T, Urabe Y, Nishide D, Kataura H. Continuous separation of metallic and semiconducting carbon nanotubes using agarose gel. Appl Phys Express 2009;2(125002):1–3.

[31] Arnold M, Stupp S, Hersam M. Enrichment of single-walled carbon nanotubes by diameter in density gradients. Nano Lett 2005;5:713–8.

[32] Ghosh S, Bachilo S, Weisman R. Advanced sorting of single-walled carbon nanotubes by nonlinear density-gradient ultracentrifugation. Nat Nanotechnol 2010;5:443–50.

[33] Green A, Duch M, Hersam M. Isolation of single-walled carbon nanotube enantiomers by density differentiation. Nano Res 2009;2:69–77.

[34] Peng X, Komatsu N, Bhattacharya S, Shimawaki T, Aonuma S, Kimura T, et al. Optically active single-walled carbon nanotubes. Nat Nanotechnol 2007;2:361–5.

[35] Liu H, Nishide D, Tanaka T, Kataura H. Large-scale single-chirality separation of single-wall carbon nanotubes by simple gel chromatography. Nat Commun 2011;2(309):1–8.

[36] Kawai M, Kyakuno H, Suzuki T, Igarashi T, Suzuki H, Okazaki T, et al. Single chirality extraction of single-wall carbon nanotubes for the encapsulation of organic molecules. J Am Chem Soc 2012;134:9545–8.

[37] LeMieux M, Roberts M, Barman S, Jin Y, Kim J, Bao Z. Self-sorted, aligned nanotube networks for thin-film transistors. Science 2008;321:101–4.

[38] Collins P, Arnold M, Avouris P. Engineering carbon nanotubes and nanotube circuits using electrical breakdown. Science 2001;292:706–9.

[39] Jin S, Dunham S, Song J, Xie X, Kim J, Lu C, et al. Using nanoscale thermocapillary flow to create arrays of pure semiconducting single-walled carbon nanotubes. Nat Nanotechnol 2013;8:335–47.

[40] Ding L, Tselev A, Wang J, Yuan D, Chu H, McNicholas T, et al. Selective growth of well aligned semiconducting single-walled carbon nanotubes. Nano Lett 2009;9:800–5.

[41] Zhang G, Qi P, Wang X, Lu Y, Li X, Tu R, et al. Selective etching of metallic carbon nanotubes by gas-phase reaction. Science 2006;314:974–7.

[42] Sato Y, Yanagi K, Miyata Y, Suenaga K, Kataura H, Iijima S. Chiral angle distribution for separated single-walled carbon nanotubes. Nano Lett 2008;8:3151–4.

[43] Haroz E, Rice W, Lu B, Ghosh S, Hauge R, Weisman R, et al. Enrichment of armchair carbon nanotubes via density-gradient ultracentrifugation: Raman spectroscopy evidence. ACS Nano 2010;4:1955–62.

[44] Blum C, Sturzl N, Hennrich F, Lebedkin S, Heeg S, Dumlich H, et al. Selective bundling of zigzag single-walled carbon nanotubes. ACS Nano 2011;5:2847–54.

[45] Green A, Hersam M. Colored semitransparent conductive coatings consisting of monodisperse metallic single-walled carbon nanotubes. Nano Lett 2008;8:1417–22.

[46] Zhang L, Tu X, Welsher K, Wang X, Zheng M, Dai H. Optical characterization and electronic devices of nearly pure (10,5) single-walled carbon nanotubes. J Am Chem Soc 2009;131: 2454–5.

[47] Richwood D, Hames BD. Preparative centrifugation: a practical approach. Oxford: Oxford University Press; 1992.

[48] Yanagi K, Miyata Y, Liu Z, Suenaga K, Okada S, Kataura H. Influence of aromatic environment on the physical properties of beta-carotene. J Phys Chem C 2010;114:2524–30.

[49] Niyogi S, Densmore C, Doon S. Electrolyte tuning of surfactant interfacial behavior for enhanced density-based separations of single-walled carbon nanotubes. J Am Chem Soc 2008;131:1144–53.

[50] Green A, Hersam M. Processing and properties of highly enriched double-wall carbon nanotubes. Nat Nanotechnol 2008;4:64–70.

[51] Yanagi K, Iitsuka T, Fujii S, Kataura H. Separation of metallic and semiconducting carbon nanotubes by using sucrose as a gradient medium. J Phys Chem C 2008;112:18889–188894.

[52] Miyata Y, Suzuki M, Fujihara M, Asada Y, Kitaura R, Shinohara H. Solution-phase extraction of ultrathin inner shells from double-wall carbon nanotubes. ACS Nano 2010;9:5807–12.

[53] Crochet J, Clemens M, Hertel T. Quantum yield heterogeneities of aqueous single wall carbon nanotube suspensions. J Am Chem Soc 2007;129:8058–9.

[54] Fagan J, Huh J, Simpson J, Blackburn J, Holt J, Larsen B, et al. Separation of empty and water-filled single-wall carbon nanotubes. ACS Nano 2011;5:3943–53.

[55] Cambré S, Wenseleers W. Separation and diameter-sorting of empty (end-capped) and water-filled (open) carbon nanotubes by density gradient ultracentrifugation. Angew Chem Int Ed 2011;50:2764–8.

[56] Rickwood D. Preparative centrifugation. Oxford: Oxford University Press; 1991.

[57] Islam M, Rojas E, Bergey D, Jojnson A, Yodh A. High weight fraction surfactant solubilization of single-wall carbon nanotubes in water. Nano Lett 2003;3:269–73.

[58] Arnold M, Suntivich J, Stupp S, Hersam M. Hydrocynamic characterization of surfactant encapsulated carbon nanotubes using an analytical ultracentrifuge. ACS Nano 2008;2:2291–300.

[59] Nair N, Kim W, Braatz R, Strano M. Dynamics of surfactant-suspended single-walled carbon nanotubes in a centrifugal field. Langmuir 2008;24:1790–5.

[60] Yurekli K, Mitchell C, Krishnamoorti R. Small-angle neutron scattering from surfactant-assisted aqueous dispersions of carbon nanotubes. J Am Chem Soc 2004;126:9902–3.
[61] Richard C, Balavoine F, Schultz P, Ebbesen TW, Mioskowski C. Supramolecular self-assembly lipid derivatives on carbon nanotubes. Science 2003;300:775–8.
[62] Tummala N, Striolo A. SDS surfactants on carbon nanotubes: aggregate morphology. ACS Nano 2009;3:595–602.
[63] Fagan JA, Zheng M, Rastogi V, Simpson J, Khripin C, Batista C, et al. Analyzing surfactant structures on length and chirality resolved (6,5) single-wall carbon nanotubes by analytical ultracentrifugation. ACS Nano 2013;7:3373–87.
[64] Wenseleers W, Vlasov I, Goovaerts E, Obraztsova E, Lobach A, Bouwen A. Efficient isolation and solubilization of pristine single-walled nanotubes in bile salt micelles. Adv Func Mater 2004;14:1105–12.
[65] Iijima S, Ichihashi T. Single-shell carbon nanotubes of 1 nm diameter. Nature 1993;363:603–5.
[66] Farkas E, Elizabeth Anderson M, Chen Z, Rinzler AG. Length sorting cut single wall carbon nanotubes by high performance liquid chromatography. Chem Phys Lett 2002;363:111–6.
[67] Heller D, Mayrhofer R, Baik S, Grinkova Y, Usrey M, Strano M. Concomitant length and diameter separation of single wall carbon nanotubes. J Am Chem Soc 2004;126:14567–73.
[68] Wakizaka Y, Nakayama K, Yokoyama M, Kanematsu Y, Sakurai Y. Separation method of single-wall carbon nanotube. Japan Patent 2008-055375. 2006. Application number: 2006-237840.
[69] Tanaka T, Jin H, Miyata Y, Kataura H. High-yield separation of metallic and semiconducting single-wall carbon nanotubes by agarose gel electrophoresis. Appl Phys Express 2008; 1(114001):1–3.
[70] Ihara K, Endoh H, Saito T, Nihey F. Separation of metallic and semiconducting single-wall carbon nanotube solution by vertical electric field. J Phys Chem C 2011;115:22827–32.
[71] Tanaka T, Jin H, Miyata Y, Fujii S, Suga H, Naitoh Y, et al. Simple and scalable gel-based separation of metallic and semiconducting carbon nanotubes. Nano Lett 2009;9:1497–500.
[72] Liu H, Tanaka T, Urabe Y, Kataura H. High-efficiency single-chirality separation of carbon nanotubes using temperature-controlled gel chromatography. Nano Lett 2013;13:1996–2003.
[73] Hirano A, Tanaka T, Kataura H. Thermodynamic determination of the metal/semiconductor separation of carbon nanotubes using hydrogels. ACS Nano 2012;6:10195–205.

Chapter 4

Preparation of Graphene with Large Area

Yong Cheol Shin*, Mildred S. Dresselhaus†, Jing Kong****

**Department of Materials Science and Engineering, Massachusetts Institute of Technology, 77 Massachusetts Avenue, Cambridge, MA 02139, USA*

†*Department of Physics, Massachusetts Institute of Technology, 77 Massachusetts Avenue, Cambridge, MA 02139, USA*

***Department of Electrical Engineering and Computer Science, Massachusetts Institute of Technology, 77 Massachusetts Avenue, Cambridge, MA 02139, USA*

4.1 INTRODUCTION

In this chapter, the synthesis of graphene by chemical vapour deposition (CVD) process will be discussed. CVD has been widely used in the microelectronics industry as a thin-film deposition method because this process can take advantage of low production cost, good film quality, scalability and good step coverage on a substrate with a complex structure. The deposition is achieved through the adsorption of vapour-phase precursor on a substrate, the pyrolysis of this precursor and a phase transformation to a condensed phase through nucleation and growth.

The synthesis of carbon materials has also taken advantage of the CVD process. In the 1960s, the graphitic thin film was already prepared on a transition metal substrate through the pyrolysis of carbon precursors, such as methane, ethylene and acetylene [1,2]. The metal substrate played a catalytic role to promote the decomposition of hydrocarbon sources. It may not be surprising that current CVD systems to synthesize graphene are not significantly different from this basic graphitic thin-film deposition process. As well as the synthesis of sp^3-carbon and diamond, sp^2-carbon nanomaterials synthesis has been shown to work well with transition metal substrates by CVD growth. For example, carbon nanotubes (CNTs) are mostly synthesized by CVD using nickel (Ni), cobalt (Co) or iron (Fe) nanoparticles [3,4]. By adjusting the process parameters, such as the size of the metal nanoparticles, CNTs with controlled height, diameter and number of walls can be prepared for large area growth. Based on these historical approaches, it can be understood why the synthesis of a new form

Carbon Nanotubes and Graphene. DOI: 10.1016/B978-0-08-098232-8.00004-8

of sp^2-carbon nanomaterial, graphene, was also attempted, based on the CVD process using transition metal substrates, although the synthesis of single-layer graphite had already been realised earlier in the 1960s [5]. In terms of the process development of graphene synthesis, the predominant parameter studied in early works was the substrate material because it turned out that the growth mechanism of graphene appeared to be strongly dependent on the type of substrate material that was exploited. With respect to the relative carbon solubility in the substrate at a typical growth temperature as high as 1000 °C, where hydrocarbon precursor can be decomposed, graphene could be formed, either as segregated carbon atoms that diffused out from the substrate bulk upon cooling or as the deposition of carbon species on the surface of the substrate. Particularly, the discovery of copper (Cu) with a negligible carbon solubility as a substrate material for graphene growth has been remarkable. Although monolayer to few-layer graphene (FLG) could be obtained by growth on a polycrystalline Ni thin film by an atmospheric pressure CVD (APCVD) process [6,7], Li *et al.* developed the process of single-layer graphene (SLG) growth over a bulk Cu foil by low-pressure CVD (LPCVD) [8]. As the process becomes self-limiting under low pressure, uniform SLG layer could be synthesized on the substrate over a large area. This opened up a route for growth of graphene that can be potentially applicable on an industrial level, beyond just a theoretical interest. After this early demonstration, significant efforts have been made toward the synthesis of graphene with improved quality and optimization of the transfer process.

In this chapter, such efforts will be described in some detail, mainly covering the Cu foil-mediated CVD process because this is the most widely utilized process to prepare SLG in large area at the present time. First, the general growth mechanism and its interactions with various CVD process parameters and substrate materials will be presented to help readers understand the behaviours that are happening during the graphene growth (Sec. 4.2). Subsequently, recently pending topics in the graphene growth process, the achievement of large domain size of graphene (Sec. 4.3) and bilayer graphene (BLG) growth (Sec. 4.4) will be discussed. Finally, the transfer process, becoming more important than before in the preparation of graphene in large area, will be described (Sec. 4.5). In summary, this chapter describes the efforts to prepare graphene with a comparable quality to mechanically exfoliated graphene, minimizing intrinsic defects and to transfer graphene onto a suitable substrate without the deterioration of quality and minimizing extrinsic defects.

4.2 GRAPHENE GROWTH ON METAL SUBSTRATES BY CVD

4.2.1 Process parameters

Graphene is a unique material by itself as a form of a monatomic layer. To understand how graphene forms, however, its synthesis process by CVD can be regarded as a deposition process of a *very* thin film through a simple pyrolysis of a hydrocarbon gas. This may allow us to model the graphene growth by CVD by adapting to the growth mechanism used in the thin-film deposition. In this manner, it can be understood how the process parameters affect the morphology of the resulting graphene. This helps in preparing graphene in large area with material properties closer to the theoretical values and those achieved in mechanically exfoliated graphene.

As can be seen in Figure 4.1, the graphene growth process by CVD typically consists of four steps: heating, annealing, growth and cooling steps. For SLG growth, a bulk Cu foil is commonly used as a catalyst substrate. First, in the heating step, the substrate temperature is ramped up to 1000 °C to facilitate the dissociation of the hydrocarbon precursor, such as methane (CH_4). In the subsequent annealing step, the substrate removes a native oxide on the surface and develops larger grain sizes and sometimes smoother surfaces at the elevated temperature. Hydrocarbon gas is introduced into the chamber in the growth step, giving rise to the formation of graphene on the substrate surface. In particular, a uniform SLG layer can be obtained over a Cu surface, based on a self-limiting process. Finally, the supply of hydrocarbon gas is stopped and the chamber is cooled down to room temperature. For other catalysts such as Ni that has a specific solubility of carbon at the growth temperature, the control of this cooling rate can influence the thickness of graphene [6]. By using a Cu foil as a substrate and methane as a precursor, the adjustable process parameters can be the growth temperature, the flow rate of gases, the partial pressure of the methane and the ambient pressure. Considering that the thin-film deposition by CVD is achieved by a phase transformation process, i.e. nucleation and growth; these process parameters can be associated with the process kinetics. When the

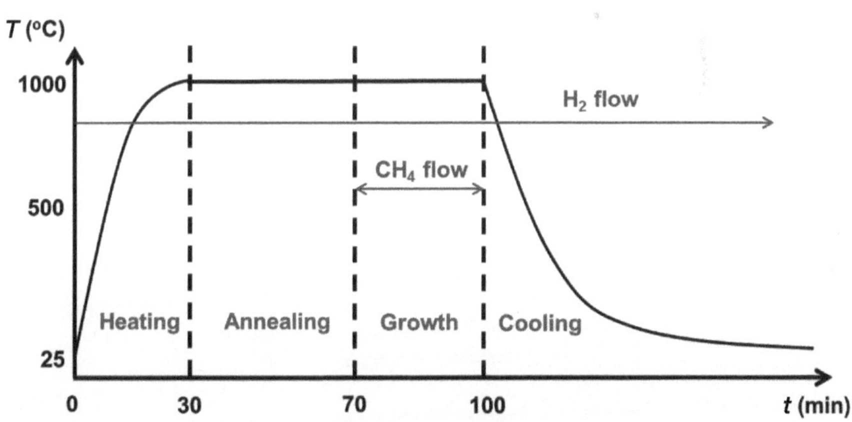

FIGURE 4.1 Typical schematic flow of the graphene growth process by CVD. The process consists of a heating, annealing, growth and cooling step. In the annealing step, the grain growth and the cleaning of the substrate are achieved. Hydrocarbon precursors are provided in the subsequent growth step.

graphene growth is assumed to be purely carried out by a surface reaction, ignoring the segregation of supersaturated carbon from the bulk substrate, either the mass-transport regime or the surface-reaction regime dominates the kinetics, depending on the competition of the corresponding fluxes.

4.2.1.1 The kinetics of the graphene growth process with respect to the process parameters

As illustrated in Figure 4.2 [9], the growth step involves the following six sequential processes: (1) mass transport of gas-phase hydrocarbon molecules from the bulk region to the surface region; (2) adsorption of these molecules on the substrate surface; (3) decomposition of the molecules into active hydrogen and carbon species; (4) surface diffusion of these active species, resulting in the formation of graphene islands; (5) desorption of active hydrogen species (H*) and hydrogen gas molecules (H_2) and (6) mass transport of the gas species in (5) from the surface region to the bulk region. These processes can be divided into two kinetic regimes: the mass transport regime that involves the diffusion of gas species through the boundary layer region ((1) and (6)); and the surface reaction regime ((2)–(5)). Because these regimes have different reaction characteristics, if first-order reaction is simply assumed, the corresponding fluxes are given by,

$$F_{mass\ transport} = h_g(C_g - C_s) \tag{4.1}$$

$$F_{surface\ reaction} = K_s C_s \tag{4.2}$$

where h_g is the mass transport coefficient, C_g and C_s the concentrations of hydrocarbon (methane in this model) in the bulk region and in the surface region, respectively and K_s is the surface reaction constant. One can consider

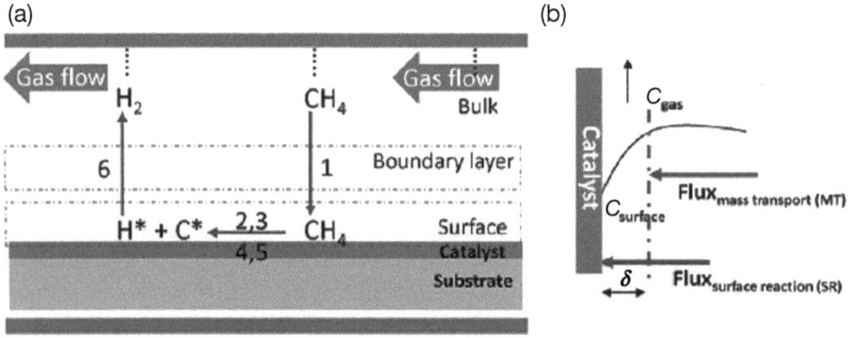

FIGURE 4.2 (a) Steps involved in a graphene growth on a substrate (Cu) where the segregation of the supersaturated carbon atoms from the bulk region is not assumed. (b) Steady-state mass transport and surface reaction fluxes where δ is the boundary layer thickness. From Ref. [9].

a steady state where both fluxes are equal to each other and the steady-state flux is then:

$$F_{mass\ transport} = F_{surface\ reaction} = F_{steady\ state} = \left[\frac{K_s h_g}{K_s + h_g}\right] C_s \tag{4.3}$$

Therefore, the kinetics of the graphene growth can be limited by the competition between K_s and h_g. In other words, when the surface reaction is fast ($K_s >> h_g$), the kinetics is dominated by mass transport of gas molecules through the boundary layer and *vice versa*. Since Equation (4.1) is derived from Fick's first law describing the diffusion of mass, h_g can be expressed as,

$$h_g = \frac{D_g}{\delta}, \quad D_g \propto \frac{1}{P}, \quad \delta \propto \frac{1}{U} \tag{4.4}$$

where D_g is the diffusion coefficient of gas molecules, δ the thickness of boundary layer formed by a gas flow, P the ambient pressure and U the flow rate of gases. Together with the observation that K_s has an exponential relationship with the substrate temperature because all the surface reactions involving adsorption, decomposition, surface diffusion and desorption are thermally activated, these allow us to control the kinetics by adjusting the process parameters. For example, when the ambient pressure is low, D_g becomes higher than the case of the atmospheric process, leading to an increase in h_g. Consequently, surface reaction becomes the rate-limiting step, and the graphene growth process appears to be self-limiting. That is the reason why SLG is mostly achieved through the Cu-mediated LPCVD process even with rough Cu foil surfaces, or varied flow rates or partial pressures of hydrocarbon are applied. In contrast, the morphology of graphene by the APCVD process shows more sensitive trends to the process parameters. For APCVD, the process kinetics is usually in the mass transport-limited regime; the process is not self-limiting, and therefore it is challenging to obtain uniform SLG. In this regime, the diffusion distance of the hydrocarbon gas molecule, or the boundary layer thickness, is an important factor. This means that the topography of the substrate surface will influence the morphology of graphene. Consequently, FLG or multilayer graphene with a preferred nucleation along the rolling lines (i.e. straight striations on a Cu-foil surface given by cold rolling process) is observed in the APCVD process [9,10]. Currently, SLG by APCVD is achieved using two strategies: (1) changing the dominant kinetics into a surface reaction-limiting process and (2) decreasing the concentration of the hydrocarbon feedstock, C_s [9,10]. In most of the cases, both strategies are applied at the same time by flowing highly diluted hydrocarbon gas with an excess of an inert carrier gas, such as argon. The increase in the flow rate by this carrier gas can reduce δ, thereby increasing h_g. The decrease in C_s also makes it harder to nucleate graphene on the substrate surface, leading to surface reaction-limiting kinetics. In addition, to exclude the effect of the topography of the substrate surface, pre-treatment of the substrate, such as by electropolishing [10] or chemomechanical polishing [11], can be applied before the CVD process to prepare a highly flat surface for the substrate.

4.2.1.2 The nucleation of graphene with respect to the process parameters

Due to its self-limiting nature, SLG can be obtained by the LPCVD process even when the process parameters are varied. However, the morphology of the graphene at its initial growth state shows more sensitive dependency on these growth parameters, as reflected in the size of a graphene domain (which is a graphene island with the same in-plane crystallographic orientation). When atomic carbons are created on the substrate surface from the dissociation of hydrocarbon precursors, a few carbon atoms attach to each other and form a nucleus. A graphene cluster made by the coalescence of these nuclei keeps growing to form a graphene domain through an adatom diffusion of carbon atoms. As time elapses these domains eventually meet and form a boundary. In contrast to the case of a typical thin-film growth, the domain boundary of as-grown graphene is assumed to be static even with further treatment, such as annealing. Since this domain boundary is an intrinsic defect that scatters carriers in electronic [12] and thermal transport [13], and degrades the mechanical strength [14] of the graphene, significant efforts were made to increase the domain size as large as possible. Achieving a large domain size is directly associated with the decrease in the nucleation density. Various strategies for achieving a large domain size will be introduced in Sec. 4.3. In this section, how process parameters are connected to the nucleation of graphene will be discussed.

As shown in Figure 4.3 [15], the nucleation density of graphene flakes can be controlled by adjusting the process parameters. If an identical state of the substrate surface is assumed, these behaviours in Figure 4.3 can be understood by applying a heterogeneous nucleation model to the graphene growth. Nucleation is a phase transformation process of supersaturated species at the expense of the addition of surface free energy by the transformation. For a heterogeneous nucleation process, the critical nucleus size (r^*) and the corresponding Gibbs free energy (ΔG^*) to achieve a critical nucleus size are given by [16],

$$r^* = -\frac{2\gamma_{AB}}{\Delta G_V}, \quad \Delta G^* = -\frac{16\pi}{3}\frac{\gamma_{AB}^3}{\Delta G_V^2}S(\theta) \tag{4.5}$$

where γ_{AB} is the surface-free energy of the phase A over the phase B (graphene over a substrate in this case), ΔG_V the change of the free energy by the phase transformation and $S(\theta)$ a shape factor of the substrate contributing to the additional promotion of the nucleation. Here, ΔG_V corresponds to the driving force of the nucleation and can be expressed as,

$$\Delta G_V \approx -RT\Omega_A^{-1}\alpha_{SS} \tag{4.6}$$

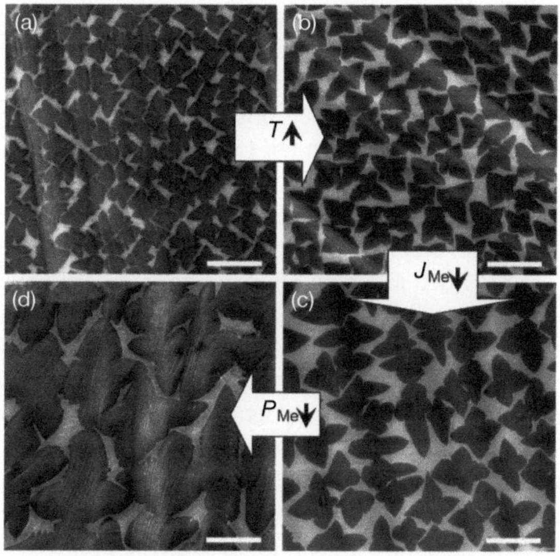

FIGURE 4.3 SEM images of partially grown graphene flakes on a Cu foil with respect to different variation of process parameters. The nucleation density obtained at a given condition (a) was decreased by (b) an increase in temperature, (c) a decrease in the flow rate of methane and (d) a decrease in the partial pressure of methane. From Ref. [15].

where R is the gas constant, T the temperature, Ω_A the molar volume of the phase A and α_{SS} the supersaturation defined by the difference of partial pressure of phase A (carbon in this case) between a given state (p_A^{act}) and the equilibrium state (p_A^{eq}).

The kinetics of the nucleation process should also be considered. The nucleation rate ($\overset{g}{N}$) is given by the attachment rate of adatoms times the concentration of clusters with the critical nucleus size (C_{r^*}),

$$\overset{g}{N} = \omega exp\left(\frac{-\Delta G_m}{RT}\right)C_{r^*} = \omega exp\left(\frac{-\Delta G_m}{RT}\right)C_0\, exp\left(\frac{-\Delta G^*}{RT}\right) \tag{4.7}$$

where ω is the vibration frequency of the atoms, ΔG_m the migration-free energy of an adatom and C_0 the total concentration of atoms. Figure 4.3(b)–(d) shows that a lower nucleation density, or a larger domain size in other words, can be achieved by increasing the substrate temperature, decreasing the flow rate of the hydrocarbon gas (methane in this case) and decreasing the partial pressure of the hydrocarbon gas, respectively. First, since this nucleation is a thermally activated process, the increase in temperature can be expected to promote the nucleation, leading to a decrease in the domain size. However, this increase in domain size shown in Figure 4.3(b) implies that the temperature appears to more actively contribute to the increase in desorption and evaporation of the decomposed carbon and hydrogen, thereby decreasing the nucleation density [17,18].

The increase in the domain size by the decrease in the flow rate in Figure 4.3(c) and the decrease in the partial pressure in Figure 4.3(d) can be accounted for by the promoted adatom attachment through surface diffusion and the suppressed supersaturation, respectively. However, this strategy to have a large domain size by decreasing the supersaturation can fail to achieve a full coverage of graphene over a substrate. Therefore, the growth step can be modified by a two-step process that aims at achieving a continuous graphene growth together with a large domain size. In the first step, a low flow rate and a low partial pressure of hydrocarbon gas are provided at a high temperature in order to nucleate graphene islands with a low density. In the subsequent growth regime, by increasing both the partial pressure and the flow rate, gaps among domains are filled up and, finally, continuous graphene with a full coverage over a substrate can be obtained. Since no further nucleation is expected as long as the temperature is fixed, the second regime only contributes to the growth of each domain by adatom attachment. This is an example showing how the quality of graphene can be controlled by adjusting the process parameters of CVD without any special treatment, thereby giving rise to the improvement in the material properties.

4.2.1.3 The role of hydrogen in the graphene growth

For a silicon thin-film deposition by the CVD process using silane (SiH_4) as a precursor, flowing hydrogen together with silane has an influence on the deposition rate of silicon by controlling the desorption rate of hydrogen decomposed from the silane [19]. This gives rise to a change in the dominant kinetics at a given growth temperature. For a graphene growth process by CVD, however, the role of hydrogen typically fed in all the steps is different to some extent. Hydrogen introduced in the heating and annealing step is related to the cleaning of the surface of the substrate. Since a Cu foil has a native oxide, hydrogen reduces this oxide in the heating and annealing step to expose the catalytic Cu surface. In the growth step, hydrogen also participates in the surface reaction beyond the role of a diluent gas for a hydrocarbon precursor. The effect of hydrogen on the graphene morphology is shown in Figure 4.4 [20]. Regardless of whether the ambient pressure was of the AP or LP type, a shape evolution of the graphene domain was observed with respect to the partial pressure of hydrogen in the growth step. Together with this shape evolution, the graphene domain size also has a dependence with the partial pressure of hydrogen as shown in Figure 4.4, and the maximum point can be understood from the dual role of hydrogen in the surface reaction; i.e. as a catalyst to dehydrogenate hydrocarbon and as an etching reagent of carbon atoms.

First-principle calculation of the graphene growth on a Cu substrate using methane as a precursor indicates that the simple direct decomposition of hydrocarbon,

$$CH_4(g) \xrightarrow[Cu]{} C(s,\ graphene) + 2H_2(g) \tag{4.8}$$

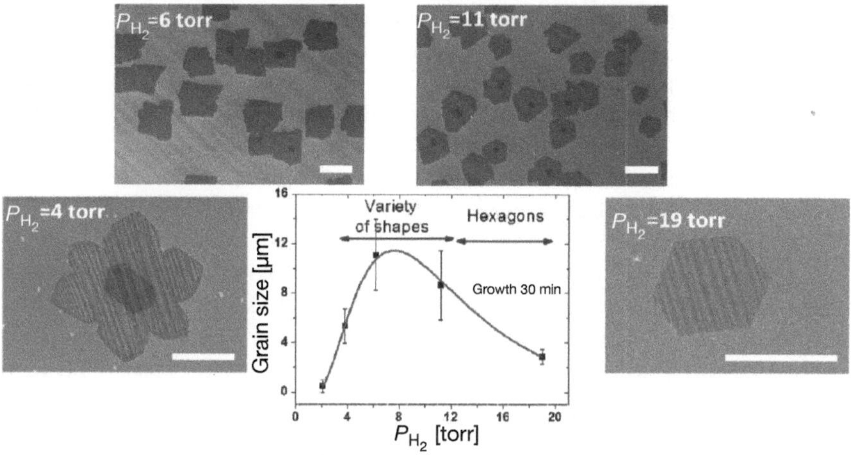

FIGURE 4.4 SEM images of partially grown graphene flakes with respect to various values of the partial pressure of hydrogen in a LPCVD graphene growth process. With the increase in the partial pressure of hydrogen, a maximum point of the domain size (grain size) of graphene was observed and a further increase in pressure led to the evolution of the domain shape to a hexagon. From Ref. [20].

is unlikely to happen because this thermal decomposition is a thermodynamically unfavourable endothermic process and the corresponding activation energy is also very high even if Cu is employed as a catalyst [21]. Therefore, one should consider more complicated reactions, including stepwise dehydrogenating reactions of hydrocarbon and hydrogen-induced reactions. According to a model [18,20], hydrogen plays the role of a catalyst to dehydrogenate methane and agglomerate one another as given by,

$$(CH_4)_{Cu} + H^*_{Cu} \Leftrightarrow (CH_3)_{Cu} + H_2 \tag{4.9}$$

$$(CH_3)_{Cu} + graphene \Leftrightarrow (graphene + C) + H_2 \tag{4.10}$$

Although the thermal dehydrogenation of methane is unfavourable, the decomposed active hydrogen atom (H^*) can provide a relatively facile route for methane dehydrogenation on the Cu surface, leading to the formation of an active hydrocarbon species. Since these active species, CH_x ($x < 4$), are unstable, they are subsequently dimerized, forming stable graphene flakes. Numerical calculations also predict that there is a threshold for the partial pressure of hydrogen (p_{H_2}) for the nucleation of graphene in the LPCVD process using Cu as a catalyst [18]. However, the further increase in p_{H_2} can lead to a decrease in the growth rate of graphene. This is due to the etching reaction of carbonaceous materials by hydrogen expressed as,

$$H_{Cu} + graphene \Leftrightarrow (graphene - C) + (CH_x)_{Cu} \tag{4.11}$$

The model accounts for this speculation by correlating the p_{H_2} with the shape evolution of the graphene islands. When a sufficiently high p_{H_2} is applied, the shape of the graphene islands appears to be hexagonal. This was attributed to the equilibrium between the graphene growth and etching, i.e. the dual role of hydrogen. Therefore, not only are other process parameters, such as the partial pressure of hydrocarbon, flow rate and temperature, but also the p_{H_2} is one of the factors to determine the size of graphene domain and the coverage of graphene over a substrate.

The preparation of SLG in large area by CVD is not as challenging as one might expect. Without a special treatment, uniform SLG over a substrate can be obtained by simply flowing methane as a hydrocarbon precursor over a commercial Cu foil as a substrate at a high temperature of around 1000 °C. Due to its self-limiting growth characteristics, there is a fairly wide process parameter window to obtain SLG on Cu by LPCVD process. However, to achieve high-quality CVD graphene material with material properties close to mechanically exfoliated graphene requires careful tuning of the growth parameters and good understanding of the growth mechanism. Better understanding of the growth of graphene is also essential for predicting behaviours when one tries to modify the growth process, e.g. by the use of plasma to decrease the process temperature [22] or the introduction of different types of carbonaceous precursors beyond methane [23,24].

4.2.2 Substrate material

As stated earlier, this chapter is mostly dealing with the method of graphene growth using Cu as a substrate, the most widely used catalyst material to grow uniform SLG in large area. Cu has negligible carbon solubility at the growth temperatures (~1000 °C) [25], so that graphene formation via segregation of carbon from Cu substrate can be mostly neglected. One can expect to employ other materials as a substrate to achieve the growth of SLG by decreasing the growth temperature to reduce the concentration of dissolved carbons. However, since the decomposition of a hydrocarbon based on hydrogen-mediated reactions for the nucleation of graphene is thermally activated, it is challenging to lower the growth temperature [17]. For these reasons, Cu is preferred as a substrate to prepare SLG in large area, but depending on applications where FLG or multilayer graphene are required, materials other than Cu can be used, with different mechanisms for graphene growth. In this section, the growth mechanism of graphene with respect to various substrate materials will be discussed. Moreover, when graphene is prepared on a metal substrate, the transfer process onto a suitable substrate is indispensable, except for the case where graphene is used as a protection layer of the metal, such as for an anti-corrosive coating [26]. The transfer process will be presented in the following in detail, but the material properties of graphene could be affected during this transfer process. In this manner, approaches to prepare graphene directly on an insulating surface, free from a further transfer processes, will be briefly introduced.

4.2.2.1 Copper

The speculation that the growth of graphene over a Cu surface is mainly a surface deposition process, because of its negligible carbon solubility, was confirmed using carbon isotopes (^{13}C and ^{12}C) alternatively during the growth process [27] (Figure 4.5(a) and (b)). As shown in Figure 4.5(c) and (d), SLG grown on a Cu foil reveals the spatially mapped G-band Raman spectra of different graphene regions, according to the types of carbon isotope ($^{12}CH_4$ and $^{13}CH_4$) introduced during the growth. This approach excluded the possibility of the segregation of carbon atoms dissolved in the bulk because these atoms would not be distributed in the order of ^{13}C and ^{12}C when diffused out of the bulk. As illustrated in Figure 4.6(c), when the same isotope tracing analysis was applied to the graphene growth on a Ni substrate, the corresponding Raman map showed a randomly distributed G-band position of FLG, regardless of the sequence of $^{13}CH_4$ and $^{12}CH_4$ flow [27]. This suggests that the Ni-mediated graphene growth follows a carbon segregation mechanism, which is in contrast

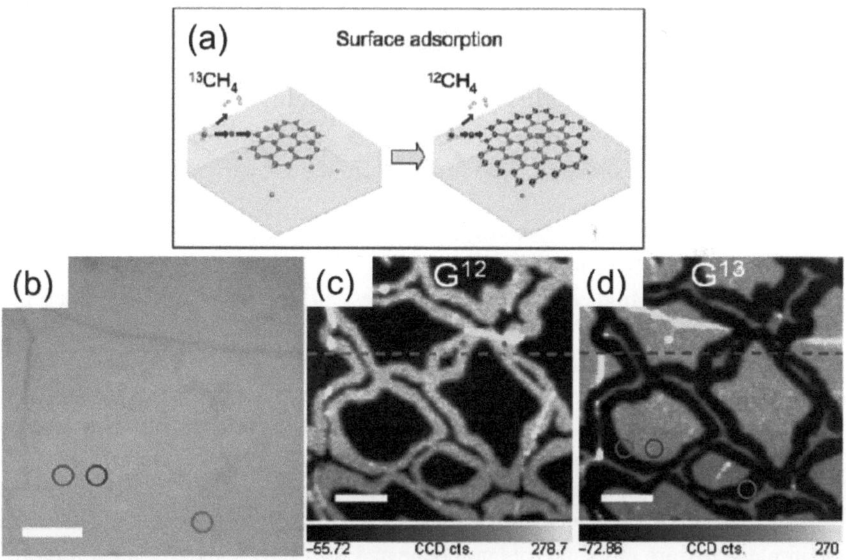

FIGURE 4.5 (a) Schematically presented graphene growth mechanism on a Cu substrate based on the surface deposition using a carbon isotope analysis. The spatially distributed carbon isotopes in the graphene sheet were expected based on the order of the sequential $^{12}CH_4$ and $^{13}CH_4$ gas flow. (b) Optical image of the transferred SLG onto a SiO_2/Si substrate where $^{12}CH_4$ and $^{13}CH_4$ were sequentially flowed over a Cu substrate in the growth step. Integrated intensity of Raman mapping of the area shown in (b) from the (c) ^{12}G-band (1560–1620 cm^{-1}) and (d) ^{13}G-band (1500–1560 cm^{-1}), which confirmed the model in (a). Scale bars are 5 μm. Adapted from Ref. [27].

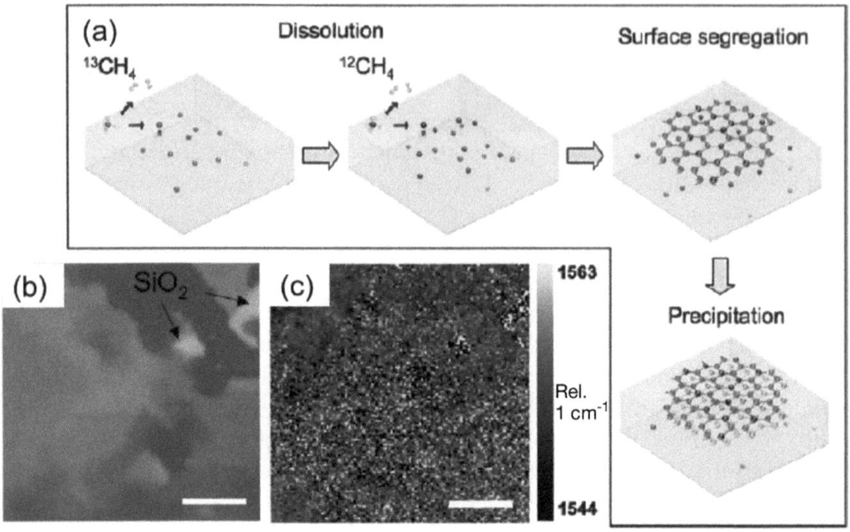

FIGURE 4.6 (a) Schematically presented graphene growth mechanism on a Ni substrate based on the segregation of dissolved carbon atoms in bulk regions using a carbon isotope analysis. Since the carbon atoms are precipitated onto the substrate surface through diffusion, randomly distributed carbon isotopes were expected, regardless of the sequence of the $^{12}CH_4$ and $^{13}CH_4$ gas flow. (b) Optical image of the transferred FLG onto a SiO_2/Si substrate where $^{12}CH_4$ and $^{13}CH_4$ were sequentially flowed over a Ni substrate in the growth step. (c) Relevant Raman mapping in terms of the position of the G-band to (b), which showed a random distribution confirming the model in (a). Scale bars are 5 μm. Adapted from Ref. [27].

to the case of the Cu substrate, and furthermore the growth result is sensitive to the cooling rate.

It is intriguing to correlate the surface state of the Cu substrate with the graphene morphology. A typically used Cu substrate is a polycrystalline foil with the purity of 99.8% or higher and with a thickness of 25 μm or thicker. Since the graphene growth on a Cu substrate starts from heterogeneous nucleation, it is hypothesized that the microstructure of the Cu substrate, such as the crystallographic orientation or the grain size of Cu, could play a role in the graphene growth. The interaction between the graphene and the Cu substrate in terms of the crystallographic orientation has been studied by examining the shape of graphene islands. The lobe shape of the graphene islands, typically observed in LPCVD and in an ultrahigh vacuum (UHV)-CVD process, was attributed to the discrepancy between the in-plane texture of the sixfold graphene and the (100)-texture of a Cu foil prepared by the rolling process [28]. The shape evolution from the lobe shape to a hexagonal shape was also explained by the degree of the interaction between the graphene and the Cu substrate; the strong interaction in the LPCVD gives rise to a lobe-shaped graphene, whereas little interplay occurs with the Cu in the APCVD process where the mass transport of precursors is more important and gives rise to a hexagonal shape [29,30]. As well as the determination of graphene islands, an electron-backscattered diffraction (EBSD) analysis showed a close packed plane of Cu, i.e. a (111) plane led to preferentially form a uniform SLG with a high growth rate. This appears to be due to the density of Cu atoms that promotes the nucleation of graphene growth as a catalyst [31].

Through examination of experimental results, it turns out that the grain boundary of Cu is generally not a preferred nucleation site for graphene growth, and in many cases, it was observed that the graphene islands were growing across the grain boundary of Cu [27]. This observation contradicts the common expectation that normally the grain boundary serves as nucleation sites. On the other hand, impurities segregated on the surface, or structural irregularities (along the rolling line) have been observed as more effective nucleation sites. It is thus important to minimize these sites because the presence of these will contribute to higher nucleation density and smaller graphene domain sizes. Although it was reported that the graphene sample grown on a Cu foil with higher purity (99.9995%) showed better performance than graphene grown from a lower-purity (99.98%) Cu foil [32], high-purity Cu foils are costly. Rather, it is much easier to clean and prepare the Cu foil with chemical treatment, high-temperature annealing and/or electrochemical or chemical–mechanical polishing before the CVD synthesis. Acetic acid has been used to remove copper oxides [33] on the foil surface. And a substantial improvement of the graphene morphology from multilayer to monolayer graphene by using an electropolished Cu foil [10] has also been reported.

For graphene CVD, bulk Cu foil has been preferred due to different reasons, such as low cost and stability against high-growth temperature. If polycrystalline thin film (300–500 nm, deposited on a substrate such as SiO_2/Si) is used, when the substrate is heated up to a temperature of around 1000 °C, dewetting in the film by surface diffusion takes place easily as the melting point of Cu is 1085 °C. Thus, for graphene growth on Cu thin films, particular care needs to be taken, such as fast heating/cooling and lower growth temperature (800–900 °C instead of 1000 °C). For example, in order to grow graphene at 1000 °C, Cu films thicker than 500 nm were used [34,35]. Or to lower the growth temperature, carbon precursors other than methane can be used, such as a liquid carbon source [36] or a solid carbon source [37]. However, the quality of the graphene grown on a Cu thin film is still poorer than the one using a Cu foil, so that the use of Cu thin film approach needs further optimization. On the other hand, in certain aspect a thin-film process is advantageous, e.g. transfer-free batch fabrication of graphene transistors has been reported [34] and direct formation of graphene at the interface between Cu thin film and the SiO_2 substrate was also implemented [38]. In this way, patterned graphene can be directly obtained via patterning the Cu thin film before the CVD growth. These are more favourable for microelectronics applications. Furthermore, single-crystalline Cu (111) thin film has been grown on basal-plane sapphire [α-Al_2O_3 (0001)] substrates and it has been shown that graphene with reduced D-band can be obtained compared with polycrystalline Cu [39].

4.2.2.2 Nickel and other metals

As shown in Figure 4.6, the main growth mechanism of graphene using Ni as a substrate is different from Cu-based growth due to the difference of carbon solubility in Cu and Ni. Because of the finite carbon solubility in Ni (1%–2%) at the growth temperature, carbon first dissolves in Ni and segregates later; thus, it has been shown that cooling rate affects the graphene morphology segregated from the Ni thin film [6]. In addition, relatively thicker graphene layers were observed along the Ni grain boundaries [6], possibly due to faster diffusion of carbon atoms along the grain boundaries. Therefore, to prepare SLG or FLG, a slow cooling rate and large Ni grain sizes are desirable. However, graphene fails to form due to suppressed segregation when the cooling rate is too low [40].

Although using CVD with these polycrystalline Ni substrates does not give rise to uniform SLG in large area, the possibility of a tunable graphene thickness by adjusting the cooling rate provides an interesting path for graphene growth with a controlled number of layers [6,41]. For example, with a fast cooling rate ($\geq$100 °C min^{-1}), more than two graphene layers over the Ni grain and further thicker graphene layers along the Ni grain boundary can be obtained, whereas fewer graphene layers, mostly consisting of one or two layers, can be obtained by applying a slow cooling rate ($\leq$25 °C min^{-1}). Particularly, it has been reported that AB-stacked BLG, in which a bandgap up to 250 meV can be created by applying a vertical electric field across the two layers [42], can be prepared via Ni-mediated CVD process [43–45]. In this respect, routes to grow graphene with other transition metals, such as cobalt (Co) [46,47] or iron (Fe) [48] as well as a Ni–X alloy (X = Cu [47,49,50], molybdenum (Mo) [51]) have also been investigated. For a binary Ni alloy system, the constituent metals contribute to controlling the number of graphene layers either by promoting a self-limiting process based on a surface deposition growth mechanism (Cu) [49] or by forming a carbide phase in the bulk with the dissolved carbon atoms (Mo) [51], resulting in the formation of uniform SLG or BLG. As compared to Cu, because these transition metals, including Ni and its alloys, have a much higher melting point (T_m of Ni, Co and Fe = 1455 °C, 1495 °C and 1538 °C, respectively) than Cu, a thin-film process is probable at a temperature as high as 1000 °C. Furthermore, wafer-size graphene growth via a segregation mechanism from these metal alloys has also been reported [47].

The epitaxial graphene growth has also been reported on (0001) surface of ruthenium (Ru) [52,53], (111) surface of iridium (Ir) [54,55], (111) surface of platinum (Pt) [56] and even (111) surface of Cu [57] under UHV conditions using an ethylene precursor. As the substrates are single crystalline, the epitaxial growth should lead to single-crystalline graphene, which is anticipated to have higher quality. By taking advantage of *in situ* surface characterization, such as low-energy electron microscopy (LEEM) during the growth, the initial nucleation and growth behaviours for each process can be explored. For the epitaxial growth of graphene on Ru (0001) by Sutter *et al.* [52], carbons decomposed from an ethylene precursor are absorbed into the Ru bulk at a temperature as high as 1150 °C. Upon slow cooling down to 825 °C, supersaturated carbon starts to be nucleated on a step edge of the substrate. This behaviour is similar to the Ni-mediated CVD growth driven by the segregation of carbon with a dynamic carbon solubility change in the bulk phase. The remarkable aspect of the epitaxial growth of graphene on Ru (0001) plane is that a lens-shaped graphene island grows further by carbon incorporation only in the 'downhill' direction onto a lower terrace of (0001) plane of Ru. This directional growth was attributed to the formation of an immobilized graphene edge by the maximized orbital overlapping between graphene and Ru on this terrace. The first layer of graphene on the Ru also showed a high coupling between graphene and Ru, as indicated by an incommensurated Moiré structure and a suppressed electrical conductivity. This graphene proceeded on the Ru surface in a stepping-down fashion and extended to 100 μm in length, which was substantially larger by two orders of magnitude than the growth on 4H-SiC (000$\bar{1}$) substrate. The second layer of graphene, which began to appear after about 80% of coverage of the first layer over Ru, turned out to be strongly decoupled from the substrate. Epitaxial graphene growth on Ir (111) appears to be different from that on the Ru (0001) substrate because the graphene growth on Ir (111) is not sensitive to whether the growth direction is 'uphill' or 'downhill' [54]. Density functional theory (DFT) calculations showed that carbon atoms are less strongly bound to Ir than to Ru [58], which could be a possible explanation for the difference. However, in both substrates, the growth rate was not linearly increased kinetically by the supersaturation of carbon adatoms [59]. This suggests that the epitaxial graphene growth may have a common growth mechanism, i.e. rate-limited by the attachment or formation of carbon clusters at the edge of graphene. Therefore, by understanding the interplay between the energy barrier to form a carbon cluster and the growth process parameters, such as temperature, epitaxial graphene growth process can be controlled in a better way [59,60].

4.2.2.3 Non-metallic substrates

As of now, the metal-mediated graphene growth by CVD can be regarded as a well-developed process insofar as high-quality graphene can be prepared in large area. However, typical applications exploiting the excellent properties of graphene require a structure where graphene is sitting on insulating substrates, such as a SiO_2/Si. This means that the transfer process of as-grown graphene from metal substrates to other substrates is essential. However, the transfer process can induce extrinsic defects and contaminants that deteriorate the inherent properties of graphene, as a result, direct growth of graphene on insulating substrates to avoid the transfer process, have also been pursued.

One widely known method is to epitaxially grow graphene on a single-crystalline SiC substrate via the sublimation process [61,62]. Since SiC is a wide bandgap semiconductor, the graphene/SiC structure itself can be utilized for electronic applications without a transfer process [63,64]. The cost of this process is relatively high since single-crystalline SiC substrates are costly and the temperature involved is usually >1200 °C. On the other hand, attempts have been made to prepare graphene on other insulating substrates, such as SiO_2/Si [43], silicon on insulator (SOI) [65], MgO [66], Al_2O_3 [67] and mica [68]. These processes may involve a modified growth process beyond a simple thermal LP-CVD process with commonly used H_2/CH_4 in the Cu foil-mediated process; e.g. molecular beam epitaxy (MBE) [68] or plasma-enhanced CVD (PECVD) [69] can be applied when other types of carbon precursors, such as alcohol [67], hexane [66] or acetylene [66], are used. However, the reported quality of graphene prepared in these ways is still inferior to those grown using metallic substrates. One possible reason could be due to the strong bonding between surface states of the non-metallic substrates and carbon species, which suppresses the surface diffusion of carbon atoms [70]. Consequently, to prepare good-quality SLG in large area directly on non-metallic substrates, alternative ways involving still a metal catalyst have been reported. In one of such methods, a Cu thin film on SiO_2/Si is used as a catalyst for the graphene growth (similar as usual graphene CVD growth), but the Cu thin film in between the graphene and the insulating substrate was gradually dewetted and evaporated in the growth process, leaving behind the graphene/insulator stack [71]. In another, it was found that when a Cu thin film (on an insulating substrate) with proper thickness (300 nm) is used, graphene can grow both on top of the Cu thin film and at the interface between the Cu and the insulating substrate. By removing the top graphene layer and the Cu thin film, the bottom graphene can be directly obtained on the initial insulating substrate [38]. It is suggested that this process takes advantage of the polycrystalline structure of the Cu thin film, and the grain boundaries of the thin film serve the role of providing the fastest diffusion paths for the carbon species.

4.3 TOWARD THE LARGE DOMAIN SIZE

As stated earlier, one intrinsic defect in the graphene prepared by most of CVD process is the domain boundary that contributes to the inferior properties than mechanically exfoliated graphene. As long as graphene is not epitaxially grown, the formation of domain boundaries in the CVD process is unavoidable because as graphene domains nucleates on the substrate, they can be of different orientations and when they meet one another, they stitch up incommensurately due to their random in-plane orientation. Since these boundaries are basically atomic imperfections of carbons [72], in-plane electronic [12] and thermal transport [13] in the graphene layer can be scattered across the boundaries and mechanical failure by in-plane tensile stress can happen along the boundaries [14]. The scattering effect by the domain boundary has already been experimentally observed in both indirect and direct ways. For continuous graphene with the different areal density of domain boundaries, when they were fabricated as field effect transistor (FET) devices, the resulting carrier mobility across the in-plane graphene layer could be described as a function of an average domain size of the graphene. As expected, an improved carrier mobility value was observed in

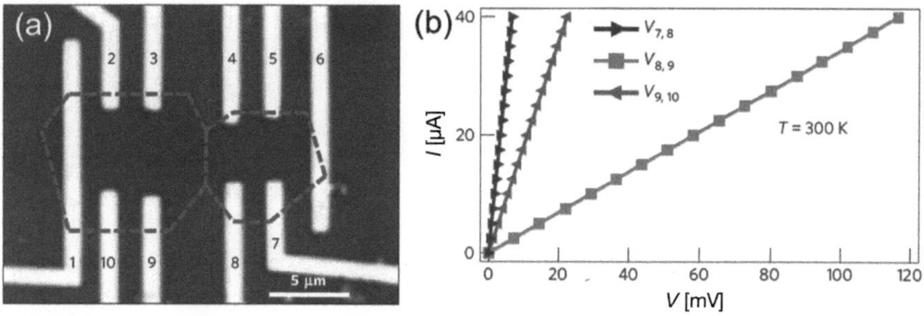

FIGURE 4.7 (a) Optical image of a device with multiple electrodes (1–10) contacting two graphene domains forming a boundary in between (indicated by dashed lines). (b) Room-temperature *I–V* curves measuring the inter-domain region (light grey (green in the web version)) across the boundary and intra-domain region (dark grey (blue in the web version) and grey (red in the web version)) inside a single domain. Highly resistive behaviour was observed in the inter-domain conduction caused by the carrier scattering across the domain boundary. From Ref. [12].

graphene with a larger domain size [73]. Direct observation of the impinging role of the domain boundary for electronic transport was provided in discontinuous graphene flakes [74]. As illustrated in Figure 4.7 [12], the resistivity obtained from four-point probe measurements showed a noticeable difference in the intra-domain and inter-domain region transport, depending on whether or not a domain boundary is involved in the conduction; while values of 2 kΩ $sq.^{-1}$ and 0.40 kΩ $sq.^{-1}$ were deduced for the resistivity from the current–voltage characteristics in the intra-domain regions, higher resistivity of 5 kΩ $sq.^{-1}$ was obtained in the inter-domain region, which confirmed the importance of scattering by domain boundaries for electronic transport. Therefore, in order to achieve high-mobility devices, it is necessary to prepare graphene with a domain size as large as possible. For Cu-mediated graphene CVD process, the size of the graphene domains is directly connected to the nucleation density during the growth. Top–down lithography has been used to facilitate the nucleation at specific sites [12,75]. With well-controlled conditions, individual hexagonal graphene flakes (a few micrometres in size) at each nucleation site can be obtained. However, even if these flakes are single crystalline and their locations are controlled, there is no guarantee for them to be of the same orientation. Thus, the continuous graphene film prepared in this way could still be polycrystalline. Another approach has been the growth of graphene domains with sub-millimetre to millimetre sizes. Such domain size is sufficiently large to fabricate microelectronic devices avoiding the involvement of a domain boundary. Recent strategies to obtain large domain sizes will be introduced in the following section that includes the control of process parameters and the modifications to the shape of the Cu substrates.

Two-step process introduced in the Sec. 4.2.1 is one example of the strategies used to achieve large domain size for graphene synthesis [15]. What has been learned through [15] is that the use of high growth temperature, low flow rate of methane feedstock and low methane partial pressure, were all attributed to producing larger domain sizes. These conditions favour low nucleation density of graphene on Cu substrates, which results in large-graphene domain sizes. On the other hand, the growth rate of graphene under these conditions is also much lower, especially when more and more of the catalytic Cu surfaces become covered with graphene. As a result, it is difficult to obtain a completely continuous graphene film using these growth conditions. In order to achieve a fully covered graphene layer over the substrate, a two-step process was proposed, where in the first step, the aforementioned growth conditions were used to obtain low nucleation density and large-graphene domain, and in the second step, higher-methane flow and partial pressure were used to enhance the graphene growth rate and capability to complete the graphene layer with reduced catalytic surface of Cu [15]. In a later report, it was also suggested that optimized flow rate of hydrogen can result in an increase in the domain size via the dual roles of hydrogen in the growth process, i.e. hydrogen can act as both the catalyst for the methane decomposition and the etching reagent of the as-grown graphene during the growth process [20]. In addition, it is anticipated that a highly clean and flat surface of the Cu foil is also important to suppress undesired graphene nucleation. Table 4.1 shows the comparison of 11 sets of growth conditions (such as Cu polishing before growth, flow rates and partial pressures CH_4, H_2, annealing time, etc.) and the corresponding domain sizes and shapes that were obtained [76]. A graphene single-crystal domain as large as $\sim$4.5 mm^2 was obtained using one set of the conditions. The experimental setup was not significantly different from the normal Cu-mediated LPCVD graphene growth [8], except for the point that the ambient pressure could be controlled by a metering valve designed to operate in parallel with a vacuum pump. By analysing the growth conditions presented in Table 4.1, it is helpful to understand the interplay between each process parameter to achieve large-graphene domain size. First, the results suggested that having a clean and flat surface is important for large-graphene domain. In this respect, electropolishing was first performed on a Cu foil and interestingly, it was claimed that high-pressure ($\sim$2 atm), long-time (7 hours) annealing under hydrogen ambient at high temperature closer to the melting point of Cu led to the remarkably flat surface of the Cu foil [76]. This long-time annealing which suppresses the nucleation of graphene was also confirmed in another APCVD growth where sub-millimetre-size graphene domain was achieved [77]. Secondly, from Table 4.1 it can be suggested that in the growth step, a high temperature as 1077 °C, and a low methane to hydrogen ratio of flow rate as 0.15–70 sccm, respectively, contributed to the largest domain size (E7 condition). Further dilution of methane failed to result in graphene nucleation, and this failure appears to be due to insufficient supersaturation. Furthermore, it can be noticed that from Table 4.1, the ambient pressure change from 8.2 to 108 Torr led to the shape evolution of the graphene domain from square shape to hexagonal shape. This, as discussed earlier, appears to be due to the increase in the partial pressure of hydrogen [20]. In another attempt to

TABLE 4.1 Eleven growth conditions (E1–E11) carried out in an experiment with different growth conditions

Sample	E1	E2	E3	E4	E5	E6	E7	E8	E9	E10	E11
Cu polishing	No	No	Yes	Yes	Yes	Yes	Yes	Yes	Yes	Yes	No
t_A (hour)	0.5	0.5	7	7	7	7	7	7	7	7	7
P_A (Torr)	8.2	8.2	1500	1500	1500	1500	1500	1500	1500	8.2	1500
$U_{H_2\_A}$ (sccm)	500	500	500	500	500	500	500	500	500	500	500
T_{A+G} (°C)	1060	1060	1060	1060	1070	1075	1077	1077	1077	1077	1077
U_{CH_4} (sccm)	5	0.15	0.15	0.15	0.15	0.15	0.15	0.3	0.3	0.15	0.15
t_G (min)	1	15	70	80	110	120	125	90	125	15	20
P_G (Torr)	8.2	8.2	8.2	108	108	108	108	108	108	108	108
U_{H_2} (sccm)	500	70	70	70	70	70	70	150	200	70	70
Domain size (mm)	~0.015	~0.03	~0.3	~0.4	~1	~1.8	~2.3	~0.1		~0.1	~0.2
Domain shape									None		

t_A, P_A, $U_{H_2\_A}$, T_{A+G}, U_{CH_4}, t_G, P_G and U_{H_2} denote to the annealing time, total pressure in the annealing step, the flow rate of hydrogen in the annealing step, the temperature in the annealing and growth step, the flow rate of methane in the growth step, growth time, total pressure in the growth step and the flow rate of hydrogen in the growth step, respectively. E7 condition that resulted in the largest domain size provides an optimal condition with a clean Cu surface, a high growth temperature and a low methane concentration. The shape evolution appears to be attributed to the partial pressure of hydrogen as discussed before. Adapted from Ref. [76].

grow graphene with large domain size, the setup was designed to allow growth parameters to be changed gradually and more independently during the growth [78]. In this work, a solid carbon source, polystyrene, was used and allowed the authors to develop a two-zone system in which the decomposition of carbon source took place at one zone (Zone A) and the graphene growth was carried out on a Cu foil at another zone (Zone B). The nucleation of graphene was controlled by adjusting the temperature of Zone A, while the growth of graphene was promoted by the gradual change of the temperature in Zone A and the flow rate of carrier gases (Ar + H_2) flowing into the Zone B. The temperature in Zone B was fixed at 1050 °C. Pulse heating in Zone A was utilized to minimize the nucleation density of graphene on copper foil, while a gradual increase in the temperature of Zone A and the flow rate of hydrogen was adapted to drive the continuous growth of the graphene grains. The partial pressure of hydrogen was fixed at a constant value to avoid etching the graphene flakes and to keep the hexagonal domain shape. The temperature increase of the Zone A contributed to break the local equilibrium at the growth front of the domain by supplying additional active carbon species while the increase in the flow rate was considered to promote the migration of the carbon species. These gradual and dynamic changes were made to overcome the saturation of the graphene growth observed in other single-zone systems [15]. This strategy can be considered as an advanced form of the earlier two-step growth. These results shown here are inspiring that the domain size of the graphene can be achieved up to millimetre sizes by adjusting the process parameters.

Large domain size of graphene was also obtained by folding the Cu foil into a special geometry. For this strategy, Li *et al.* folded a Cu foil into an envelope, shown in Figure 4.8(c), and obtained large-graphene domain size up to 0.5 mm on the inside surface of the Cu enclosure using growth conditions similar to normal LPCVD growth and slightly higher growth temperature (1035 °C) [79]. The morphology of the graphene grown on the outside surface of the enclosure was not significantly different from what is typically obtained in a LPCVD process, although the formation of FLG was observed more often. This aspect has been utilized for the growth of BLG as will be discussed in Sec. 4.4. It was found from recent studies [80] that small amount of methane leaks through the folding edge and thus graphene grows with methane on the inside surface of the Cu enclosure. As the methane concentration inside the enclosure is much lower (the model and data fitting in [80] suggest it is 1000 times lower than methane concentration outside), the nucleation of graphene on the inside surface is much delayed and the nucleation density is much lower, which favours the growth of large domains. It is also worth pointing out that the growth on the inside surface usually takes much longer due to the low growth rate (as a result of low methane concentration). Figure 4.8(a) shows an example of such single-domain flakes (scanning electron microscopic (SEM) image), and since there is no domain boundaries within the flake, a FET device fabricated with such a graphene flake showed carrier mobility values higher than 4000 $cm^2\ V^{-1}s^{-1}$ at

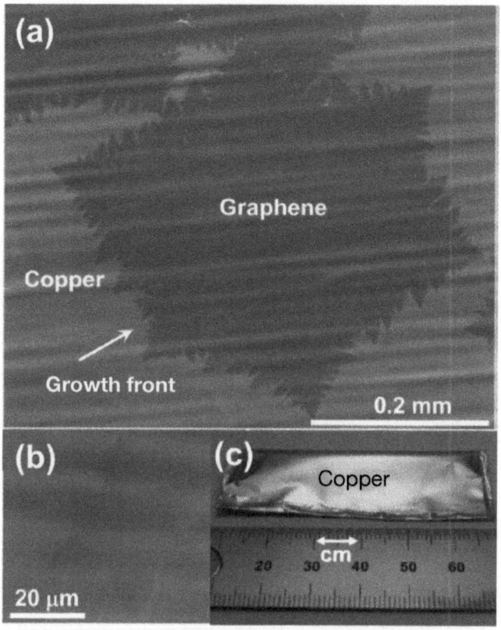

FIGURE 4.8 (a) SEM image of a single-domain graphene flake with the size of ~0.4 mm grown on a Cu foil inside the closure (c). (b) Graphene nucleus at an initial stage of the growth showing a sixfold lobe shape. (c) The Cu foil enclosure prepared for the experiment. Adapted from Ref. [79].

room temperature on SiO_2/Si [79], whereas when transferred to a hexagonal boron nitride (h-BN) substrate, the device characteristics were comparable to the quality of exfoliated graphene, showing a carrier mobility of up to 45,000 $cm^2\ V^{-1}s^{-1}$ [74]. The use of h-BN to screen substrate effects will be discussed later.

A more extremely modified form of a Cu substrate is one that has a phase change from solid to liquid. When the graphene growth was carried out at 1120 °C, which is higher than the melting point of Cu, hexagonally close packed graphene domains (also in hexagon shape) were self-aligned like a honeycomb [81]. In a solid Cu foil, the nucleation takes place in an inhomogeneous manner along the randomly distributed high surface energy sites that can facilitate the nucleation (such as impurities or structural irregularities of Cu). Even if the nucleation sites on the Cu surface can be lithographically constrained, the in-plane orientations of each graphene domain are different from one another. However, since the surface states of liquid Cu are homogeneous, regular-shaped nuclei can be distributed on the surface with a relatively lower density than on a solid Cu surface. Besides, the energetically promoted surface diffusion of carbon atoms can lead to a self-aligned distribution of each domain to minimize the surface/edge energy, i.e. in a close-packed arrangement. This approach, aiming for the achievement of a self-assembled, compact and ordered graphene structure, provides a new route to prepare graphene with good quality; however, with this approach further optimization is still necessary to increase the size of each single domain or to commensurately stitch together the close-packed domain arrays in a seamless way.

4.4 BILAYER GRAPHENE (BLG) GROWTH

One interesting graphene structure is a BLG. Even though a large percentage of the synthesis efforts have focused on the preparation of SLG, however, because SLG does not have a bandgap, its potential applications (especially in electronic devices) are limited. Opening a bandgap in graphene while maintaining the rest of its own excellent properties, is of great interest. The bandgap opening in graphene can be achieved by patterning a SLG into narrow ribbons having width <10 nm. Such a structure is called a graphene nanoribbon (GNR) [82]. Nevertheless, it has been challenging to obtain controlled GNR with atomically sharp edges; typical lithography and etching technique only give rise to imperfect edges which cause severe charge carrier scattering. As a result, the carrier mobility of GNRs is significantly lowered to only a few hundred $cm^2\ V^{-1}s^{-1}$. Another option for introducing a bandgap in graphene is to use epitaxial substrate to induce a bandgap [83], including 53 meV of bandgap opening along with h-BN substrate [84]. However, it is challenging to obtain single-crystalline domain graphene epitaxially on single-crystalline substrates (especially in large areas) and the process in this area has been limited. Apart from these methods, an alternative route is to use BLG with Bernal stacking (AB-stacking) where half of the carbon atoms in the top layer lie directly on top of the atoms in the bottom layer and the other half carbon atoms are above the centre of the hexagonal ring of the bottom layer. It has been proposed by theory [85,86] that if a vertical electric field is applied across the two graphene layers, a bandgap proportional to the field can be created. Experiments have confirmed this and a bandgap up to 250 meV have been observed by optical spectroscopy [42,87,88]. Therefore, to obtain large-area AB-stacked BLG is highly desirable for generating a tunable bandgap in graphene.

Many groups have explored a variety of methods for producing large-area BLG by CVD process. For example, it was reported that using solid carbon source, such as spin-coated polymer layer or self-assembled monolayer (SAM) on an insulating substrate, by depositing a Ni capping layer on top, after high-temperature annealing, BLG grows at the interface between the Ni layer and the substrate [45]. A two-step epitaxial growth was also carried out on Cu to achieve BLG [89]. Basically, SLG was first grown on one of the Cu foils under LPCVD first; afterwards the Cu foil with graphene was put into the furnace again for subsequent growth, but this time with another bare Cu foil upstream. It was argued that the upstream bare Cu foil help to catalyse the decomposition of hydrocarbon to generate the active species for graphene growth, and as a result, the second graphene layer grows on the downstream foil during this second growth. A similar principle was used in another report, where the authors used a long Cu stripe in the LPCVD growth and found that at the downstream end of the Cu stripe there is large percentage of graphene bilayer [90]. Other transition metal catalysts, such as Ru (0001) [52] and Ir (111) [91], have been used to achieve Bernal-stacked BLG, while methods utilizing only Ni have generated turbostratically stacked BLG [6,43]. Cu–Ni alloys have also been used to achieve BLG while Ni and Cu share the same face-centred cubic (FCC) structure with only one atomic number difference and thus form a bimetallic phase. By using an 88.00% Cu, 9.90% Ni (weight percent) commercial Cu–Ni alloy foil and LPCVD growth, BLG was obtained and a surface precipitation mechanism was suggested. It was found that higher temperature, longer exposure time, higher carbon diffusion rate, higher H_2 concentration and slower cooling rate all contributed to higher BLG coverage [92].

In all these BLG synthesis studies, in order to distinguish whether the graphene bilayer is AB-stacked or randomly stacked (i.e. turbostratic stacking), a combination of Raman spectroscopy and transmission electron microscopy (TEM) has been used to identify the stacking orientation [87]. These methods generally rely on evaluating the full width at half maximum (FWHM) of the G' (or 2D) Raman peak combined with selected area electron diffraction (SAED) patterns to verify the stacking orientation [93]. However, SAED is a time-consuming process and only examines a relatively small sample region (100 nm ~ 1 μm) and the Raman spectra do not always provide full confidence for identifying the type of bilayer stacking. By using sequential ^{12}C and ^{13}C methane during the synthesis, Fang *et al.* [94] reported an efficient and confident method to distinguish AB-stacked bilayer from random stacking via Raman spectroscopy, which as a result allows quick evaluation of the area percentage of AB-stacking in a graphene sample. This was based on the fact that there is a strong coupling between the two layers in an AB-stacked BLG, while such coupling is much weaker or absent in turbostratic BLG. For single-layer ^{12}C and ^{13}C graphene, their G' (2D) Raman peak occurs at two distinct locations (~2700 cm^{-1} for ^{12}C graphene and ~2650 cm^{-1} for ^{13}C graphene by the relationship: $\omega(^{13}C)=\sqrt{12/13}\,\omega(^{12}C)$). In BLG, if one layer is ^{12}C and the other layer is ^{13}C, for turbostratic BLG because the two layers are independent from each other, two distinct peaks are observed at G' (2D) region, while for AB-stacked BLG, the coupling between the two layers results in a broad ($\Delta\omega > 100$ cm^{-1}) peak for the G' (2D) peak [95]. Therefore, by mapping the FWHM of the G' (2D) peak for a BLG sample grown with ^{12}C as one layer and ^{13}C as the other, the area percentage of AB-stacked BLG can be quickly obtained as shown in Figure 4.9.

In some of the BLG syntheses, investigations into whether the second layer grows on top or below the first layer were carried out. These studies helped to understand the growth mechanism of the BLG. Interestingly, among these cases most of them suggest the second layer grows from underneath [94,96] except one report [97]. This difference (i.e. whether the second layer grows on top or below) could be due to the difference in growth conditions, type or arrangement

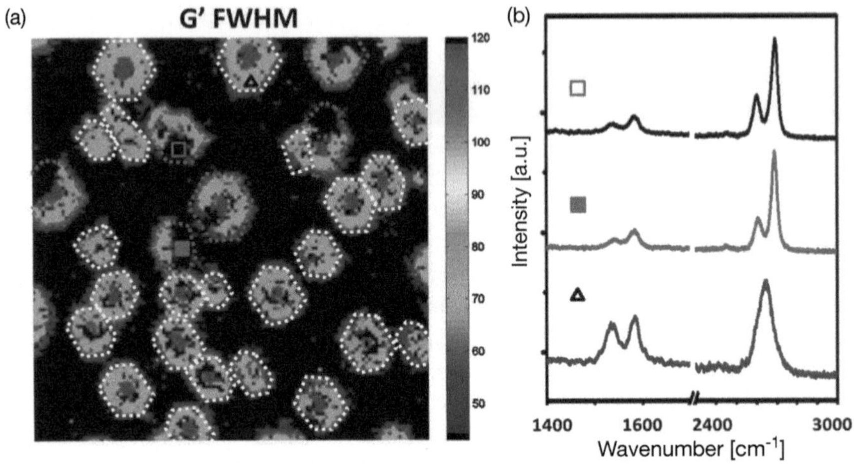

FIGURE 4.9 (a) Raman mapping of integrated FWHM of G′ band of $^{12}C/^{13}C$ BLG over an area of 60 μm by 60 μm. $^{12}C/^{13}C$ AB-stacked BLG (dotted white line) and regions containing some $^{12}C/^{13}C$ turbostratic BLG (dotted grey line (dotted red line in the web version)) are remarked. (b) Representative Raman spectra taken from (a) two turbostratic BLG regions (empty and filled squares) to show that these turbostratic BLG regions have similar spectra that can be clearly distinguished with a AB-stacked BLG region (empty triangle) in G′ band. From Ref. [94].

of the growth substrates, etc. In Refs [94,96], Cu enclosures were used for the growth of BLG. The Cu enclosure geometry was initially developed to obtain large-area (a few hundred micrometres) single-crystalline graphene domain [79]. It turns out such growth substrate geometry (and the longer growth time involved) also favours the growth of BLG. In Ref. [94], BLG was obtained from the outside surface of the Cu enclosure, whereas in another Ref. [96], BLG grown on the inside surface of the enclosure was studied. The second layer was grown underneath the first layer.

4.5 GRAPHENE TRANSFER

When graphene is prepared by a metal-mediated CVD process, the use of graphene on a metal substrate is limited, except for application as an anti-corrosive layer [12,26,98]. Therefore, more often it is required to transfer the graphene from the metal substrate onto suitable substrates, such as dielectrics for electronic applications. In order to obtain high-quality graphene material, not only it is important to optimize the graphene growth, but the transfer process is also critical to avoid deterioration of the graphene quality during transfer. Since most of the transfer processes are carried out with wet etchant, surface tension of the etchant (or water) could cause ripping or breaks in the graphene layer. Furthermore, it has been shown that the protective layer (most often poly(methyl methacrylate) (PMMA) is used) or the etchant (e.g. acids or $FeCl_3$) can leave residues (e.g. FeO_x) on graphene, or the transfer process (e.g. exposure to etchant or annealing) causes doping in graphene; thus, particular care and various strategies have been proposed to improve the transfer process. Other aspects of consideration for graphene transfer include how to make the process capable of scaling up, and integratable with the graphene synthesis process. There have been some initial developments in this area, such as roll-to-roll process for graphene transfer [99]. In this section, most of our discussions will focus on the studies carried out for transfer process optimization in which maintaining graphene quality is the top priority. There will be four sub-sections: first in Section 4.5.1, the choice of the protective layer and the corresponding results and considerations will be discussed, and then in Section 4.5.2, the impact of the target substrates' surface on the graphene quality and corresponding surface treatments that have been investigated to avoid graphene quality degradation will be described. The third sub-section 4.5.3 discusses the direct transfer of graphene onto target substrates (in which the substrate itself serves as the protective layer). The last sub-section 4.5.4 introduces transfer processes that resembles exfoliation, which allows the recycling of the metal substrates and will reduce the cost for graphene production in the future.

4.5.1 Choice of the protective layer

Since graphene is basically a single atomic layer, when it is transferred onto a target substrate, a protective layer (also referred to as supporting layer or sacrificial layer) providing mechanical robustness and handling convenience is typically used. For the mechanical exfoliation of graphene from highly ordered pyrolytic graphite (HOPG), an adhesive tape plays a supporting role. This mechanical transfer process can also be employed on the CVD-grown large-area graphene from metallic substrates. However, since it is challenging to homogeneously control the relative adhesion between the tape/graphene and graphene/metal substrate, such transfer tends to result in holes and cracks in the graphene, which are detrimental to the graphene quality as a large-area film. Therefore, instead of the mechanical exfoliation, the most straightforward transfer process is to isolate the protective layer/graphene stack by chemically etching the metal substrate and then transfer the stack to a target substrate, and finally the protective layer is removed, leaving the graphene on the target substrate. There are two basic requirements for the protective layer: first, it should be resistant to the metal etchant and second, it should be able to be removed without damaging the graphene.

Polymers have been found to be very suitable for this purpose, and the most commonly used has been the e-beam lithography resist [100] PMMA, which was also used in the transfer of CNTs [101]. In addition, polydimethylsiloxane (PDMS), which has been used in soft lithography [102] has also been applied to graphene transfer. The PDMS transfer process [7] takes advantage of the robustness and good adhesion to graphene. At first, cured PDMS is brought onto a graphene/metal stack and weakly bound to the surface. Subsequently, when this PDMS/graphene/metal stack is placed in a metal-specific etchant bath, the metal layer is selectively etched away and the undamaged PDMS/graphene stack is floated on the etchant because this etchant is mostly aqueous. This PDMS/graphene stack is subsequently scooped out of the bath, rinsed and dried. The stack can afterward be fabricated as a device by itself in which the PDMS is used as a substrate for a sensing application [103], or alternatively, PDMS can be used as a stamp, to transfer the graphene onto a target substrate [7]. The advantage of this PDMS-supported transfer is the compatibility with soft lithography. By exploiting a pre-patterned PDMS mould in the transfer process, such a micro- or nano-pattern of the mould can be reproduced in the graphene, and then the patterned graphene can be stamped to a target substrate [104]. Another advantage is that the mould can be reused. However, for monolayer graphene grown on Cu foils, from our experiments it has been found that when using PDMS layer, having a complete film transferred without any breaks or tears is quite challenging. Therefore, it might be more suitable for multilayer graphene transfer. In addition, in order to let the PDMS adhere well to the as-grown graphene on Cu foils, the PDMS should not be hard-baked, but this tends to leave residues on graphene (similar to PMMA). At present, the PDMS transfer is not used as widely as the PMMA transfer. Nevertheless, it is a very useful alternative method offering the capability of direct patterning particularly for applications involving soft-lithography.

The PMMA-supported method is the most widely used transfer process at present for graphene prepared by CVD [6]. The use of PMMA as a supporting layer originates from the transfer process developed for CNTs from their growth substrates [101]. PMMA also has a good adhesion to graphene, and therefore its use provides more ambivalent effects. Graphene over a large area can be more uniformly intact with the PMMA supporting layer with less ripping of the layer. On the other hand, PMMA cannot be easily peeled off once graphene is on the other substrates. Figure 4.10 is a schematic diagram illustrating the step-by-step process of PMMA transfer. First, a PMMA layer with hundreds of nanometres up to 1 μm in thickness is spun onto an as-grown graphene/metal stack. The thickness of the PMMA layer can be adjusted by changing the spinning rate and the concentration of PMMA solution. To evaporate the solvent in this PMMA layer, the PMMA/graphene/metal stack is baked at an elevated temperature. Subsequently, this stack is placed into a metal-specific etchant bath to etch away the metal substrate. When graphene is grown on a Cu foil by CVD, because the entire surface of the Cu foil is covered with graphene, the graphene grown on the back side of the Cu foil needs to be removed by oxygen plasma before the etching step [9]. Otherwise, after the Cu is etched, the broken pieces of graphene from the back side (which are not protected by the PMMA layer) will float in the solution and later adhere to the (front side) graphene/PMMA layer. The etching typically takes 20 min to 1 hour depending on the metal layer thickness and the type of an etchant. After that, the PMMA/graphene stack

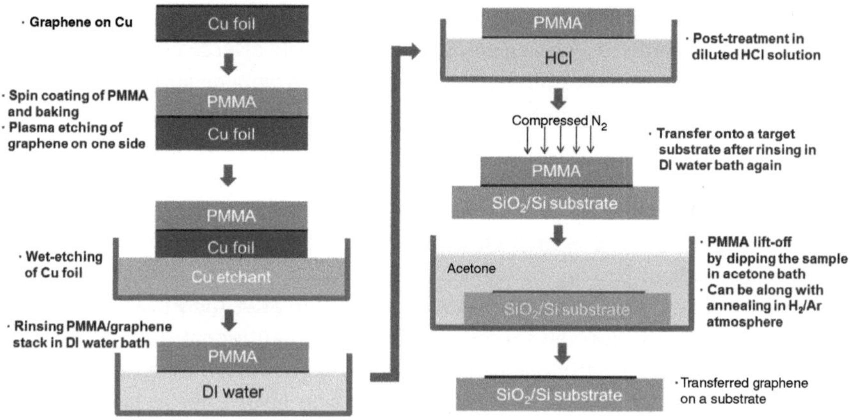

FIGURE 4.10 General PMMA-supported graphene transfer process onto an arbitrary substrate (SiO_2/Si in this case). At first, PMMA is spun on an as-grown graphene/Cu stack and cured. Subsequently, Cu is selectively wet-etched in a Cu-specific etchant bath. After thoroughly rinsing with DI water, the post-treatment using HCl to remove by-products created by the etching is followed. After another rinsing with DI water, the PMMA/graphene stack is brought onto a target substrate. Finally, by lifting off the PMMA in acetone, the graphene prepared onto the target substrate is available for further study or use.

will float on the etchant. Although the etching time is determined by the thickness of the metal substrate and the aggressiveness of the etchant, over-etching is quite often carried out (especially with Cu etchants, such as $FeCl_3$, $Fe(NO_3)_3$ or $(NH_4)_2S_2O_8$), and it has not been noticed that there is a clear degradation on the graphene quality for excessive etching time. After the metal is etched away, the PMMA/graphene stack is scooped from the etchant and brought to a deionized (DI) water bath to be rinsed multiple times for an hour up to overnight soaking. As some etchants (especially $FeCl_3$, $Fe(NO_3)_3$) could leave residues on the graphene, one additional treatment with hydrochloric acid (HCl), is sometimes used in between the DI water rinsing steps [9]. After these steps, the PMMA/graphene stack floating on the DI water bath is scooped by the desired target substrate. To remove any possibly trapped water between the PMMA/graphene and the substrate, compressed nitrogen (N_2) gas is gently blown perpendicular to the PMMA/graphene/substrate stack, pushing out the water droplets through the sides of the PMMA/graphene layer. After this step, the stack is dried to let the PMMA/graphene form a conformal adhesion on the substrate surface. Finally, the PMMA layer is removed by solvents such as acetone, or by annealing under elevated temperature (350–500 °C) and Ar/H_2 gas ambient, or by a combination of both. Process described here are the most basic steps for graphene transfer using PMMA protective layer. In order to avoid any degradation and maintain the graphene quality as much as possible during the transfer, various approaches were used to improve the transfer procedures. For example, for the graphene surface that is in contact with the target substrate, it is desirable to avoid carrying over any contaminants or impurities from the etching process. Liang *et al.* reported that a modified RCA cleaning treatment could effectively remove metal nanoparticles and organic contaminants created on graphene surface during the transfer process, resulting in better performance in the fabricated devices [105]. This modified RCA process is different from a typical RCA cleaning to clean Si wafers [106], only involving two cleaning steps using diluted H_2O_2/HCl and H_2O_2/NH_4O_4 solutions in order. On the other hand, for the other side of the graphene layer where it is in contact with PMMA, how to remove the PMMA completely without leaving residues has been another effort of research during the last few years. For instance, it was reported that after the PMMA/graphene layer is transferred onto the target substrate, if an additional drop of PMMA solution was applied and coated on the PMMA/graphene/substrate stack, the new drop of solution will help the original PMMA layer to have partial or full redissolution, which will be helpful obtaining a more conformal adhesion of the graphene to the target substrate. In this way, it was found that there are much reduced cracks in the transferred graphene layer [107]. To remove the PMMA, as an acetone-only treatment fails to fully remove the PMMA, annealing at an elevated temperature higher than 360 °C [108] for a few hours under a hydrogen/argon atmosphere has been used to follow the acetone treatment [9,109–111]. This thermal treatment has been widely used to yield a relatively clean surface of the graphene with minimal PMMA residues. However, recent results also pointed out some issues regarding this post-annealing. When fabricating devices with exfoliated graphene on SiO_2/Si and investigating whether thermal annealing can effectively remove PMMA (e-beam lithography resist) residues, it was found that thermal treatment above 300 °C in vacuum was required to effectively remove resist residues on graphene surfaces; however, annealing at temperatures >300 °C will concomitantly bring graphene in close contact with SiO_2 substrates and induce increased coupling between them, which leads to heavy hole doping and severe degradation of carrier mobility in graphene devices [112]. The second issue is that this annealing treatment cannot completely remove the PMMA layer. Through a TEM study with a suspended graphene sample placed on a TEM grid, it was found that the PMMA layer is decomposed in a two-step scheme [113]. The PMMA facing the air is easily decomposed at a temperature as low as 160 °C while the PMMA facing the graphene (within the same PMMA coating) appears to be much harder to decompose. Furthermore, the TEM examination reveals that annealing cannot completely remove the PMMA. Higher-temperature annealing (>250 °C) does not really yield a much cleaner surface compared with lower (200 °C) temperature annealing, but increases the risk of graphene structural damage. It was found that the cleanliness of the graphene surface remains far from satisfactory even after annealing at up to 700 °C in a TEM vacuum chamber. It is suggested that although thermal annealing can help the graphene to have a clean surface by decomposing the PMMA layer, this treatment also adversely promotes the formation of local PMMA residues that have a strong bonding with graphene. For these reasons, non-polymer materials, such as evaporated Au thin film, have been used as a supporting layer [114]. Although in Ref. [114] Au thin film was used because of the incompatibility of PMMA fabrication with the target substrate instead for the purpose of removing residues, it appears that the quality of the resulting graphene (as reflected from the performance of the devices) was reasonably good. Nevertheless, systematic studies should be carried out to compare this type of transfer with the PMMA transfer regarding damage to the graphene or residue issue. Obviously, the cost is a lot more than using PMMA for the transfer.

4.5.2 The effect of target substrates on graphene quality

In terms of the quality of graphene, the interface between graphene and the target substrate is also of interest. Even if the surface of the graphene that makes contacts with the target substrate is very clean, the coupling between the graphene and the substrate can lead to deterioration of graphene properties. For electronic applications, a typical dielectric substrate is a SiO_2/Si. As mentioned earlier, the annealing treatment turned out to promote the coupling and hole doping of the graphene layer by the target substrate. This coupling was attributed to adsorbates, such as water trapped during the transfer [115]. Even a thorough N_2 blowing in the transfer step cannot completely eliminate the water molecules at the interface because the surface group, silanol (SiOH), on the SiO_2 is hydrophilic [116]. An inhomogeneous distribution of these polar groups/adsorbates results in the formation of local charge puddles that act as scattering centres for charge carriers. The corrugated surface morphology of SiO_2 on the atomic scale also gives rise to local charge impurities distribution and local strains are induced in the graphene by the formation of ripples. One solution to cope with local charge puddles is to screen the SiO_2 surface by coating it with hydrophobic materials or functional groups before the graphene transfer, such as surface treatment using hexamethyldisilazane (HMDS) [116] or SAMs [117], such as octyltrichlorosilane (OTS) or octadecyltrichlorosilane (ODTS). Suppressed p-doping and enhanced electronic properties for the fabricated FET devices were reported as the result of hydrophobic surface treatment. The carrier mobility increased with increasing alkyl chain length (HMDS $<$ OTS $<$ ODTS), where graphene on the ODTS SAM-treated surface showed more than a threefold increase in the mobility, compared to an untreated sample [117]. The longer chain length was understood to be more effective in screening the charge impurities on the SiO_2 surface. Particularly, when the SiO_2 surface was treated by the octadecyltrimethoxysilane (OTMS) SAM with a closely packed structure and an even longer inert alkyl chain, the resulting carrier mobility increased up to 47,000 $cm^2\ V^{-1}s^{-1}$, which is almost one order of magnitude greater than that for the untreated sample [118].

Besides the surface treatments with the organic molecules or functional groups, one could insert another material in between the graphene and the SiO_2/Si substrate. For such cases, h-BN has been identified as the superior substrate for graphene electronic devices. It has large bandgap (5.97 eV; thus, it is insulating), very small lattice mismatch with graphene (1.7%), atomic surface flatness, and an inert surface free from dangling bonds or charge impurities. All these characteristics make h-BN the best counterpart substrate for graphene [119,120]. (Other two-dimensional materials were also under investigation, e.g. mica has also been used as a common imaging substrate due to its flat surface; however, molecular films of water as thin as 0.4 nm have been observed, which could contribute to charge doping of graphene [121].) With mechanically exfoliated h-BN as substrate, it was also shown that CVD-grown graphene flakes with single-crystal domain showed comparable performance (in terms of carrier mobility) as those reported for mechanically exfoliated graphene [75], thus pointing out the importance of substrate for the graphene FET devices. At present, these excellent graphene devices with h-BN substrates were made with mechanically exfoliated h-BN flakes. There have also been considerable efforts in the CVD synthesis of large-area h-BN films (e.g. [122,123]); comparable-quality h-BN films have not been reliably grown with CVD yet.

4.5.3 Direct transfer of graphene onto target substrates

Most graphene transfer has been carried out with the PMMA technique shown in Figure 4.10. This method has several drawbacks: the complete removal of PMMA residues is challenging as mentioned previously; and most flexible substrates either dissolve in acetone or cannot withstand the annealing temperature; thus, graphene can only be transferred to a limited number of flexible substrates. In order to overcome this, methods that directly transfer graphene onto the target substrates were developed. Regan *et al.* used a drop of isopropanol (IPA) to assist a TEM grid to adhere to graphene/Cu. After the Cu is etched away, graphene is directly transferred to the TEM grid substrate [124]. O'Hern *et al.* used pressing to let the polycarbonate (PC) membrane substrate adhere to the graphene/Cu before etching away the Cu [125]. Verma *et al.* used hot press lamination process to adhere polyethylene terephthalate (PET) substrates to the graphene/Cu and thus transferred CVD graphene to PET substrates [126]. Han *et al.* also used lamination but with poly(ethylene-*co*-vinyl acetate) as a glue for the PET substrates to adhere to graphene/Cu in order to carry out the transfer [127]. Martins *et al.* utilized this lamination technique to transfer onto various different flexible substrates. These substrates are both the target substrate and the protective layer during the transfer [128]. A wide range of substrates were investigated, including polytetrafluoroethylene (PTFE) filter membranes, polyvinyl chloride (PVC), cellulose nitrate/cellulose acetate (CN/CA) filter membranes, PC, paraffin, PET, paper and even cloth. By comparing the properties of these substrates, two critical factors to ensure a successful transfer on *bare* substrates were identified: the substrate's hydrophobicity and good contact between the substrate and graphene. For substrates that do not satisfy those requirements, PMMA was used as a surface modifier or glue to ensure successful transfer. Therefore, the general methodology developed in this work can be applied to suit the need for CVD-graphene on flexible substrates.

4.5.4 Non-destructive exfoliation transfer process

One practical requirement is that the production cost of the graphene synthesis by CVD cannot be excessively high. Presently, SLG with a carrier mobility showing three orders of magnitude can be prepared with a CVD system that simply consists of a stationary thermal furnace, a mechanical pump, gas flow lines for CH_4 and H_2 and a metal substrate, such as a Ni thin film or a Cu foil. However, the post-transfer process can increase the cost significantly because in most of the present working procedures, the metal substrate is wasted by etching, i.e. it is treated as an expendable. Moreover, the disposal of the chemical etchants containing the metal wastes creates additional cost and is environmentally hazardous by itself. For these reasons, the most favourable strategy is an elimination of the transfer process by directly growing graphene on an arbitrary substrate. As discussed earlier, however, it is still challenging, on one hand, to obtain graphene with comparable quality to the one available by established metal-mediated CVD process. For these reasons, efforts have been made to develop a transfer process not requiring the etching of the metal substrate and allowing its reuse to grow graphene repeatedly. Firstly, a dry transfer process using a polyimide (PI)/Au thin film [129] or a thermal release tape [130] could enable the mechanical exfoliation of graphene from the substrate. In contrast to the mechanical exfoliation from HOPG, since this process isolates single to few layers grown on the substrate, a homogeneous force distribution is required when graphene is detached from the substrate and transferred on a target substrate. If this detachment process is not well controlled, the transferred graphene is prone to have physical damage, such as due to cracks or holes. A remarkable transfer process without sacrificing the metal substrate is to electrochemically exfoliate the graphene by generating hydrogen in between the graphene and the metal substrate [131,132]. This bubbling transfer process was already demonstrated through an experiment that takes advantage of preparing graphene with a millimetre-size domain on a Pt foil [131]. After PMMA is spun on such a graphene/Pt foil stack, an electrochemical cell that consists of PMMA/graphene/Pt foil as a cathode, another Pt foil as an anode and 1 M NaOH as an electrolyte is constructed. When 1 A of constant current was maintained across the electrodes, hydrogen bubbles started to generate on the surface of the Pt as a result of the water reduction reaction and these bubbles separated the PMMA/graphene layer from the Pt substrate. Although Pt is not an inexpensive material, by exploiting this transfer process, the Pt substrate can be repeatedly used to prepare graphene with a millimetre-scale domain size. The feasibility of this transfer process together with a process for the continuous growth of large-area graphene on a Cu foil was also demonstrated [132]. Therefore, the bubble transfer process is expected to be available soon as a low-cost alternative to the established chemical etching-based transfer process.

4.6 CONCLUDING REMARKS

In this chapter, the preparation of graphene in large area was discussed, mostly based on the CVD process, covering from the general growth mechanism and its interaction with various process parameters to issues, such as the achievement of large domain size and BLG growth. For the CVD growth, monolayer graphene growth in large area using Cu foil substrate is now the most widely used process, but the synthesis using other metallic substrates has also been actively investigated. There are still many open questions regarding the graphene growth yet to be answered and addressed, e.g. how and when the nucleation occurs or how to achieve uniform BLG or FLG, etc. In terms of the graphene quality, although graphene with single-crystalline domains up to millimetre size has been obtained, to reliably achieve graphene with quality comparable to exfoliated ones and on a large scale is still a task yet to be accomplished. This will require not only optimizing both the graphene growth and the transfer process, but ideally having large-area high-quality CVD-grown h-BN to serve as a superior substrate as well. In terms of graphene transfer, although existing efforts have addressed many issues and challenges, there has not been much development in removing the wrinkles in the CVD graphene. Wrinkles have been observed ubiquitously in the CVD graphene (and CVD h-BN) layers, likely due to the negative thermal expansion coefficient of graphene (and h-BN) and the transfer process. It has been shown that wrinkles cause scattering of carriers (which reduce mobility) and the carbon atoms along the wrinkle are found to be less stable due to curvature effect. It would be highly desirable to have wrinkle-free graphene for future applications.

REFERENCES

[1] Karu AE, Beer M. Pyrolytic formation of highly crystalline graphite films. J Appl Phys 1966;37:2179–81.

[2] Presland AE, Walker PL. Growth of single-crystal graphite by pyrolysis of acetylene over metals. Carbon 1969;7:1–4.

[3] Li WZ, Xie SS, Qian LX, Chang BH, Zou BS, Zhou WY, et al. Large-scale synthesis of aligned carbon nanotubes. Science 1996;274:1701–3.

[4] Ren ZF, Huang ZP, Xu JW, Wang JH, Bush P, Siegal MP, et al. Synthesis of large arrays of well-aligned carbon nanotubes on glass. Science 1998;282:1105–7.

[5] May JW. Platinum surface leed rings. Surf Sci 1969;17:267–70.

[6] Reina A, Jia XT, Ho J, Nezich D, Son HB, Bulovic V, et al. Large area, few-layer graphene films on arbitrary substrates by chemical vapor deposition. Nano Lett 2009;9:30–5.

[7] Kim KS, Zhao Y, Jang H, Lee SY, Kim JM, Kim KS, et al. Large-scale pattern growth of graphene films for stretchable transparent electrodes. Nature 2009;457:706–10.

[8] Li X, Cai W, An J, Kim S, Nah J, Yang D, et al. Large-area synthesis of high-quality and uniform graphene films on copper foils. Science 2009;324:1312–4.

[9] Bhaviripudi S, Jia XT, Dresselhaus MS, Kong J. Role of kinetic factors in chemical vapor deposition synthesis of uniform large area graphene using copper catalyst. Nano Lett 2010;10:4128–33.
[10] Luo Z, Lu Y, Singer DW, Berck ME, Somers LA, Goldsmith BR, et al. Effect of substrate roughness and feedstock concentration on growth of wafer-scale graphene at atmospheric pressure. Chem Mater 2011;23:1441–7.
[11] Han GH, Guenes F, Bae JJ, Kim ES, Chae SJ, Shin H-J, et al. Influence of copper morphology in forming nucleation seeds for graphene growth. Nano Lett 2011;11:4144–8.
[12] Yu Q, Jauregui LA, Wu W, Colby R, Tian J, Su Z, et al. Control and characterization of individual grains and grain boundaries in graphene grown by chemical vapour deposition. Nat Mater 2011;10:443–9.
[13] Bagri A, Kim S-P, Ruoff RS, Shenoy VB. Thermal transport across twin grain boundaries in polycrystalline graphene from nonequilibrium molecular dynamics simulations. Nano Lett 2011;11:3917–21.
[14] Grantab R, Shenoy VB, Ruoff RS. Anomalous strength characteristics of tilt grain boundaries in graphene. Science 2010;330:946–8.
[15] Li X, Magnuson CW, Venugopal A, An J, Suk JW, Han B, et al. Graphene films with large domain size by a two-step chemical vapor deposition process. Nano Lett 2010;10:4328–34.
[16] Porter DA, Easterling KE. Phase Transformations in Metals and Alloys. Nelson Thornes; 1992.
[17] Kim H, Mattevi C, Calvo MR, Oberg JC, Artiglia L, Agnoli S, et al. Activation energy paths for graphene nucleation and growth on Cu. ACS Nano 2012;6:3614–23.
[18] Mehdipour H, Ostrikov KK. Kinetics of low-pressure, low-temperature graphene growth: toward single-layer single-crystalline structure. ACS Nano 2012;6:10276–86.
[19] Plummer JD, Deal MD, Griffin PB. Silicon VLSI Technology: Fundamentals, Practice and Modeling. Prentice Hall; 2000.
[20] Vlassiouk I, Regmi M, Fulvio P, Dai S, Datskos P, Eres G, et al. Role of hydrogen in chemical vapor deposition growth of large single-crystal graphene. ACS Nano 2011;5:6069–76.
[21] Zhang W, Wu P, Li Z, Yang J. First-principles thermodynamics of graphene growth on Cu surfaces. J Phys Chem C 2011;115(36):17782–7.
[22] Terasawa T-o, Saiki K. Growth of graphene on Cu by plasma enhanced chemical vapor deposition. Carbon 2012;50(3):869–74.
[23] Yao Y, Li Z, Lin Z, Moon K-S, Agar J, Wong C. Controlled growth of multilayer, few-layer, and single-layer graphene on metal substrates. J Phys Chem C 2011;115(13):5232–8.
[24] Li Z, Wu P, Wang C, Fan X, Zhang W, Zhai X, et al. Low-temperature growth of graphene by chemical vapor deposition using solid and liquid carbon sources. ACS Nano 2011;5(4):3385–90.
[25] Lopez GA, Mittemeijer E. The solubility of C in solid Cu. Scripta Mater 2004;51(1):1–5.
[26] Prasai D, Tuberquia JC, Harl RR, Jennings GK, Bolotin KI. Graphene: corrosion-inhibiting coating. ACS Nano 2012;6(2):1102–8.
[27] Li X, Cai W, Colombo L, Ruoff RS. Evolution of graphene growth on Ni and Cu by carbon isotope labeling. Nano Lett 2009;9(12):4268–72.
[28] Wofford JM, Nie S, McCarty KF, Bartelt NC, Dubon OD. Graphene islands on Cu foils: the interplay between shape, orientation, and defects. Nano Lett 2010;10(12):4890–6.
[29] Robertson AW, Warner JH. Hexagonal single crystal domains of few-layer graphene on copper foils. Nano Lett 2011;11(3):1182–9.
[30] Wu B, Geng D, Guo Y, Huang L, Xue Y, Zheng J, et al. Equiangular hexagon-shape-controlled synthesis of graphene on copper surface. Adv Mater 2011;23(31):3522–5.

[31] Wood JD, Schmucker SW, Lyons AS, Pop E, Lyding JW. Effects of polycrystalline Cu substrate on graphene growth by chemical vapor deposition. Nano Lett 2011;11(11):4547–54.

[32] Fan L, Zou J, Li Z, Li X, Wang K, Wei J, et al. Topology evolution of graphene in chemical vapor deposition, a combined theoretical/experimental approach toward shape control of graphene domains. Nanotechnology 2012;23(11):5605.

[33] Chavez KL, Hess DW. A novel method of etching copper oxide using acetic acid. J Electrochem Soc 2001;148(11):G640–3.

[34] Levendorf MP, Ruiz-Vargas CS, Garg S, Park J. Transfer-free batch fabrication of single layer graphene transistors. Nano Lett 2009;9(12):4479–83.

[35] Tao L, Lee J, Chou H, Holt M, Ruoff RS, Akinwande D. Synthesis of high quality monolayer graphene at reduced temperature on hydrogen-enriched evaporated copper (111) films. ACS Nano 2012;6(3):2319–25.

[36] Reddy KM, Gledhill AD, Chen C-H, Drexler JM, Padture NP. High quality, transferrable graphene grown on single crystal Cu(111) thin films on basal-plane sapphire. Applied Physics Letters 2011;98(11).

[37] Zhang B, Lee WH, Piner R, Kholmanov I, Wu Y, Li H, et al. Low-temperature chemical vapor deposition growth of graphene from toluene on electropolished copper foils. ACS Nano 2012;6(3):2471–6.

[38] Sun Z, Yan Z, Yao J, Beitler E, Zhu Y, Tour JM. Growth of graphene from solid carbon sources. Nature 2010;468(7323):549–52.

[39] Su C-Y, Lu A-Y, Wu C-Y, Li Y-T, Liu K-K, Zhang W, et al. Direct formation of wafer scale graphene thin layers on insulating substrates by chemical vapor deposition. Nano Lett 2011;11(9):3612–6.

[40] Yu Q, Lian J, Siriponglert S, Li H, Chen YP, Pei S-S. Graphene segregated on Ni surfaces and transferred to insulators. Appl Phys Lett 2008;93(11):3103.

[41] Reina A, Thiele S, Jia X, Bhaviripudi S, Dresselhaus MS, Schaefer JA, et al. Growth of large-area single- and bi-layer graphene by controlled carbon precipitation on polycrystalline Ni surfaces. Nano Res 2009;2(6):509–16.

[42] Zhang Y, Tang T-T, Girit C, Hao Z, Martin MC, Zettl A, et al. Direct observation of a widely tunable bandgap in bilayer graphene. Nature 2009;459(7248):820–3.

[43] Yan Z, Peng Z, Sun Z, Yao J, Zhu Y, Liu Z, et al. Growth of bilayer graphene on insulating substrates. ACS Nano 2011;5(10):8187–92.

[44] Umair A, Raza H. Controlled synthesis of bilayer graphene on nickel. Nanoscale Res Lett 2012;7(1):437–1437.

[45] Peng Z, Yan Z, Sun Z, Tour JM. Direct growth of bilayer graphene on SiO_2 substrates by carbon diffusion through nickel. ACS Nano 2011;5(10):8241–7.

[46] Ramon ME, Gupta A, Corbet C, Ferrer DA, Movva HCP, Carpenter G, et al. CMOS-compatible synthesis of large-area, high-mobility graphene by chemical vapor deposition of acetylene on cobalt thin films. ACS Nano 2011;5(9):7198–204.

[47] Liu N, Fu L, Dai B, Yan K, Liu X, Zhao R, et al. Universal segregation growth approach to wafer-size graphene from non-noble metals. Nano Lett 2011;11(1):297–303.

[48] John R, Ashokreddy A, Vijayan C, Pradeep T. Single- and few-layer graphene growth on stainless steel substrates by direct thermal chemical vapor deposition. Nanotechnology 2011;22(16):5701.

[49] Liu X, Fu L, Liu N, Gao T, Zhang Y, Liao L, et al. Segregation growth of graphene on Cu-Ni alloy for precise layer control. J Phys Chem C 2011;115(24):11976–82.

[50] Chen S, Cai W, Piner RD, Suk JW, Wu Y, Ren Y, et al. Synthesis and characterization of large-area graphene and graphite films on commercial Cu-Ni alloy foils. Nano Lett 2011;11(9):3519–25.

[51] Dai B, Fu L, Zou Z, Wang M, Xu H, Wang S, et al. Rational design of a binary metal alloy for chemical vapour deposition growth of uniform single-layer graphene. Nat Commun 2011;2:522.
[52] Sutter PW, Flege J-I, Sutter EA. Epitaxial graphene on ruthenium. Nat Mater 2008;7(5):406–11.
[53] Preobrajenski AB, Ng ML, Vinogradov AS, Martensson N. Controlling graphene corrugation on lattice-mismatched substrates. Phys Rev B 2008;78(7):3401.
[54] Coraux J, N'Diaye AT, Busse C, Michely T. Structural coherency of graphene on Ir(111). Nano Lett 2008;8(2):565–70.
[55] Loginova E, Nie S, Thuermer K, Bartelt NC, McCarty KF. Defects of graphene on Ir(111): rotational domains and ridges. Phys Rev B 2009;80(8):5430.
[56] Sutter P, Sadowski JT, Sutter E. Graphene on Pt(111): growth and substrate interaction. Phys Rev B 2009;80(24):5411.
[57] Gao L, Guest JR, Guisinger NP. Epitaxial graphene on Cu(111). Nano Lett 2010;10(9):3512–6.
[58] N'Diaye AT, Bleikamp S, Feibelman PJ, Michely T. Two-dimensional Ir cluster lattice on a graphene moire on Ir(111). Phys Rev Lett 2006;97(21):5501.
[59] Loginova E, Bartelt NC, Feibelman PJ, McCarty KF. Factors influencing graphene growth on metal surfaces. New J Phys 2009;11:063046.
[60] Zangwill A, Vvedensky DD. Novel growth mechanism of epitaxial graphene on metals. Nano Lett 2011;11(5):2092–5.
[61] Vanbommel AJ, Crombeen JE, Vantooren A. Leed and Auger-electron observations of Sic (0001) surface. Surf Sci 1975;48(2):463–72.
[62] Emtsev KV, Bostwick A, Horn K, Jobst J, Kellogg GL, Ley L, et al. Towards wafer-size graphene layers by atmospheric pressure graphitization of silicon carbide. Nat Mater 2009;8(3):203–7.
[63] Lin YM, Dimitrakopoulos C, Jenkins KA, Farmer DB, Chiu HY, Grill A, et al. 100-GHz transistors from wafer-scale epitaxial graphene. Science 2010;327(5966):662.
[64] Lin YM, Valdes-Garcia A, Han SJ, Farmer DB, Meric I, Sun YN, et al. Wafer-scale graphene integrated circuit. Science 2011;332(6035):1294–7.
[65] Hong G, Wu Q-H, Ren J, Lee S-T. Mechanism of non-metal catalytic growth of graphene on silicon. Appl Phys Lett 2012;100(23):1604.
[66] Rummeli MH, Bachmatiuk A, Scott A, Borrnert F, Warner JH, Hoffman V, et al. Direct low-temperature nanographene CVD synthesis over a dielectric insulator. ACS Nano 2010;4(7):4206–10.
[67] Miyasaka Y, Nakamura A, Temmyo J. Graphite thin films consisting of nanograins of multilayer graphene on sapphire substrates directly grown by alcohol chemical vapor deposition. Jpn J Appl Phys 2011;50(4). DH12-1–DH-12-4.
[68] Lippert G, Dabrowski J, Lemme M, Marcus C, Seifarth O, Lupina G. Direct graphene growth on insulator. Phys Stat Sol B 2011;248(11):2619–22.
[69] Malesevic A, Vitchev R, Schouteden K, Volodin A, Zhang L, Van Tendeloo G, et al. Synthesis of few-layer graphene via microwave plasma-enhanced chemical vapour deposition. Nanotechnology 2008;19(30):5604.
[70] Sun J, Cole MT, Lindvall N, Teo KBK, Yurgens A. Noncatalytic chemical vapor deposition of graphene on high-temperature substrates for transparent electrodes. Appl Phys Lett 2012;100(2):2102.
[71] Ismach A, Druzgalski C, Penwell S, Schwartzberg A, Zheng M, Javey A, et al. Direct chemical vapor deposition of graphene on dielectric surfaces. Nano Lett 2010;10(5):1542–8.

[72] Huang PY, Ruiz-Vargas CS, van der Zande AM, Whitney WS, Levendorf MP, Kevek JW, et al. Grains and grain boundaries in single-layer graphene atomic patchwork quilts. Nature 2011;469(7330):389–92.

[73] Petrone N, Dean CR, Meric I, van der Zande AM, Huang PY, Wang L, et al. Chemical vapor deposition-derived graphene with electrical performance of exfoliated graphene. Nano Lett 2012;12(6):2751–6.

[74] Tsen AW, Brown L, Levendorf MP, Ghahari F, Huang PY, Havener RW, et al. Tailoring electrical transport across grain boundaries in polycrystalline graphene. Science 2012;336(6085):1143–6.

[75] Wu W, Jauregui LA, Su Z, Liu Z, Bao J, Chen YP, et al. Growth of single crystal graphene arrays by locally controlling nucleation on polycrystalline Cu using chemical vapor deposition. Adv Mater 2011;23(42):4898–903.

[76] Yan Z, Lin J, Peng Z, Sun Z, Zhu Y, Li L, et al. Toward the synthesis of wafer-scale single-crystal graphene on copper foils. ACS Nano 2012;6(10):9110–7.

[77] Wang H, Wang G, Bao P, Yang S, Zhu W, Xie X, et al. Controllable synthesis of submillimeter single-crystal monolayer graphene domains on copper foils by suppressing nucleation. J Am Chem Soc 2012;134(8):3627–30.

[78] Wu T, Ding G, Shen H, Wang H, Sun L, Jiang D, et al. Triggering the continuous growth of graphene toward millimeter-sized grains. Adv Funct Mater 2013;23:198.

[79] Li X, Magnuson CW, Venugopal A, Tromp RM, Hannon JB, Vogel EM, et al. Large-area graphene single crystals grown by low-pressure chemical vapor deposition of methane on copper. J Am Chem Soc 2011;133(9):2816–9.

[80] Fang W, Hsu AL, Song Y, Birdwell AG, Amani M, Dubey M, et al. Asymmetric growth of bilayer graphene on copper enclosures using low pressure chemical vapor deposition. (submitted to ACS Nano).

[81] Geng D, Wu B, Guo Y, Huang L, Xue Y, Chen J, et al. Uniform hexagonal graphene flakes and films grown on liquid copper surface. Proc Natl Acad Sci USA 2012;109(21):7992–6.

[82] Han MY, Oezyilmaz B, Zhang Y, Kim P. Energy band-gap engineering of graphene nanoribbons. Phys Rev Lett 2007;98(20):6805.

[83] Zhou SY, Gweon GH, Fedorov AV, First PN, De Heer WA, Lee DH, et al. Substrate-induced bandgap opening in epitaxial graphene. Nat Mater 2007;6(10):770–5.

[84] Giovannetti G, Khomyakov PA, Brocks G, Kelly PJ, van den Brink J. Substrate-induced band gap in graphene on hexagonal boron nitride: ab initio density functional calculations. Phys Rev B 2007;76(7):3103.

[85] McCann E, Fal'ko VI. Landau-level degeneracy and quantum hall effect in a graphite bilayer. Phys Rev Lett 2006;96(8):6805.

[86] McCann E. Asymmetry gap in the electronic band structure of bilayer graphene. Phys Rev B 2006;74(16):1403.

[87] Mak KF, Lui CH, Shan J, Heinz TF. Observation of an electric-field-induced band gap in bilayer graphene by infrared spectroscopy. Phys Rev Lett 2009;102(25):6405.

[88] Oostinga JB, Heersche HB, Liu X, Morpurgo AF, Vandersypen LMK. Gate-induced insulating state in bilayer graphene devices. Nat Mater 2008;7(2):151–7.

[89] Yan K, Peng H, Zhou Y, Li H, Liu Z. Formation of bilayer Bernal graphene: layer-by-layer epitaxy via chemical vapor deposition. Nano Lett 2011;11(3):1106–10.

[90] Liu L, Zhou H, Cheng R, Yu WJ, Liu Y, Chen Y, et al. High-yield chemical vapor deposition growth of high-quality large-area AB-stacked bilayer graphene. ACS Nano 2012;6(9):8241–9.

[91] Nie S, Walter AL, Bartelt NC, Starodub E, Bostwick A, Rotenberg E, et al. Growth from below: graphene bilayers on Ir(111). ACS Nano 2011;5(3):2298–306.

[92] Wu Y, Chou H, Ji H, Wu Q, Chen S, Jiang W, et al. Growth mechanism and controlled synthesis of AB-stacked bilayer graphene on Cu-Ni alloy foils. ACS Nano 2012;6(9):7731–8.
[93] Kim K, Coh S, Tan LZ, Regan W, Yuk JM, Chatterjee E, et al. Raman spectroscopy study of rotated double-layer graphene: misorientation-angle dependence of electronic structure. Phys Rev Lett 2012;108(24):6103.
[94] Fang W, Hsu AL, Caudillo R, Song Y, Birdwell AG, Zakar E, et al. Rapid identification of stacking orientation in isotopically labeled chemical-vapor grown bilayer graphene by Raman spectroscopy. Nano Lett 2013;13(4):1541–8.
[95] Araujo PT, Mafra DL, Sato K, Saito R, Kong J, Dresselhaus MS. Unraveling the interlayer-related phonon self-energy renormalization in bilayer graphene. Scientific Rep 2012;2:1017.
[96] Li Q, Chou H, Zhong J-H, Liu J-Y, Dolocan A, Zhang J, et al. Growth of adlayer graphene on Cu studied by carbon isotope labeling. Nano Lett 2013;13(2):486–90.
[97] Kalbac M, Frank O, Kavan L. The control of graphene double-layer formation in copper-catalyzed chemical vapor deposition. Carbon 2012;50(10):3682–7.
[98] Chen SS, Brown L, Levendorf M, Cai WW, Ju SY, Edgeworth J, et al. Oxidation resistance of graphene-coated Cu and Cu/Ni alloy. ACS Nano 2011;5(2):1321–7.
[99] Bae S, Kim H, Lee Y, Xu X, Park J-S, Zheng Y, et al. Roll-to-roll production of 30-inch graphene films for transparent electrodes. Nat Nanotechnol 2010;5(8):574–8.
[100] Hatzakis M. Electron resists for microcircuit and mask production. J Electrochem Soc 1969;116(7):1033–7.
[101] Jiao L, Fan B, Xian X, Wu Z, Zhang J, Liu Z. Creation of nanostructures with poly(methyl methacrylate)-mediated nanotransfer printing. J Am Chem Soc 2008;130(38):12612–3.
[102] Lee JN, Park C, Whitesides GM. Solvent compatibility of poly(dimethylsiloxane)-based microfluidic devices. Anal Chem 2003;75(23):6544–54.
[103] Wang YY, Burke PJ. A large-area and contamination-free graphene transistor for liquid-gated sensing applications. Appl Phys Lett 2013;103(5):2103.
[104] Kang SJ, Kim B, Kim KS, Zhao Y, Chen Z, Lee GH, et al. Inking elastomeric stamps with micro-patterned, single layer graphene to create high-performance OFETs. Adv Mater 2011;23(31):3531–5.
[105] Liang X, Sperling BA, Calizo I, Cheng G, Hacker CA, Zhang Q, et al. Toward clean and crackless transfer of graphene. ACS Nano 2011;5(11):9144–53.
[106] Kern W. Handbook of semiconductor wafer cleaning technology: science, technology, and applications. Noyes Publications; 1993.
[107] Li XS, Zhu YW, Cai WW, Borysiak M, Han BY, Chen D, et al. Transfer of large-area graphene films for high-performance transparent conductive electrodes. Nano Lett 2009;9(12):4359–63.
[108] Kashiwagi T, Inaba A, Brown JE, Hatada K, Kitayama T, Masuda E. Effects of weak linkages on the thermal and oxidative-degradation of poly(methyl methacrylates). Macromolecules 1986;19(8):2160–8.
[109] Kedzierski J, Hsu P-L, Reina A, Kong J, Healey P, Wyatt P, et al. Graphene-on-insulator transistors made using C on Ni chemical-vapor deposition. IEEE Electron Device Lett 2009;30(7):745–7.
[110] Lin Y-C, Jin C, Lee J-C, Jen S-F, Suenaga K, Chiu P-W. Clean transfer of graphene for isolation and suspension. ACS Nano 2011;5(3):2362–8.
[111] He KT, Wood JD, Doidge GP, Pop E, Lyding JW. Scanning tunneling microscopy study and nanomanipulation of graphene-coated water on mica. Nano Lett 2012;12(6):2665–72.
[112] Cheng Z, Zhou Q, Wang C, Li Q, Wang C, Fang Y. Toward intrinsic graphene surfaces: a systematic study on thermal annealing and wet-chemical treatment of SiO_2-supported graphene devices. Nano Lett 2011;11(2):767–71.

[113] Lin Y-C, Lu C-C, Yeh C-H, Jin C, Suenaga K, Chiu P-W. Graphene annealing: how clean can it be? Nano Lett 2012;12(1):414–9.

[114] Hsu C-L, Lin C-T, Huang J-H, Chu C-W, Wei K-H, Li L-J. Layer-by-layer graphene/TCNQ stacked films as conducting anodes for organic solar cells. ACS Nano 2012;6(6):5031–9.

[115] Nagashio K, Yamashita T, Nishimura T, Kita K, Toriumi A. Electrical transport properties of graphene on SiO_2 with specific surface structures. J Appl Phys 2011;110(2):4513.

[116] Lafkioti M, Krauss B, Lohmann T, Zschieschang U, Klauk H, v Klitzing K, et al. Graphene on a hydrophobic substrate: doping reduction and hysteresis suppression under ambient conditions. Nano Lett 2010;10(4):1149–53.

[117] Lee WH, Park J, Kim Y, Kim KS, Hong BH, Cho K. Control of graphene field-effect transistors by interfacial hydrophobic self-assembled monolayers. Adv Mater 2011;23(30):3460–4.

[118] Wang X, Xu J-B, Wang C, Du J, Xie W. High-performance graphene devices on SiO_2/Si substrate modified by highly ordered self-assembled monolayers. Adv Mater 2011;23(21):2464–8.

[119] Dean CR, Young AF, Meric I, Lee C, Wang L, Sorgenfrei S, et al. Boron nitride substrates for high-quality graphene electronics. Nat Nanotechnol 2010;5(10):722–6.

[120] Wang QH, Jin Z, Kim KK, Hilmer AJ, Paulus GLC, Shih C-J, et al. Understanding and controlling the substrate effect on graphene electron-transfer chemistry via reactivity imprint lithography. Nat Chem 2012;4(9):724–32.

[121] Shim J, Lui CH, Ko TY, Yu Y-J, Kim P, Heinz TF, et al. Water-gated charge doping of graphene induced by mica substrates. Nano Lett 2012;12(2):648–54.

[122] Song L, Ci L, Lu H, Sorokin PB, Jin C, Ni J, et al. Large scale growth and characterization of atomic hexagonal boron nitride layers. Nano Lett 2010;10(8):3209–15.

[123] Kim KK, Hsu A, Jia X, Kim SM, Shi Y, Hofmann M, et al. Synthesis of monolayer hexagonal boron nitride on Cu foil using chemical vapor deposition. Nano Lett 2012;12(1):161–6.

[124] Regan W, Alem N, Aleman B, Geng B, Girit C, Maserati L, et al. A direct transfer of layer-area graphene. Appl Phys Lett 2010;96(11):3102.

[125] O'Hern SC, Stewart CA, Boutilier MSH, Idrobo J-C, Bhaviripudi S, Das SK, et al. Selective molecular transport through intrinsic defects in a single layer of CVD graphene. ACS Nano 2012;6(11):10130–8.

[126] Verma VP, Das S, Lahiri I, Choi W. Large-area graphene on polymer film for flexible and transparent anode in field emission device. Appl Phys Lett 2010;96(20):3108.

[127] Han GH, Shin H-J, Kim ES, Chae SJ, Choi J-Y, Lee YH. Poly(ethylene co-vinyl acetate)-assisted one-step transfer of ultra-large graphene. Nano 2011;6(1):59–65.

[128] Martins LGP, Song Y, Zeng T, Dresselhaus MS, Kong J, Araujo PT. Direct transfer of graphene onto flexible substrates. Proc Natl Acad Sci USA 2013;110(44):17762–7.

[129] Unarunotai S, Murata Y, Chialvo CE, Kim H-s, MacLaren S, Mason N, et al. Transfer of graphene layers grown on SiC wafers to other substrates and their integration into field effect transistors. Appl Phys Lett 2009;95(20):2101.

[130] Caldwell JD, Anderson TJ, Culbertson JC, Jernigan GG, Hobart KD, Kub FJ, et al. Technique for the dry transfer of epitaxial graphene onto arbitrary substrates. ACS Nano 2010;4(2):1108–14.

[131] Gao L, Ren W, Xu H, Jin L, Wang Z, Ma T, et al. Repeated growth and bubbling transfer of graphene with millimetre-size single-crystal grains using platinum. Nat Commun 2012;3:699.

[132] Wang Y, Zheng Y, Xu X, Dubuisson E, Bao Q, Lu J, et al. Electrochemical delamination of CVD-grown graphene film: toward the recyclable use of copper catalyst. ACS Nano 2011;5(12):9927–33.

Chapter 5

Optical Properties of Carbon Nanotubes

Riichiro Saito*, Mildred S. Dresselhaus†, ‡

**Department of Physics, Tohoku University, Sendai 980-8578, Japan*

†*Department of Electrical Engineering and Computer Science, Massachusetts Institute of Technology, 77 Massachusetts Avenue, Cambridge, MA 02139, USA*

‡*Department of Physics, Massachusetts Institute of Technology, 77 Massachusetts Avenue, Cambridge, MA 02139, USA*

5.1 INTRODUCTION

In this chapter, we discuss the optical properties of single-walled carbon nanotubes (SWCNTs). When light (or a photon) is incident on a SWCNT, a valence (2p) or core (1s, 2s) electron is excited to the conduction or higher lying unoccupied energy bands and the electron becomes a photo-excited electron. This phenomenon is called optical, X-ray or ultra-violet (UV) absorption, depending on the energy of the photon, and the electronic interaction of the photo-absorption process is generally an electric–dipole interaction (Figure 5.1(a)). The originally occupied state of the photo-excited electron becomes unoccupied upon optical absorption in a valence or core energy band, and the unoccupied state behaves like a particle with a positive charge, which is called a hole. A bound state is also formed by the Coulomb interaction between the photo-excited electron and the hole that is left behind, which is called a valence or core exciton [1].

The photo-excited electron (a hole) can be scattered by emitting phonons by exciton–phonon or electron–phonon interactions and these excitations relax to the energy bottom (top) of the conduction (valence) band. Then after a time on the order of 10 ps–1 ns, the photo-excited electron and the hole (the exciton) are recombined by emitting a photon by spontaneous emission of a photon, which is known as photoluminescence (PL, Figure 5.1(b), left). Each of these processes, including photo absorption, relaxation and photo emission, is called an optical process. Here we will discuss the optical properties of a SWCNT in terms of a sequence of optical processes [2].

A photon can be emitted not only by PL but also by light scattering processes. When the incident light changes its momentum and/or the energy in the presence of a material, we say that the light is scattered by the material. Scattered light is essential, e.g. to recognize the colour of a material. The scattering phenomenon

Carbon Nanotubes and Graphene. DOI: 10.1016/B978-0-08-098232-8.00005-X

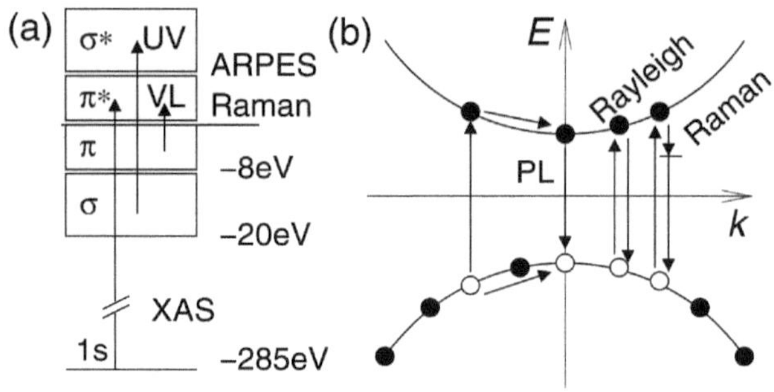

FIGURE 5.1 (a) XAS occurs from the 1s core state to unoccupied states. UV (VL) absorption occurs from the σ (π) energy band to the σ^* (π^*) band. ARPES and Raman spectroscopy are probed by UV and VL, respectively. (b) Optical processes of PL (left), Rayleigh and Raman scattering of light (middle and right). Here solid and open circles correspond to electrons and holes, respectively.

is expressed by consecutive optical processes such as optical absorption, scattering of a photo-excited electron and photon emission (Figure 5.1(b), right). The difference between the PL and light scattering phenomena is the characteristic time over which the scattering event occurs, which is 1000 times faster (10 fs–1 ps) for light scattering than that for PL (10 ps–1 ns). There are two kinds of scattering processes, the elastic and inelastic scattering of light, which are called, respectively, Rayleigh and Raman scattering (Figure 5.1(b), right) [3].

In Rayleigh scattering, because of the vibrating dipole moment that is induced in a material by the incident light, an electromagnetic wave (a photon) is generated by the vibrating dipole moment and the scattered photon propagates in a *different* direction from that of the original incident photon without losing any energy. Since this process is generally fast (10 fs), the Rayleigh spectra has a broad spectral width ($\sim$100 cm^{-1} = 11.6 meV) due to the uncertainty relation between time and energy. In Raman scattering, on the other hand, which is the inelastic scattering of light, the photo-excited electron loses or gains energy, respectively, by generating or absorbing some elementary excitations such as phonons or by exciting another exciton, by exciton–phonon or exciton–exciton interactions, respectively. In this case, the scattered light loses or gains the energy of the elementary excitation, which we call, respectively, the Stokes and anti-Stokes shifts of light, from which we can get information about the phonon or the electron–phonon interaction [2].

Here, in this chapter, we discuss the optical processes of SWCNTs for the energy range of visible light (VL). Before going into the main part of the chapter, we briefly overview the electronic structure and optical process in the overall energy region. The electronic structure of a SWCNT consists of carbon 1s orbitals, sp^2-hybridized σ orbitals and π orbitals (Figure 5.1(a)) [4]. The optical transition energies from the 1s, σ and π orbitals are, respectively, around 285 eV, 5–20 eV, below 5 eV so that X-ray absorption (XAS), UV and VL absorption are relevant to measure the absorption coefficient of the light [3].

In XAS, we can measure the chemical shifts and polarization dependence of light coming from the XAS spectra from a carbon atom, from which we can determine the change of the chemical bond between two C atoms in a SWCNT, such as in attached amorphous carbon, or between sp^3 carbon or other chemical bonding with other atoms (see, e.g. Ref. [5]). UV spectra are mainly used for angle-resolved photo-electron emission spectroscopy (ARPES) in which the kinetic energy of a photo-excited electron is directly measured (for graphene, see Ref. [6]). From ARPES, we can obtain the electronic energy dispersion relation of the valence electrons. However, in the case of SWCNTs, since an isolated single SWCNT is not sufficient to get a signal from ARPES, we do not have ARPES data of an individual SWCNT. Hereafter, we mainly discuss Raman spectroscopy of SWCNTs measured by VL in which an exciton (a bounded pair of a photo-excited electron and a hole) picture is essential for nanotubes as discussed in subsequent sections.

Since a SWCNT is a one-dimensional material, a photo-excited electron and a hole are bound to one another in the cylindrical shell of a SWCNT, and thus the electron–electron interaction between them on a SWCNT becomes large [7]. In fact the exciton binding energy becomes up to 1 eV for smaller-diameter ($\leq$0.6 nm) nanotubes (see, e.g. Refs [8,9]). Thus, the exciton for SWCNTs exists even at room temperature while we can observe excitons only at low temperature for conventional semiconductors. Thus, when we consider the Raman process of a nanotube, the optical transition energy corresponds to excitation energy of an exciton. Because of the symmetry and structure of a SWCNT, the exciton has a special character which appears only in a SWCNT.

In Sec. 5.2, we review the character of the exciton and how to calculate the exciton energy as a function of the chirality of nanotubes. In Sec. 5.3, we show the so-called exciton Kataura plot and then we discuss the modification of the transition energy by the surrounding materials, which is called the environmental effect [10]. Since the Coulomb interaction between a photo-excited electron and a hole is modified by changing the dielectric constant outside (or inside) of a SWCNT, it is important to discuss the environmental effect when assigning the structure of a SWCNT from the optical transition energy. In Sec. 5.5, we will discuss the exciton effect in Raman spectroscopy [2].

5.2 EXCITON ENERGY CALCULATION

5.2.1 Many-body effect in an exciton

Here we briefly introduce the special characteristics of an exciton in a SWCNT and we then show calculational methods and results. There is an attractive Coulomb interaction between a photo-excited electron and a hole called an exciton binding energy, E_{bd} which lowers the optical transition energy. On the other hand, there is a repulsive Coulomb interaction between a photo-excited electron (or a hole) and many valence electrons, which pushes the energy state of the

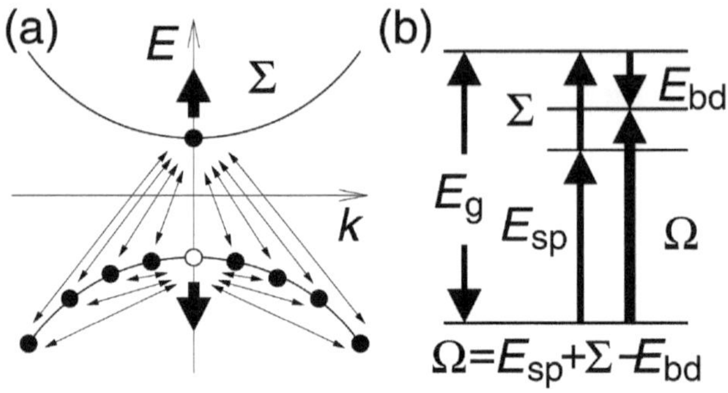

FIGURE 5.2 (a) Self-energy of the Coulomb interaction for a photo-excited electron and a hole with many valence electrons. There is a repulsive (attractive) Coulomb interaction between a photo-excited electron (a hole) and many valence electrons and this interaction pushes the energy state of the photo-excited electron (hole) higher (lower). (b) The exciton energy Ω is a sum of (1) single particle energy E_{sp}, (2) the self-energy Σ and (3) exciton binding energy between the photo-excited electron and a hole E_{bd}; $\Omega = E_{sp} + \Sigma + E_{bd}$. The sum of $\Sigma + E_{bd}$ is many-body effect.

photo-excited electron (the hole) higher (lower) and this energy is called the self-energy of the Coulomb interaction, Σ (Figure 5.2(a)) [7,11,12]. The sum of the exciton binding energy and the self-energy, $\Sigma + E_{bd}$ is called the many-body effect. When we add the single particle excitation energy, E_{sp}, which is expressed by the energy difference between the conduction and valence energy bands, to the many-body effect, we get the transition energy of an exciton, Ω (Figure 5.2(b)) [8]. When we discuss exciton levels in a conventional semiconductor material, the exciton energy levels lie within the energy gap below the bottom of the conduction band. In this case, we implicitly take account of the self-energy term for the energy gap of the semiconductor, $E_g = E_{sp} + \Sigma$. In the case of SWCNTs, if we assume that there is no consideration of the self-energy in the tight-binding parameters, the exciton-energy levels appear at higher energies than the bottom of the conduction band, since the self-energy term (positive) is larger than the exciton binding energy (defined to have a positive value), and thus the many-body effect term becomes a positive value [7].

On the other hand, if we implicitly include a self-energy term in the tight-binding parameters, we just subtract E_{bd} from E_g for obtaining Ω. In fact, when we consider the nearest-neighbour transfer integral for π electrons, which is denoted by γ_0 for graphite, the value of γ_0 is in the range 2.5–2.7 eV [4]. When we try to fit the transition energies of the resonance Raman spectra (RRS) for many isolated nanotubes, having an average diameter of 1.4 nm, by a simple tight-binding calculation, the fitted γ_0 was found to have the relatively large value of 2.89 eV [13]. In the early papers of single nanotube Raman spectroscopy, we did not calculate exciton energies. Thus, the many-body effect can be included by increasing the γ_0 value, which still works for a limited region of SWCNT diameters or transition energies. However, it does not work well for describing the transition energies over a wide region of diameters or energies, but rather needs inclusion of the concept of excitons.

5.2.2 Dark and bright exciton states

One of the important concepts of an exciton in a SWCNT is the symmetry requirement of spin and sub-lattice (or valleys). Here the sub-lattices mean that there are two sub-lattices, A and B, in a SWCNT because of the two lattice sites for the carbon atoms in graphene [4]. The valleys mean that there are two regions of the wavevector K in the two-dimensional (2D) Brillouin zone, around two inequivalent hexagonal corners, the K and K' points, where an optical transition occurs.

The spin of an exciton is defined by the spin state of a photo-excited electron and the spin state of a hole (Figure 5.3). The spin of a hole is defined by an opposite sign to the spin of an electron that would occupy the hole state. For example, if an up-spin electron is excited from a valence state to a conduction state, the spin of the hole is defined by a down spin (Figure 5.3). Here it is noted that an optical dipole transition does not change the direction of the spin and that the spin of the photo-excited electron is an up spin, too. Then a spin singlet $S = 0$ exciton is formed with an up-spin electron and a down-spin hole, which is consistent with an optically allowed dipole transition. On the other hand, if there is a magnetic impurity present, this impurity can change the spin direction of the photo-excited electron, and the exciton is then transformed to be a triplet exciton ($S = 1$). The triplet exciton, as shown in Figure 5.3, is optically forbidden, and thus no photo emission occurs for the triplet exciton. Such an exciton is called a dark exciton. Triplet excitons can be generated by the exchange of electrons (or holes) between two excitons by an electron–electron interaction, too.

It is very important to state that dark exciton states exist even in the case of the singlet exciton when we consider the symmetry of the exciton wavefunction. Here we consider the C_2 rotational symmetry in a SWCNT in which the C_2 axis is perpendicular to the nanotube axis and goes through the centre of a C–C bond

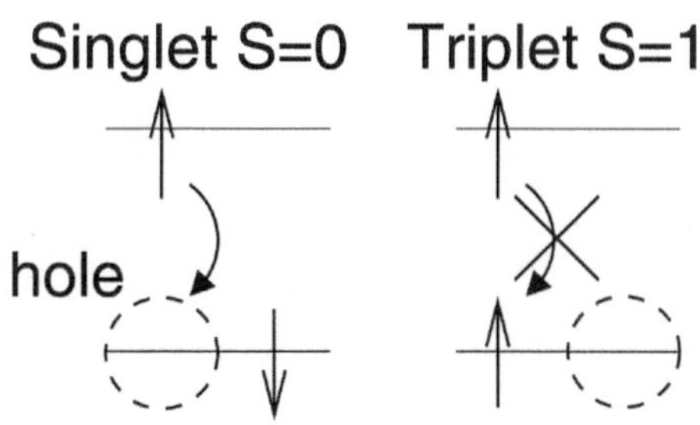

FIGURE 5.3 Spin singlet exciton (left) and triplet exciton (right). The spin of a hole is defined by the opposite direction of spin as the electron had when it occupied that site. The triplet exciton is optically forbidden and no photo emission occurs and thus results in dark excitons.

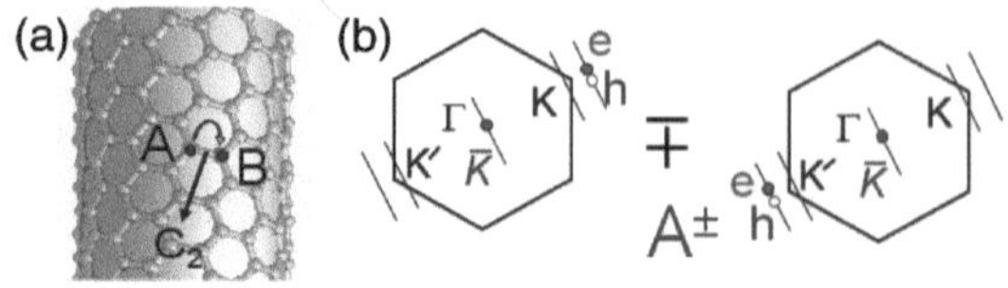

FIGURE 5.4 (a) The C_2 rotation that is perpendicular to the nanotube axis and exchanges the A and B atoms. (b) Even A^+ and odd A^- exciton wavefunctions in which an electron–hole pair near the K point is coupled with an electron–hole pair near the K′ point anti-symmetrically (− of ± for A^+) or symmetrically (+ of ± for A^-). A^+ becomes a bright exciton while A^- becomes a dark exciton.

(Figure 5.4(a)). For the C_2 rotation, the A and B sub-lattices exchange carbon atoms with each other in real space, while the one-dimensional k vector changes from k to $-k$ in k space.

In general, when there is a symmetry operation R that does not change the Hamiltonian H (or in terms of group theory, R commutes with H, $[H, R] = 0$), any eigenfunction of H should also be an eigenfunction of R (or an irreducible representation of a group which has R). If we consider a simple point group that has only two symmetry operations or symmetry operators, $\{E, C_2\}$ (E is an identity operator), all wavefunctions should be either symmetric or anti-symmetric (even or odd functions, Ψ_+ or Ψ_-) for C_2. Such wavefunctions can be obtained for a given eigenfunction of H, Ψ by

$$\Psi_\pm = (E \pm C_2)\Psi, \; C_2\Psi_+ = \Psi_+ \text{ and } C_2\Psi_- = -\Psi_- \tag{5.1}$$

in which ($E \pm C_2$) are known as projection operators for each irreducible representation. The exciton wavefunction Ψ in Equation (5.1) is given by a linear combination of products of electron (conduction band, c) and hole (valence band, v) wavefunctions, $\Psi_C(k)\Psi_V(k)$,

$$\Psi = \sum_k Z(k)\Psi_c(k)\Psi_v(k) \tag{5.2}$$

where $Z(k)$ is a Fourier transform of the envelop of the exciton wavefunction, which will be solved by the Bethe–Salpeter equation (see Sec. 5.2.3). Further, the wavefunctions $\Psi_C(k)$ and $\Psi_v(k)$ are expressed by a linear combination of the Bloch functions $\Phi_i^s(k,r)$, ($i = c$ or v, and $s =$ A or B sub-lattices)

$$\begin{aligned} \Psi_c(k) &= C_c^{\mathrm{A}}(k)\Phi_c^{\mathrm{A}}(k,r) - C_c^{\mathrm{B}}(k)\Phi_c^{\mathrm{B}}(k,r) \\ \Psi_v(k) &= C_v^{\mathrm{A}}(k)\Phi_v^{\mathrm{A}}(k,r) + C_v^{\mathrm{B}}(k)\Phi_v^{\mathrm{B}}(k,r) \end{aligned} \tag{5.3}$$

where $C_i^s(k)$, ($i = c$ or v, $s =$ A or B) are coefficients for the Bloch orbitals by the tight-binding calculation [4]. For the C_2 rotation of the wavefunction, the labels A and B are exchanged with each other, and k is changed to $-k'$ in the 2D Brillouin zone. Thus, the Bloch orbitals are transformed by a C_2 rotation as

$$C_2\Phi_i^{\mathrm{A}}(k,r) = \Phi_i^{\mathrm{B}*}(k,r) \quad \text{and} \quad C_2\Phi_i^{\mathrm{B}}(k,r) = \Phi_i^{\mathrm{A}*}(k,r), \quad (i = c \text{ and } v) \tag{5.4}$$

As for the coefficients of the Bloch orbitals, $C_i^s(k)$, we can use the expression of wavefunction coefficients in Equation (5.3) [14] around the high-symmetry points $\mathbf{K} = (0, -4\pi/(3a))$ and $\mathbf{K}' = (0, +4\pi/(3a))$ in which the wavevector $\mathbf{k} = (k_x, k_y)$ is measured from $\mathbf{K}(\mathbf{K}')$;

$$\begin{aligned} C_c^{\mathrm{A}}(\mathbf{k}) &= \frac{1}{\sqrt{2}}, \quad C_c^{\mathrm{B}}(\mathbf{k}) = \frac{\mp k_y + ik_x}{\sqrt{2}k} \\ C_v^{\mathrm{A}}(\mathbf{k}) &= \frac{1}{\sqrt{2}}, \quad C_v^{\mathrm{B}}(\mathbf{k}) = \frac{\pm k_y - ik_x}{\sqrt{2}k} \end{aligned} \tag{5.5}$$

where $-$ (+) of $\mp$ for $C_c^{\mathrm{B}}(\mathbf{k})$ and + ($-$) of $\pm$ for $C_v^{\mathrm{B}}(\mathbf{k})$ are taken for the $\mathbf{K}(\mathbf{K}')$ points.

Applying the C_2 symmetry operation to the wavefunction coefficients, we obtain the following relationship:

$$\begin{aligned} C_2 C_c^{\mathrm{A}}(k) &= -C_c^{\mathrm{B}}(-k)^*, \quad C_2 C_v^{\mathrm{B}}(k) = -C_c^{\mathrm{B}}(-k)^*, \\ C_2 C_v^{\mathrm{A}}(k) &= C_v^{\mathrm{B}}(-k)^*, \quad C_2 C_v^{\mathrm{B}}(k) = C_v^{\mathrm{B}}(-k)^* \end{aligned} \tag{5.6}$$

where k ($-k$) is measured with respect to the K (K′) point. We then get the following relations,

$$C_2 \Psi_c(k) = -\Psi_c(-k)^* \quad \text{and} \quad C_2 \Psi_v(k) = \Psi_v(-k)^* \tag{5.7}$$

Thus, $\Psi_\pm$ in Equation (5.1),

$$\psi_\pm = \sum_k Z(k)\left(\Psi_c(k)\Psi_v(k) \mp \Psi_c(k)^* \Psi_v(k)^*\right) \tag{5.8}$$

In Figure 5.4(b), we show the $\Psi_\pm$, which is a linear combination of the two exciton pair (e and h) wavefunctions at k near the K point and at $-k$ near the K′ point. It is noted that the centre of momentum of an exciton is given by $k_e - k_h$, and thus in both cases the centre of momentum exists at the Γ point at the centre of the Brillouin zone.

When we only consider $2p_z$ orbitals on the xy plane, we show that the matrix element of the dipole transition vanishes for the odd function for C_2, Ψ_- at any k point before taking the sum over k in Equation (5.8). The matrix elements are proportional to $\langle \Psi_f | \nabla | \Psi_i \rangle$, where Ψ_i and Ψ_f are the wavefunctions for the initial and final states, respectively. When we take the z axis in the direction perpendicular to the cylindrical surface of the SWCNT (see Figure 5.4), and thus the same as the C_2 axis), $\partial/\partial z$ of ∇ is an even operator for C_2. Since Ψ_i is an even function, $\langle \Psi_f | \nabla | \Psi_i \rangle$ vanishes when $\Psi_f = \Psi_-$. It is noted here that $\partial/\partial x$ or $\partial/\partial y$ of ∇ is an odd function of x or y, and the product of two $2p_z$ is an even function of x or y, and therefore the atomic matrix element for $\partial/\partial x$ or $\partial/\partial y$ always vanishes; $\langle 2p_z | \partial/\partial x | 2p_z \rangle = \langle 2p_z | \partial/\partial x | 2p_z \rangle = 0$. Thus, the odd function Ψ_- corresponds to dark exciton states which can neither absorb nor emit a photon, while the even function Ψ_+ is a bright exciton. Since the shape of the wavefunctions is different for Ψ_+ and Ψ_-, the exciton energy levels are not degenerate but are singlet exciton states, denoted by A1 and A2.

There are other exciton states which are also dark exciton states and are doubly degenerate (E excitons). In the case of an E exciton, when an electron exists near the K point, a hole must also exist near the K' point or vice versa. Even though the electron and hole are separated in k space, the electron and hole can be bounded in real space. Since the electron and hole have different wavevectors, optical dipole transitions between them are not possible, and thus excitons with E symmetry are dark excitons. The centre of momentum of E excitons exists near the K point in contrast to the A^+ and A^- excitons which have their centre of momentum near the Γ point and have different symmetry properties as discussed in the previous sections. E-exciton states appear as intermediate states when the electron or hole of the A^+ and A^- excitons is scattered from the K point to the K′ point (intervalley scattering) by exciton–phonon interaction.

5.2.3 Bethe–Salpeter equation

Next we briefly explain how to calculate the exciton energy Ω_n and the wavefunction Ψ^n [7,8,11,15].[1] Since the exciton wavefunction is localized in real space by the exciton binding energy, the wavevector of an electron ($\mathbf{k}_c$) or a hole ($\mathbf{k}_v$) is no longer a good quantum number. Thus, the exciton wavefunction Ψ_n for the n-th exciton energy Ω_n is given by a linear combination of Bloch functions at many $\mathbf{k}_c$ and $\mathbf{k}_v$ wavevectors with appropriate weight functions Z(k) in Equation (5.8). The Z(k) is calculated so as to minimize the energy of the exciton and an eigenvalue equation for mixing different wavevectors through the Coulomb interaction is obtained by the so-called Bethe–Salpeter equation [11]. The Bethe–Salpeter equation is solved as an eigenvalue problem of a large matrix in which the basis functions of the matrix are one-electron wavefunctions specified by the wavenumbers $\mathbf{k}_c$ and $\mathbf{k}_v$

$$\sum_{\mathbf{k}_c\mathbf{k}_v}\left\{[E(\mathbf{k}_c)-E(\mathbf{k}_v)]\delta_{\mathbf{k}_c^{'}\mathbf{k}_c}\delta_{\mathbf{k}_v^{'}\mathbf{k}_v}+K(\mathbf{k}_c^{'}\mathbf{k}_v^{'},\mathbf{k}_c\mathbf{k}_v)\right\}\Psi^n(\mathbf{k}_c\mathbf{k}_v)=\Omega_n\Psi^n(\mathbf{k}_c^{'}\mathbf{k}_v^{'}) \quad (5.9)$$

where $E(\mathbf{k}_c)$ and $E(\mathbf{k}_v)$ in the diagonal matrix element of Equation (5.9) are the quasi-electron and quasi-hole energies, respectively.[2] Ω_n is the energy of the n-th excitation, and $\Psi^n(\mathbf{k}_c\mathbf{k}_v)$ are the components of the excitonic wavefunctions.

1. If the reader wants to know more precise information that would be useful for using Sec. 5.2.3, please refer to previous references (review articles or a book) [1,2,3,16].

2. 'Quasi-particle' means that we add a Coulomb interaction to the one-particle energy and that the particle has a finite lifetime.

The off-diagonal term of Equation (5.9), $K(\mathbf{k}_c^{'}\mathbf{k}_v^{'},\mathbf{k}_c\mathbf{k}_v)$ is given by a sum of direct and exchange interaction matrices, K^{d} and K^{x},

$$K(\mathbf{k}_c^{'}\mathbf{k}_v^{'},\mathbf{k}_c\mathbf{k}_v) = -K^{\mathrm{d}}(\mathbf{k}_c^{'}\mathbf{k}_v^{'},\mathbf{k}_c\mathbf{k}_v) + 2\delta_{\mathrm{S}}K^{\mathrm{x}}(\mathbf{k}_c^{'}\mathbf{k}_v^{'},\mathbf{k}_c\mathbf{k}_v) \qquad (5.10)$$

with $\delta_{\mathrm{S}} = 1$ for spin-singlet and 0 for spin-triplet states. Here, K^{d} and K^{x} are given by the following two-centre integrals on **r** and **r'**, [17]

$$\begin{aligned} K^{\mathrm{d}}(\mathbf{k}_c^{'}\mathbf{k}_v^{'},\mathbf{k}_c\mathbf{k}_v) &\equiv W(\mathbf{k}_c^{'}\mathbf{k}_c,\mathbf{k}_v^{'}\mathbf{k}_v) \\ &= \int d\mathbf{r}' d\mathbf{r}\, \psi^*_{\mathbf{k}_c^{'}}(\mathbf{r}')\psi_{\mathbf{k}_c}(\mathbf{r}')w(\mathbf{r}',\mathbf{r})\psi_{\mathbf{k}_v^{'}}(\mathbf{r})\psi^*_{\mathbf{k}_v}(\mathbf{r}) \\ K^{\mathrm{x}}(\mathbf{k}_c^{'}\mathbf{k}_v^{'},\mathbf{k}_c\mathbf{k}_v) &= \int d\mathbf{r}' d\mathbf{r}\, \psi^*_{\mathbf{k}_c^{'}}(\mathbf{r}')\psi_{\mathbf{k}_v^{'}}(\mathbf{r}')v(\mathbf{r}',\mathbf{r})\psi_{\mathbf{k}_c}(\mathrm{r})\psi^*_{\mathbf{k}_v}(\mathrm{r}) \end{aligned} \qquad (5.11)$$

where w and v are the screened and bare Coulomb potentials, respectively and ψ the quasi-particle eigenfunction as discussed in subsequent sections.

As discussed in Sec. 5.2.1, the quasi-particle energies are the sum of the single-particle energy [$\varepsilon(\mathbf{k})$] and self-energy ($\Sigma(\mathbf{k})$),

$$E(\mathbf{k}_i) = \varepsilon(\mathbf{k}_i) + \Sigma(\mathbf{k}_i), (i = c, v) \qquad (5.12)$$

where $\Sigma(\mathbf{k})$ is expressed by

$$\begin{aligned} \Sigma(\mathbf{k}_c) &= -\sum_{\mathrm{q}} W(\mathbf{k}_c(\mathbf{k}+\mathrm{q})_v, (\mathbf{k}+\mathrm{q})_v\mathbf{k}_c) \\ \Sigma(\mathbf{k}_v) &= -\sum_{\mathrm{q}} W(\mathbf{k}_v(\mathbf{k}+\mathrm{q})_v, (\mathbf{k}+\mathrm{q})_v\mathbf{k}_v) \end{aligned} \qquad (5.13)$$

In order to obtain the matrix elements of Equation (5.9), the single-particle Bloch wavefunction $\psi_{\mathrm{k}}(\mathbf{r})$ and screening potential W are obtained by either a first principles calculation [11] or an extended tight-binding wavefunction and a random-phase approximation (RPA) calculation [7,8]. In the RPA, the static screened Coulomb interaction is expressed by $W = V/\kappa\varepsilon(\mathbf{q})$, in which the screening of π electrons is expressed by the dielectric function $\varepsilon(\mathbf{q}) = 1 + v(\mathbf{q})\Pi(\mathbf{q})$. Here $\Pi(\mathbf{q})$ is a polarization function given by the RPA and the remaining screening of the core electrons or surrounding materials is expressed by the dielectric constant κ. Further for the unscreened Coulomb potential V, the so-called Ohno potential is commonly used for the π orbitals [18].

$$v(|\mathbf{R}_{u's'} - \mathbf{R}_{0s}|) = \frac{U}{\sqrt{\left(\frac{4\,\varepsilon_0}{e^2}U|\mathbf{R}_{us} - \mathbf{R}_{0s'}|\right)^2 + 1}} \qquad (5.14)$$

where U is the maximum energy to place two electrons on a single site where $|\mathbf{R}_{us} - \mathbf{R}_{s'}| = 0$ and this maximum energy is taken as $U \equiv U_{\pi_a\pi_a\pi_a\pi_a} = 11.3\,\mathrm{eV}$ for π orbitals[3] [18].

5.3 THE CALCULATED EXCITON ENERGIES

5.3.1 The exciton Kataura plot

The optical transition energies for (n, m) SWCNTs are plotted in the so-called Kataura plot as a function of nanotube diameter d_t. Kataura et al. first used this plot for explaining the optical-absorption spectra of a SWCNT bundle sample [19]. Later the Kataura plot [20] was frequently used for obtaining the transition energies of each (n, m) SWCNT in resonance Raman excitation profiles [21], resonance Rayleigh spectra [22] or in PL spectra [23]. For a given (n, m) index, there are some high-intensity optical transition energies from the i-th valence sub-band to the i-th conduction band, denoted by E_{ii}. Because of the periodic boundary conditions around the circumferential direction of a SWCNT, the wavevectors in the circumferential direction are discrete while the wavevectors along the nanotube axis become continuous [4,24]. When we plot possible k vectors in the 2D Brillouin zone, the k vectors lie on line segments aligned periodically in the direction of nanotube axis direction, and these special line segments are known as *cutting lines* [25,26]. The energy sub-bands are defined for each cutting line and the i-th sub-band is labelled by the distance of each cutting line from the K (K′) point in the 2D Brillouin zone.

There are several versions of the Kataura plot even within the author's group: (1) when using a simple tight-binding method in which the trigonal warping effect is included [20], (2) when the extended tight-binding method is used, in which the curvature effect is included [27], (3) when the exciton Kataura plot is used, in which the exciton energy is calculated [10] and (4) finally when the exciton Kataura plot is used with the environmental effect [10]. If the reader wants the most updated data of the transition energy, visit 'Exciton Kataura Plot Page' managed by the authors.[4]

In Figure 5.5, the transition energies E_{ii} are plotted versus inverse diameter $1/d_t$ in the range of $0.67 < d_t < 3.0$ nm [10]. Here the static dielectric constant $\kappa = 2.22$ is used for solving the Bethe–Salpeter equation [8], which reproduces the experimental transition energies used for a SWCNT well for a SWCNT bundle sample with $d_t = 1.4$ nm. The calculated results also reproduce well experimental data of single SWCNTs for a larger diameter by Michel et al. [28] in which (n, m) are measured by electron diffraction from the lower energies, E_{11}^{S}, E_{22}^{S}, E_{11}^{M}, E_{44}^{S}, E_{55}^{S}, E_{22}^{M} and so on, where the labels S and M denote semiconducting and metallic SWCNT, respectively (see Figure 5.5). The data points connected by solid lines and with numbers shown in Figure 5.5 are the so-called *family numbers*, $2n + m$ for (n, m), in which $2n + m$ = constant gives similar E_{ii} values, since SWCNTs with the same $2n + m$ family number have similar

---

3. The Coulomb interaction of two π electrons does not diverge even at $|\mathbf{R}_{us} - \mathbf{R}_{s'}| = 0$ if we consider the wavefunctions in the integration of the Coulomb interaction. In the integration, the Jacobian term ($\propto |\mathbf{R}_{us} - \mathbf{R}_{s'}|^2$) prevents the divergence of the integrand function.

4. See the web page at http://flex.phys.tohoku.ac.jp/eii/

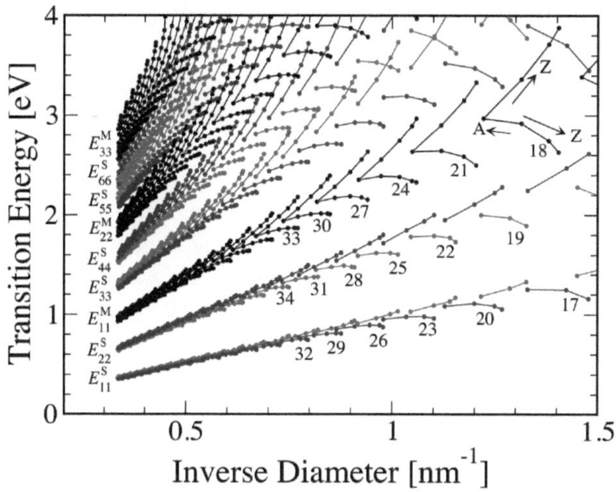

FIGURE 5.5 The bright exciton energy Kataura plot as a function of inverse diameter $1/d_t$ for a static dielectric constant $\kappa = 2.22$. The exciton energies E_{ii} shown here are up to E^S_{66} and E^M_{33}, where S and M denote, respectively, semiconductor and metallic SWCNTs. The $2n + m$ family patterns are clearly seen for smaller-diameter SWCNTs. The arrows with A (Z) symbols correspond to the near armchair (zigzag) SWCNTs.

SWCNT diameters to one another. If $2n + m$ is a multiple of 3, then the SWCNTs are metallic nanotubes, while others are semiconducting SWCNTs. When we divide $2n + m$ by 3, the remainder, mod($2n + m$,3), can be 0, 1 and 2. As for semiconducting SWCNTs, there are two cases 1 and 2 for mod($2n + m$,3), which we call, respectively, Type I and II semiconductors SWCNTs. Depending on Type I and II, the family pattern within the same E^S_{ii} appear on opposite sides of each other. This behaviour comes from the fact that the cutting line position in the 2D Brillouin zone relative to the K point appears on the opposite side for Type I and II semiconducting SWCNTs and this phenomenon is understood by the so-called *trigonal warping effect* of the energy sub-band of SWCNTs [20]. In the case of metallic SWCNTs, in which mod($2n + m$,3) = 0, two cutting lines appear at the same distance from the K point and each cutting line gives an exciton energy which splits into two exciton energies.

5.4 EXCITON ENVIRONMENTAL EFFECT

The calculated exciton energies in the Kataura plot can be used for the assignment of (n, m) by comparing theoretical prediction with experimental observation of the transition energies by PL, Rayleigh and Raman spectroscopy. In the case of experiments, however, the observed E_{ii} values for a (n, m) changes up to 100 meV depending on the surrounding materials which are used as surfactants for dissolving SWCNTs into aqueous solution [29]. Even for air-suspended SWCNTs, the effect of neighbouring SWCNTs in a SWCNT bundle or the presence of oxygen molecules can modify the E_{ii} values, which we call the exciton environmental effect. This effect causes a modification of the many-body term in an exciton in which the Coulomb interaction is screened by the surrounding materials or the materials in the hollow core of SWCNTs.

A simple way to take account of the environmental effect is to change the dielectric constant κ depending on the surrounding materials. However, a single value of κ is not sufficient even for explaining the observed E_{ii} values with different i's within an (n, m) because the size of an exciton wavefunction for a given (n, m) is different for different E_{ii}'s. Thus, the E_{ii} exciton states feel the surrounding materials differently from the electric field which goes out from a SWCNT (see Figure 5.6) [30]. In solving the Bethe–Salpeter equation, κ

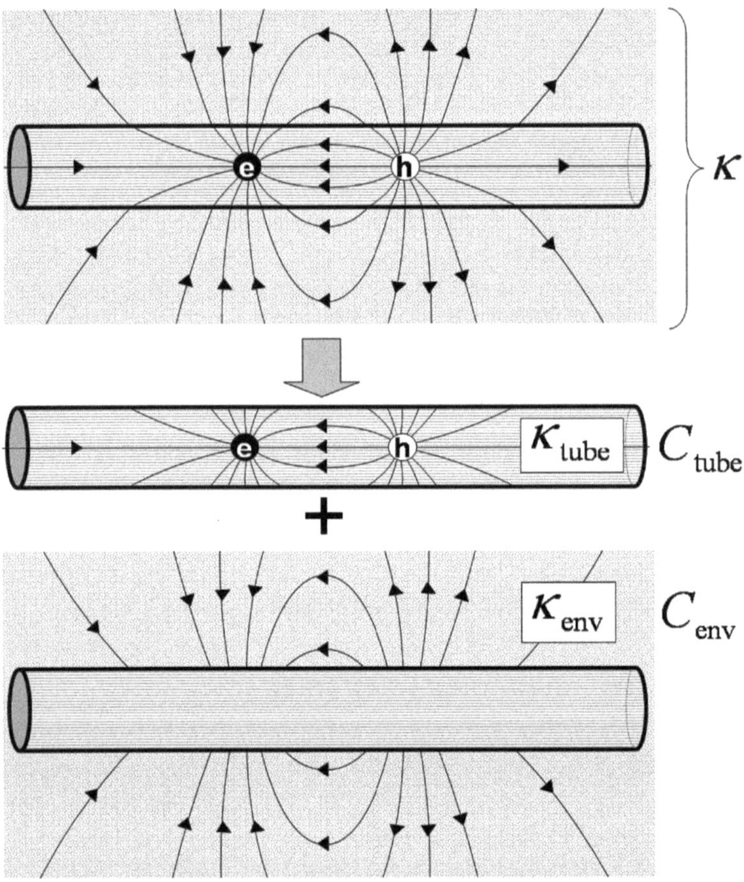

FIGURE 5.6 A schematic of the connection between the net dielectric constant κ and the dielectric constant of the surrounding material κ_{env} and that of the nanotube itself κ_{tube} [30].

denotes both the screening of σ and the core electrons of the SWCNTs and that of the surrounding materials. Thus, we can rewrite κ as

$$\frac{1}{\kappa} = \frac{C_{\text{tube}}}{\kappa_{\text{tube}}} + \frac{C_{\text{env}}}{\kappa_{\text{env}}} \tag{5.15}$$

where κ_{tube} is the dielectric constant within a nanotube except for the π electrons,[5] and C_{tube} and C_{env} are coefficients for the tube itself and for the environment outside of the nanotube, respectively. Then κ in Equation (5.15) is understood by analogy to the serial connection of two capacitors with different dielectric constants, in which the electric flux lines pass through the two dielectric materials (Figure 5.6). In order to understand the environmental effect, Perebeinos et al. proposed that the exciton binding energy E_{bd} of a SWCNT can be scaled by a single parameter, α as

$$E_{\text{bd}} \propto d_{\text{t}}^{\alpha-2} \mu^{\alpha-1} \kappa^{-\alpha} \tag{5.16}$$

where μ is the effective reduced mass of an exciton since the Coulomb interaction does not have a finite range for the acting interaction [18] and the energy can be scaled by changing the length or mass etc. If we take $\alpha = 1.40$ in Equation (5.16), the observed E_{ii}^{S} values for semiconducting SWCNTs (S-SWCNTs) reproduce theoretical prediction well for a large region of d_{t}. However, if we apply the scaling formula to the metallic SWCNTs (M-SWCNTs), it does not work since the screening of π electrons in M-SWCNTs is very large compared with the screening by σ and by the environment denoted by κ. Thus, the exciton size of M-SWCNTs is much larger than that of S-SWCNTs and the exciton binding energy of M-SWCNTs is much smaller than that of S-SWCNTs. If we consider the difference between M-SWCNTs and S-SWCNTs, we should modify κ depending on the metallicity, diameter, etc.

In order to investigate the environmental effect for each M-SWCNT and S-SWCNT, we first fit the κ value in the Bethe–Salpeter equation to reproduce the experimental E_{ii} value for each case. Then we plot the optimized κ as the function

$$\kappa \propto p^{A} (1/d_{\text{t}})^{B} (1/\mu)^{C} \tag{5.17}$$

where the integer p corresponds to the ratio of the distances of the cutting lines for each E_{ii} transition from the K-point in 2D Brillouin zone of graphene [20], and $p = 1,2,3,4,5$ stands for E_{11}^{S}, E_{22}^{S}, E_{11}^{M}, E_{33}^{S} and E_{44}^{S}, respectively. We get two fitting functions of κ for S-SWCNTs and M-SWCNTs as shown in Figure 5.7. Here we adopt the parameters of Equation (5.17); $A = 1.0$, $B = 1.6$ and $C = 0.8$.

5. The screening of π electrons is considered by the polarization function $\Pi(q)$ within the RPA in the Bethe–Salpeter equation.

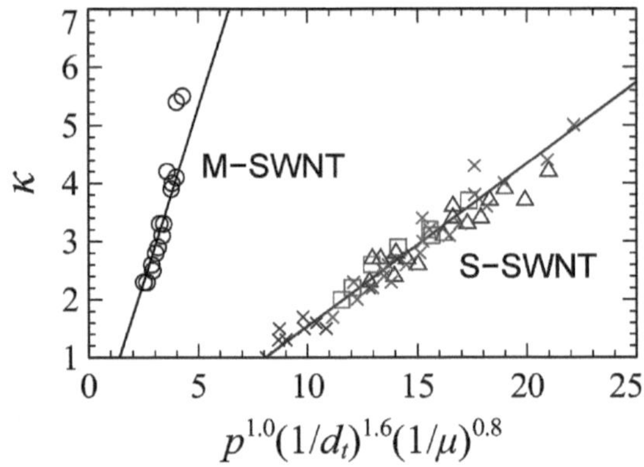

FIGURE 5.7 The fitted κ as a function of (p, d_t, μ). Two different functions are needed, one for M-SWCNTs (circle) and one for S-SWCNTs (others), respectively. The data plotted here are based on the alcohol chemical vapour deposition sample. The following symbols are used: E_{11}(○), E_{22}(×), E_{33}(Δ) and E_{44}(□) [10].

The two fitting functions are further merged into one function if the fitted κ is replotted as the following functional form

$$\kappa = C_\kappa \left\{ p^a (1/d_t)^b (1/\ell_k)^c \right\} \tag{5.18}$$

The variable ℓ_k is defined by the width of the exciton wavefunction in k space. When the exciton states are spatially localized, many k wavevectors should be mixed with one another for representing the exciton wavefunction in Equation (5.8), and thus the ℓ_k becomes large. The parameters (a, b, c) that are thus determined are common for all different samples so as to both optimize the correlation between κ and (p, d_t, ℓ_k), and to minimize differences of the E_{ii}'s between theory and experiment. The optimized values of (a, b, c) are found to be (0.80 ± 0.10, 1.60 ± 0.10, 0.40 ± 0.05), respectively.

In Figure 5.8, a series of results are shown for the fitted κ function for different experimental E_{ii} data including (a) super growth (SG) [31,32], (b) alcohol catalyst chemical vapour deposition (ACCVD) [32,33] and (c) high-pressure carbon monooxide (HiPCO) SWCNTs wrapped by sodium dodecyl sulphate (SDS) [34]. For each sample, the κ functions can be unified for S-SWCNT and M-SWCNT transitions (E_{ii}^{S} and E_{ii}^{M}) for the same parameters. Further, this unification between the lower (E_{11}^{S}) and higher transitions ($E_{33}^{\mathrm{S}}, E_{44}^{\mathrm{S}}$) was not available in the earlier work [32]. In Figure 5.8(d), the data for three different kinds of samples are merged into one figure and it is clear only that the slope for each data set is different from one another. Further, if we divide the slopes of the ACCVD and HiPCO data by the slope of the SG, we can get unified κ values, as shown in Figure 5.8(e). Thus, all the environmental information which is special for each sample can be expressed by one parameter of this relative slope, e.g. for the ACCVD sample

$$\tilde{C}_\kappa(\mathrm{ACCVD}) \equiv C_\kappa(\mathrm{ACCVD}) / C_\kappa(\mathrm{SG}) \tag{5.19}$$

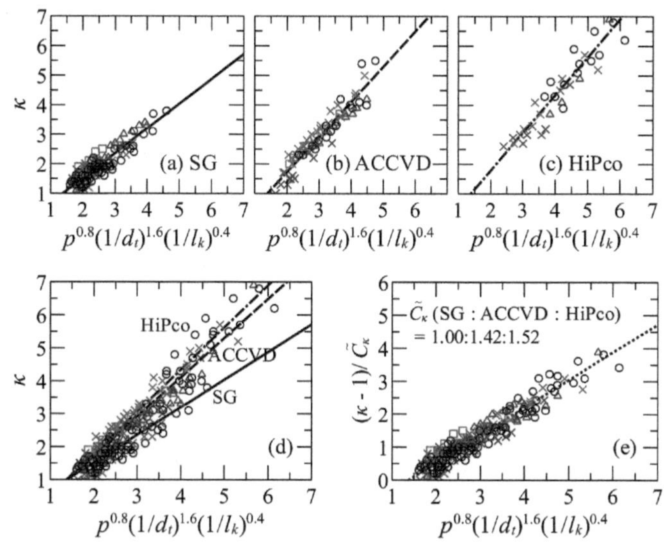

FIGURE 5.8 The κ function for: (a) SG, (b) ACCVD and (c) HiPCO samples. (d) The three samples of (a), (b) and (c) are plotted on the same figure with a fitted slope C_κ for each sample. (e) All the κ functions collapse on to a single line after dividing each function with the corresponding $\tilde{C}_\kappa$. The following symbols are used: $E_{11}(\bigcirc)$, $E_{22}(\times)$, $E_{33}(\Delta)$ and $E_{44}(\square)$ [10].

The values of $\tilde{C}_\kappa$ are now obtained for many different samples and different measurements (PL and Raman) [10]. These data are being updated and kept on the web page (http://flex.phys.tohoku.ac.jp/eii, Exciton Kataura plot pages).

Finally we will give a useful formula for directly obtaining the energy shift of E_{ii} by the environment. Even if we get $\tilde{C}_\kappa$ for a given environment, it is not useful if we do not have the calculated E_{ii}^{env} values to compare with the experiment. Here we define the energy shift $\delta E_{ii}^{\mathrm{env}}$ for a given environment as the difference of that E_{ii} from that for the SG sample: E_{ii}^{SG} [10],

$$\delta E_{ii}^{\mathrm{env}} \equiv E_{ii}^{\mathrm{SG}} - E_{ii}^{\mathrm{env}} = \tilde{C}_\kappa \left[A + B\left(\frac{p}{d_t}\right) + C\left(\frac{p}{d_t}\right)^2 \right] \tag{5.20}$$

where $A = -42.80 \pm 1.26$ meV, $B = 46.34 \pm 1.32$ meVnm^{-1} and $C = -7.47 \pm 0.65$ meVnm^{-2}, are common parameters for all types of environments. Once one gets the list of calculated value of E_{ii}^{SG} from the web page as above, you can get E_{ii}^{env}. The function in Equation (5.20) will give all possible experimental E_{ii} values from PL and RRS within 50 meV [10]. If you have an unknown environmental data set, we can fit the $\tilde{C}_\kappa$ value for that data set with previous experimental data for E_{ii} coming from many previous measurements and we can also update the reference exciton Kataura plot web page.

5.5 EXCITON EFFECT IN RAMAN SPECTROSCOPY

5.5.1 (n,m) assignment from resonance Raman spectra

Finally, we discuss the exciton effect in Raman spectroscopy. If the intermediate states of optical processes of Raman (or Rayleigh) scattering (Figure 5.1) are real electronic states,[6] then the Raman intensity is enhanced (say up to 10^3 times), which we call RRS. In the case of SWCNTs, if the resonance energy is one of the E_{ii}, the enhancement effect is much more significant than for a simple RRS transition since the joint density of states at E_{ii} becomes singular in a one-dimensional material and this effect is known as *the van Hove singularity*. In fact, the density of states at the energy minimum, E_{min} for a simple parabolic energy dispersion in one dimension behaves like $1/\sqrt{E - E_{min}}$, which diverges at $E = E_{min}$. In the case of an exciton, since the exciton wavefunction is spatially localized, the exciton–photon matrix element is much larger (100 times) than the electron–photon matrix element. Thus, a strong absorption occurs for the exciton state which makes the resonance Raman intensity very large. The exciton energy also has an energy dispersion as a function of the centre of momentum, K, of the exciton. Thus, a similar van Hove singularity occurs at $K = 0$, although the exciton with non-zero momentum can not be excited by a dipole transition.

When we can change the energy of the incident light, E_L, by changing the laser excitation energy and measure the Raman intensity as a function of E_L (Raman excitation profile, Figure 5.9(a)), we get information about the E_{ii} for many (n, m), from which we can make an experimental Kataura plot [34] (Figure 5.9(b)). By combining this information with the exciton Kataura plot which was discussed in Sec. 5.3.1, the information about E_{ii} can be used for assigning the (n, m) values for SWCNTs using Raman spectroscopy [36].

5.5.2 Exciton–exciton interaction and electronic Raman spectra

When the power of the incident light becomes strong (e.g. more than 10^9 W cm^{-2}), many excitons are excited along a SWCNT. Since an exciton can diffuse along the SWCNT, two excitons on a SWCNT can be scattered by each other by the Coulomb interaction. In this scattering process, one exciton loses energy to the ground state and another exciton gains the energy at a higher exciton energy, which is known as *an Auger process*. The Auger process is one of the most

---

6. Here 'real' electronic states are electronic states of the energy band. In general, the intermediate states of the optical processes occurring in Raman and Rayleigh scattering can be 'virtual' processes. A virtual state is not an existing state but rather is a state given by a linear combination of the excited states. In perturbation theory, we consider all possible excited states with the matrix elements of the perturbed Hamiltonian in a higher-order correction. When we take the summation on all excited states as a state, the state thus obtained is called a virtual state. In the case of Raman scattering, the electron can be excited to a virtual state at $E = E_L$ by an incident photon. If the photon energy is matched to one of the excited states, the major contribution of the coefficient of the excited states to the virtual state is the resonant excited state.

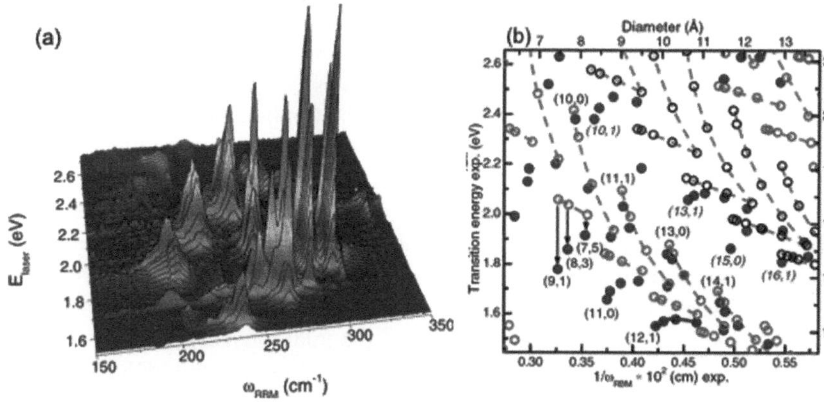

FIGURE 5.9 (a) RBM Raman intensity of HiPCO SWCNTs dispersed in an SDS aqueous solution [23], as a function of incident laser energy E_L and the RBM frequency, which is inversely proportional to the nanotube diameter $1/d_t$ [34]. (b) Experimental Kataura plot (filled circles), E_{ii} versus ω_{RBM} data obtained by Telg et al. [35]. Open circles come from calculation, and grey and black circles indicate calculated optical transitions from semiconducting (E_{22}^{S} and E_{33}^{S}) and metallic (E_{11}^{M}) tubes, respectively.

important non-radiative decay processes of the exciton in the case of SWCNTs. A bright exciton can also decay into a dark exciton state by emitting a phonon which is about 10–50 meV smaller than that of the bright exciton state. Thus, at low temperature below 50 K, most excitons lie in dark exciton states, from which we do not expect PL. The existence of dark exciton states and Auger processes make the quantum efficiency of an exciton in SWCNTs quite low ($\sim$1%). In order to improve the quantum efficiency, we must separate excitons on a SWCNT, perhaps making use of the fact that the dark exciton states should have a higher energy than the bright exciton states. Although there are many challenges for increasing the quantum efficiency of SWCNTs, we should wait for some breakthrough to occur for solving this problem.

The exciton–exciton interaction appears in M-SWCNTs, too. In the case of M-SWCNTs, the exciton which is generated by the incident light with an energy E_L is relaxed to E_{ii} by losing the energy $E_L - E_{ii}$. This energy is used for exciting the exciton for the metallic linear energy sub-band of M-SWCNTs by the exciton–exciton interaction. Farhat et al. have observed a new Raman feature exclusively in M-SWCNTs which is ascribed to the electronic Raman scattering (ERS) [37]. A broad ERS feature (at $\sim$500 cm^{-1}) is observed in the energy region between the radial breathing mode (RBM) and the G band and shows no phonon Raman feature by the following arguments: (1) in comparison to the phonon spectral width ($\sim$1–50 cm^{-1}), the ERS width is much broader ($\sim$500 cm^{-1}) and has a smaller peak intensity ($I_{ERS} \approx 0.6 I_G$); (2) the energy of the inelastic scattered light in the phonon Raman spectra is changed by changing the laser excitation energy E_L, while the ERS peak position does not change but keeps constant at E_{ii}^{M}; (3) the ERS feature is suppressed by changing the Fermi energy, which indicates that the origin of this spectrum comes from exciton excitations in the linear band of M-SWCNTs by the Coulomb interaction. Hasdeo et al. recently calculated the relevant optical process in which the Coulomb interaction is essential for generating the exciton at the linear energy band [38]. What is surprising about their calculation is their result for the direct Coulomb interaction at the wavevector $q = 0$, which is generally considered to be a maximum, becomes absent for the present exciton–exciton interaction because of the symmetry of the two sub-lattices [38]. This vanishing intensity arises because the two direct Coulomb interactions from the A and B sites give an opposite sign to each other which cancel the total intensity. Thus, a second-order exciton–exciton interaction is essential for explaining the ERS effect.

Since the ERS spectra has a broad ESR feature and does not change its energy position with changing E_{L}, the broad spectra of ERS and the phonon Raman spectra may appear in the same energy region if we select E_{L} around the sum of $E_{ii} + E_{\mathrm{ph}}$. Then the scattering amplitude of the ERS and phonon Raman features interfere with each other, which is the origin of the asymmetric lineshape of the phonon Raman peaks and is known as the Breit–Wigner–Fano (BWF) lineshape;

$$I(\omega) = I_0 \frac{\left[1 + (\omega - \omega_0)/q_{\mathrm{BWF}}\Gamma\right]^2}{1 + \left[(\omega - \omega_0)/\Gamma\right]^2} \tag{5.21}$$

where $1/q_{\mathrm{BWF}}$, Γ and ω_0 are, respectively, the asymmetry factor, the spectral width and the central frequency of the discrete phonon. If $1/q_{\mathrm{BWF}} = 0$, we get a symmetric Lorentzian function. Brown et al. observed a BWF lineshape for the G-band Raman spectra in M-SWCNTs as a function of the diameter and found that the asymmetry factor $1/q_{\mathrm{BWF}}$ depends on the SWCNT diameter d_{t} [39]. In the ERS measurement $1/q_{\mathrm{BWF}}$ depends not only on d_{t} but also on the relative position of the ERS feature to the G-band phonon [37].

In Figure 5.10(a) we plot the calculated result for the (E_{L}) dependence of the Raman intensity as a function of the scattered photon energy $(h\omega_{\mathrm{s}})$ [38], which is directly compared with the experimental results of the ERS measurement (Figure 5.10(b) [37]). Here we calculate the E_{L} dependence of the Raman intensity near $M_{22}^{\mathrm{L}} = 2.08$ eV of the (23,14) M-SWCNT. The ERS appearing at $M_{22}^{\mathrm{L}} = 2.08$ eV has a very broad spectral width (FWHM$^{\mathrm{ERS}} \approx 50$ meV) with a peak intensity almost comparable to that of the RBM. At $E_{\mathrm{L}} = 2.07$ eV, the ERS spectrum starts to appear and modifies the RBM and the G-band lineshapes. According to Fano [40], the asymmetry factor $|1/q_{\mathrm{BWF}}|$ is proportional to the coupling constant between the continuum spectrum and the discrete spectrum. The asymmetry factor $|1/q_{\mathrm{BWF}}|$ reaches a maximum at $E_{\mathrm{L}} - M_{22}^{\mathrm{L}} = 40$ meV because the intensity and the peak position of the ERS allow the ERS to have a very strong overlap with the G band (not shown here) [38].

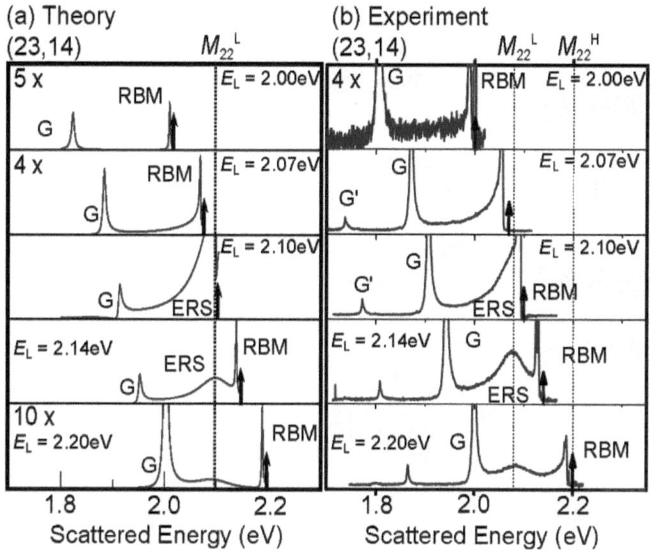

FIGURE 5.10 (a) Calculated result [38] and (b) experimental results (adapted from Ref. [37]) of Raman intensity versus scattered photon energy $(\hbar\omega_s)$ for a (23,14) M-SWCNT where we have the calculated $M_{22}^{L} = 2.10$ eV and the experimental $M_{22}^{L} = 2.08$ eV. The laser excitation energies E_L are taken as 2.00, 2.07, 2.10, 2.14 and 2.20 eV.

5.6 SUMMARY

In summary, we have presented an overview of the optical properties of SWCNTs especially on the subject of the exciton photo-physics of SWCNTs. Because of the large exciton binding energy, SWCNTs provide unique optical phenomena in which the exciton plays an essential role in describing the dynamics of the photo-excited carriers. The existence of the dark exciton makes the quantum efficiency for photon emission low. Since the symmetry of the SWCNT lattice is essential for the appearance of the dark exciton, some symmetry breaking treatment for SWCNTs would be highly beneficial for improving the quantum efficiency of SWCNTs and for their potential application in optical devices. However, if we introduce too much symmetry breaking, such sites will become a non-radiative centre for PL. Thus, many trial and error efforts will be needed for changing the symmetry and will require experimental effects combined with some predictions by theory. An exciton is a promising confined particle for carrying energy from one location to another place. If we make functionalized SWCNTs with some attached molecules, we can perhaps make a unique energy conversion system by using light, which can be applied to solar cells or artificial photosynthesis systems. Since the transition energy can be changed easily by changing the (n, m) value, we can use SWCNTs for communication devices or as a single photon device once we solve the problem of increasing the quantum efficiency. There are many ways to investigate exciton properties. Controlling the exciton dynamics and lattice dynamics with exciton–phonon interactions is likely to become a highly active research area in the near future with particular emphasis given to the optical properties at a picosecond time scale and vibrational frequencies in the terahertz range.

ACKNOWLEDGEMENTS

R. S. acknowledges MEXT grants Nos. 25286005 and 25107005. M.S.D. acknowledges MURI N00014-09-1-1063.

REFERENCES

[1] Dresselhaus MS, Dresselhaus G, Saito R, Jorio A. Raman spectroscopy of carbon nanotubes. Phys Rep 2005;409:47–99.

[2] Saito R, Hofmann M, Dresselhaus G, Jorio A, Dresselhaus MS. Raman spectroscopy of graphene and carbon nanotubes. Adv Phys 2011;60:413–550.

[3] Jorio A, Dresselhaus MS, Saito R, Dresselhaus G. Raman spectroscopy in graphene related systems. Weinheim, Germany: Wiley-VCH Verlag GmbH & Co KGaA; 2010.

[4] Saito R, Dresselhaus G, Dresselhaus MS. Physical properties of carbon nanotubes. London: Imperial College Press; 1998.

[5] Chowdhury MT, Saito R, Dresselhaus MS. Polarization dependence of X-ray absorption spectra in graphene. Phys Rev B 2012;85:115410.

[6] Castro-Neto AH, Guinea F, Peres NMR, Novoselov KS, Geim AK. The electronic properties of graphene. Rev Mod Phys 2009;81:109–62.

[7] Ando T. Excitons in carbon nanotubes. J Phys Soc Jpn 1997;66:1066–73.

[8] Jiang J, Saito R, Samsonidze GG, Jorio A, Chou SG, Dresselhaus G, et al. Chirality dependence of the exciton effects in single-wall carbon nanotubes. Phys Rev B 2007;75:035407.

[9] Jiang J, Saito R, Sato K, Park JS, Samsonidze GG, Jorio A, et al. Exciton-photon, exciton-phonon matrix elements and resonance Raman intensity. Phys Rev B 2007;75:035405.

[10] Nugraha ART, Saito R, Sato K, Araujo PT, Jorio A, Dresselhaus MS. Dielectric constant model for environmental effects on the exciton energies of single wall carbon nanotubes. Appl Phys Lett 2010;97:091905.

[11] Spataru CD, Ismail-Beigi S, Benedict LX, Louie SG. Excitonic effects and optical spectra of single-walled carbon nanotubes. Phys Rev Lett 2004;92:077402.

[12] Wang F, Dukovic G, Brus LE, Heinz TF. The optical resonances in carbon nanotubes arise from excitons. Science 2005;308:838–41.

[13] Jorio A, Saito R, Hafner JH, Lieber CM, Hunter M, McClure T, et al. Structural (*n*, *m*) determination of isolated single-wall carbon nanotubes by resonant Raman scattering. Phys Rev Lett 2001;86:1118–21.

[14] Grüneis A, Saito R, Samsonidze GG, Kimura T, Pimenta MA, Jorio A, et al. Inhomogeneous optical absorption around the K point in graphite and carbon nanotubes. Phys Rev B 2003;67:165402.

[15] Ando T. Theory of electronic states and transport in carbon nanotubes. J Phys Soc Jpn 2005;74:777–817.

[16] M. S. Dresselhaus, G. Dresselhaus, R. Saito, A. Jorio, Exciton photophysics of carbon nanotubes. In : S.R., Leone, J.T., Groves, R.F., Ismagilov, G., Richmond, (Eds.), Annual reviews of physical chemistry chemical physics, Vol. 58, Annual Reviews, Palo Alto, CA, 2007, pp. 719-747.

[17] Rohlfing M, Louie SG. Electron-hole excitations and optical spectra from first principles. Phys Rev B 2000;62:4927.

[18] Perebeinos V, Tersoff J, Avouris P. Scaling of excitons in carbon nanotubes. Phys Rev Lett 2004;92:257402.

[19] Kataura H, Kumazawa Y, Maniwa Y, Umezu I, Suzuki S, Ohtsuka Y, et al. Optical properties of single-wall carbon nanotubes. Synth Metals 1999;103:2555–8.

[20] Saito R, Dresselhaus G, Dresselhaus MS. Trigonal warping effect of carbon nanotubes. Phys Rev B 2000;61:2981–90.

[21] Jorio A, Souza Filho AG, Dresselhaus G, Dresselhaus MS, Saito R, Hafner JH, et al. Joint density of electronic states for one isolated single-wall carbon nanotube studied by resonant Raman scattering. Phys Rev B 2001;63:245416.

[22] Sfeir MY, Beetz T, Wang F, Huang L, Huang XMH, Huang M, et al. Optical spectroscopy of individual single walled carbon nanotubes of defined chiral structure. Science 2006;312:554–6.

[23] O'Connell MJ, Bachilo SM, Huffman XB, Moore VC, Strano MS, Haroz EH, et al. Band gap fluorescence from individual single walled carbon nanotubes. Science 2002;297:593–6.

[24] Saito R, Fujita M, Dresselhaus G, Dresselhaus MS. Electronic structures of carbon fibers based on C_{60}. Phys Rev B 1992;46:1804–11.

[25] Samsonidze GG, Saito R, Jorio A, Pimenta MA, Souza Filho AG, Grüneis A, et al. The concept of cutting lines in carbon nanotube science. J Nanosci Nanotechnol 2003;3:431–58.

[26] Saito R, Sato K, Oyama Y, Jiang J, Samsonidze GG, Dresselhaus G, et al. Cutting lines near the fermi energy of single-wall carbon nanotubes. Phys Rev B 2005;72:153413.

[27] Samsonidze GG, Saito R, Kobayashi N, Grüneis A, Jiang J, Jorio A, et al. Family behavior of the optical transition energies in single-wall carbon nanotubes of smaller diameters. Appl Phys Lett 2004;85:5703–5.

[28] Michel T, Paillet M, Meyer JC, Popov VN, Henrard L, Sauvajol J-L. E_{33} and E_{44} optical transitions in semiconducting single-walled carbon nanotubes: electron diffraction and Raman experiments. Phys Rev B 2007;75(15):155432.

[29] Ohno Y, Iwasaki S, Murakami Y, Kishimoto S, Maruyama S, Mizutani T. Chirality-dependent environmental effects in photoluminescence of single-walled carbon nanotubes. Phys Rev B 2006;73(23):235427. URL http://link.aps.org/abstract/PRB/v73/e235427.

[30] Miyauchi Y, Saito R, Sato K, Ohno Y, Iwasaki S, Mizutani T, et al. Dependence of exciton transition energy of single-walled carbon nanotubes on surrounding dieletcric materials. Chem Phys Lett 2007;442:394–9.

[31] Araujo PT, Jorio A. The role of environmental effects on the optical transition energies and radial breathing mode frequency of single wall carbon nanotubes. Phys Stat Sol B 2008;245:2201–4.

[32] Araujo P, Jorio A, Dresselhaus MS, Sato K, Saito R. Diameter dependence of the dielectric constant for the excitonic transition energy of single-wall carbon nanotubes. Phys Rev Lett 2009;103:146802.

[33] Araujo PT, Doorn SK, Kilina S, Tretiak S, Einarsson E, Maruyama S, et al. Third and fourth optical transitions in semiconducting carbon nanotubes. Phys Rev Lett 2007;98:067401.

[34] Fantini C, Jorio A, Souza M, Strano MS, Dresselhaus MS, Pimenta MA. Optical transition energies for carbon nanotubes from resonant Raman spectroscopy: environment and temperature effects. Phys Rev Lett 2004;93:147406.

[35] Telg H, Maultzsch J, Reich S, Hennrich F, Thomsen C. Chirality distribution and transition energies of carbon nanotubes. Phys Rev Lett 2004;93:177401.

[36] Jorio A, Pimenta MA, Souza Filho AG, Saito R, Dresselhaus G, Dresselhaus MS. Characterizing carbon nanotube samples with resonance Raman scattering. New J Phys 2003;5: 139.1–139.17.

[37] Farhat H, Berciaud S, Kalbac M, Saito R, Heinz TF, Dresselhaus MS, et al. Observation of electronic Raman scattering in metallic carbon nanotubes. Phys Rev Lett 2011;107:157401.

[38] Hasdeo EH, Nugraha ART, Sato K, Saito R, Dresselhaus MS. Origin of electronic Raman scattering and the Fano resonance in metallic carbon nanotubes. Phys Rev B 2013;88(11):115107.

[39] Brown SDM, Jorio A, Corio P, Dresselhaus MS, Dresselhaus G, Saito R, et al. Origin of the Breit–Wigner–Fano lineshape of the tangential G–band feature of metallic carbon nanotubes. Phys Rev B 2001;63:155414.

[40] Fano U. Effects of configuration interaction on intensities and phase shifts. Phys Rev 1961;124:1866.

Chapter 6

Phonon Structures and Raman Effect of Carbon Nanotubes and Graphene

Hans Kuzmany
Universität Wien, Fakultät für Physik, Strudlhofgasse 4, A-1090 Wien, Austria

6.1 INTRODUCTION

Raman scattering (RS) is a key analytical tool for carbon nanophases, in particular for carbon nanotubes (CNTs) and graphene. After the discovery of multiwalled carbon nanotubes (MWCNTs) by Iijima in 1991 [1] and of single-walled carbon nanotubes (SWCNTs) by Iijima et al. [2] and Bethune et al. [3] by transmission electron microscopy, RS immediately started to play a dominant role in studying details of their structures and geometries [4]. Even before that RS has been widely used to characterize carbon phases with respect to their defect structure. It was the so-called D line (defect-induced Raman line) which was used to check the crystallite size L in graphite and low-quality sp^2 carbon systems from the Tuinstra–Koenig relation [5]

$$I(D)/I(G) = C(\lambda)/L \tag{6.1}$$

Here $I(D)$ and $I(G)$ are integrated intensities of the D line and of the graphitic line (G line), respectively and C is a constant which depends on the light wavelength λ used for excitation. For example, $C(514\ \text{nm}) = 0.44$ nm if L is given in nanometres. In a good approximation C scales as λ^4. A detailed analysis of graphitic materials using the $I(D)/I(G)$ ratio was reported in Ref. [6] and, very recently, $I(D)/I(G)$ and $I(D')/I(G)$ ratios were used to characterize defects in graphene [7].

The reason for the widespread use of RS to study the sp^2 carbon phases is the strong signal, originating from a general resonance enhancement, the simultaneous probing of phonon structures and electronic properties, and the dramatic development of instrumentation for RS. Accordingly several review articles were published already on this or related topics [8–15].

RS is a spectroscopic technique that allows studying quasi-particles and their excitations. In our case this is limited, with one exception, to a study of the electronic structure and the excitation of phonons. Properties that can be

Carbon Nanotubes and Graphene. DOI: 10.1016/B978-0-08-098232-8.00006-1

analysed are diameter and chirality of the tubes, functionalization, tube–tube interaction, interaction with environment, doping, orientation, defects, etc., all dependent on temperature and pressure or strain. With respect to basic physics, information can be obtained on scattering rates, electron–phonon coupling and thermal and mechanical properties, such as stiffness and phonon relaxation. Electronic structure comes in particular from resonance excitation and the phonon structure from Raman shift. In the following sections particular attention will be paid to the resonance enhancement of the scattering, to dispersive Raman lines and to RS from $q \neq 0$ modes.

This review starts with descriptions of RS as it is used for SWCNTs and graphene with particular emphasis on the double resonance Raman process and new developments in instrumentation. The next section presents a summary on phonons and phonon dispersions from a theoretical point of view. In the third section, an explicit review on RS from SWCNT is presented where the radial breathing mode (RBM), the D line, the G line and the 2D line are discussed, including their response to doping. A section on tube filling and double-walled carbon nanotubes (DWCNTs) follows. The review on RS from CNTs ends with a special section on selected experiments such as tip-enhanced RS (TERS), electronic RS, linear chains inside the tubes etc. The last two sections discuss RS from graphene, including few-layer graphene (FLG), and an analysis of overtones and combination modes in SWCNTs and graphene.

6.2 THE RAMAN PROCESS WITH PARTICULAR EMPHASIS ON SP² CARBON PHASES

In a first-order Raman process an electron (hole) is first excited by an electron–photon interaction. In the excited state electron–phonon (hole–phonon) interaction leads to the creation (Stokes process) or absorption (anti-Stokes process of a phonon which drives the system into a new state with one more (or one less) phonon). From this state the electron (hole) recombines to the initial electronic ground state. In the overall process energy and momentum must be conserved as described by

$$\hbar\omega_1 = \hbar\omega_2 \pm \hbar\omega_\mathrm{p}, \qquad \hbar k_1 = \hbar k_2 \pm \hbar q \tag{6.2}$$

where k_1, k_2 and ω_1, ω_2 are the wavevectors and frequencies of the incident and scattered light, respectively. q and ω_p are wavevector and frequency of the phonon. Since the momentum of the photon is very small, the maximum q vector which can contribute to the first-order scattering process is $q_{\max} \approx 2k_1 \ll q_{\mathrm{BZ}}$ where q_{BZ} is the wavevector for the edge of the Brillouin zone. Therefore, it is usually claimed that only $q \approx 0$ phonons or Γ-point phonons can contribute to RS.

6.2.1 Quantum mechanics of Raman scattering

For a first-order Raman process the most relevant part of the electron–photon interaction is the $H^A \propto pA$ term of the interaction Hamiltonian where p and A are the electron momentum and the vector potential of the radiation, respectively [16]. RS must be analysed in this case in second-order time-dependent perturbation theory. For electronic RS the Hamiltonian H^{AA} is relevant which can be used in first-order perturbation theory [17]. The Feynman diagrams for the two processes are depicted in Figure 6.1(a). Full drawn lines are electron propagators (→) or hole propagators (←). The propagators for photons and phonons are as indicated. The diagram for the phonon RS leads to the description of the Raman process according to

$$(K_{2f,10})_{\rho,\sigma} = \sum_{i,f'} \frac{\langle \omega_2, f \mid H_\rho^A \mid 0, f' \rangle \langle 0, f' \mid H_{\mathrm{ep}} \mid 0, i \rangle \langle 0, i \mid H_\sigma^A \mid \omega_1, 0 \rangle}{(\hbar\omega_1 - E_{\mathrm{i}}^{\mathrm{e}} - \mathrm{i}\gamma)(\hbar\omega_2 - E_{\mathrm{f'}}^{\mathrm{e}} - \mathrm{i}\gamma)} \tag{6.3}$$

The notation given here follows the description in Ref. [18]. $|0\rangle$ is the initial (electronic) state. $|i\rangle$ and $|f'\rangle$ are intermediate (electronic and phonon) states and $|f\rangle$ the final (electronic and phonon) state. The sum runs over all intermediate electronic states in $E_{\mathrm{i}}^{\mathrm{e}}$ and $E_{\mathrm{f'}}^{\mathrm{e}}$. H_ρ^A and H_σ^A are the ρ, σ components of the electron–photon interaction operator. H_{ep} is the electron–phonon coupling operator. $\hbar\omega_1$ and $\hbar\omega_2$ are the energies of the incident and scattered light, which

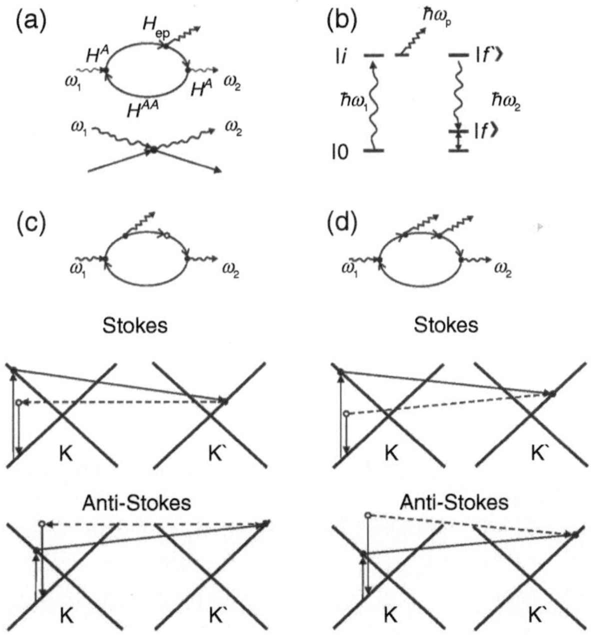

FIGURE 6.1 RS processes for graphene. (a) Feynman diagram for H^A and for H^{AA}. (b) Energy diagram for phonon scattering. (c) Second-order scattering with one impurity and one phonon. The vector diagram shows the corresponding double resonance scattering; (d) same as (c) but for two-phonon scattering. The dashed lines are non-resonant transitions.

differ by the energy of the scattered phonon $\hbar\omega_p$. Eventually the Raman intensity is obtained from $I \propto |K_{2f,10}|^2$. If at least one of the intermediate states is an eigenstate the denominator becomes resonant and the process is strongly enhanced. As it is immediately evident from the formula, resonance conditions can be fulfilled for the incident or for the scattered light, the latter occurring always for higher energies than the former. Figure 6.1(b) has the energy level diagram for the case of Stokes scattering. The final state $|f\rangle$ is higher than the initial state by the energy of the created phonon. The electronic component of this state is equal to the initial state. A similar diagram and description holds for the anti-Stokes process where one phonon is absorbed. For the explicit calculation of the scattering process all possible time orderings of the scattering events must be considered as it is a general rule in quantum mechanics.

In the case of solids the sum over the intermediate states is usually replaced by an integral over the energy and the relevant states are described by the joint density of states, g_{jds}, (JDOS). In addition it is often assumed that the matrix elements do not depend strongly on the energy as compared to the JDOS and can therefore be removed from the integral. This is in particular so if the JDOS is jammed with Van Hove singularities such as in SWCNTs. In this case the expression for $K_{2f,10}$ can be written in a simpler form as

$$K_{2f,10} = M_f M_{f0} M_0 \int \frac{g_{\mathrm{jds}}(E)\mathrm{d}E}{(\hbar\omega_1 - E - \mathrm{i}\gamma)(\hbar\omega_1 - (\hbar\omega_p + E) - \mathrm{i}\gamma)} \tag{6.4}$$

where the explicit matrix elements from Equation 6.3 are replaced by the symbol M and considered energy independent. Within the tight-binding approximation the density of states (DOS) for achiral tubes can be written in an analytic form. For a (n, n) armchair tubes this is

$$\mathrm{DOS}^{AC}(E,n) = \sum_s \frac{E / \pi a \gamma_0^2}{\sin(k_s(E)a/2)[\cos(s\pi/n) - 2\cos(k_s(E)a/2)]} \tag{6.5}$$

s runs over all bands. a and γ_0 are graphene lattice constant and π-electron transfer integral, respectively. Due to the symmetry of valence-band DOS and conduction-band DOS this function multiplied by a factor two can be used in a good approximation in Equation 6.4 for $g_{\mathrm{jds}}(E)$.

Whether a phonon mode gives a Raman signal at all depends on the symmetry of the vibration. The latter are correlated to the Raman tensors $\alpha_{\rho\sigma}$. The Raman tensors are, e.g., tabulated in Refs [10,16] and allow to obtain the scattering intensity as a function of light polarization from $I = |e_i \alpha e_s|^2$ where e_i and e_s are the polarization vectors of the incident and scattered light, respectively.

6.2.2 Double resonance Raman scattering

For the Raman processes in SWCNTs and graphene second-order processes are of great importance, particularly if they are resonance enhanced. Such processes can consist of two-phonon inelastic scattering or one inelastic and one elastic scattering process. Examples are depicted in Figure 6.1(c) and (d). The lower part of Figure 6.1(c) and (d) depicts the scattering processes as they were first suggested by C. Thomsen and S. Reich to explain the dispersion of the D line and of the 2D line in graphite and graphene [20]. The crossed lines represent the Dirac cone. Figure 6.1(c) shows the second-order one-phonon scattering. An electron is excited in resonance from the valence band to the conduction band. Then, a K-point phonon scatters it in a Stokes (anti-Stokes) process to the next nearest K' point in another resonance process. From there the electron has to get back to the original Dirac cone by an elastic scattering mediated by an impurity. This process is non-resonant but allows the electron to recombine with the hole. The elastic scattering is indicated in the figure as a horizontal arrow. Figure 6.1(d) has the similar processes where the elastic scattering is replaced by a second phonon. In this case no impurity is needed. Both graphs in Figure 6.1(c) and (d) describe the inter-valley backward scattering process. In principle the scattering process can start at any point on the circumference of the Dirac cone at K and can end up at any point of the Dirac cone at K', as long as resonance conditions are satisfied. Since in the scattering process two resonant transitions are involved, it is called 'double resonance Raman scattering' (DRRS). A detailed theoretical description of the DRRS is given in Ref. [21].

There is also the possibility for an intra-valley DRRS as it will be discussed later.

The processes described above are quite common. One important consequence of the DRRS is the shift of the Raman line with shifting laser excitation energy. This is evident, since the wavevector of the inter-valley phonon changes with laser energy with a concomitant change of the phonon frequency. A Raman line which shifts with laser excitation energy is called 'dispersive'. Besides the above-described double resonance scattering process systematic sample inhomogeneities can be an other source for dispersive Raman lines [22,23].

6.2.3 Calculation of scattering cross sections

Figure 6.2 shows calculated scattering cross sections for selected phonon modes of SWCNTs. Cross sections for the RBM and for the G line were evaluated on the basis of tight-binding and extended tight-binding models. Matrix elements for the electron–photon interaction H^A and for the electron–phonon interaction H_{ep} were evaluated separately as a function of tube diameter [19]. As it was demonstrated from two-photon absorption experiments [24] optical excitation in SWCNTs is dominated by excitonic transitions rather than by transitions to extended band states. This holds at least for the low-energy transitions E_{11}^S and E_{22}^S in the semiconducting (SC) tubes. In Ref. [19] the evaluated cross sections and Raman intensities are compared explicitly for excitonic and band-to-band transition. The values plotted in Figure 6.2 are RBM cross sections for E_{22} transitions and the corresponding exciton transitions. As shown the overall

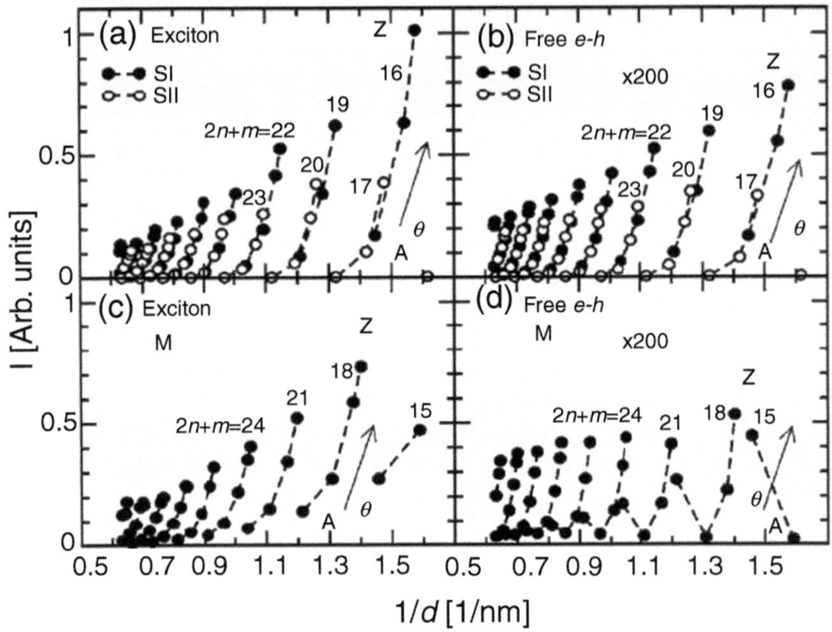

FIGURE 6.2 Calculated Raman intensities per length for RBM modes in sSWCNT and mSWCNT for excitonic transitions (a), (c) and for band-to-band transitions (b) and (d); open and filled symbols in (a) and (b) are for family II and family I, respectively. The numbers are branched with equal $2n + m$. The arrows indicate decreasing chiral angle, from armchair to zigzag. From Ref. [19], copyright with the American Physical Society. Permission for reproduction from November 2012, Ref. #21759.

behaviour with respect to dependence on chiral angle and tube diameter is very similar. However, the absolute values for the cross sections are a factor 200 larger for the excitonic excitation. Also, the diameter dependence for the excitonic excitation is noticeably stronger. For the metallic (M) tubes the Raman intensity has a minimum for decreasing chiral angle in the case of transition to the band states. For the G line in SC tubes Raman intensities from longitudinal optic (LO) and transverse optic (TO) modes show different behaviour. Both, in the excitonic and in the extended band model the LO intensities increase with decreasing chirality while they decrease for the TO modes (not shown). Also, the TO mode intensities are about a factor 10 smaller as compared to the LO intensities. Similar results were reported in Ref. [25] on the level of symmetry adapted non-orthogonal tight-binding (noTB) calculations, albeit only for band-to-band transitions.

6.2.4 Raman instrumentation

Instrumentation for RS progressed remarkably in recent years with respect to measuring resonance profiles, very high spatial resolution, Raman imaging and substantial increase in luminosity of spectrometers. Gas lasers and solid-state-based lasers are now available over the whole spectral range of visible light; quite a few of them are tunable. By using special filters Fourier transform (FT) RS with infrared (IR) excitation is now possible with commercial instruments down to 20 cm^{-1} mode frequency.

Triple monochromators are available with resolution better than 1 cm^{-1}. Single monochromators are used in combination with notch filters or edge filters (Rayleigh filters). Such instruments have a strongly enhanced luminosity and allow, e.g., an easy recording of RS from individual tubes. Rayleigh filter systems are available for Raman shifts down to 10 cm^{-1}. In confocal Raman systems the light is focused by a microscope and unwanted stray light is removed by a pinhole and with a spatial filter. Such instruments yield spatial resolution of around 1 μm and allow depth profiling. They are often used for imaging. High-resolution streamline imaging allows for spatial resolution of 0.5 μm. Up to almost 2000 spectra per minute can be recorded.

In order to obtain spatial resolution below the diffraction limit near-field RS is available. There was a breakthrough in this technique when TERS became available where the tip of an atomic force microscope (AFM) is illuminated from the bottom and scattered light is collected from the near-field region. The tip must approach the sample to a distance of less than 30 nm. Spatial resolution of 15 nm is possible. If the tip is gold-coated, additional enhancement comes from surface-enhanced Raman effect (SERS). Tip-enhanced Raman spectrometers are now commercially available in combination with an AFM which allows automatic control for the correct distance between sample and tip.

As far as time resolution is concerned, picosecond and femtosecond Raman systems are available.

6.3 PHONONS IN SWCNTS AND GRAPHENE

Graphene is a rather simple two-dimensional solid, which has been analysed with respect to vibrational properties on the level of tight binding or density functional theory (DFT). The most prominent vibration appears as the G line in the Raman spectra. It is tangential to the graphene plane and has the signature of a C = C stretch vibration. From the calculated results for graphene the zone-folding (ZF) procedure allows to obtain phonon structures of CNTs. This works except for the RBM, which is the second prominent vibration in SWCNTs. The RBM has almost pure radial character. Alternatively to ZF, noTB or DFT can be applied directly to evaluate the properties of SWCNTs. For the DFT calculations various codes have been used, such as VASP (Vienna *ab initio* Simulation Package), SIESTA (Spanish Initiative for Electronic Simulations with Thousands of Atoms), PWSCF (Plane-Wave Self-Consistent Field) or ABINIT.

6.3.1 Phonons in graphene

Experiments and theory of phonons in graphene are discussed in subsequent sections, neglecting that some of the experimental results were obtained from

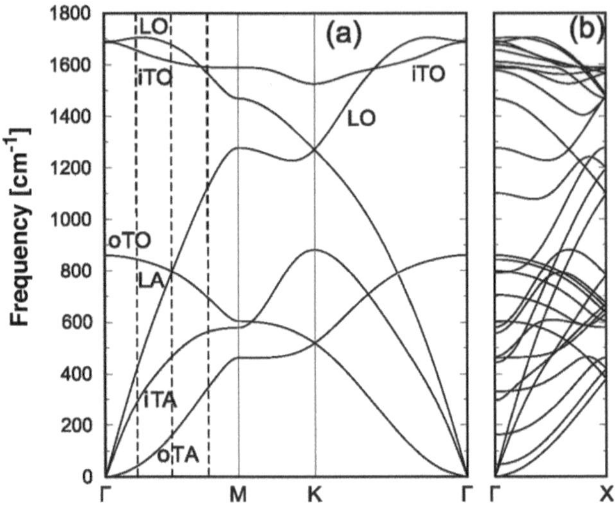

FIGURE 6.3 Phonon dispersion for graphene from a DFT calculation (a) and phonon dispersion for a (4, 4) SWCNT as obtained from ZF (b). For the armchair tubes ZF proceeds along Γ − M. The dashed lines in (a) mark the folding vectors for the (4, 4) tubes. Courtesy of A. Rubio.

graphite. Early work on the phonon dispersion comes from 1995 as evaluated from the VASP code [26] and from 1999 the SIESTA code [27], respectively. Figure 6.3(a) shows results depicting the three acoustic and the three optical modes for directions Γ to M, M to K and K to Γ.

At the Γ point the LO and the in-plane TO (iTO) phonons are degenerated, of symmetry E_{2g} and Raman active. Away from this point they split into non-degenerated modes [30]. The third out-of-plane mode (oTO) is of B_{2g} symmetry and neither Raman nor IR active. The acoustic modes are of u symmetry. The LO mode shows a dip at the Γ point (overbending), and the iTO mode shows a dip at the K point. Both structures come from a strong electron–phonon coupling and are the signature of a Kohn anomaly. In several experiments such as X-ray RS [30], reflection electron energy loss spectroscopy or high-resolution electron energy loss spectroscopy agreement was obtained with DFT calculations. Also, extended tight-binding calculations in the form of noTB [31] yield good agreement with experiments except for a requested downscaling of the frequencies by a factor 0.9. A recent detailed study including Coulomb correlation of the many-body system (GW approximation) revealed an even deeper dip of the iTO mode at the K point extending right away to the frequency of the LO phonon [21]. This was confirmed experimentally from inelastic X-ray scattering in Ref. [32].

To obtain a direct presentation of the phonon dispersion force constants can be used. A set of force constants that was adapted to represent the results from DFT calculations is given in Ref. [33].

6.3.2 Phonons in SWCNTs

A good review on the theory of phonons in SWCNTs is given in Ref. [34]. The easiest way to derive frequencies and dispersions is ZF of the graphene dispersion of Figure 6.3(a). This is demonstrated for a (4,4) armchair tube in Figure 6.3(b). Whereever the dashed lines in Figure 6.3(a) cut a branch of the phonon dispersion the nanotube has a Γ point phonon. For a general (n, n) armchair tube the Brillouin zone has to be cut at n wavevectors. The total number of the so received $q = 0$ phonons is rather large. They are defined by the number $N_C = 2N$ of carbon atoms per unit cell, where N is the number of benzene rings in the cell. For a (10, 10) tube this yields, e.g., 180 vibrational degrees of freedom. An explicit evaluation by ZF for a larger set of tubes was reported in Ref. [35].

The explicit application of DFT calculation allows evaluation of phonon frequencies and phonon dispersion if the unit cell of the tube is not too large. Several 100 atoms per supercell can be handled with good computers. This means such calculations are limited in general to armchair and zigzag tubes. Using the VASP code, the G-line frequencies were calculated as function of tube diameter [36] for mainly zigzag tubes as shown in Figure 6.4. According to this the Raman active G line has six components, two each with symmetries A_{1g}, E_{1g} and E_{2g}, respectively. In the graph of the figure an abbreviated symbol for the assignment is used. For comparison the figure has also the G-line frequencies as obtained from ZF the DFT-calculated graphene dispersion. The difference

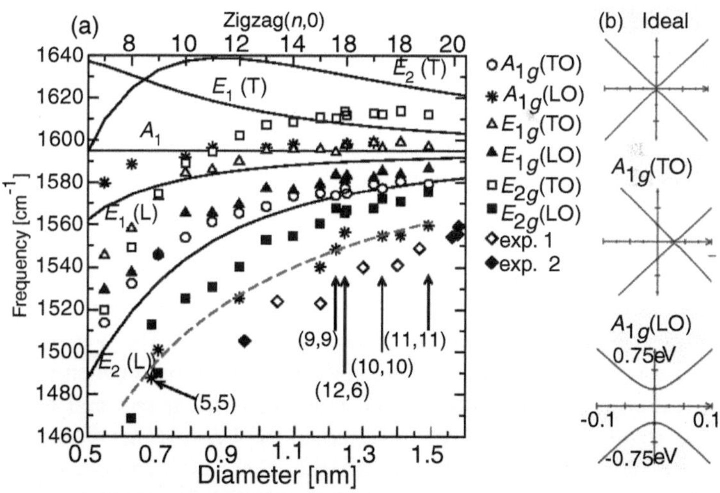

FIGURE 6.4 G-line frequencies for six tangential modes versus tube diameter as calculated from VASP (symbols from the legend). Open and filled diamonds are experimental results for M tubes [28,29]. The dashed line (dashed red line in web version) is an empirical fit. Arrows point to LO modes from additional M tubes. Full lines are results from ZF. (a) Electron–phonon coupling scheme for A_{1g}(LO) and A_{1g}(TO) modes (b). Courtesy of G. Kresse.

between the results from ZF and from *ab initio* calculations is remarkable. First of all there is only one A_{1g} line in the former, since LO and TO modes are degenerated for graphite at Γ. Second the frequencies from ZF are generally higher than those from DFT calculation. This is natural, since curvature reduces force constants due to reduction of $\pi-\pi$ overlap and in ZF curvature is not considered. Similarly, deviations between ZF and *ab initio* calculations are more dramatic for tubes with smaller diameter.

The A_{1g}(LO) mode is almost independent of curvature unless the diameters d are really very small. This reflects the fact that there is no curvature along the tube axis. In contrast, transverse to the tube axis curvature is increasing with decreasing tube diameter, which is reflected in the rather strong dependence of mode frequency on d for the A_{1g}(TO) mode. The most remarkable result from the calculation is the strong reduction of the A_{1g}(LO) frequencies for M tubes. For the 1-nm tube the reduction is 60 cm^{-1}. The red dashed line is a fit from $\omega_{\mathrm{M}}^{\mathrm{LO}} = 1619 - 866 / d$. The reason for this downshift was found to be a particular strong electron–phonon coupling of this mode. The mechanism is indicated in Figure 6.4(b). Whereas the A_{1g}(TO) mode shifts the Dirac point without changing the energy, the motion of the A_{1g}(LO) leads to a substantial change of the electronic states. The calculation also demonstrates that this electron–phonon coupling strongly increases with decreasing tube diameter. This behaviour is reminiscent to electronically induced Peierls distortion like in a Kohn anomaly. This Kohn anomaly was demonstrated in several subsequent papers for SWCNTs [37,38] and graphite [39]. In a more extended calculation VASP was used to calculate phonon frequencies and dispersions for all Raman active tangential and radial modes [40].

The disadvantage of the full *ab initio* calculations has been avoided by extended tight-binding calculations such as the noTB procedure. With this technique curvature can be considered and frequencies as well as dispersions for all phonon modes can be calculated [31]. After proper parameterization and rescaling the calculated frequencies versus tube diameter very good agreement with the DFT calculations is obtained.

Another possibility to avoid expensive calculation on the DFT level is the use of helical unit cells. As it was shown in Ref. [41] each nanotube belongs to a particular line group which means a space group in only one dimension. The tubes can be constructed from only two atoms (helical unit cell) if screw axis operation is used to replicate the structure. However, exactly $gcd(n,m)$ helices ($gcd(n,m)$ means greatest common divisor of (n,m)) are needed to generate the tube, all with the same screw axis. Since the helical unit cell is so small the helical Brillouin zone is very large, equal to $N_C/gcd(N,m)$. There are only $6 \times gcd(n,m)$ branches of phonons for the helical zone but from this all phonon branches for the standard zone can be calculated [10]. This technique was successfully applied in Refs [14,34] to calculate phonon dispersions in small and medium large diameter SWCNT with arbitrary chirality. For the example of a (16, 1) tube $gcd(16,1) = 1$. Thus, there are only 6 phonon branches in the helical zone while there are 1092 branches for the 364 carbons per conventional cell.

6.3.3 Approximate relations for phonon frequencies

Since the DFT and extended tight-binding calculations imply much effort, in several papers parameterized versions of the results are given which represent the *ab initio* calculations very well. They are useful for comparison with experiments or for further calculations. For example, the results for the phonon frequencies from the noTB calculations can be written in a parameterized form as [31]

$$\omega_{\mathrm{RBM}} = \frac{a_2}{(d/2)^{n_2}} + \frac{a_3}{(d/2)^{n_3}} \cos 3\theta \tag{6.6}$$

The values for a_2, a_3, n_2, n_3 are 1300.28, -149.83, 1.00692, 2.348, respectively. θ is the chiral angle of the tubes. Similar parameterized relations were obtained for the various G mode components.

Another useful parameterization for RBM frequencies was given in Ref. [42] for SC and M tubes as

$$\omega_{\mathrm{RBM}} = A/d + B + [C + B\cos^2 3\theta]/d^2 \tag{6.7}$$

Parameter values for SC(M) tubes are 227(227) nm, 7.3(11.8) cm^{-1}, $-1.1(-2.7)$ cm^{-1} nm^2, $-0.9(-2.7)$ cm^{-1} nm^2 for *A*, *B*, *C*, *D*, respectively. Both equations become relevant for small-diameter tubes. For very narrow tubes a DFT-corrected value d_{DFT} for the tube diameter should be used [43]. A parameterized version of this value is [44] $1/d_{\mathrm{DFT}} = 1/d - 0.005/d^2 - 0.0013/d^4$ where d is the geometrical tube diameter.

Finally, if the full dispersion of the modes is requested a useful relation, at least for the RBM, is [45]

$$\omega(q_r, q_\phi) = \{\omega_{\mathrm{k}} + \omega_1 q_r [1 - 0.06\cos(3q_\phi)]\}\left(1 - \frac{Cq_r}{d^2}\right) \tag{6.8}$$

q_r and q_ϕ are phonon wavevectors in polar coordinates, counted from the K point. ω_1, ω_{k} and C are parameters.

6.4 RAMAN SCATTERING FROM SWCNT

In spite of the very large number of $q = 0$ modes the Raman spectrum of SWCNTs is rather simple. One reason is the fact that only few of the modes are Raman active. The number and species of the Raman active modes are independent of the tube size and given as [10]

$\Gamma = 2A_{1g} + 3E_{1g} + 3E_{2g}$ zigzag tubes (8 modes)
$\Gamma = 2A_{1g} + 2E_{1g} + 4E_{2g}$ armchair tubes (8 modes)
$\Gamma = 3A_1 + 5E_1 + 6E_2$ chiral tubes (14 modes)

The total symmetric modes (A modes) can only be observed for parallel orientation of incident and scattered light polarization. *E* modes may be observed either for parallel or for crossed polarization [10]. The Raman spectra become

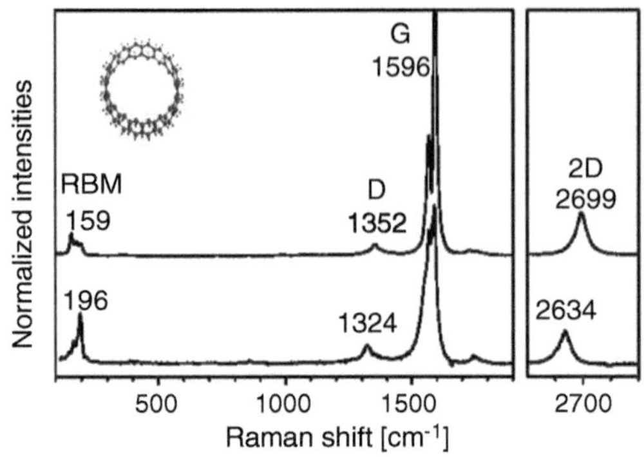

FIGURE 6.5 Raman spectra for SWCNT as excited by a 647-nm laser (lower spectrum) and a 488-nm laser (upper spectrum). The letters RBM, D, G and 2D assign most important lines. Numbers give line positions in cm^{-1}. The inset depicts the normal coordinates for the RBM, which are almost perpendicular to the tube surface.

even more simple as most of the modes are weak scatterers. Figure 6.5 depicts two spectra as recorded in the most common frequency range with two different laser lines.

Four Raman lines dominate: the RBM, the defect-induced line (D line), the G line and the overtone of the D line (2D line). The latter is often also assigned as G' line. Clearly, all Raman lines disperse. For the G line dispersion is only seen from the first spectral moment versus laser energy. In the following, we will discuss the properties of the lines in detail.

6.4.1 The radial breathing mode

The RBM cannot be derived from ZF, since the ZF works with a planar graphene lattice. It needs special treatment as, e.g., provided from a force constant model by Jishi et al. [35] or from *ab initio* calculations [27,40,43,46]. As long as the curvature of the tubes is not too large, the RBM frequency scales with inverse tube diameter d described by the empirical formula

$$\omega_{\mathrm{RBM}} = C_1 / d + C_2 \tag{6.9}$$

where C_1 and C_2 are constants. C_1 is considered as more or less intrinsic whereas C_2 takes care for the environment of the tubes such as bundling, tension, surfactants or substrate effects. From the *ab initio* calculations in Ref. [46] values for C_1 were 236 and 232 nm cm^{-1} (for d in nanometres) for isolated armchair and zigzag tubes, respectively. From experiments a large variety of values for C_1 were suggested [10] all in the range from around 220 to 260 nm cm^{-1}. A good value could be 227. Similarly for C_2, where values from zero to more than 20 cm^{-1} are reported. Thus, from the Raman frequencies a good estimation of the tube diameters can indeed be obtained but the frequencies do not allow to obtain diameters enough accurate to derive chiralities. To obtain the

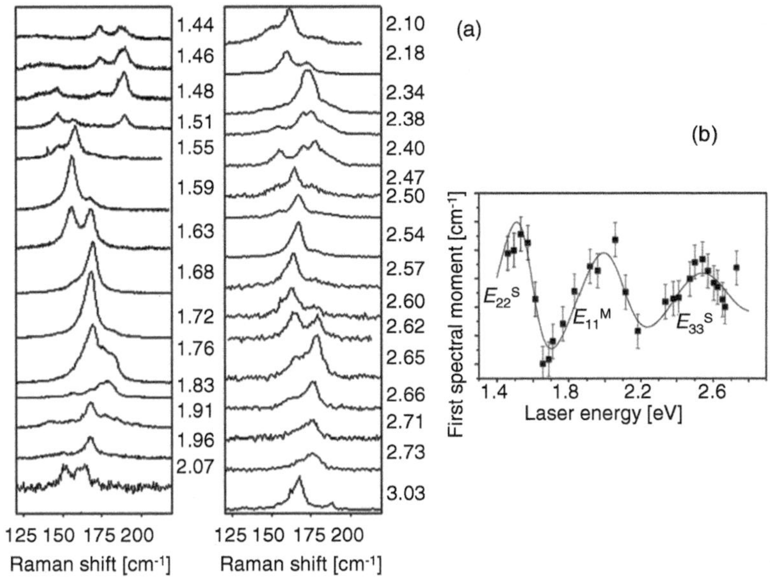

FIGURE 6.6 (a) Raman response for RBM mode as excited with 30 different lasers of energies as indicated in electronvolts. (b) First spectral moment of the measured RBM response (black squares) and calculated response using Equation 6.4 with a tight-binding JDOS and an adjusted distribution of tube diameters. Courtesy of M. Hulman.

latter it needs to combine RBM mode frequencies and resonance transitions as discussed in subsequent sections. However, as long as the $1/d$ law holds a linear relation between RBM frequency and transition energies (Kataura plot) is valid.

As depicted from Figure 6.5 the pattern of the RBM response depends strongly on the energy of the exciting laser. Figure 6.6 shows a set of spectra for tubes with an average diameter of 1.5 nm as excited with 30 different lasers. The peaks in the pattern correspond to the various tubes in resonance. However, response from individual tubes is overlapping which does not allow any identification of the peaks. The overall appearance of the spectra suggests an oscillatory behaviour. This was explicitly demonstrated in Refs [47] and [11]. The right part of the figure depicts the first spectral moments (experiment) and a calculated response. The oscillatory dispersion is a consequence of photo-selective resonance scattering. With increasing laser energy, successively increasing optical transitions E_{ii}^{S} and E_{ii}^{M} become relevant. Concomitant to the selection of the transitions, the tube diameters determine the resonance energy, starting from larger and proceeding to smaller diameters for increasing laser energies.

To identify the peak positions in the RBM spectra it needs to go to smaller-diameter tubes. Standard materials for such studies are high-pressure carbon mono-oxide (HiPco) tubes with an average diameter of around 1 nm. For such tubes the resonance cross section can be determined for the individual peaks by recording spectra with a large set of lasers. This means for each of the individual

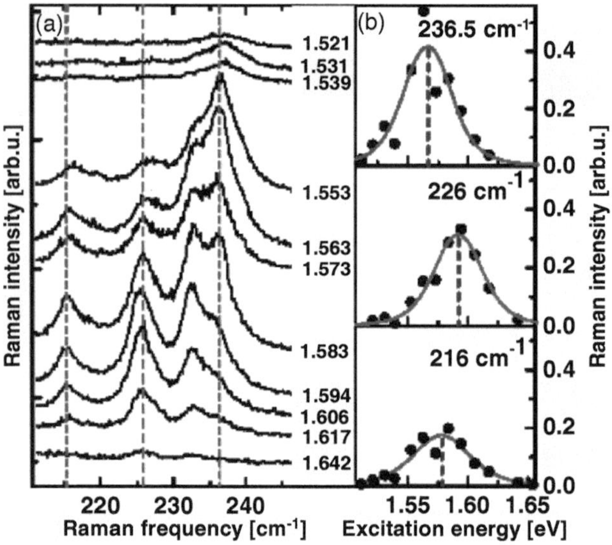

FIGURE 6.7 Raman spectra of HiPco tubes with $d \approx 1$ nm as excited with 11 different lasers lines as indicated in electronvolts (a). Tubes were measured as dispersed by a surfactant to avoid bundling. (b) Cross sections for Raman lines selected by the vertical dashed lines in (a). Courtesy of H. Telg.

peaks the RBM frequency and the transition energy can be determined experimentally. Chirality assignment performed in this way is superior to assignments from, e.g., luminescence as response from SC and M tubes can be studied [48]. Figure 6.7 depicts Raman spectra from HiPco tubes.

The Raman peaks exhibit resonance behaviour with the exciting laser. This, together with the measured frequency, allows to assign the peaks to tube chiralities and in turn to obtain a corresponding Kataura plot as discussed in subsequent sections.

As the tubes diameters become smaller deviations of E_{ii} from the simple Kataura plot become relevant. The deviations exhibit family behaviour. Family behaviour originates from the possible intersections of the cutting lines along the $\Gamma - \mathrm{K}$ distance [49]. For SC tubes family I and family II are characterized as $(2n + m)\mathrm{mod}3 = 1$ and $(2n + m)\mathrm{mod}3 = 2$ (or -1), respectively. Family behaviour for the deviation of the optical transitions from linearity in $1/d$ is systematic. For the E_{22}^{S} transition SC family I tubes have systematically lower transition energies than linear whereas SC family II tubes deviate upwards. For E_{11}^{S} and E_{33}^{S} it is the other way round and continues to alternate for higher transitions. Family behaviour for M tubes is defined by $d_{\mathrm{R}} = gcd(n, m)$ for family I and $d_{\mathrm{R}} = 3gcd(n, m)$ for family II where $d_{\mathrm{R}} = gcd(2n + m, 2m + n)$. M tubes were so far observed to have reduced transition energies as compared to the $1/d$ dependence. With decreasing chiral angle within one branch ($2n + m$ constant), i.e. proceeding from armchair-like to zigzag-like tubes, deviations become stronger. Such deviations were observed in several works [42,48, 50].

The influence of curvature on transition energies was certainly a challenge for calculations beyond simple tight binding. Successful results were reported from noTB calculations [51], and from extended tight-binding calculations [52]. Such calculations do not yet provide agreement with the experiments as they were based on single particle excitation. Improvement was obtained by introducing a logarithmic correction term to take care for many-body effects such as excitons [53], and a chirality correction. The corrections are most conveniently used as suggested in Ref. [54] where the transition energies were calculated from

$$E_{ii}(p,d) = p\frac{a}{d} + p\frac{ab\log cd/p}{d} + \beta_p\frac{\cos 3\theta}{d^2} \tag{6.10}$$

$p = 1, 2, 3, 4, 5$ is a transition number index for $E_{11}^{\mathrm{S}}, E_{22}^{\mathrm{S}}, E_{11}^{\mathrm{M}}, E_{33}^{\mathrm{S}}, E_{44}^{\mathrm{S}}$, respectively. $a = 1.049$ eV nm, $b = 0.456$ and $c = 0.812$ nm^{-1} are fitting constants. β_p are family-dependent parameters given for family I(II) and $p = 1, 2, 3, 4, 5$ as 0.05(−0.07), −0.19(0.14), −0.19(), 0.42(−0.42), −0.4(0.4), respectively. With these corrections good agreement over the whole range of transition energies and chiralities up to $p = 3$ is obtained [42,54]. Figure 6.8 shows a correspondingly modified Kataura plot. Open symbols are as calculated with the noTB theory [51] and corrected using Equation 6.10. Experimental results were averaged from Refs [55] and [48].

For transitions with $p > 3$ an additional correction term of the form $0.305/d$ must be added to Equation 6.10. This is a consequence of a loss of excitonic effects, which are expected to be relevant only for the lower energy transitions [54].

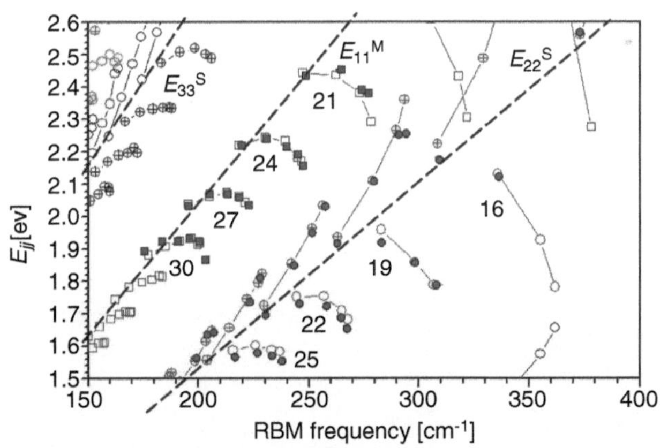

FIGURE 6.8 Popov–Kataura plot for thin SWCNTs. Shown are transition energies for E_{22}^{S}, E_{11}^{M} and E_{33}^{S}. Open symbols are as calculated from the noTB model. Squares: M tubes, circles: SC tubes family I, crossed circles: SC tubes family II. Full symbols are averaged values for HiPco results. Dashed lines indicate simple tight-binding behaviour. Numbers are values of $2n + m$ for the various branches. Courtesy of R. Pfeiffer.

Results from RBM measurements are often plotted as two-dimensional Raman maps where x and y axes are the RBM frequency and the exciting laser, respectively, while the Raman intensity is plotted on a colour scale. Alternatively, three-dimensional plots are used as in Ref. [55].

Due to the strong resonance enhancement of the scattering cross section there were several successful reports to measure the RBM from individual SWCNTs by resonant confocal RS [54,56–58]. In these cases the tubes were spread out in a very dilute concentration on a SiO_2/Si substrate and first localized by AFM. With high-resolution scanning by a tunable laser even the JDOS could be analysed. Figure 6.9 depicts results for an anti-Stokes and a Stokes Raman cross section.

The asymmetry of the cross section originates from incoming and outgoing resonance. The full drawn line was calculated from Equation 6.4 with a g_{jds} of the form

$$g_{\text{jds}}(E) = \text{Re}\left[\sum_i \frac{a_{\text{CC}}E}{d\gamma_0\sqrt{(E-E_{ii}-\text{i}\gamma)(E-E_{ii}+\text{i}\gamma)}}\right] \tag{6.11}$$

which approaches the tight-binding and ZF calculation for all tubes. a_{CC} = 0.143 nm, E_{ii} = 1.655 ± 0.003 eV, γ_0 = 2.9 eV and γ = 8 meV are the nearest-neighbour carbon–carbon distance, the position of the Van Hove singularity under consideration, the tight-binding transfer integral and the width of the Van Hove singularities, respectively. Since the selected tube was metallic, the

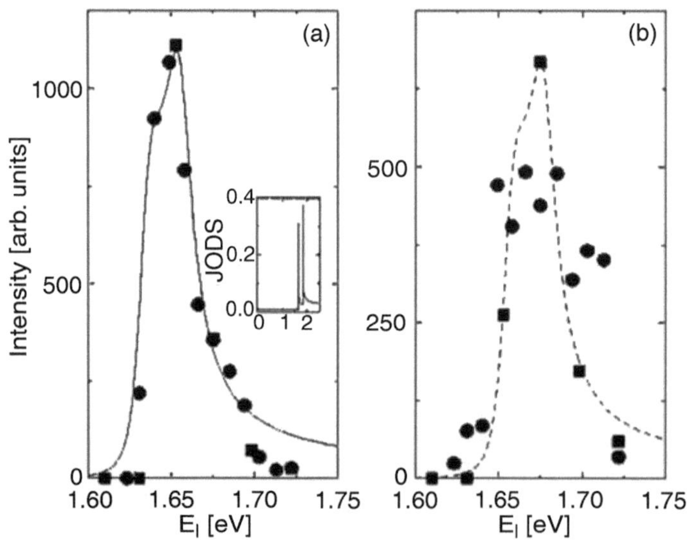

FIGURE 6.9 Raman cross sections for an isolated SWCNT deposited on a Si/SiO_2 substrate. (a) Anti-Stokes; (b) Stokes response after excitation with a tunable Ti:Sapphire laser. Inset: g_{jds} for a (18, 0) SWCNT at E_{11}^{M}. Courtesy of A. Jorio.

Van Hove singularity is split from trigonal warping with peaks at 1.655 and 1.831 eV, respectively. From the value of the RBM frequency and from the position of the resonance the tube was assigned to (18, 0). Only the resonance from the lower energy component of the split g_{jds} could be observed.

6.4.2 Raman scattering from D line and 2D line

The D line and the 2D line in SWCNTs exhibit a quasi-linear dispersion very similar to graphene and graphite. The slopes are around 50 and 100 cm^{-1} eV, respectively, for standard-diameter tubes. Detailed studies showed a weak oscillation on top of the linear increase for the D and 2D frequencies [59,60]. In contrast, the energy dependence for the G line is completely flat. Figure 6.10 depicts the dispersion for tubes with 1.45-nm diameter. After subtraction of the oscillation the dispersion for the D line and for the 2D line was 51.8 and 103 cm^{-1} eV, respectively. The origin of the D and 2D line in SWCNTs is the same as it was discussed in the introduction for graphite. This means DRRS including defects for the D line and two-phonon DRRS for the 2D line. The oscillation is again from photoselective resonance scattering and thus not intrinsic. However, an intrinsic dispersion can be observed for selected lasers if one excites individual tubes at different transitions E_{ii} [61].

The quantitative description of the dispersion effect is not as straight as in graphene, since only a selection of states on the Dirac cone is available due to

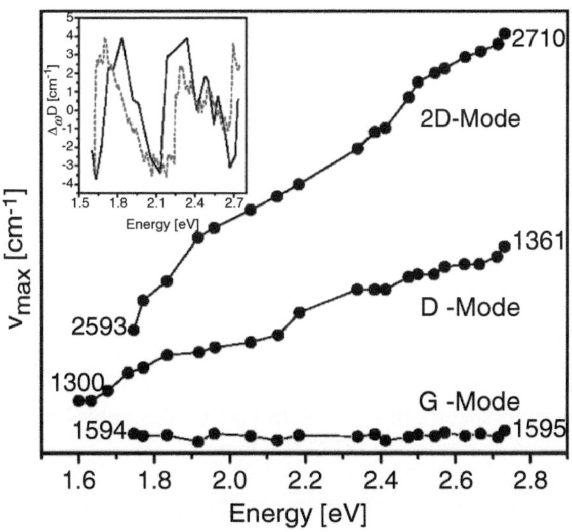

FIGURE 6.10 Dispersion of D and 2D lines as compared to the dispersion of G line. The numbers indicate lowest and highest observed frequency. Inset: oscillatory part after subtraction of a linear background. Experiment (black) and calculated (dashed (red dashed in the web version)). Courtesy of J. Kürti.

quantum confinement. For the double resonance the transition from K to K' must match the Van Hove singularities. One thus needs to satisfy in this case triple resonance conditions. Calculations, where this triple resonance was considered, describe the linear dispersion and the superimposed oscillations [59,60] very well as it can be seen from the insert of Figure 6.10.

The line positions of the D and 2D lines were found to depend on the tube diameter according to Ref. [62]. For standard tube diameters this dependence is given by

$$\omega_{\mathrm{D}} = 1354.8 - 16.5/d, \qquad \omega_{2\mathrm{D}} = 2708.1 - 35.4/d \tag{6.12}$$

if the spectra are excited with a green laser of 2.41 eV. For $d \rightarrow \infty$ both relations extrapolate to the values for graphite ($\omega_{\mathrm{D,G}} = 1355\ \mathrm{cm}^{-1}$, $\omega_{\mathrm{2D,G}} = 2710\ \mathrm{cm}^{-1}$).

The D line has a counterpart if the double resonance process is established by an intra-valley scattering. In this case the Raman line appears at 1620 cm^{-1} and is known as D' line.

6.4.3 Raman scattering from G mode

The G mode yields the strongest Raman line in the SWCNT spectrum. In contrast to graphite, it is split into several components. In experimental work the components are often assigned as G^+ and G^- or ω_{G}^{+} and ω_{G}^{-}. In detail they are components of the LO and iTO branch of graphene as they result from ZF. Line pattern as excited with various lasers are depicted in Figure 6.11(a). The spectra are correlated to a Kataura plot in Figure 6.11(b) as calculated from Equation 6.10. For changing the laser energy from higher to lower values the line positions and line pattern remain at first almost constant and resonance occurs at the E_{33}^{S} transition. As soon as the energy reaches the E_{11}^{M} transition at around 2 eV, the G^+ peak is strongly reduced and a broad line appears at lower frequencies with a Breit–Wigner–Fano (BWF) lineshape. This line represents the response from the LO mode in M tubes.

The transition to the resonance from the M tubes depends strongly on the tube diameter. For 1-nm-diameter HiPco tubes it occurs already in the green spectral range [63]. Also, the position of the BWF line downshifts with decreasing tube diameter [64]. On further reducing the laser energy the resonance for the M tubes is lost; the spectra become weak until at around 1.6 eV the E_{22}^{S} transitions becomes resonant.

From recording spectra of tubes with various diameters, an empirical diameter dependence was reported [28] of the form

$$\omega_{\mathrm{G,S}}^{\mathrm{TO}} = 1591 - 47.7/d^2, \qquad \omega_{\mathrm{G,M}}^{\mathrm{LO}} = 1591 - 79.5/d^2 \tag{6.13}$$

Doping SWCNTs with electron donors such as Li, K or Rb (n doping [65–67]) or $\mathrm{FeCl_3}$ (p doping [68]) is another procedure where G-line and RBM-line response play an important role. The doping process is rather an intercalation

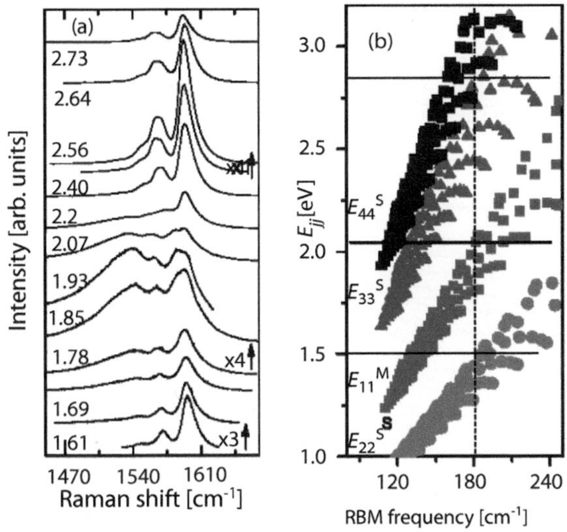

FIGURE 6.11 Raman signal for the G line in SWCNTs as excited with various lasers indicated in electronvolt (a) and Kataura plot as calculated from Equation 6.10 for tubes with standard diameters of ≈1.4 nm (b). The perpendicular line in (b) indicates the average RBM frequency of the sample. Horizontal lines are guides for the eye. Courtesy of T. Pichler.

with a concomitant charge transfer from the inserted atom to the tube. Experiments where the Raman response is measured in situ with the doping process are quite common. For the SC tubes doping leads in a first stage to a decrease in the RBM intensity while the G line remains unchanged. In contrast, for excitation with red lasers where the M tubes are addressed, immediate reduction of the G-line intensity is observed [65]. This is due to a rapid filling of the states at the Fermi level. As a consequence the E_{11}^{M} transition gets quickly Pauli blocked. For stronger doping the G line of the SC tubes shifts upwards, broadens and loses intensity [66,67]. Eventually, it further broadens and shifts downwards from about 1595 to 1560 cm^{-1}. In the highly doped state the G line adopts a BWF shape. The upshift and downshift of the G-mode frequency are described in terms of bond length change as it used to be done for graphite. However, quantitative understanding has not yet been obtained.

A particular elegant procedure for the doping makes use of electrochemical (EC) charging the tubes. Depending on their potential versus a reference electrode the tubes can be positively or negatively charged. This means p and n doping can be reversibly controlled [69,70]. In this case either anions or cations interact strongly with the tubes and provide a charge transfer. Figure 6.12 depicts RS as excited with a red laser after EC doping. The decrease in the response from the M tubes (line on the left) is evident for p and n doping. The downshift of LO mode (ω_G^+, line on the right) is also evident but it amounts only about 5 cm^{-1} as compared to the 35 cm^{-1} downshift after heavy alkali metal doping. A good review on Raman response of EC-doped SWCNTs is given in Ref. [71].

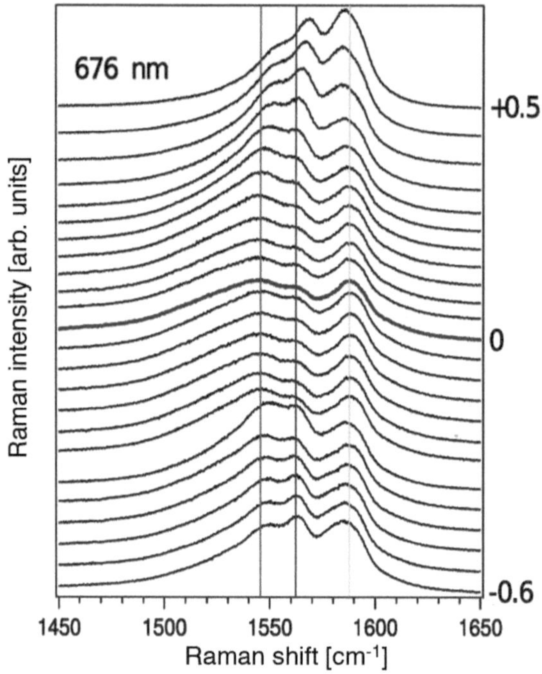

FIGURE 6.12 G line response for excitation with a red laser of SWCNT doped electrochemically in a 0.2 M $LiClO_4$/acetonitrile electrolyte. The potential was shifted in steps of 0.05 V from −0.6 V to +0.5 V versus the Ag electrode. Perpendicular lines mark the positions of the LO mode for M tubes and the positions for TO and LO modes of SC tubes. Spectra are displaced for clarity. Courtesy of M. Kalbac.

Using the EC technique it was recently possible to dope individual tubes and record their Raman spectra [38,72].

6.5 RAMAN SCATTERING OF SWCNT FUNCTIONALIZED BY FILLING

SWCNTs are not just special with respect to their 1D nature. As important is their hollow inside which was shown to provide a nano-sized template for growing 1D materials and to allow for chemical reactions confined to one dimension. Originally tube filling was performed with C_{60} fullerenes from where the structures were given the name 'peapods'. This name is now generally used for the filled tubes. RS plays an important role for the analysis and understanding of the peapod structures and for the reactions inside the tubes.

6.5.1 Raman scattering from peapods

A large number of different molecules and crystalline materials were filled into the tubes so far. A good review about various filling experiments can be found

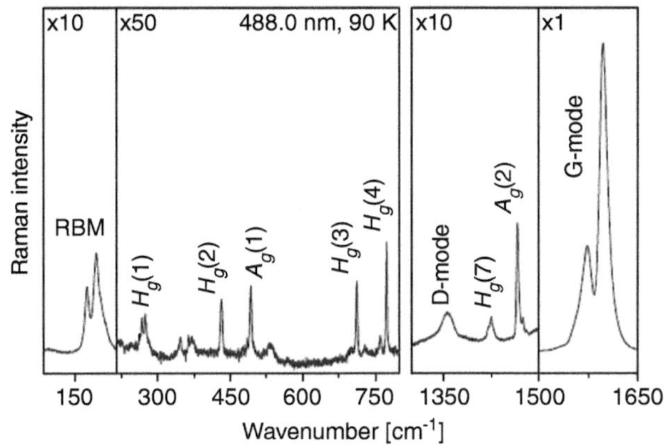

FIGURE 6.13 Raman spectrum of SWCNTs filled with C_{60} as excited with a blue laser. The sharp lines with the symmetry species assign the response from the C_{60} cage. Broad lines are from the nanotubes. Courtesy of T. Pichler.

in Ref. [73] and for organic molecules in Ref. [74]. After removing the caps of the tubes by oxidation they can be filled at elevated temperatures either from the gas phase of the filler material, from solution, or from the melt. To yield good information about the material inside the tubes from RS, resonance excitation is important. This is, e.g., the case for β-carotene [75] or fullerene peapods.

Most prominent fillers are the C_{60} molecules. RS for such peapods shows the full spectrum of the C_{60} molecule on top of the spectrum of the SWCNT, as shown in Figure 6.13. The ratio between the integrated intensity of the $A_g(2)$ mode from the fullerenes and the G mode from the SWCNTs can be used as a measure for the quality of filling [76]. For good filling and excitation with a blue laser this ratio should be larger than 5%.

Peapods can be doped either electrochemically [77] or by intercalation of electron donors [78]. Figure 6.14 compares RS from doped peapods with spectra from several other structures. Figure 6.14(a) depicts the response from pristine tubes in comparison to heavily intercalated peapods for excitation with three different lasers. The selected spectral range covers the nanotube G line and the fullerene pentagonal pinch mode $A_g(2)$. The G line response is drawn off scale to make the pentagonal pinch mode response better visible. The distance between the frequency position for the neutral C_{60} (right arrow) and the frequency where it appears after heavy doping (left arrow) is 42 cm^{-1}, which indicates a −6 charge state for the fullerene. Looking at the Raman signal in the spectral range of the radial and intermediate frequency (IMF) modes as depicted in Figure 6.14(b), the spectrum from the doped peapods (centre) has two additional lines assigned as P and located at 360 and 620 cm^{-1}. These lines are not observed in the spectrum of Rb_6C_{60} (bottom). They are rather a signature of polymeric C_{60} as demonstrated in the top spectrum of Figure 6.13(b). Here the additional lines are marked as P as well. This spectrum was recorded for RbC_{60},

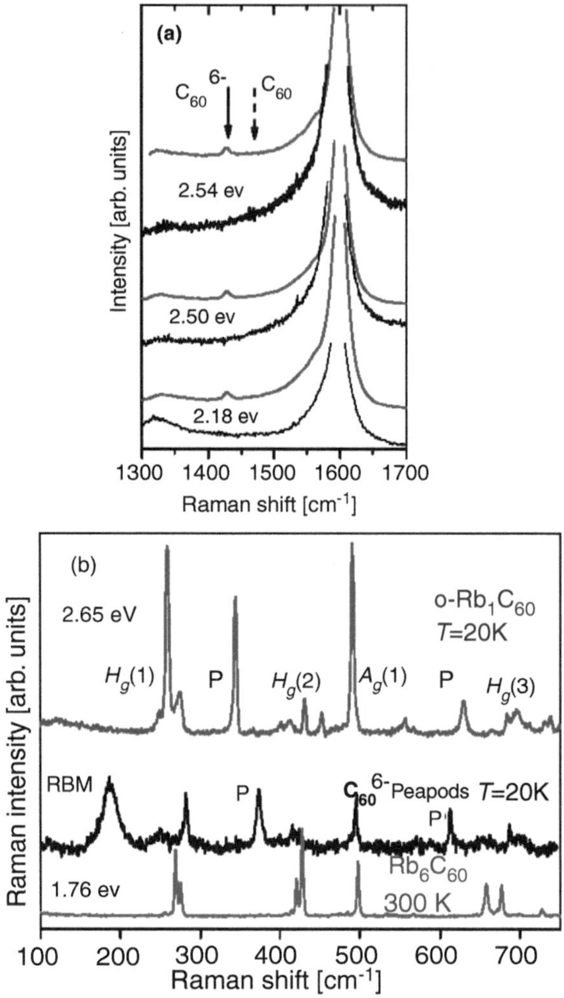

FIGURE 6.14 (a) Raman response in the frequency range of the G line for heavily doped peapods and unfilled references. Right arrow: frequency position for pentagonal pinch mode in neutral C_{60}; left arrow: pentagonal pinch mode frequency for C_{60}^{6-}. (b) Raman spectra of Rb_6C_{60} (bottom), doped peapods (centre) and polymeric Rb_1C_{60} (top). Courtesy of T. Pichler.

which is known to be a polymer. Thus, we learn that there is a polymerized phase of C_{60}^{6-} inside the SWCNT [78]. In addition some of the lines from the charged fullerenes exhibit a BWF lineshape showing that they are in a metallic state. In a more refined work, it was demonstrated from an analysis of the Raman spectra that doping the C_{60} molecule can lead to a continuous charge transfer [79] in contrast to the line-phase doping observed for crystalline A_xC_{60}. Interestingly EC p doping of peapods yields a strong enhancement of the Raman response from the fullerene (anodic enhancement), instead of a reduction [71].

6.5.2 Raman scattering from double-walled carbon nanotubes

Much attention has been paid to chemical reactions inside the tubes, in particular to the growth of DWCNTs. DWCNTs were first grown by S. Bandow et al. [80] by heating peapods in vacuum for several hours at 1250 °C. This has now become a standard procedure to prepare DWCNTs [81]. Alternatively DWCNT can be grown directly from a chemical vapour deposition (CVD) process for properly selected growth parameters [82,83]. A review on growth and properties of DWCNTs is available in Ref. [84].

Raman spectra of DWCNTs show the response from the inner and the outer tubes as depicted in Figure 6.15 (top) while the response from the fullerenes in the peapod precursor (centre) is continuously decreasing during the transformation process.

The appearance of the large number of sharp lines in the spectrum of the DWCNTs is unconventional as the lines are almost a factor of 10 more narrow than RBM lines from conventional tubes [85]. At low temperatures line widths reduce down to 0.40 cm^{-1}. This means the lifetime of the RBM phonons is very long (quality factor $Q = \nu/\delta\nu \approx 500$) and not terminated by defect scattering. In other words, the inner tubes are highly defect-free as they were grown in a clean room with nano dimensions.

In addition to the extremely small line width the number of lines in the cluster is much larger than the expected number of tubes in the corresponding diameter range. Exciting DWCNT with different lasers reveals several clusters. From measuring the resonance cross section for the individual peaks in the cluster the set of lines turned out to have the same resonance and therefore to originate from the same inner tube. This tube is accommodated in different outer

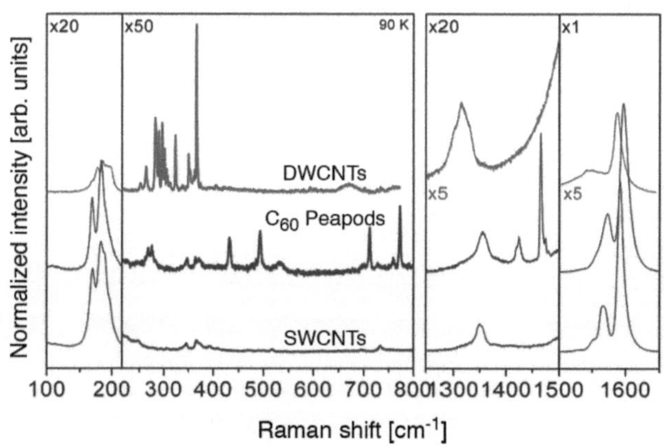

FIGURE 6.15 Raman spectra for SWCNTs (bottom), peapods (centre) and DWCNT (top) as excited with 514, 488 and 614 nm lasers, respectively. The response from the fullerenes in the spectrum at the centre converted into a set of narrow and clustered lines around 300 cm^{-1} originating from the inner tube RBM modes. Courtesy of R. Pfeiffer.

tubes and therefore appears with different RBM frequency. This is depicted in Figure 6.16. It shows two sets of clustered lines corresponding to (6, 5) and (6, 4) inner tubes in different outer tubes. Each cluster is spread out over more than 20 cm^{-1} and contains about 10 lines. The lower part of the figure depicts the cross section response on a false colour scale, plotted on top of the Popov–Kataura plot from Figure 6.8. Each peak in the cluster shows almost the same resonance profile and each cluster starts off at a position in the Raman map which corresponds to a tube from HiPco material (green-filled circles). The rest of the members of the cluster are upshifted in frequency [86,87]. Thus, the lines in the cluster represent 'pair spectra'. Each line corresponds to a pair of the same inner tube in different outer tubes.

Quantitative evidence for this interpretation was derived from a comparison of the observed line positions with a calculation of frequency shift due to intertube interaction [88]. As expected this interaction depends strongly on the wall-to-wall distance and can shift RBM frequencies by up to 30 cm^{-1}. Figure 6.17

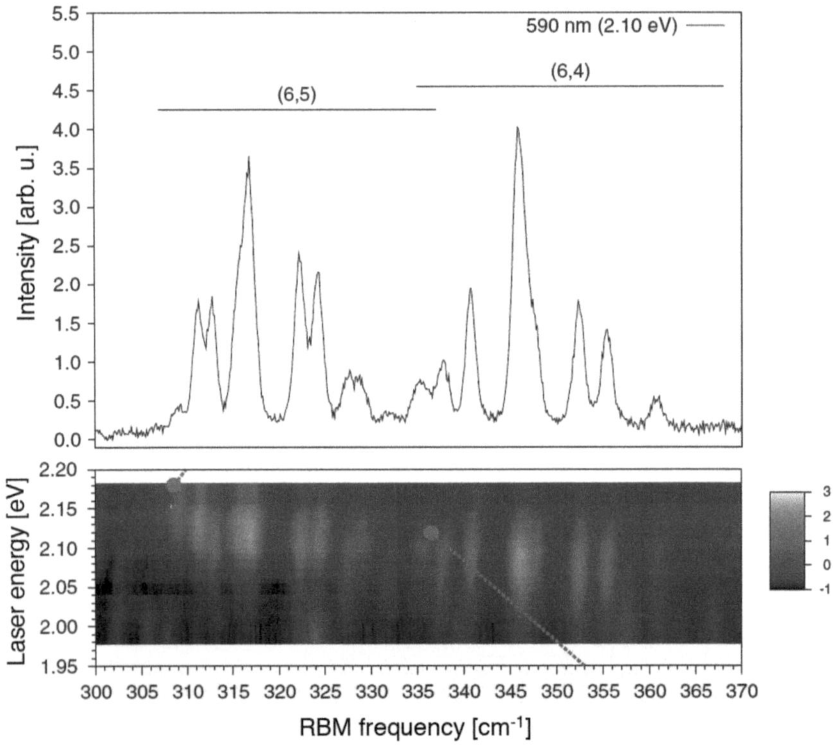

FIGURE 6.16 Raman response for two cluster of lines assigned to (6, 5) and (6, 4) tubes on a blown up scale, for 590-nm laser excitation (upper part). Selection of the corresponding Raman map with (6, 5) and (6,4) tubes from HiPco samples as filled circles (green-filled circles in the web version) (lower part). Courtesy of R. Pfeiffer.

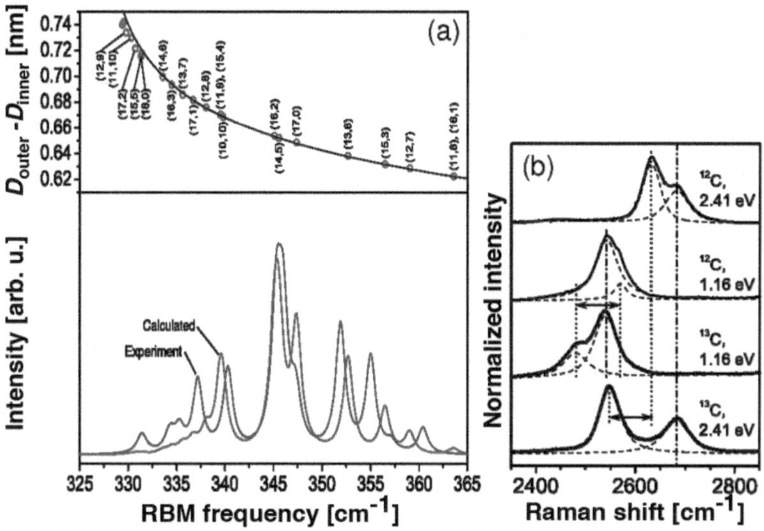

FIGURE 6.17 (a) Upper: calculated RBM frequencies for a (6, 4) SWCNT in various outer tubes plotted versus diameter difference between inner and outer tubes. Lower: experimental results in comparison to the calculation. Each tube frequency in the upper part of the figure was dressed with a Lorentzian line. Courtesy of R. Pfeiffer. (b) Spectra of 2D line excited with different lasers and for two different isotope substitutions for the inner tubes, as indicated. Dashed lines are Lorentzian curve fits. Dash-dotted lines are guides for the eye. Double arrows indicate isotope shift. Courtesy of F. Simon.

has results from a calculation using a simple reduced mass approximation for the tubes and a van der Waals potential for the wall-to-wall interaction [86]. The upper part of the figure depicts the calculated RBM frequencies for the various pairs of the (6, 4) tube. The intensity profile for the calculated response in the lower part of the figure was obtained by assuming a Gaussian distribution for the probability to observe a particular inner tube–outer tube pair.

The G line and the second-order Raman spectra from inner tubes in DWCNTs are not so easy to detect, since they overlap with the response from the outer tubes. A general help to detect such overlapping lines is the use of ^{13}C isotope-labelled inner tubes [45,89,90] as they can be prepared from a ^{13}C fullerene peapod. In this way the G line from the inner tubes is downshifted from the overlapping outer tube LO and TO components [89]. Similarly for the D mode. Since the diameter dependence of the D-mode frequency is rather weak and the D-mode line profile is structured [60], the contributions from inner and outer shell tubes can only be separated if the inner tubes are isotope labelled.

Isotope labelling was particularly useful for the study of the 2D lines in DWCNTs. Even though in this case response from the inner and outer tubes can be recognized without isotope labelling [91] an unexpected behaviour in the response of this line could only be clarified [45] by comparing DWCNTs with inner tubes grown from ^{13}C and ^{12}C. This is demonstrated in

Figure 6.17(b). The top spectrum is for excitation with 2.41 eV. The high- and low-frequency components are from the outer and inner tubes, respectively, as expected from Equation 6.12. For excitation with a laser at 1.16 eV there are also two components but now the higher-frequency component is from the inner tubes and the lower-frequency component is from the outer tubes. This can be evidenced from experiments with isotope-substituted inner tubes as depicted in spectrum 3 from top. Whereas the lower-frequency component remains unchanged the higher-frequency component shifts down by hundred wavenumbers due to the enhanced tube mass. To control the isotope-induced downshift the spectrum at the bottom depicts the response from the isotope labelled tubes when the tubes are excited with a green laser. Now, as compared to the spectrum on top, the high-frequency peak remains unchanged and the low-frequency peak is downshifted. As a consequence the low-diameter inner tubes do exhibit not only a lower frequency as one might expect from Equation 6.12 but also the slope of the linearized dispersion changed from 104 to 60 eV cm^{-1}. This behaviour was also documented from a DFT calculation [45].

RS also allows to study the dynamics of the transition from the peapods to the DWCNTs [92]. Both, the vanishing of the response from the fullerenes as well as the increasing response from the inner tubes can be analysed. Evaluating the integrated intensities for the various Raman lines as a function of annealing time yields time constants for the various processes. Figure 6.18 depicts examples for the transformation at 1250 °C. Time constants for (7, 2) and (8, 3) tubes were 16 and 38 min, respectively. The time constant for the decay of the linear chain response in X-ray diffraction (XRD) relates well to the time constants for the growth of the DWCNTs. On the other hand, the time constant for the vanishing of the response from C_{60} is noticeably smaller compared to the values for the growth of the inner tubes. This means there is an intermediate Raman dark phase. The variation of the time constants for the inner tube growth is a consequence of the effort to generate these tubes from the fullerene. Fusion of C_{60} leads immediately to a (5, 5) armchair tube. The further away the resulting diameter of the inner tube is, the more steps are needed to convert the fullerene into the appropriate inner tube. This is demonstrated in the insert of Figure 6.18. Time constant is minimum for best match of the inner tube diameter to the (5, 5) tube and increases with increasing diameter difference. The results support models for inner tube growth by fullerene coalescence [93,94].

To record RS from individual DWCNTs was problematic, since suspending individual tubes in a solvent was difficult. However, very recently such experiments were possible [95,96]. In Ref. [95] a DWCNT (12, 8)@(14, 16) was analysed. Chiral indices were determined from transmission electron microscopy (TEM) and electron diffraction. The observed Raman frequencies were 133 and 186 cm^{-1} for the inner and outer tubes, respectively. According to the enhanced interaction due to the large diameters of the tubes modes were strongly coupled

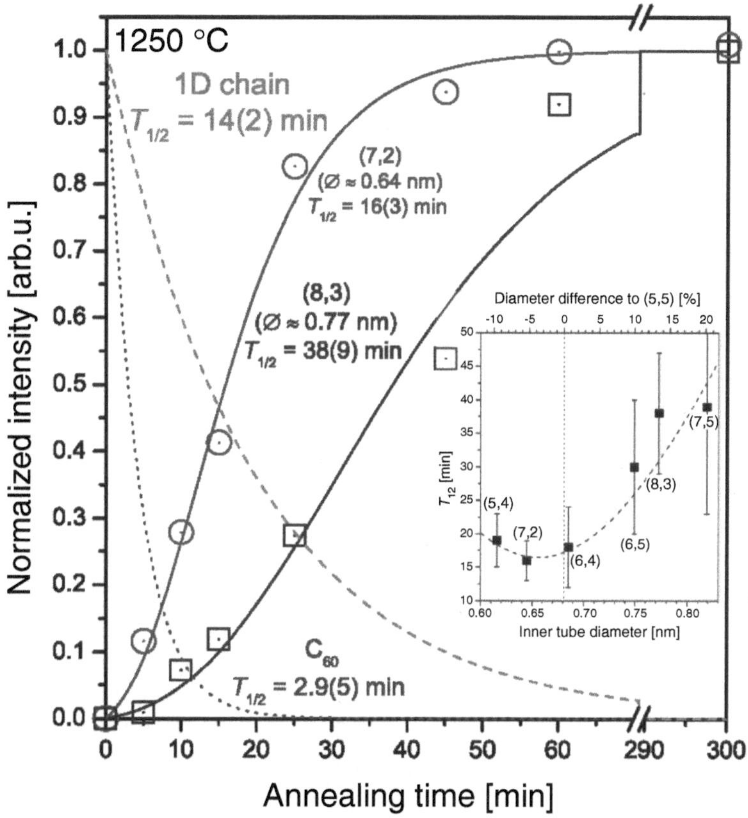

FIGURE 6.18 Normalized Raman line intensity versus transformation time for C_{60} peapods. Circles (red circles in the web version) and squares (blue squares in the web version) are for (7, 2) and (8, 3) inner tubes, respectively. Full-drawn lines are fits with a Gompertz-like function. Dotted and dashed lines represent the vanishing of the fullerene pentagonal pinch mode and of the linear chain peak from XRD. Inset: growth time for the various inner tubes. Top scale: diameter difference to (5, 5) for the listed inner tubes. Courtesy of R. Pfeiffer.

and exhibited the features of a coupled breathing like modes. Therefore, the $1/d$ law could not be applied in the same form for the two tubes.

Among the various possibilities to grow tubes from other peapods the use of metallocenes provided the best yield. For this particular feedstock only rather low temperatures of the order of 500 °C are needed to convert the metallocene peapods into DWCNTs. The most-stable metallocene is ferrocene ($FeCp_2 = Fe(CH_5)_2$). First DWCNTs grown from $FeCp_2$ were reported in Ref. [97] with several subsequent works in Refs [98,99]. As shown in detail in Ref. [98] the RBM response for C_{60}- and $FeCp_2$-grown DWCNT is identical as far as the positions of the lines in the pair spectra is concerned. However, they are characteristically different concerning the first spectral moments. This is demonstrated in Figure 6.19. It depicts pair spectra for (6, 5) and (6, 4) cluster. Line

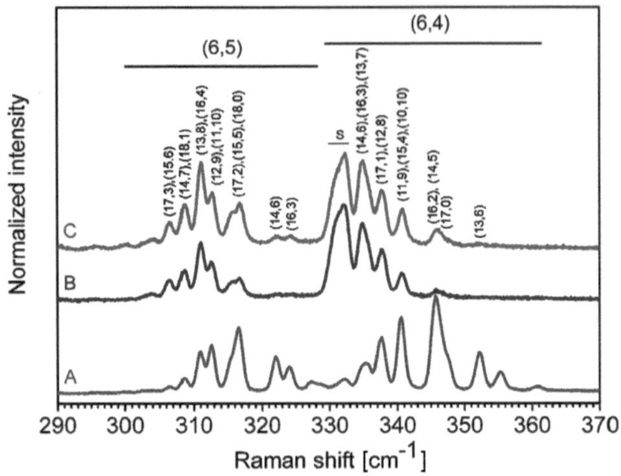

FIGURE 6.19 RBM frequency cluster for (6, 5) and (6, 4) inner tubes in DWCNTs grown under different conditions. (A) DWCNT grown from C_{60} peapods, (B and C) DWCNT grown from $FeCp_2$ peapods for transformation at 600 °C (B) and at 1150 °C (C). The small numbers in parentheses assign the outer tubes. S is the response of (6, 4) in several large outer tubes. Courtesy of R. Pfeiffer.

positions are exactly the same for all spectra shown. In particular, for the $FeCp_2$-grown DWCNTs the transformation temperatures has almost no influence on the line pattern. On the other hand, comparing spectrum A with B and C reveals that the centre of gravity is upshifted for the fullerene-grown DWCNT.

This means for the latter on average the distance between the walls is smaller. Consequently it was concluded that the process for the growth of DWCNTs from $FeCp_2$ is characteristically different. TEM analysis demonstrated indeed that iron carbide (Fe_3C) particles are grown inside the tubes. This suggests a catalytic growth process where the iron carbide particles first dissolve the $FeCp_2$ and subsequently, after supersaturation is reached, release carbon atoms, which form the inner tubes.

6.5.3 Raman scattering from tubes with ultrahigh curvature

There was considerable interest to grow SWCNT with very small diameter corresponding to very high curvature. Such tubes are expected to not only yield strong deviations for the electronic and vibrational properties from tight-binding-derived structures but also exhibit enhanced electron–phonon coupling and concomitant challenging possibilities for superconductivity. Since the very high curvature tubes are highly unstable due to reduced π–π overlap, they must be grown in a template such as zeolites [100,101] or SWCNTs [102,103]. RS for ultrahigh curvature tubes was described in several papers [101,102,104,105]. In order to obtain ultrahigh curvature tubes inside a master tube small-diameter tubes must be used as starting material. As a consequence, molecules noticeably smaller than C_{60} must be used as filler. Ferrocene with a molecular dimension of $\approx 0.37 \times 0.34$ nm turned out to be appropriate for filling into the 1-nm-diameter HiPco tubes. From molecular dynamical calculations possible filling was shown for tubes with diameters down around 0.94 nm ((12, 0) or (7, 7)), corresponding to an inner tube diameter of around 0.3 nm. Successful filling and transformation to DWCNTs resulted in a set of ultrahigh curvature inner tubes with chiralities down to (5, 0) tubes and DFT diameters $d_{DFT} = 0.404$ nm ($d = d_{TB} = 0.390$ nm). RBM frequencies were as high as 570 cm^{-1}. Figure 6.20 depicts RBM spectra for such tubes as excited with various red lasers. Resonance behaviour is clearly shown for Raman line at 470 cm^{-1} and for the inner tube mode at 360 cm^{-1}. From resonance Raman excitation electronic transition energies and RBM frequencies were determined for all observed inner tubes and allowed to obtain a Kataura plot with dramatic deviations from standard behaviour [102]. For tubes with diameters as small as 0.5 nm and below corrections to the RBM frequencies and even to the tube diameters are necessary as described in Sec. 6.3.

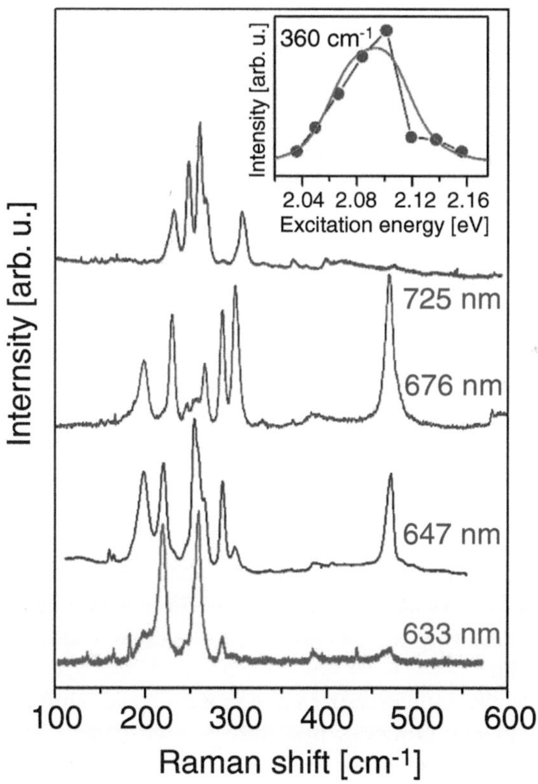

FIGURE 6.20 RBM Raman spectra for DWCNTs grown from $FeCp_2$@HiPco. Excitation was for four different lasers. The strong line at 470 cm^{-1} is the resonant response from one special inner tube. Inset: resonance behaviour for another inner tube. Courtesy of W. Plank.

Even smaller diameters were observed for tubes grown in the pores of a single crystal AIPO-11 zeolite [101]. Tubes with 0.3 nm diameter were detected by RS with a RBM of 760 cm^{-1}. The tubes were assigned to (2, 2) armchair species. The G^+ line was at around 1580 cm^{-1} and thus in the expected frequency region if they were considered as SC. In contrast, the G^- response was strongly downshifted to 1200 and 1400 cm^{-1} for the M and SC configuration of the (2, 2) tube, respectively.

6.6 SPECIAL RAMAN EXPERIMENTS

6.6.1 Effects of temperature, pressure and atomic substitution

Temperature-dependent RS is generally characterized by an increase in line width with increasing temperature due to a decay of optical phonons into two acoustic modes. In this case the full-width half maximum (FWHM) line width Γ follows from

$$\Gamma(\omega_p, T) = \gamma + \Gamma(\omega_p, 0)\left(1 + \frac{2}{\exp(\hbar\omega_p / 2k_B T) - 1}\right) \tag{6.14}$$

where γ, ω_p and T are the extrinsic line width, the phonon frequency and the temperature, respectively. In the case of SWCNTs γ is usually rather large which leads to a weak relative temperature dependence. For SWCNT bundles, the increase in line width for the RBM line was only 5 cm^{-1}, from 5.5 to 10.5 cm^{-1} for temperatures between 10 K and 500 K. From a fit $\Gamma(\omega_p, 0) = 1$ and $\gamma = 4.5$ cm^{-1} was obtained [106]. There are indeed many reasons for inhomogeneous broadening such as tube–tube interaction, strain, overlapping response from different tubes, impurities, interaction with electrons, etc.

Measurements on individual tubes might be more reliable [28]. Even at low temperatures line width for the RBM are still broad, of the order of 3 cm^{-1}. The observed line widths decrease with decreasing tube diameter. This was interpreted as originating from a confinement-induced reduction of phonons available for the interaction process. Large scattering of observed values for the line widths still indicates a contribution from inhomogeneous broadening.

A best choice to study intrinsic line width behaviour would be an analysis of inner tubes in DWCNTs. As discussed above, in this case the low temperature line width for the RBM goes down to 0.4 cm^{-1}, which could be close to the real intrinsic line width.

Under high hydrostatic pressure phonon frequencies shift upwards. At least for the high-frequency G modes this shift is almost linear in pressure as observed from the shift of the Raman lines [107,108]. This dependence can be described by $\omega_G(P) = \omega_G(0) + 0.5P$ where $\omega_G(0)$ and P are the various G-mode frequency components at zero pressure and the applied pressure in gigaPascal, respectively.

Substitutional doping of SWCNTs with nitrogen or phosphorus has led to other special Raman experiments. A review on the possibilities to n-doped SWCNT by substitution carbon with nitrogen or phosphorus is presented in Ref. [109]. RS after nitrogen doping was in particular discussed in Ref. [110]. Nitrogen-doped tubes (N-SWCNTs) were grown using a floating catalyst aerosol CVD method with CO as a carbon source and ammonia as a nitrogen source. RS from the RBM indicates reduced diameters for N-SWCNTs grown under the same conditions as unsubstituted tubes. The D line remains unchanged with the substitution process, which is surprising, since scattering centres are created. On the other hand, the LO mode for the M tubes disappears very similarly to intercalation doping or EC doping. This indicates an increase in free carrier concentration as expected. The 2D line shifts to lower frequencies with doping, which is partly due to the average decrease in tube diameters according to Equation 6.12.

6.6.2 Special Raman lines

Special heat treatment of DWCNTs leads to new structure in the tubes represented by new Raman lines between G line and 2D line. If the tubes are heated to rather high temperatures around 1600 °C, a Raman line appears around 1800 cm^{-1}. It was suggested to result from a coalescence of two DWCNTs with a resulting larger DWCNT and a linear chain inside [111–113]. Therefore, new line was called 'coalescence-induced mode' (CIM). However, most recently such modes were also observed after heat treatment of adamantane-filled tubes for temperatures as low as 500 °C where tube coalescence is very unlikely [114]. Figure 6.21 shows Raman spectra for CVD-grown DWCNTs after high-vacuum heat treatment at 1460 °C. The new CIM lines at 1855 cm^{-1} are resonant-enhanced and often exhibit satellites. Details about the origin of the low-temperature-grown CIM modes are still under investigation.

6.6.3 Tip-enhanced Raman scattering

The strong Raman response of the G line makes it an appropriate signal for nanoscale imaging of individual tubes. Likewise, if the defect concentration is high enough, the D line can be used. The recently developed TERS combined with the SERS effect allows to follow shape and structures of the tubes along its extension with a resolution of down to 15 nm [115,116]. Figure 6.22 depicts images and the RS of tubes recorded with 0.5 mW of a circularly polarized 633-nm laser [117]. The sample was kept at a distance of 2 nm from the tip. An M tube was selected by the laser. Before any treatment the tube was well located on the substrate from the G-band image. The D-band image of the same tube revealed defect scattering at one position along the tube. After generation of additional defects by very localized laser radiation new bright spots appeared in the D-band image. The two spectra in Figure 6.22(d) show that at position B

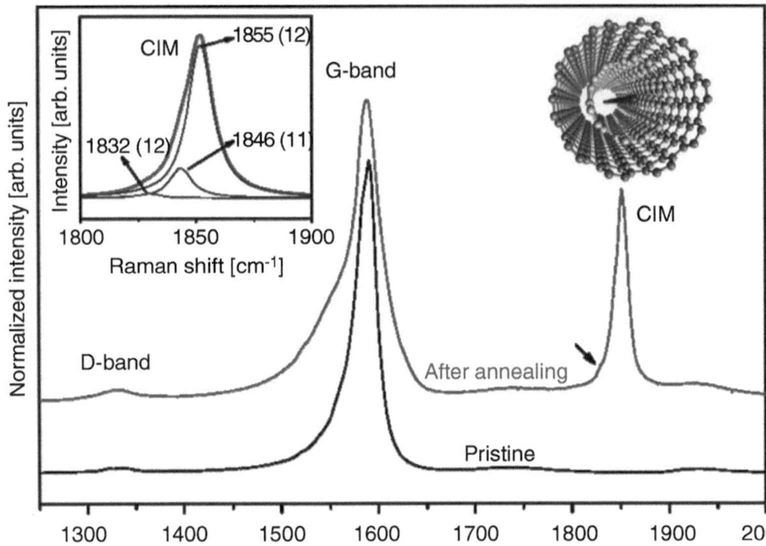

FIGURE 6.21 Raman spectra from DWCNTs excited with a 567-nm laser as grown and after annealing at 1460 °C. The new line at 1855 cm^{-1} is typical for a CC triple bond (carbyne) where the carbons have adopted a sp hybridization. The left and right insets depict a lineshape analysis and a DWCNT with an encaged carbyne chain, respectively. Courtesy of L. Shi.

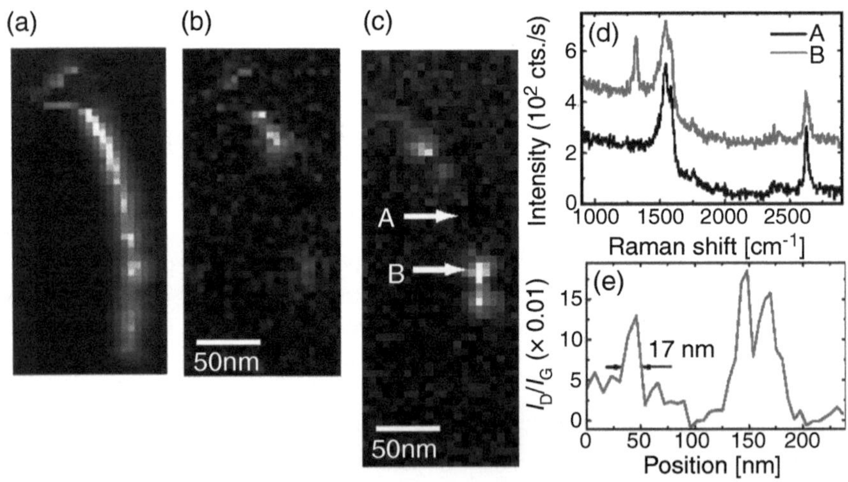

FIGURE 6.22 Tip-enhanced Raman spectra recorded along a SWCNT. (a) Image of a tube constructed from the G line. (b) Image constructed from the D line. (c) Image constructed from the D line after introduction of defects at B. (d) Raman spectra in the high-frequency region recorded at position A (lower spectrum, black) and B (upper spectrum, grey (red in the web version)) after defect generation. (e) I_D/I_G values as recorded along the tube. Courtesy of A. Hartschuh.

new D-band scattering was created. The scan over the corresponding part of the tube revealed a double dot scattering centre Figure 6.22(e). From the observed width of 17 nm and the known instrument resolution of 15 nm the spatial extent of the D-band scattering was estimated to be 2 nm.

6.6.4 Electronic Raman scattering

Electronic RS has been known and intensively studied in highly degenerate semiconductors where the Fermi energy is shifted into the conduction band or in metals (superconductors). Since the M tubes have free carriers, it was challenging to look for such inelastic scattering (see Feynman graph in Figure 6.1). Electronic RS usually appears as a broad background continuum. Recently signatures in the Raman response from individual M tubes (mSWCNTs) were reported [118], which were ascribed to electronic RS. A corresponding theoretical description was given in Ref. [119]. In the experiments a rather broad line with FWHM of about 620 cm^{-1} was observed for excitation with a red laser. The energy of the Raman light corresponded to a transition energy of an mSWCNT. Depending on the tube diameter this was E_{22}^{M-} or E_{22}^{M+}, or, for thinner tubes at $E_{11}^{M-,+}$ where + and − correspond to the splitting due to the trigonal-warped Dirac cone. In all cases the Raman line position for the scattered light, measured in electronvolts, was independent of the incident light. In other words, the response showed a strong dispersion in terms of Raman shift. For a (23, 14) tube the line was observed at ≈500 cm^{-1} for excitation with 2.14 eV. The emission energy was 2.08 eV, independent of excitation energy and well coinciding with the E_{22}^{M-} transition energy for the (23, 14) tube. For a thinner tube with chiral indices (15, 6) excitation with a laser around 1.70 eV yielded a similar result with Raman signatures at constant energies of 1.62 and 1.76 eV for the E_{11}^{M-} and E_{11}^{M+} transition, respectively.

The observation of a broad peak rather than a background continuum is due to the resonance effect for the scattered light. After excitation of the system to any, even virtual, state a continuum of electronic excitations is possible between the Dirac cone states of the M tubes. However, for a given excitation only those transitions are seen which lead to a scattered light energy equal to a resonant transition. Support for this interpretation came from electrochemically shifting the Fermi level upwards or downwards. In this case Pauli blocking hinders the resonance transition for the scattered light and the resonance-enhanced electronic RS disappears.

6.6.5 Orientation-dependent Raman scattering

The strong geometrical and electronic anisotropy render the Raman response strongly dependent on light polarization. Parallel orientation (VV) or perpendicular orientation (VH) for the polarization of incident and scattered light is used. The scattering intensity is obtained from the Raman tensor α as described in Sec. 6.2.1. In SWCNTs the point group symmetries are either D_q or D_{qh} with $q > 3$. Thus, only the corresponding tensors are relevant. Since the trace and the anisotropy of the tensors are invariant and independent of orientation total symmetric modes A_{1g} (A_1) can be selected from non-total symmetric modes E_{ig} (E_i), $i = 1, 2$ even for unoriented samples if a VV and a VH scattering geometry is used. More efficient analyses are, however, obtained from well-aligned SWC-NTs or for RS from individual tubes. In this case two fundamental problems must be considered. Resonance conditions in the scattering experiments and any tube–tube interaction, e.g., in SWCNT bundles can relax or even change group theoretical selection rules. In addition, the special shape of the nanotubes gives rise to strong depolarization effects if the light polarization is perpendicular to the tube axis. This is known as 'antenna effect' and leads to an additional suppression of optical transitions for this geometry [120].

For highly aligned SWCNTs spectra are often recorded for VV and VH orientation while the tubes are rotated around an axis parallel to the incident light [121,122]. In such experiments the observed Raman intensity is plotted in a polar diagram as a function of the angle ϕ between orientation direction of the tubes and polarization of the incident light. For VV scattering geometry dependence like $\cos^4\phi$ is obtained as expected from the product of two dipole interaction events, each one for the incident and scattering photon. Corrections for the exact response may occur from not perfectly aligned tubes. For the VH geometry dependence like $\sin^2\phi\cos^2\phi$ is expected. Such functional dependencies were observed repeatedly.

In more refined experiments polarization-dependent measurements were obtained from individual tubes with tube axis rotating by an angle ϕ from the incident light polarization [28,123]. Both VV and VH orientations were used and clearly showed the $\cos^4\phi$ and $\sin^2\phi\cos^2\phi$ law for the scattering. For VV scattering with light polarization parallel to the tube axis only total symmetric modes are expected to appear in the spectrum. However, in the experiment *A* and *E* modes are observed, which was explained from relaxation of selection rules due to resonance conditions. In contrast, for VH scattering the total symmetric modes are well suppressed and the *E* modes appear enhanced, at least as compared to the VV scattering. In this way in Ref. [28] five out of the six components of the G line in SC tubes were identified as *A* modes at 1578, 1592 cm^{-1}; E_1 modes at 1566, 1592 cm^{-1} and one E_2 mode at 1555 cm^{-1}.

6.6.6 Raman scattering from semiconductor-metal separated tubes

Recently separation of mSWCNTs and semiconducting (sSWCNTs) tubes became possible [124]. In several Raman experiments use was made from this preparation technique. A problem is related to the strong resonance enhancement of, e.g., SC tubes in the metallic fraction. Even for very good separation such tubes give a noticeable signal. From a detailed analysis it was shown in Ref. [125] that the separation process leads to an enhanced concentration of armchair or near-armchair tubes in the metallic fraction.

Very good evidence for the separation can be obtained from DWCNT grown from semiconducting and metallic fractions after separation [126]. Since one observes in this case cluster of lines for one inner tube in different outer tubes, the clusters are significantly different if the outer tubes are SC or M. In fact the clusters from SWCNT@mSWCNT and SWCNT@sSWCNT are complementary and add to the cluster for SWCNT@uSWCNT where uSWCNT stands for unsorted SWCNTs.

6.7 RAMAN SCATTERING FROM GRAPHENE

Graphene, the monolayer of graphite, is the most recent member of the carbon nanophases. Exceptional excitement about this material came from the unusual electronic properties discovered in 2004 by Novoselov et al. [127]. Since then, graphene was studied in numerous transport experiments. However, RS soon became a key experimental tool for the characterization of the material as well.

6.7.1 Graphene and few-layer graphene

In the early experiments graphene was obtained from exfoliation of graphite. In this process single-layer graphenes (SLGs) can be obtained but usually bilayer graphene (BLG) and even FLG were produced as well. Optical microscopy or AFM techniques have been used to check the number of layers. Both techniques are inconvenient as for the optical analysis one needs a special substrate consisting of a 90- or 300-nm SiO_2 layer on a Si waver to allow application of the ratio of colour difference method (RCD) [128]. AFM itself is a delicate procedure and needs good experience. In contrast, RS turned out to be a convenient method as the spectra change characteristically with the number of layers [129–131]. After this discovery numerous experiments were reported where RS was used to characterize SLG, BLG or, more general, FLG. The RS of SLG resembles very much the spectrum of graphite with the characteristic D line, G line and the corresponding overtones. For the understanding of the D line and of the overtones DRRS was successfully applied. Interestingly in graphene the 2D line is usually stronger than the G line. Spectra became particularly interesting as controlled FLG became available. In some sense BLG is the analogue to DWCNTs.

According to the increased unit cell the electronic spectrum and the phonon dispersion for FLG split into two or more components. Even though at very low excitation energies the Dirac nature of the band structure is retained [132], for optical and resonance Raman experiments the parabolic part of the split bands are relevant. For these the higher valence band and the lower conduction band touch each other at the K point. Particular attention has been paid to the behaviour of the 2D band as it changes most characteristically with the number of layers. Figure 6.23 has an example as described in Ref. [130].

For SLG the 2D band has only one component of Lorentzian type. Two-layer graphene has four components, according to the increased unit cell and

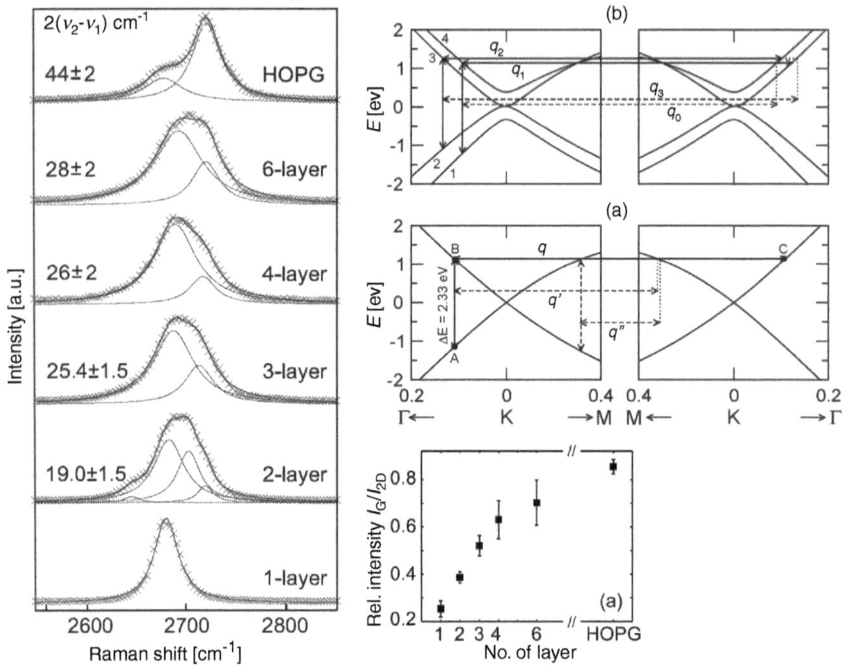

FIGURE 6.23 Raman response of the 2D line in FLG (left) and electronic structure from DFT calculation with the scattering processes for double resonance processes indicated for monolayer and BLG (right, top). The dashed lines in the left panel are fitted Lorentzian components. The numbers on the left side are frequency differences between peaks for two Lorentzian fits. The lower part of the right panel shows the ratio I_G/I_{2D} of the integrated intensities as a function of layer number. Courtesy of L. Wirtz.

the interaction between the layers. Two of the components are weak, as shown in Figure 6.23 (left panel). With increasing layer number the experiments can be well fitted with only two Lorentzians with an almost constant intensity, in spite of the increasing scattering volume. In contrast, the G line increases and broadens with increasing layer number. This means the ratio I_G/I_{2D} depicted in Figure 6.23 (right, bottom) is a good measure for the number of layers in FLG. However, as described in next sections, attention must be paid to the fact that any charging of the graphene layers modifies I_G/I_{2D}.

The double resonance process for SLG depicted in Figure 6.23, right panel (top, lower graph), is as explained in Sec. 6.2.2. Scattering vectors are drawn horizontally for simplicity. The alternative double resonance scattering possibilities assigned as q' and q'' are considered as weak and therefore not observed. In BLG we expect four lines according to the scattering diagram of Figure 6.23, right panel (top, upper graph). Again double resonance is only strong for the transitions two to three and one to four with wavevectors q_2 and q_1. The other indicated transitions are weak. The peak positions for the two innermost peaks in Figure 6.23, left panel are 2683 and 2702 cm^{-1}, respectively.

FLG has additional interesting vibrational properties as now interlayer shear modes and compressional modes, so-called 'z modes', are possible. With increasing number of layers the number of such modes increases [133], like the modes in a linear chain. Eventually they transform into the corresponding dispersive bands in graphite. According to the high masses of the involved layers and the weak interlayer coupling such frequencies have low energy in the range of 30–40 cm^{-1} but could recently be observed by RS [134]. The corresponding peaks in Raman spectra were given the name 'C lines', pointing to their origin from interlayer coupling. The shear mode in BLG is Raman active and of E_g symmetry. With increasing layer number n more modes, notably all double degenerated, exist and can be Raman and/or IR active. Figure 6.24 depicts some results for Stokes and anti-Stokes scattering. From Figure 6.23 the dependence of the C-peak position on layer number is evident. From the $n-1$ possible shear modes only the mode with the highest frequency is observed.

The frequencies for the interlayer shear modes can be derived from a linear chain model using the interlayer force constant α as a parameter. In n-layer graphene the highest frequency is Raman active. In this mode, neighbouring layers move in opposite directions. The frequency can be obtained from

$$\omega_C(n) = \sqrt{\frac{\alpha}{2\pi^2 c^2 \mu}}\sqrt{1+\cos(\pi / n)} \tag{6.15}$$

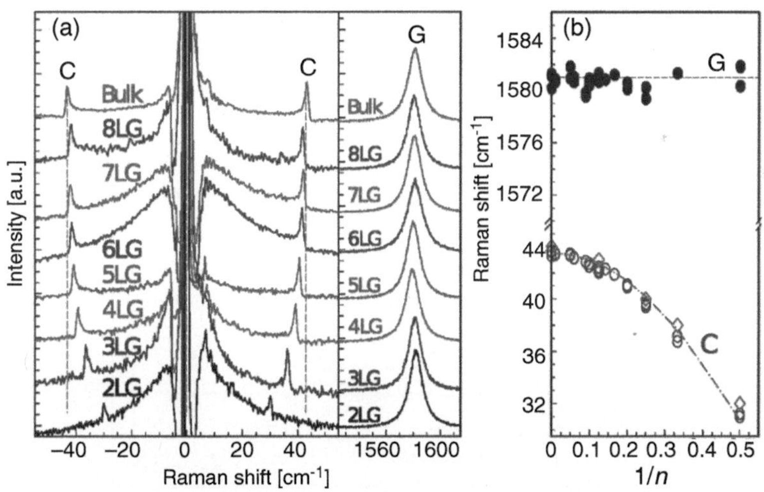

FIGURE 6.24 Stokes and anti-Stokes low-energy Raman spectra of FLG showing the shift of the shear mode peak C with layer number and the unshifted G line in contrast (a). Dashed lines are guides for the eye. (b) Depicts G-line and C-line frequencies versus inverse number of layers. The dash-dotted line is from Equation 6.15; diamonds are from DFT calculations (b). Spectra were recorded for suspended FLG using a single monochromator with three BragGrate notch filters and 541-nm laser excitation. Courtesy of A. Ferrari.

where $\mu = 7.6 \times 10^{-25}$ kg nm^{-2} is the layer mass per unit area and c is the speed of light. From a fit of the experiments $\alpha = 12.8 \times 10^{18}$ N m^{-3} is obtained. Results from DFT calculations, as presented in Ref. [134], are in very good agreement with the empirical results described in previous sections.

The shape of the C line is asymmetric and can be well fitted with a BWF profile. This indicates quantum interference in the scattering process with continuum excitations. Best candidate for such excitations are free electrons in the parabolic bands which touch at the K point but also electrons in the Dirac cones which are superimposed at the K point [132].

Another possibility to check on the number of layers in FLG was recently suggested from two-phonon RS where one phonon is the compressional mode in z direction (ZO' phonon) and the other is the intra-valley scattered LO phonon close to the Γ point [135]. The latter was suggested to contribute in a Stokes process, whereas the former contributes in an anti-Stokes process. The Raman response from these combination modes was called N lines. Spectra are depicted in Figure 6.25. Since the compressional mode does not exist in SLG, N lines appear only for number of layers $n > 1$.

The N lines are about a factor 100 weaker than the G line but, as evident from Figure 6.25, exhibit a well-expressed dependence on layer number. Layer numbers up to $n = 11$ could be identified.

There is an equivalent N-type mode on the high-energy side of the G line which originates from the Stokes scattering process of the ZO' compressional

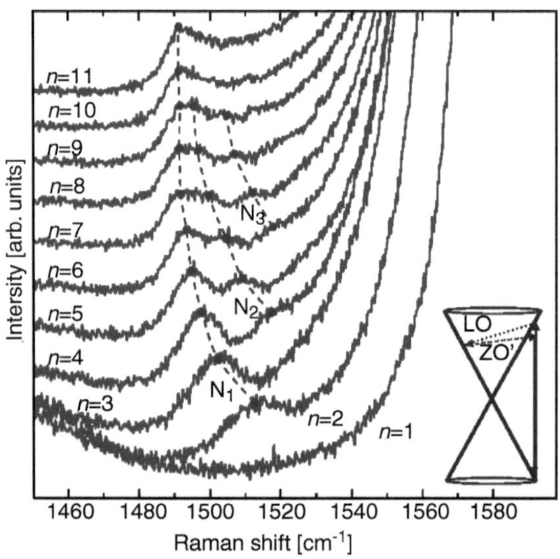

FIGURE 6.25 RS from FLG in the spectral range just below the G line for excitation with 633-nm laser. The number of layers is indicated as n. Capital N_i assign the overtone contributions of the compressional modes. Dashed lines are guides to the eye. Inset: suggested intra-valley double resonance scattering process for a selected time order. Courtesy of J. Maultzsch.

modes and may be called N$^+$ line (LOZO'$^+$ in the original work). Dependence on number of layers is quite similar. The anti-Stokes and the Stokes contributions to the Raman spectra are symmetric to the D' line at 1616 cm^{-1}. From calculations N and N$^+$ are expected to be dispersive in a highly non-linear manner.

BLG can have other interesting structures as they were recently studied [136,137]. After exfoliation from graphite, backfolding may incidentally happen for some monolayers. In this case BLG is created where the layers are arbitrarily stacked in contrast to the normal AB Bernal stacking. Depending on the angle of backfolding commensurate or incommensurate structures may appear. They can be observed as Moire pattern in TEM [136,138]. In some sense this has again analogy to DWCNT [139]. Such structures have a very interesting D-line response. Even if the starting graphene single layer and the part of the flake which was not backfolded have no D line, indicating high-quality graphene, the bilayer part of the flake can have two or even three types of D lines [136] as shown in Figure 6.26. One type is broad and dispersive, one is strongly resonant in the ultraviolet but not dispersive and the third is small and only observed with the ultraviolet laser. The lines were given the names I bands, in particular $I_{envelope}$, I_{fixed} and I_{new}. The line position of I_{fixed} varies with folding orientation.

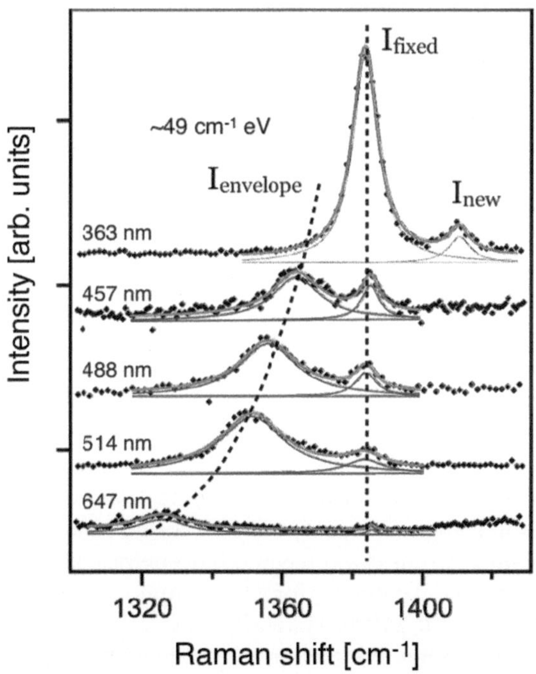

FIGURE 6.26 Raman spectra from backfolded graphene as excited with various lasers. Dashed lines are guides for the eye. They demonstrate the dispersive and the non-dispersive character for $I_{envelope}$ and for I_{fixed}, respectively. Full-drawn lines are Lorentzian fits. From Ref. [136], copyright with the American Physical Society. Permission for reproduction from November 2012, Ref. #21759.

The dispersion of $I_{envelope}$ corresponds well to the dispersion of 50 eV cm^{-1} known for defect scattering.

The origin of the new lines is traced back to a DRRS process where the periodic potential induced by one layer into the other plays the role of an impurity potential. This periodic potential replaces the random potential of the defects and allows for elastic backscattering. For non-periodic defects many vectors in the reciprocal space can contribute to compensate the phonon wavevector from the inter-valley phonon scattering and these vectors change with the exciting laser. In contrast, for the periodic perturbations only selected Fourier components can compensate the phonon scattering and these components do not depend on the exciting laser. They only depend on the geometry of the static interlayer perturbation. Therefore, the Raman line does not disperse but different mutual rotations of the two lattices yield different lines at 1384, 1382, 1371 and 1395, respectively, as observed for different samples. If many different individually unresolved Fourier components contribute, they may all contribute and resonance enhancement selects some of them leading to the dispersive envelope line.

6.7.2 Special Raman experiments with graphene: electronic Raman scattering and edge state scattering

Like in the case of SWCNT graphene can be doped with electrons or holes. This can be done chemically by adding donor or acceptor molecules or, more conveniently, by field effect. The latter technique has the advantage of allowing immediate undoping or change from n doping to p doping while Raman spectra can be recorded in situ. Conventional field doping with a back gate [140] and EC doping with a top gate using a solid electrolyte at the gate [141] were demonstrated. In both cases Raman response was studied as a function of carrier concentration. EC doping has the advantage of allowing for higher electron or hole concentrations due to the very narrow Debye layer developing on the boundary between electrolyte and graphene. In Ref. [141] the solid electrolyte polyethylene oxide + $LiClO_4$ was used as a gate on top of a graphene monolayer. Carrier concentration and current were controlled by a gate voltage connected to a platinum electrode on top of the solid electrolyte and by source and drain connected to the graphene. Raman spectra were recorded with a confocal spectrometer for excitation with a 514.5-nm laser. The G line was found to shift upwards in a slightly asymmetric way for holes and electrons. For the former a shift of up to 30 cm^{-1} was recorded. In contrast, the 2D line expired an upshift for p doping but eventually, for strong doping, a downshift for electron doping. Importantly, the $I(2D)/I(G)$ intensity ratio was found to depend strongly on the carrier concentration. For the observed range of concentrations between nominally zero and 4×10^{13} cm^{-2} it changes by almost a factor 6. This suggests to use this ratio for a determination of the doping concentration.

The changes of G-line and 2D-line frequencies have different origin. In the mSWCNTs and in graphene a strong electron–phonon coupling lead to a downshift of the phonon energy at the Γ point and at the K point due to the Kohn anomaly as already discussed. When the Fermi level is shifted away from the Dirac point the downshift due to the electron–phonon interaction is not effective any more and the line stiffens. The upshift of the G line goes along with a line narrowing by about a factor 2, since decay channels close due to Pauli blockade. For the behaviour of the 2D line changes in the bond length due to doping were suggested and qualitatively demonstrated from a DFT calculation.

In another, more classical way of doping carbon atoms are substituted by nitrogen [142]. The doping process can be studied by RS from the G band and 2D band. Simple graphitic like substitution leads to only weak changes of the bands. Clustered substitution where the N atoms adopt a pyrrolic or a pyridinic configuration yields an enhanced D-line response and a downshift of the 2D line. The G line appears broadened due to an overlap with the D' line. Recently there were also Raman studies reported for B doping of graphene [143].

There are strong efforts to induce a gap into graphene. One possibility is to cut very narrow ribbons from a graphene sheet and open a gap in this way by quantum confinement. However, gap opening was not very efficient so far. Only gaps up to 400 meV could be obtained but the generated edge states are interesting from several points of view. For example, the electronic structure for armchair edges and zigzag edges is significantly different. The latter are formed from the same type of graphene sub-lattice and thus exhibit a flat band at the K point where the charge density is strongly localized to the edge atoms. In contrast, the former edges have alternating contributions from the two different sub-lattices and flat band states are absent. RS is an excellent tool to check on the nature of these states.

Since edge states are defects, D-band intensities will be enhanced if the laser is focused on the edge of a ribbon. Imaging indeed showed in general that the D-band image is very bright at the edges whereas it is uniform and weak over the ribbon [6,144]. This was surprising, in detail, as RS from edge states is subjected to strict selection rules as far as the orientation of light polarization and chirality of the edge are concerned [145]. A zigzag edge cannot produce D-band scattering, since the only possible orientation of the elastic backscattering is perpendicular to the edge and therefore cannot satisfy a KK' scattering condition [146,147]. In contrast, for an armchair edge backscattering is possible. Good evidence for this difference was recently obtained from Raman experiments of graphene flakes with holes. The holes were either round which means consisting of mixture of armchair and zigzag edges or hexagonal consisting of only zigzag edges [144]. Raman intensity ratios between D band and G band were a factor 30 smaller for the latter flakes as compared to the flakes with round holes. In the described experiments circularly polarized light was used to avoid influence of polarization selection rules. Interestingly, the limitations by selection rules for possible DRRS are not valid for the intra-valley double resonance as it happens for the D' band. Indeed, the D' response is not dependent on the chirality of the edge [146].

Another possibility to open a gap in graphene is hydrogenation. This process can indeed open a large gap as it converts part of the sp^2 bonds to sp^3 bonds turning the graphene into something like a two-dimensional diamond [148].

6.8 SECOND-ORDER RAMAN SPECTRA AND COMBINATION MODES IN SWCNTS AND GRAPHENE

The D lines (D and D' line) and the 2D line are most prominent members of second-order spectra in SWCNTs and graphene as they were already discussed in the previous sections. Also, the above-discussed N line is a combination mode. There are, however, additional Raman lines particularly in the spectral range between RBM lines and G line and between 1600 and 2900 cm^{-1}, which are due to overtones and combination modes. Since in the latter spectral range combinations modes of acoustic and optical phonons dominate, it has been called 'LOLA region'. To a large extent the modes appear likewise in SWCNTs, graphene and graphite, which suggests to review these features together in a special section. In most cases DRRS gives a good explanation for the spectra. Assignment of the lines to special combination modes is not straightforward and often different assignments were reported for the same Raman feature.

6.8.1 Intermediate frequency modes, M lines, iTOLA line and overtone lines in CNT

In the spectral region between RBM and D line several weak and broad Raman features appear. They are classified as 'intermediate frequency modes' (IMF modes) [149] and originate from zone-folded optic and acoustic combination modes of graphene. Some of them are dispersive as, e.g. a mode around 950 cm^{-1} with a dispersion of 100 cm^{-1} eV. Other IMF modes, e.g. at 700, 860, 1070 cm^{-1} are not dispersive.

On the high-frequency side of the G line a double-peaked line is observed around 1730 and 1750 cm^{-1}. From their appearance the lines were named M^+ and M^- [150]. In graphite similar lines are observed but upshifted by 20 cm^{-1}. The M^+ line is almost independent of the laser used for excitation whereas the M^- line shows a weak dispersion of 26 cm^{-1} eV. This behaviour is depicted in Figure 6.27(a). The origin of these lines is assigned to an overtone of the oTO-derived mode of graphene at 867 cm^{-1}. This is concluded from a comparison of calculated and observed dispersion of the lines as depicted in Figure 6.27(b). Intra-valley and inter-valley scattering processes are possible, corresponding to either $|q| \approx 0$ or $|q| \approx 2|k|$. Since only the $|q| \approx 2|k|$ wavevector shifts with light energy this scattering process is assigned to the M^- line (inter-valley scattering) and the non-dispersive M^+ line is assigned to the intra-valley scattering. The upshift of the M lines for graphite is considered as a consequence of zero curvature in this material.

In SWCNT bundles an additional broad line is observed around 1950 cm^{-1}, which is assigned to the combination mode of iTO mode and longitudinal

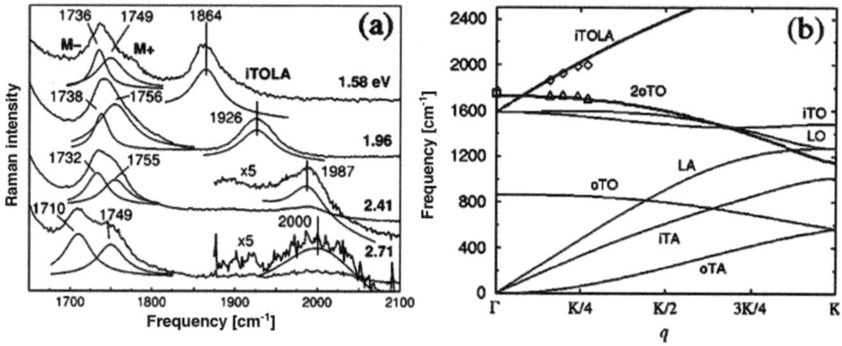

FIGURE 6.27 Selected spectral regions of Raman response for SWCNTs. (a) Shows the weak shift of the M^- line (left) and the very strong shift of the iTOLA line (right) with decreasing laser energy. (b) Depicts the phonon dispersion for graphene fundamental modes and for combination and overtone modes. The diamonds and triangles are results from (a) after transformation to q space using Equation 6.16. From Ref. [150], copyright with the American Physical Society. Permission for reproduction from November 2012, Ref. #21759.

acoustic (LA) mode [150]. This line disperses strongly with 120 cm^{-1} eV which originates from the acoustic mode contribution. The assignment is again obtained from comparing the dispersion with the dispersion of various combination modes for the graphite phonons as depicted in Figure 6.27(b). After acid treatment of the tubes additional lines were observed [151].

In addition to the above-described lines from $q \neq 0$ also simple overtone lines at $q = 0$ are observed at 3180 cm^{-1} (2G line) and at 2900 cm^{-1} (2LO line) [13].

6.8.2 Raman scattering from combination modes in graphene

At the high-frequency side of the G band the D' band is observed at 1620 cm^{-1}. It results from intra-valley DRRS of the LO phonon, involving electronic states at the same K point and using a defect for backscattering. This is in contrast to the D line that uses inter-valley scattered iTO phonons involving electronic states at different K points. As a consequence the D line scattering probes phonons with large wavevector q, close to the K point whereas D' probes small wavevectors close to the Γ point. The D' has a weak dispersion of only 10 cm^{-1} eV.

M bands like in SWCNTs are only observed for FLG with $n > 1$ and are assigned to intra-valley scattered overtone of the oTO mode. With increasing layer number they shift upwards in frequency and eventually reach the values for graphite [152]. Interestingly, in twisted BLG (incommensurate stacking) the M modes disappear.

In the combination mode range between 1800 and 2100 cm^{-1} observation of three lines is reported but assignment is not yet agreed. Various combination modes such as LA, TA and LO are in discussion [153,154]. The observed Raman lines are highly dispersive, of the order of 200 cm^{-1} eV indicating their origin from DRRS.

The shape of the 2D band in FLG graphene with large values for n was observed to depend strongly on the stacking behaviour. AB stacking yields only one line whereas turbostratic stacking yields two peaks [6], which are separated as much as 40 cm^{-1}.

Finally, at 2950 cm^{-1} a Raman line was observed for the combination mode D + G.

In the case of graphite and graphene, the DRRS has been used successfully to analyse dispersion for $q \neq 0$ phonons. This is at least possible for phonons close to the Γ point and close to the K point. The requested q vector for the phonons is obtained from the laser energy E_{laser} used for excitation [155] according to

$$E_{\text{laser}} = (\sqrt{3}/2)\gamma_0 ak \tag{6.16}$$

with $q = 2k$ expressing the condition for double resonance. a is the graphene lattice constant. The equation is derived from the special form of the band structure of graphene (Dirac cone at the K point) as depicted in Figure 6.1. The above relation holds for intra-valley as well as for inter-valley scattering if in the latter the phonon wavevector is measured from the K point. For q vectors close to the Γ point and close to the K point intra-valley and inter-valley scattering must be considered, respectively. This method was first successfully applied for the analysis of phonon dispersion of graphite [156] but is now often used to check the graphene phonon dispersion.

ACKNOWLEDGEMENTS

Work supported by the FWF in Austria (P 14893). I am grateful to my colleagues for the supply of original data and to A. Stangl for the assistance during preparation of the figures. Valuable discussions with Prof. J. Kürti, Dr. A. Grüneis, Dr. C. Kramberger and Prof. T. Pichler, are gratefully acknowledged.

REFERENCES

[1] Iijima S. Helical microtubules of graphitic carbon. Nature 1991;354:56.
[2] Iijima S, Ichihashi T. Single-shell carbon nanotubes of 1-nm diameter. Nature 1993;363: 603.
[3] Bethune DS, Kiang CH, DeVries MS, Gorman G, Savoy R, Beyers R. Cobalt-catalysed growth of carbon nanotubes with single-atomic-layer walls. Nature 1993;363:605.
[4] Thess A, Lee R, Nikolaev P, Dai H, Petit P, Robert J, et al. Crystalline Ropes of Metallic Carbon Nanotubes. Science 1996;273:483.
[5] Tuinstra F, Koenig JL. Raman Spectrum of Graphite. J Chem Phys 1970;53:1126.
[6] Pimenta MA, Dresselhaus G, Dresselhaus MS, Cancado LG, Jorio A, Saito R. Studying disorder in graphite-based systems by Raman spectroscopy. Phys Chem Chem Phys 2007;9:1276.
[7] Eckmann A, Felten A, Mishchenko A, Britnell L, Krupke R, Novoselov KS, et al. Probing the Nature of Defects in Graphene by Raman Spectroscopy. Nano Lett 2012;12:3925.
[8] Saito R, Hofmann M, Dresselhaus G, Jorio A, Dresselhaus MS. Raman spectroscopy of graphene and carbon nanotubes. Adv Phys 2011;60:413.

[9] Jorio A, Saito G, Dresselhaus M, Dresselhaus M. Raman spectroscopy in graphene related systems. Wiley-VCH; 2010.

[10] S. Reich, C. Thomsen, and M. J., Carbon nanotubes, basic concepts and physical properties (Wiley-VCH, 2004), Chapters 7 and 8, p. 115–133.

[11] Kuzmany H, Hulman M, Pfeiffer R, Simon F. Raman scattering of carbon nanotubes. NATO Science Series, Mathematics, Physics & Chemistry 2006;II/222:89.

[12] Dresselhaus MS, Jorio A, Saito R. Characterizing Graphene, Graphite, and Carbon Nanotubes by Raman Spectroscopy. Ann Rev Condens Matter Phys 2010;1:89.

[13] Dresselhaus M, Dresselhaus G, Saito R, Jorio A. Raman spectroscopy of carbon nanotubes. Phys Rep 2005;409:47.

[14] Koltai J, Zólyomi V, Kürti J. Phonon dispersion of small diameter semiconducting chiral carbon nanotubes – a theoretical study. Phys Stat Sol B 2008;245:2137.

[15] Kramberger C, Pichler T. Advances in carbon nanotubes. Pan Stanford Publishing; 2012. Chapter 4.

[16] Kuzmany H. Solid-state spectroscopy, an introduction. Springer Heidelberg; 2009.

[17] Pinczuk A, Burstein E. Fundamentals of Inelastic Light Scattering in Semiconductors and Insulators. Topics Appl Phys 1975;8:23.

[18] Martin RM, Falicov LM. Light scattering in solids I. Springer, Berlin; 1983. Chapter 3, p. 79.

[19] Jiang J, Saito R, Sato K, Park JS, Samsonidze GG, Jorio A, et al. Exciton-photon, exciton-phonon matrix elements, and resonant Raman intensity of single-wall carbon nanotubes. Phys Rev B 2007;75:035405.

[20] Thomsen C, Reich S. Double-resonant Raman scattering in graphite. Phys Rev Lett 2000;85:5214.

[21] Venezuela P, Lazzeri M, Mauri F. Theory of double-resonant Raman spectra in graphene: Intensity and line shape of defect-induced and two-phonon bands. Phys Rev B 2011;84: 035433.

[22] Abstreiter G, Cardona M, Pinczuk A. Light Scattering by Free Carrier Excitations in Semiconductors. Topics Appl Phys 1984;54:5.

[23] Kuzmany H. The Particle in the Box Model For Resonance Raman-Scattering in Polyacetylene. Pure Appl Chem 1985;57:235.

[24] Wang F, Dukovic G, Brus LE, Heinz TF. The Optical Resonances in Carbon Nanotubes Arise from Excitons. Science 2005;308:838.

[25] Popov VN, Henrard L, Lambin P. Electron-phonon and electron-photon interactions and resonant Raman scattering from the radial-breathing mode of single-walled carbon nanotubes. Phys Rev B 2005;72:035436.

[26] Kresse G, Furtmüller J, Hafner J. Ab initio force constant approach to phonon dispersion relations of diamond and graphite. Europhys Lett 1995;32:729.

[27] Sanchez-Portal D, Artacho E, Soler JM, Rubio A, Ordejon P. Ab initio structural, elastic, and vibrational properties of carbon nanotubes. Phys Rev B 1999;59:12678.

[28] Jorio A, Souza AG, Dresselhaus G, Dresselhaus MS, Swan AK, Unlu MS, et al. G-band resonant Raman study of 62 isolated single-wall carbon nanotubes. Phys Rev B 2002;65:155412.

[29] Kuzmany H, Plank W, Hulman M, Kramberger C, Grüneis A, Pichler T, et al. Determination of SWCNT diameters from the Raman response of the radial breathing mode. Eur Phys J B 2001;22:307.

[30] Maultzsch J, Reich S, Thomsen C, Requardt H, Ordejon P. Phonon dispersion in graphite. Phys Rev Lett 2004;92:075501.

[31] Popov VN, Lambin P. Radius and chirality dependence of the radial breathing mode and the G-band phonon modes of single-walled carbon nanotubes. Phys Rev B 2006;73:085407.

[32] Grüneis A, Serrano J, Bosak A, Lazzeri M, Molodtsov SL, Wirtz L, et al. Phonon surface mapping of graphite: Disentangling quasi-degenerate phonon dispersions. Phys Rev B 2009;80:085423.
[33] Grüneis A. PhD thesis, Tohoku University (2004).
[34] Zólyomi V, Koltai J, Kürti J, Kuzmany H. Phonons of single-walled carbon nanotubes. Research Signpost; 2008. Chapter 7, 297-332.
[35] Jishi RA, Venkataraman L, Dresselhaus MS, Dresselhaus G. Phonon modes in carbon nanotubules. Chem Phys Lett 1993;209:77.
[36] Dubay O, Kresse G, Kuzmany H. Phonon Softening in Metallic Nanotubes by a Peierls-like Mechanism. Phys Rev Lett 2002;88:235506.
[37] Piscanec S, Lazzeri M, Robertson J, Ferrari AC, Mauri F. Optical phonons in carbon nanotubes: Kohn anomalies, Peierls distortions, and dynamic effects. Phys Rev B 2007;75:035427.
[38] Farhat H, Sasaki K, Kalbac M, Hofmann M, Saito R, Dresselhaus MS, et al. Softening of the Radial Breathing Mode in Metallic Carbon Nanotubes. Phys Rev Lett 2009;102:126804.
[39] Piscanec S, Lazzeri M, Mauri F, Ferrari AC, Robertson J. Kohn anomalies and electron-phonon interactions in graphite. Phys Rev Lett 2004;93:185503.
[40] Dubay O, Kresse G. Accurate density functional calculations for the phonon dispersion relations of graphite layer and carbon nanotubes. Phys Rev B 2003;67:035401.
[41] Damnjanovic M, Milosevic I, Vukovic T, Sredanovic R. Full symmetry, optical activity, and potentials of single-wall and multiwall nanotubes. Phys Rev B 1999;60:2728.
[42] Jorio A, Fantini C, Pimenta MA, Capaz RB, Samsonidze GG, Dresselhaus G, et al. Resonance Raman spectroscopy (n,m)-dependent effects in small-diameter single-wall carbon nanotubes. Phys Rev B 2005;71:075401.
[43] Kürti J, Zólyomi V, Kertesz M, Sun G. The geometry and the radial breathing mode of carbon nanotubes: beyond the ideal behaviour. New J Phys 2003;5:125.
[44] Kramberger C, Pfeiffer R, Kuzmany H, Zolyomi V, Kuerti J. Assignment of chiral vectors in carbon nanotubes. Phys Rev B 2003;68:235404.
[45] Simon F, Zólyomi V, Pfeiffer R, Kuzmany H, Koltai J, Kürti J. Unusual Raman dispersion for D and 2D lines in high-curvature single-walled carbon nanotubes revealed by C-13 isotope substitution. Phys Rev B 2010;81:125434.
[46] Kürti J, Kresse G, Kuzmany H. First-principles calculations of the radial breathing mode of single-wall carbon nanotubes. Phys Rev B 1998;58:R8869.
[47] Milnera M, Kuerti J, Hulman M, Kuzmany H. Periodic resonance excitation and intertube interaction from quasicontinuous distributed helicities in single-wall carbon nanotubes. Phys Rev Lett 2000;84:1324.
[48] Telg H, Maultzsch J, Reich S, Hennrich F, Thomsen C. Chirality Distribution and Transition Energies of Carbon Nanotubes. Phys Rev Lett 2004;93:177401.
[49] Samsonidze GG, Saito R, Jorio A, Pimenta MA, Souza AG, Gruneis A, et al. The concept of cutting lines in carbon nanotube science. J Nanosci Nanotechnnol 2003;3:431.
[50] Maultzsch J, Telg H, Reich S, Thomsen C. Radial breathing mode of single-walled carbon nanotubes: Optical transition energies and chiral-index assignment. Phys Rev B 2005; 72:205438.
[51] Popov VN. Curvature effects on the structural, electronic and optical properties of isolated single-walled carbon nanotubes within a symmetry-adapted. New J Phys 2004;6:17.
[52] Samsonidze GG, Saito R, Kobayashi N, Grüneis A, Jiang J, Jorio A, et al. Family behavior of the optical transition energies in single-wall carbon nanotubes of smaller diameters. Appl Phys Lett 2004;85:5703.
[53] Kane CL, Mele EJ. Electron Interactions and Scaling Relations for Optical Excitations in Carbon Nanotubes. Phys Rev Lett 2004;93:197402.

[54] Araujo PT, Doorn SK, Kilina S, Tretiak S, Einarsson E, Maruyama S, et al. Third and fourth optical transitions in semiconducting carbon nanotubes. Phys Rev Lett 2007;98:067401.
[55] Fantini C, Jorio A, Souza M, Strano MS, Dresselhaus MS, Pimenta MA. Optical transition energies for carbon nanotubes from resonant Raman spectroscopy: Environment and temperature effects. Phys Rev Lett 2004;93:147406.
[56] Kneipp K, Kneipp H, Corio P, Brown SDM, Shafer K, Motz J, et al. Surface-enhanced and normal Stokes and anti-Stokes Raman spectroscopy of single-walled carbon nanotubes. Phys Rev Lett 2000;84:3470.
[57] Jorio A, Souza AG, Dresselhaus G, Dresselhaus MS, Saito R, Hafner JH, et al. Joint density of electronic states for one isolated single-wall carbon nanotube studied by resonant Raman scattering. Phys Rev B 2001;63:245416.
[58] Jorio A, Saito R, Hafner JH, Lieber CM, Hunter M, McClure T, et al. Structural *(n,m)* Determination of Isolated Single-Wall Carbon Nanotubes by Resonant Raman Scattering. Phys Rev Lett 2001;86:1118.
[59] Kürti J, Zólyomi V, Grueneis A, Kuzmany H. Double resonant Raman phenomena enhanced by van Hove singularities in single-wall carbon nanotubes. Phys Rev B 2002;65:165433.
[60] Zólyomi V, Kürti J, Grüneis A, Kuzmany H. Origin of the Fine Structure of the Raman D Band in Single-Wall Carbon Nanotubes. Phys Rev Lett 2003;90:157401.
[61] Maultzsch J, Reich S, Schlecht U, Thomsen C. High-energy phonon branches of an individual metallic carbon nanotube. Phys Rev Lett 2003;91:087402.
[62] Souza Filho AG, Jorio A, Samsonidze GG, Dresselhaus G, Pimenta MA, Dresselhaus MS, et al. Competing spring constant versus double resonance effects on the properties of dispersive modes in isolated single-wall carbon nanotubes. Phys Rev B 2003;67:035427.
[63] Kukovecz A, Kramberger C, Georgakilas V, Prato M, Kuzmany H. A detailed Raman study on thin single-wall carbon nanotubes prepared by the HiPCO process. Eur Phys J B 2002;28:223.
[64] Brown SDM, Jorio A, Corio P, Dresselhaus MS, Dresselhaus G, Saito R, et al. Origin of the Breit-Wigner-Fano lineshape of the tangential G-band feature of metallic carbon nanotubes. Phys Rev B 2001;63:155414.
[65] Bendiab N, Anglaret E, Bantignies JL, Zahab A, Sauvajol JL, Petit P, Mathis C, et al. Stoichiometry dependence of the Raman spectrum of alkali-doped single-wall carbon nanotubes. Phys Rev B 2001;64:245424.
[66] Chen G, Furtado CA, Bandow S, Iijima S, Eklund PC. Anomalous contraction of the C-C bond length in semiconducting carbon nanotubes observed during Cs doping. Phys Rev B 2005;71:045408.
[67] Chen G, Furtado CA, Kim UJ, Eklund PC. Alkali-metal-doping dynamics and anomalous lattice contraction of individual debundled carbon nanotubes. Phys Rev B 2005;72:155406.
[68] Kukovecz A, Pichler T, Pfeiffer R, Kuzmany H. Diameter selective charge transfer in p- and n-doped single wall carbon nanotubes synthesized by the HiPCO method. Chem Commun 2002;1730–1.
[69] Kavan L, Frank O, Green AA, Hersam MC, Koltai J, Zólyomi V, et al. In situ Raman spectroelectrochemistry of single-walled carbon nanotubes: Investigation of materials enriched with (6,5) tubes. J Phys Chem C 2008;112:14179.
[70] Kalbac M, Kavan L, Dunsch L, Dresselhaus MS. Development of the tangential mode in the Raman spectra of SWCNT bundles during electrochemical charging. Nano Lett 2008;8:1257.
[71] Kavan L, Dunsch L. Electrochemistry of carbon nanotubes. Carbon Nanotubes 2008;111:567.
[72] Park JS, Sasaki K, Saito R, Izumida W, Kalbac M, Farhat H, et al. Fermi energy dependence of the G-band resonance Raman spectra of single-wall carbon nanotubes. Phys Rev B 2009;80:081402.

[73] Sloan J, Monthioux M. Filled carbon nanotubes: (x@CNTs). Wiley; 2012. Chapter 5.
[74] Takenobu T, Takano T, Shiraishi M, Murakami Y, Ata M, Kataura H, et al. Stable and controlled amphoteric doping by encapsulation of organic molecules inside carbon nanotubes. Nat Mater 2003;2:683.
[75] Yanagi K, Miyata Y, Kataura H. Highly stabilized beta-carotene in carbon nanotubes. Adv Mater 2006;18:437.
[76] Kuzmany H, Pfeiffer R, Kramberger C, Pichler T, Liu X, Knupfer M, et al. Analysis of the concentration of C-60 fullerenes in single wall carbon nanotubes. Appl Phys A 2003;76:449.
[77] Kalbac M, Kavan L, Zukalova M, Dunsch L. Two positions of potassium in chemically doped C-60 peapods: An in situ spectroelectrochemical study. J Phys Chem B 2004;108:6275.
[78] Pichler T, Kuzmany H, Kataura H, Achiba Y. Metallic polymers of C-60 inside single-walled carbon nanotubes. Phys Rev Lett 2001;87:267401.
[79] Pichler T, Kukovecz A, Kuzmany H, Kataura H, Achiba Y. Quasicontinuous electron and hole doping of C-60 peapods. Phys Rev B 2003;67:125416.
[80] Bandow S, Takizawa M, Hirahara K, Yudasaka M, Iijima S. Raman scattering study of double-wall carbon nanotubes derived from the chains of fullerenes in single-wall carbon nanotubes. Chem Phys Lett 2001;337:48.
[81] Kataura H, Maniwa Y, Kodama T, Kikuchi K, Hirahara K, Iijima S, et al. Fullerene-peapods: synthesis, structure, and Raman spectroscopy. AIP Conf Proc 2001;591:251.
[82] Hafner JH, Bronikowski MJ, Azamian BR, Nikolaev P, Rinzler AG, Colbert DT, et al. Catalytic growth of single-wall carbon nanotubes from metal particles. Chem Phys Lett 1998;296:195.
[83] Endo M, Muramatsu H, Hayashi T, Kim YA, Terrones M, Dresselhaus NS. Nanotechnology: 'buckypaper' from coaxial nanotubes. Nature 2005;433:476.
[84] Pfeiffer R, Pichler T, Kim YA, Kuzmany H. Double-wall carbon nanotubes. Topics Appl Phys 2008;111:495.
[85] Pfeiffer R, Kuzmany H, Kramberger C, Schaman C, Pichler T, Kataura H, et al. Unusual high degree of unperturbed environment in the interior of single-wall carbon nanotubes. Phys Rev Lett 2003;90:225501.
[86] Pfeiffer R, Simon F, Kuzmany H, Popov VN. Fine structure of the radial breathing mode of double-wall carbon nanotubes. Phys Rev B 2005;72:161404.
[87] Simon F, Kukovecz A, Kramberger C, Pfeiffer R, Hasi F, Kuzmany H, et al. Diameter selective reaction processes of single-wall carbon nanotubes. Phys Rev B 2005;71:165439.
[88] Pfeiffer R, Kramberger C, Simon F, Kuzmany H, Popov VN, Kataura H. Interaction between concentric tubes in DWCNTs. Eur Phys J B 2004;42:345.
[89] Simon F, Kramberger C, Pfeiffer R, Kuzmany H, Zólyomi V, Kürti J, et al. Isotope Engineering of Carbon Nanotube Systems. Phys Rev Lett 2005;95:017401.
[90] Zólyomi V, Simon F, Rusznyak A, Pfeiffer R, Peterlik H, Kuzmany H, et al. Inhomogeneity of ^{13}C isotope distribution in isotope engineered carbon nanotubes: Experiment and theory. Phys Rev B 2007;75:195419.
[91] Pfeiffer R, Kuzmany H, Simon F, Bokova SN, Obraztsova E. Resonance Raman scattering from phonon overtones in double-wall carbon nanotubes. Phys Rev B 2005;71:155409.
[92] Pfeiffer R, Holzweber M, Peterlik H, Kuzmany H, Liu Z, Suenaga K, et al. Dynamics of carbon nanotube growth from fullerenes. Nano Lett 2007;7:2428.
[93] Han SW, Yoon M, Berber S, Park N, Osawa E, Ihm J, et al. Microscopic mechanism of fullerene fusion. Phys Rev B 2004;70:113402.
[94] Kim YH, Lee IH, Chang KJ, Lee S. Dynamics of fullerene coalescence. Phys Rev Lett 2003;90:065501.

[95] Levshov D, Than TX, Arenal R, Popov VN, Parret R, Paillet M, et al. Experimental Evidence of a Mechanical Coupling between Layers in an Individual Double-Walled Carbon Nanotube. Nano Lett 2011;11:4800.
[96] Bouilly D, Cabana J, Meunier F, Desjardins-Carriere M, Lapointe F, Gagnon P, et al. Wall-Selective Probing of Double-Walled Carbon Nanotubes Using Covalent Functionalization. ACS Nano 2011;5:4927.
[97] Guan LH, Shi ZJ, Li MX, Gu ZN. Ferrocene-filled single-walled carbon nanotubes. Carbon 2005;43:2780.
[98] Shiozawa H, Pichler T, Grueneis A, Pfeiffer R, Kuzmany H, Liu Z, et al. A catalytic reaction inside a single-walled carbon nanotube. Adv Mater 2008;20:1443.
[99] Li YF, Hatakeyama R, Kaneko T, Okada T. Nano sized magnetic particles with diameters less than 1nm encapsulated-in single-walled carbon nanotubes. Jpn J Appl Phys Lett 2006;45:L428.
[100] Tang ZK, Sun HD, Wang J, Chen J, Li G. Mono-sized single-wall carbon nanotubes formed in channels of $AlPO_4$-5 single crystal. Appl Phys Lett 1998;73:2287.
[101] Tang ZK, Zhai JP, Tong YY, Hu XJ, Saito R, Feng YJ. Resonant Raman scattering of the smallest single-walled carbon nanotubes. Phys Rev Lett 2008;101:047402.
[102] Plank W, Pfeiffer R, Schaman C, Kuzmany H, Calvarasi M, Zerbetto F, et al. Electronic structure of carbon nanotubes with ultrahigh curvature. ACS Nano 2010;4:4515.
[103] Plank W, Pfeiffer R, Scharman C, Wirtz L, Calvaresi M, Zerbetto F, et al. Electronic structure and radial breathing mode for carbon nanotubes with ultra-high curvature. Phys Stat Sol B 2010;247:2774.
[104] Jorio A, Souza AG, Dresselhaus G, Dresselhaus MS, Righi A, Matinaga FM, et al. Raman studies on 0.4 nm diameter single wall carbon nanotubes. Chem Phys Lett 2002;351:27.
[105] Hulman M, Kuzmany H, Dubay O, Kresse G, Li L, Tang ZK, et al. Raman spectroscopy of single wall carbon nanotubes grown in zeolite crystals. Carbon 2004;42:1071.
[106] Iliev MN, Litvinchuk AP, Arepalli S, Nikolaev P, Scott CD. Fine structure of the low-frequency Raman phonon bands of single-wall carbon nanotubes. Chem Phys Lett 2000;316:217.
[107] Venkateswaran UD, Masica DL, Sumanasekera GU, Furtado CA, Kim UJ, Eklund PC. Diameter dependent wall deformations during the compression of a carbon nanotube bundle. Phys Rev B 2003;68:241406(R).
[108] Loa I. Raman spectroscopy on carbon nanotubes at high pressure. J Raman Spectrosc 2003;34:611.
[109] Ayala P, Arenal R, Loiseau A, Rubio A, Pichler T. The physical and chemical properties of heteronanotubes. Rev Mod Phys 2010;82:1843.
[110] Susi T, Kaskela A, Zhu Z, Ayala P, Arenal R, Tian Y, et al. Nitrogen-Doped Single-Walled Carbon Nanotube Thin Films Exhibiting Anomalous Sheet Resistances. Chem Mater 2011;23:2201.
[111] Endo M, Kim YA, Hayashi T, Muramatsu H, Terrones M, Saito R, et al. Nanotube coalescence-inducing mode: A novel vibrational mode in carbon systems. Small 2006;2:1031.
[112] Fantini C, Cruz E, Jorio A, Terrones M, Terrones H, Van Lier G, et al. Resonance Raman study of linear carbon chains formed by the heat treatment of double-wall carbon nanotubes. Phys Rev B 2006;73:193408.
[113] Shi L, Sheng L, Yu L, An K, Ando Y, Zhao X. Ultra-thin double-walled carbon nanotubes: A novel nanocontainer for preparing atomic wires. Nano Res 2011;4:759.
[114] Zhang J, Feng Y, Ishiwata H, Miyata Y, Kitaura R, Dahl JEP, et al. Synthesis and Transformation of Linear Adamantane Assemblies inside Carbon Nanotubes. ACS Nano 2012;6: 8674.

[115] Soares JS, Barboza APM, Araujo PT, Barbosa Neto NM, Nakabayashi D, Shadmi N, et al. Modulating the Electronic Properties along Carbon Nanotubes via Tube-Substrate Interaction. Nano Lett 2010;10:5043.

[116] Hartschuh A. Tip-Enhanced Near-Field Optical Microscopy. Angew Chem Int Ed 2008; 47:8178.

[117] Georgi C, Hartschuh A. Tip-enhanced Raman spectroscopic imaging of localized defects in carbon nanotubes. Appl Phys Lett 2010;97:143117.

[118] Farhat H, Berciaud S, Kalbac M, Saito R, Heinz TF, Dresselhaus MS, et al. Observation of electronic Raman scattering in metallic carbon nanotubes. Phys Rev Lett 2011;107:157401.

[119] Kashuba O, Fal'ko VI. Selection rules for Raman-active electronic excitations in carbon nanotubes. Phys Rev B 2012;85:113401.

[120] Fantini C, Pimenta MA, Dantas MSS, Ugarte D, Rao AM, Jorio A, et al. Micro-Raman investigation of aligned single-wall carbon nanotubes. Phys Rev B 2001;63:161405.

[121] Hwang J, Gommans HH, Ugawa A, Tashiro H, Haggenmueller R, Winey KI, et al. Polarized spectroscopy of aligned single-wall carbon nanotubes. Phys Rev B 2000;62:R13310.

[122] Kramberger C. On the polarization dependent Raman spectra of aligned carbon nanotubes. Appl Phys A 2012;20:324.

[123] Jorio A, Pimenta MA, Souza AG, Samsonidze GG, Swan AK, Unlu MS, et al. Resonance Raman spectra of carbon nanotubes by cross-polarized light. Phys Rev Lett 2003;90:107403.

[124] Yanagi K, Miyata Y, Kataura H. Optical and Conductive Characteristics of Metallic Single-Walled Carbon Nanotubes with Three Basic Colors; Cyan, Magenta, and Yellow. Appl Phys Express 2008;1:034003.

[125] Haroz EH, Rice WD, Lu B, Ghosh S, Hauge RH, Weisman RB, et al. Enrichment of Armchair Carbon Nanotubes via Density Gradient Ultracentrifugation: Raman Spectroscopy Evidence. ACS Nano 2010;4:1955.

[126] Simon F, Kramberger C, Peterlik H, Pfeiffer R, De Blauwe K, Pichler T, et al. Raman response from double-wall carbon nanotubes based on metallicity selected host SWCNTs. Phys Stat Sol B 2010;247:2880.

[127] Novoselov KS, Geim AK, Morozov SV, Jiang D, Zhang Y, Dubonos SV, et al. Electric field effect in atomically thin carbon films. Science 2004;306:666.

[128] Chen Y-F, Liu D, Wang Z-G, Li P-J, Hao X, Cheng K, et al. Rapid Determination of the Thickness of Graphene Using the Ratio of Color Difference. J Phys Chem C 2011;115:6690.

[129] Ferrari AC, Meyer JC, Scardaci V, Casiraghi C, Lazzeri M, Mauri F, et al. Raman spectrum of graphene and graphene layers. Phys Rev Lett 2006;97:187401.

[130] Graf D, Molitor F, Ensslin K, Stampfer C, Jungen A, Hierold C. Spatially resolved raman spectroscopy of single- and few-layer graphene. Nano Lett 2007;7:238.

[131] Gupta A, Chen G, Joshi P, Tadigadapa S, Eklund PC. Raman scattering from high-frequency phonons in supported n-graphene layer films. Nano Lett 2006;6:2667.

[132] Mayorov AS, Elias DC, Mucha-Kruczynski M, Gorbachev RV, Tudorovskiy T, Zhukov A, et al. Interaction-Driven Spectrum Reconstruction in Bilayer Graphene. Science 2011;333:860.

[133] Michel KH, Verberck B. Theory of rigid-plane phonon modes in layered crystals. Phys Rev B 2012;85:094303.

[134] Tan PH, Han WP, Zhao WJ, Wu ZH, Chang K, et al. The shear mode of multilayer graphene. Nat Mater 2012;11:294.

[135] Herziger F, May P, Maultzsch J. Layer-number determination in graphene by out-of-plane phonons. Phys Rev B 2012;85:235447.

[136] Gupta AK, Tang YJ, Crespi VH, Eklund PC. Nondispersive Raman D band activated by well-ordered interlayer interactions in rotationally stacked bilayer graphene. Phys Rev B 2010;82:241406.

[137] Carozo V, Almeida CM, Ferreira EHM, Cancado LG, Achete CA, Jorio A. Raman Signature of Graphene Superlattices. Nano Lett 2011;11:4527.

[138] Righi A, Costa SD, Chacham H, Fantini C, Venezuela P, Magnuson C, et al. Graphene Moire patterns observed by umklapp double-resonance Raman scattering. Phys Rev B 2011; 84:241409.

[139] Tison Y, Giusca CE, Sloan J, Silva SRP. Registry-Induced Electronic Superstructure in Double-Walled Carbon Nanotubes, Associated with the Interaction between Two Graphene-Like Monolayers. ACS Nano 2008;2:2113.

[140] Yan J, Zhang Y, Kim P, Pinczuk A. Electric field effect tuning of electron-phonon coupling in graphene. Phys Rev Lett 2007;98:166802.

[141] Das A, Pisana S, Chakraborty B, Piscanec S, Saha SK, Waghmare UV, et al. Monitoring dopants by Raman scattering in an electrochemically top-gated graphene transistor. Nat Nanotechnol 2008;3:210.

[142] Podila R, Chacoacuten-Torres J, Spear JT, Pichler T, Ayala P, Rao AM. Spectroscopic investigation of nitrogen doped graphene. Appl Phys Lett 2012;101:123108.

[143] Kim YA, Fujisawa K, Muramatsu H, Hayashi T, Endo M, Fujimori T, et al. Raman Spectroscopy of Boron-Doped Single-Layer Graphene. ACS Nano 2012;6:6293.

[144] Krauss B, Nemes-Incze P, Skakalova V, Biro LP, von Klitzing K, Smet JH. Raman Scattering at Pure Graphene Zigzag Edges. Nano Lett 2010;10:4544.

[145] Saito R, Furukawa M, Dresselhaus G, Dresselhaus MS. Raman spectra of graphene ribbons. J Phys C Condens Matter 2010;22:334203.

[146] Cancado LG, Pimenta MA, Neves BRA, Medeiros-Ribeiro G, Enoki T, Kobayashi Y, et al. Anisotropy of the Raman spectra of nanographite ribbons. Phys Rev Lett 2004;93:047403.

[147] Cancado LG, Pimenta MA, Neves BRA, Dantas MSS, Jorio A. Influence of the atomic structure on the Raman spectra of graphite edges. Phys Rev Lett 2004;93:247401.

[148] Elias DC, Nair RR, Mohiuddin TMG, Morozov SV, Blake P, Halsall MP, et al. Control of Graphene's Properties by Reversible Hydrogenation: Evidence for Graphane. Science 2009;323:610.

[149] Fantini C, Jorio A, Souza M, Ladeira LO, Souza AG, Saito R, et al. One-dimensional character of combination modes in the resonance Raman scattering of carbon nanotubes. Phys Rev Lett 2004;93:087401.

[150] Brar VW, Samsonidze GG, Dresselhaus MS, Dresselhaus G, Saito R, Swan AK, et al. Second-order harmonic and combination modes in graphite, single-wall carbon nanotube bundles, and isolated single-wall carbon nanotubes. Phys Rev B 2002;66:155418.

[151] Ellis AV. Second-order overtone and combination modes in the LOLA region of acid treated double-walled carbon nanotubes. J Chem Phys 2006;125:121103.

[152] Cong C, Yu T, Saito R, Dresselhaus GF, Dresselhaus MS. Second-Order Overtone and Combination Raman Modes of Graphene Layers in the Range of 1690–2150 cm^{-1}. ACS Nano 2011;5:1600.

[153] Cong C, Yu T, Sato K, Shang J, Saito R, Dresselhaus GF, et al. Raman Characterization of ABA- and ABC-Stacked Trilayer Graphene. ACS Nano 2011;5:8760.

[154] Mafra DL, Kong J, Sato K, Saito R, Dresselhaus MS, Araujo PT. Using the G' Raman Cross-Section To Understand the Phonon Dynamics in Bilayer Graphene Systems. Nano Lett 2012;12:2883.

[155] Saito R, Jorio A, Souza AG, Dresselhaus G, Dresselhaus MS, Pimenta MA. Probing phonon dispersion relations of graphite by double resonance Raman scattering. Phys Rev Lett 2002;88:027401.

[156] Grüneis A, Saito R, Kimura T, Cancado LG, Pimenta MA, Jorio A, et al. Determination of two-dimensional phonon dispersion relation of graphite by Raman spectroscopy. Phys Rev B 2002;65:155405.

Chapter 7

Transport Properties of Carbon Nanotubes and Graphene

Daniela Dragoman*, Mircea Dragoman†
**Physics Department, University of Bucharest, P.O. Box MG-11, 077125 Bucharest, Romania*
†National Institute for Research and Development in Microtechnology (IMT), P.O. Box 38-160, 023573 Bucharest, Romania

Graphene is the first two-dimensional (2D) crystalline material, and is in fact a graphite monolayer with a thickness of 0.34 nm [1,2]. Graphene consists of carbon atoms in the sp^2 hybridization state, each atom being covalently bonded to three others and thus forming a honeycomb lattice consisting of two intertwined triangular sub-lattices. Graphene is retrieved in many carbon-based materials, such as graphite and carbon nanotubes (CNTs). Single-walled carbon nanotubes (SWCNTs) can be understood as one rolled-up graphene sheet, while in multi-walled carbon nanotubes (MWCNTs) multiple sheets of graphene are rolled up forming multi-wall cylinders.

7.1 CONDUCTION PROPERTIES OF GRAPHENE AND CARBON NANOTUBES

The unique properties of graphene, summarized in Table 7.1, are direct consequences of its structure. The strong and rigid σ bonds formed by hybridization of the nearest planar sp^2 hybrid orbitals in graphene (see Figure 7.1(a)) determine the extraordinary strength and flexibility of this material, characterized by an exceptionally high breaking strength and Young's modulus [3]. On the other hand, since the σ bonds are localized, the conduction properties of graphene are determined by the π and π^* bonds formed by hybridization of the remaining p_z atomic orbitals of nearest C atoms, as shown in Figure 7.1(b). Because the p_z orbitals are normal to the (x,y) graphene plane, the electric conductivity in the graphene plane is good, the charge carriers behaving as a 2D electron gas.

7.1.1 Conduction properties of graphene

The two interpenetrating triangular sub-lattices of the honeycomb lattice contribute each to the wavefunction of charge carriers, which becomes a spinor in graphene rather than a scalar, as in common semiconductor materials.

Carbon Nanotubes and Graphene. DOI: 10.1016/B978-0-08-098232-8.00007-3

TABLE 7.1 Graphene properties

Parameter	Value at room temperature
Mean-free-path	400 nm, even 1 μm for graphene deposited on hexagonal BN
Mobility	40,000 $cm^2 V^{-1}s^{-1}$ (intrinsic), but could attain 100,000 $cm^2 V^{-1}s^{-1}$ in suspended graphene or in graphene deposited on hexagonal BN
Effective mass of carriers	0 near the Dirac point
Thermal conductivity	5000 $W m^{-1}K^{-1}$, better than in many metals
Young's modulus	1.5 TPa, 10 times greater than in steel

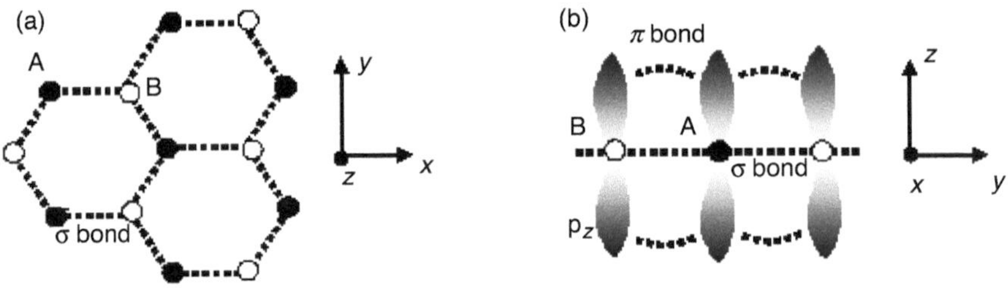

FIGURE 7.1 (a) Graphene honeycomb lattice, consisting of interpenetrating A and B triangular sub-lattices and (b) the π bonds.

Tight-binding calculations that account only for interactions between nearest neighbours lead to an energy dispersion of the form [2]

$$E(\mathbf{k}) = \pm t\sqrt{3 + 2\cos(\sqrt{3}k_y a) + 4\cos(\sqrt{3}k_y a/2)\cos(3k_x a/2)} \tag{7.1}$$

where a = 0.142 nm is the distance between two C atoms and t = 2.75 eV the hopping energy. The plus and minus signs in Equation (7.1) are associated with the upper (π^*) and lower (π) bands, referred to also as the electron and hole states, respectively. The dispersion relation of graphene is displayed in Figure 7.2, from which it follows that at the so-called Dirac points K and K' the energy bandgap vanishes.

In fact, near the Dirac points, i.e. at low energies, the dispersion relation becomes $E = \pm\hbar|\mathbf{k}|v_F$, where $v_F = 3ta/2\hbar \cong c/300$ is the Fermi velocity and $\mathbf{k} = (k_x, k_y)$ the wavevector of charge carriers relative to Dirac points. This linear dispersion relation corresponds to a vanishing effective mass of charge carriers. Unlike in common semiconductors, where the transport properties of electrons and holes differ because of their different effective masses, in graphene the transport properties are the same for electron or hole states since the dispersion relation is perfectly symmetric around the Dirac point.

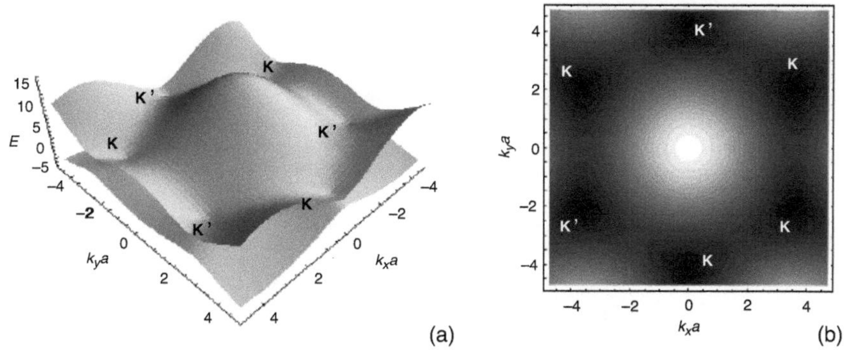

FIGURE 7.2 (a) Dispersion relation and graphene and (b) its contour plot.

Summarizing, the charge carriers in graphene are described, at low energies, by a mass- less Dirac-like equation, in contrast with common semiconductors or metals, in which the electrons and holes obey the Schrödinger equation. This difference in the transport equation of charge carriers is at the origin of the unique conduction properties of graphene. Note that the two components of the spinorial wavefunction in graphene refer to the contributions of the two sub-lattices to electron or hole states, whereas in the relativistic Dirac equation for massive particles the two spinor components refer to spin-up and spin-down charge carriers.

7.1.1.1 Electrical doping and ambipolar transport in graphene

If we deposit graphene on a SiO_2 layer grown thermally over a doped silicon substrate with a thickness $t = 300$ nm, the Si substrate, which has a metal-like behaviour, acts as a back gate (BG) that modulates the carrier density n via the relation [4]

$$n = \varepsilon_0 \varepsilon_d V_G / te \tag{7.2}$$

where ε_0 and ε_d are the dielectric permittivities of air and SiO_2, respectively and V_G the gate voltage. This gate modulation shifts the Fermi energy level in graphene, initially located at the Dirac point, where $n = 0$. The Fermi energy shift, illustrated in Figure 7.3, shows that electrical gating, which dictates the transport type (n or p) in graphene device, is similar to chemical doping, typically used for changing the transport type in semiconductor devices. The transport is ambipolar, i.e. it is performed by electrons or holes, depending on the gate voltage.

It must be emphasized that, although the density of states vanishes at the Dirac point, the conductivity has a minimum finite value at this point, the origin and extent of which are still under debate; there are several theories attempting to explain this phenomenon [5].

The Fermi energy shift around the Dirac point can be modelled by introducing a gate-dependent potential energy V in the Dirac-like equation, the linear dispersion relation becoming

$$E = V \pm \hbar |\mathbf{k}| v_F \tag{7.3}$$

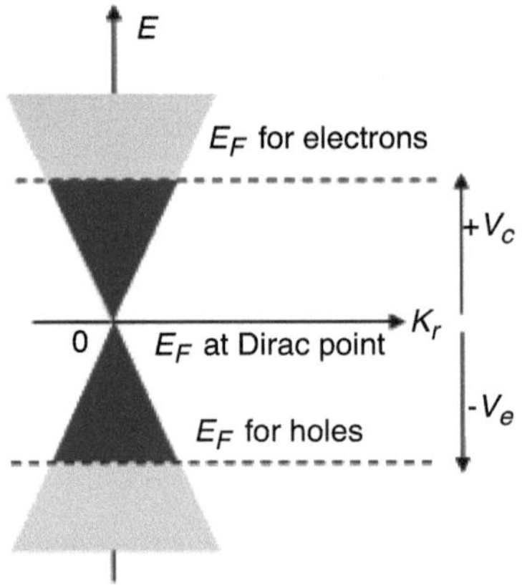

FIGURE 7.3 Ambipolar transport in graphene induced by electrical gating.

The ambipolar transport characteristics and the lack of a bandgap at Dirac points lead to the so-called Klein paradox [6], which in graphene implies that at an interface between two regions with potential energies $V_1 > 0$ and $V_2 < 0$ (see Figure 7.4) charge carrier transport requires complete transformation of electron states into hole states. Experimental studies of Klein paradox in *n–p* junctions in graphene can offer insights into the experimentally unverifiable Klein paradox for massive relativistic particles; in particular, the Klein paradox at a step-like barrier is invalidated [7].

7.1.1.2 Mobility in graphene

The mobility in graphene depends strongly on the substrate on which it is deposited. This parameter dictates the ultimate performances of graphene-based devices, in particular the performances of field-effect transistors (FETs). The mobility could be quite high in graphene FETs with a top-gate (TG) configuration and a high *k* dielectric isolating the metal gate from the graphene channel.

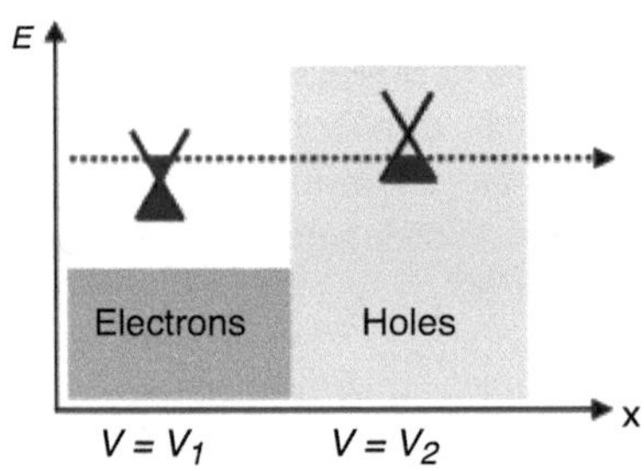

FIGURE 7.4 Klein paradox at an *n–p* junction in graphene.

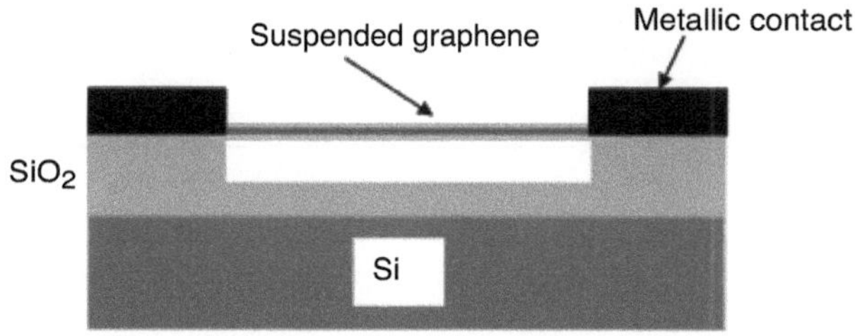

FIGURE 7.5 Contacted suspended graphene sheet.

Its maximum value reaches 23,600 $cm^2V^{-1}s^{-1}$ for an Al_2O_3 high k dielectric with a thickness of 38 nm and a relative electrical permittivity of $\varepsilon_r = 9.1$ [8]. A further increase in mobility can be achieved by using substrate-less graphene or suspended graphene (see Figure 7.5), which reduces dramatically the scattering of carriers with the substrate. In the latter cases the mobility becomes very high, around 200,000 $cm^2V^{-1}s^{-1}$ [9]. Another way to increase the mobility in graphene is to encapsulate it in hexagonal boron nitride (BN) because both materials are lattice matched [10]. In this way, mobilities higher than 100,000 $cm^2V^{-1}s^{-1}$ and mean-free paths longer than 1 μm are observed at room temperature. The present performances of electron beam nanolithography allow fabrication of ballistic graphene devices with a mean free path of 400 nm or even 1 μm at room temperature, paving the way towards ultrafast electronic devices.

7.1.2 Conduction properties of carbon nanotubes

Carbon nanotubes (CNTs) are rolled-up graphene sheets. Nowadays, CNTs can be separated into metallic and semiconducting, both having a purity of about 99%. CNT production is a mature technology and the CNT can be positioned in the prescribed location in almost any geometry (vertical, horizontal) using various technologies.

As in graphene, the structure of CNTs dictates their mechanical and electrical properties, which are summarized in Table 7.2. The mechanical properties have similar values as in graphene, since they are determined by the same planar bonds in the graphene sheet, but the conduction properties are different.

As in nanowires, by rolling the graphene sheet the motion of charge carriers in CNTs are confined in the circumferential direction, which implies that the wavevector component along this direction takes discrete values. As a result, the charge carriers in graphene propagate only the axial direction of the CNT, the one-dimensional (1D) dispersion relation in CNTs consisting of lines through the 2D dispersion relation of graphene. These lines correspond to the discrete values of the circumferential wavevector component and are determined by the rolling direction of the graphene sheet. If at least one of these lines passes through a Dirac point, the CNT is metallic, as shown in Figure 7.6(a) (has no

TABLE 7.2 CNT properties

Parameter	Value and units	Observations
Current density	$>10^9$ A cm^{-2}	1000 times larger than the current density in copper
Mobility	10,000–50,000 cm^2 V^{-1}s^{-1}	Lower than in graphene
Mean-free path (ballistic transport)	300–700 nm in semiconducting CNT 1000–3000 nm in metallic CNT	Measured at room temperature At least three times larger than in the best semiconducting heterostructures, but comparable to graphene
Conductance in ballistic transport	$G = 4e^2/h = 155$ μS	$1/G = 6.5$ kΩ
Thermal conductivity	6600 W m^{-1}K^{-1}	
Young's modulus	1 TPa	10 times stronger than in steel, comparable to graphene

bandgap for that circumferential wavevector component, or mode), and is semiconducting otherwise (see Figure 7.6(b)).

The bandgap of semiconducting CNTs is related to their diameter d through [11]

$$E_g(\text{eV}) \cong 0.9 / d(\text{nm}) \tag{7.4}$$

for a Fermi velocity of $v_F = 8 \times 10^5$ ms^{-1}. From Equation (7.4) it follows that by changing the diameter from a fraction of a nanometre to several nanometres, the bandgap of semiconducting CNTs can be engineered at desired values in

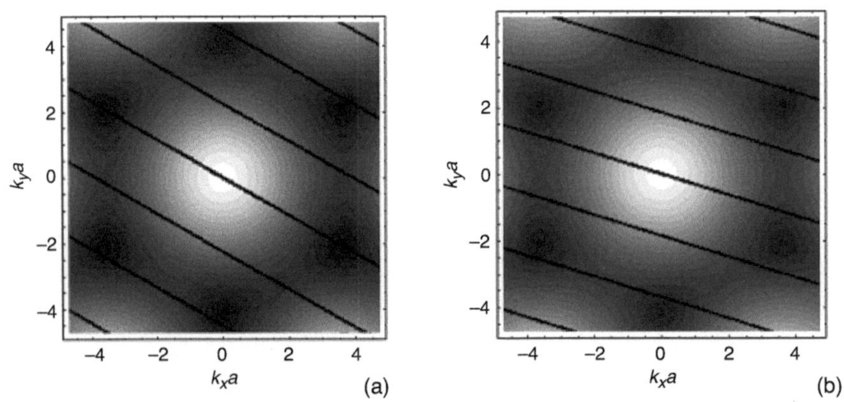

FIGURE 7.6 Dispersion relation in (a) metallic CNTs and (b) semiconducting CNTs.

the 20 meV–2 eV range. This unique property is a direct consequence of the dispersion relation and cannot be found in any other semiconductor, although bandgap engineering is at the core of modern semiconducting device. It should be noted that, in semiconducting CNTs the electron and hole masses are finite and identical, unlike in common semiconductors, since they originate from the symmetric dispersion relation of graphene around the Dirac point.

Besides its diameter dependence, the CNT bandgap can be tuned by chemical, mechanical, electrical or magnetic functionalization [11]. Analytical expressions for the main electrical parameters of metallic and semiconducting CNTs, such as density of states, effective mass and mobility, can be found in Ref. [12].

7.2 ELECTRONIC DEVICES BASED ON GRAPHENE AND CARBON NANOTUBES

Due to the Klein paradox in graphene, it is not possible to modulate the transmission coefficient T of ballistic graphene-based devices through electrostatic gates, since at normal incidence T is always unity. This distinctive behaviour, which is not desirable in nanodevices, can be overcome by devising structures involving oblique incidence or magnetic fields.

Several devices have been proposed based on oblique incidence on a gated region, as that illustrated in Figure 7.7(a). In this case, the dependence of the transmission coefficient on an applied voltage V_D across the device can be tuned by the applied gate voltage (see Figure 7.7(b) for two values of V_2). In the region with $T=0$ charge carrier propagation is forbidden since the wavenumber in gated graphene becomes imaginary, the result being the appearance of a negative differential resistance region in the current–voltage characteristics [13]. Alternatively, low transmission can be obtained in periodically gated structures due to the formation of minibands [14,15].

In the presence of magnetic fields B normal to the graphene plane the dispersion relation changes in both graphene and CNTs due to the formation of

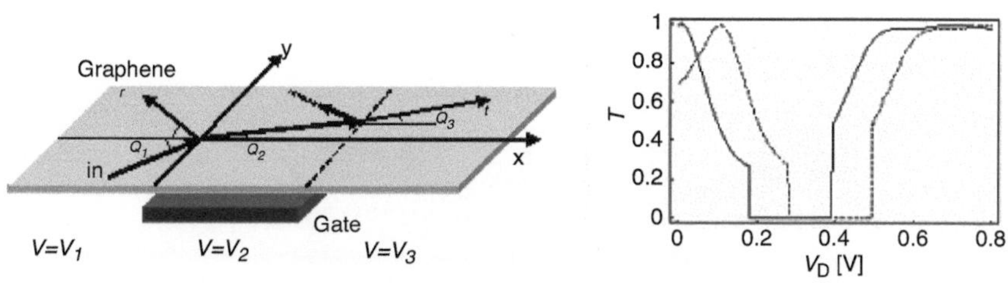

(a) (b)

FIGURE 7.7 (a) Charge carrier transport through a gated graphene region and (b) a typical voltage dependence of the transmission coefficient for different V_G values.

Landau levels. This is the only known mechanism of energy discretization in 2D graphene, the discrete levels being given by [2]

$$E_n - V = \pm v_F \sqrt{2eB\hbar\,|n|} \tag{7.5}$$

with n an integer. Relation (7.5) is the same in graphene and CNTs, but is much more difficult to evidence in the last case due to the small diameter of CNTs [16,17]. Expression (7.5) differs from the Landau levels $E_n = \hbar(eB/m_{\text{eff}})(n+1/2)$ for charge carriers with effective mass m_{eff} obeying the Schrödinger equation, and leads to quantized values of the Hall conductivity,

$$\sigma_{xy} = (n+1/2)(4e^2/h) \tag{7.6}$$

The discretized energy levels in magnetic fields could be used to implement devices such as electron spectrometers [18].

7.2.1 Field-effect transistors based on graphene

Although many graphene-based devices have been proposed, there are few actually implemented. The most common device based on graphene (or CNT) is the FET. A graphene FET is depicted in Figure 7.8.

In such a FET, the graphene (monolayer or bilayer) channel is modulated by one or two gates, and can have various configurations. In the BG configuration, the substrate is a doped Si, and it acts as a metal-like BG able to modulate the carrier density in graphene channels. On the other hand, if the substrate is of high-resistivity Si or SiC, a TG is needed to modulate the carrier density in graphene. However, a combination of TG and BG is possible in some graphene transistors, but the doped Si substrate is lossy and therefore the transistor cannot work at very high frequencies. The oxide used for the isolation of metal gates is usually SiO_2, but oxides such as Al_2O_3 or HfO_2 can be encountered also, especially for TG. For TG the thickness of oxides ranges between few nanometres and 100 nm. Comprehensive reviews about graphene transistors can be found in Refs [19] and [20]. Although graphene FETs have reached cutoff frequencies of 300 GHz [21], there is a lot of room for improvements of graphene FET

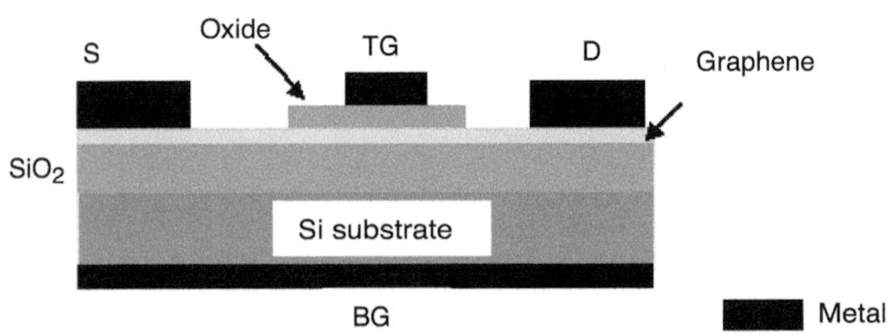

FIGURE 7.8 General configuration of a graphene FET containing both TG and BG.

performances. Enhanced performances are expected specifically from TG configurations, since doped substrates are lossy in microwaves and millimetre waves [22,23]. In particular, the graphene contact resistance is the main obstacle against outstanding performances of graphene FETs.

To be more specific, let us consider a graphene FET with an insulating substrate, i.e. in the TG configuration. We denote the gate length and width with L and W, respectively. In the quantum capacitance limit regime, when the gate capacitance C_G is much larger than the quantum capacitance C_Q, the channel voltage has the simplified expression [23]

$$V_{\mathrm{ch}} = (x / L)V_{\mathrm{D}} \tag{7.7}$$

where x is the position along the channel. In this case, the drain current is given by [23]

$$I_{\mathrm{D}}(V_{\mathrm{D}}, V_{\mathrm{G}}) = e(W / L)\mu^{*}V_{\mathrm{D}}[n_0 + \gamma(V_{\mathrm{G}}^2 + V_{\mathrm{D}}^2 / 3 - V_{\mathrm{D}}V_{\mathrm{G}})] \tag{7.8}$$

where $n_0 \cong 10^{11}\,\mathrm{cm}^{-2}$ is the concentration of impurities, $\mu^{*} = \mu[1 + (\mu V_{\mathrm{D}} / v_{\mathrm{sat}}L)]$ with μ the mobility of the graphene FET and $\gamma = e^2 / \pi\hbar^2 v_{\mathrm{F}}^2$ constant. The saturation velocity is taken as $v_{\mathrm{sat}} = v_{\mathrm{F}} / 2$, where v_{F} is the Fermi velocity. If the mobility is not known, it can be extracted from [24]

$$\mu = \frac{1}{en}\frac{\partial I_{\mathrm{D}} / W}{\partial V_{\mathrm{D}} / L} \tag{7.9}$$

where n is the gate-dependent carrier concentration given in Equation (7.2).

For $L = W = 1$ μm and $\mu = 300\ \mathrm{cm^2Vs^{-1}}$, the drain-voltage characteristic $I_{\mathrm{D}} - V_{\mathrm{D}}$ for several gate voltages V_{G} is depicted in Figure 7.9(a). These characteristics have various shapes. At V_{G} = 0.5 V, as shown separately in Figure 7.9(b), the cubic dependence of the drain current on the drain voltage is evident and the $I_{\mathrm{D}} - V_{\mathrm{D}}$ curve is divided into three regions: (I) sublinear; (II) saturation, for which the Dirac point is found at $V_{\mathrm{D}} = V_{\mathrm{G}}$ and (III) superlinear. However, as the gate voltage increases, only the first two regions of the $I_{\mathrm{D}} - V_{\mathrm{D}}$ dependence are apparent in Figure 7.9(a), but the superlinear region exists for all curves, at higher drain voltages. The superlinear region appears in graphene FETs due to the lack of a bandgap, while in materials with bandgaps greater than 0.35 eV, as in CNTs or common semiconductors, the saturation region extends indefinitely [19].

The $I_{\mathrm{D}} - V_{\mathrm{G}}$ dependence, displayed in Figure 7.10 for $V_{\mathrm{D}} = 0.2$ V, shows that the drain current is not zero even if the gate voltage is zero – a characteristic typical for graphene physics and related to the minimum conductivity. The minimum of the $I_{\mathrm{D}}(V_{\mathrm{G}})$ curve is the Dirac point, for which $V_{\mathrm{G}} = V_{\mathrm{G,\,Dirac}}$. If $V_{\mathrm{G}} > V_{\mathrm{G,\,Dirac}}$ we have n-type conductivity, while for $V_{\mathrm{G}} < V_{\mathrm{G,\,Dirac}}$ we have p-type conductivity. The $I_{\mathrm{D}} - V_{\mathrm{G}}$ curve has a typical 'V'-shape in graphene FET measurements.

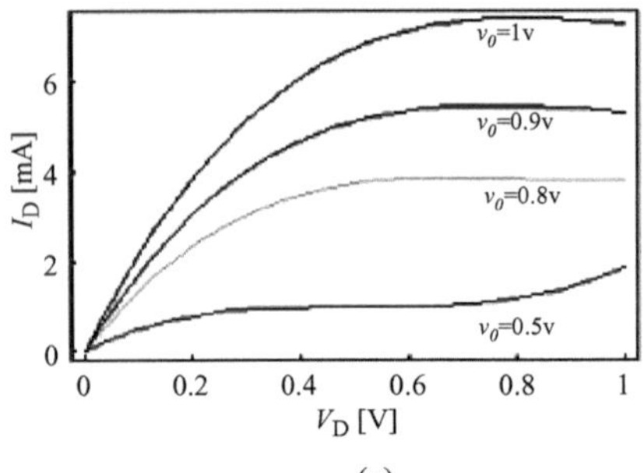

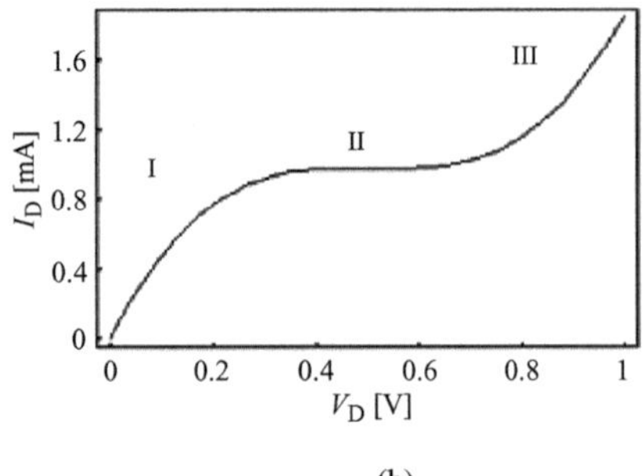

FIGURE 7.9 (a) $I_D - V_D$ characteristics for various gate voltages and (b) close-up for $V_G = 0.5$ V.

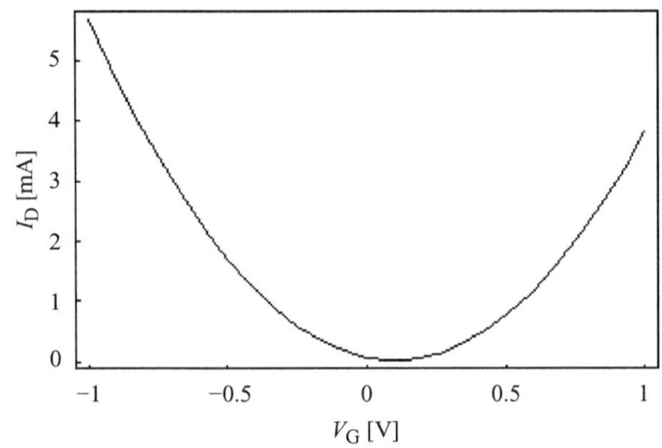

FIGURE 7.10 $I_D - V_G$ dependence for $V_D = 0.2$ V.

The transconductance is a key parameter of any transistor and is defined as $g_m = \partial I_D / \partial V_G$. This parameter is directly related to the cutoff frequency of the transistor:

$$f_c = 1/(2\pi g_m C_G) \tag{7.10}$$

where C_G is the gate capacitance.

The graphene/metal contact is the real problem of graphene transistors because its value cannot be decreased under certain limits. Metals with high workfunctions are preferred and produce low resistance in graphene because of their closeness to the workfunction of graphene (−4.5 eV) [25]. In this respect, Ti/Pd/Au displays a contact resistance of 750 Ω μm, corresponding to a contact resistivity of 2×10^{-6} Ω cm^2, which is close to the state-of-the-art value for the contact resistance in GaAs technology [26]. A very high cutoff frequency (see Equation (7.10)) implies a high g_m and a low length gate L In this respect, the self-aligned gate technology has produced the highest cutoff frequencies in graphene FETs, in the range 300–700 GHz [27]. In particular, GaN nanowires were used as gates with widths in the 45–100 nm range. To further increase the graphene-FET performances, the FETs must work in the quantum capacitance limit, which implies higher oxide capacitances, i.e. a thickness of the oxide of few nanometres even for high-permittivity dielectrics. Recently, such a graphene FET was fabricated using Y_2O_3 (relative permittivity of 10) as gate dielectric with the thickness of 3.9 nm [28]. In this case, the oxide capacitance is 2.3 μF cm^{-2}, near the quantum capacitance.

7.2.2 Field-effect transistors based on carbon nanotubes

Semiconducting CNT FETs do not encounter the same conceptual problems as graphene due to their finite bandgaps given by $E_g = 0.9 / d(nm)$. CNT FETs are able to work as graphene FETs, in TG, BG or a combination of TG and BG configurations (for a review, see Ref. [19]). In principle, the best performances of CNT FETs are obtained when the semiconducting CNTs forming a thin film are arranged in parallel using dielectrophoresis. Operating frequencies of 80 GHz are reported [29]. However, TG single CNT transistors (see Figure 7.11) show impressive results. Very recently, a 9-nm gate length CNT FET with HfO_2 with a thickness of 3 nm as dielectric was fabricated [30]. This transistor has outperformed the best Si devices due to its impressive diameter-normalized current density of 2.41 mA μm^{-1}, which is four times larger than the state-of-the-art value. In many cases, the contact between the CNT and metals is Schottky, and thus we have a Schottky CNT FET [31]. If the potential at the top of the barrier is proportional to the gate voltage: $E_{top} = -eV_G + e^2nC_e$, where n is the non-degenerate carrier concentration, we have a simple $I_D(V_D, V_G)$ dependence, which is given by [32]:

$$I_D(V_D, V_G) = e[8m_{eff}e(V_G - V_T) / \hbar^2\pi^2(1 + C_q / C_e)]^{1/2}(\mu/L)V_D(1 + V_D / V_w)^{-1} \tag{7.11}$$

where $C_e = 2\pi\varepsilon_0\varepsilon_r / \ln[(2t + d)/d]$, $C_q = 2e^2 / hv_F$, $V_T = -E_F / e$ is the threshold voltage and $V_C = v_{sat}L/\mu$, with $v_{sat} = \pi\hbar n / 4m_{eff}$ and $m_{eff} / m_0 = 0.008 / d(\text{nm})$

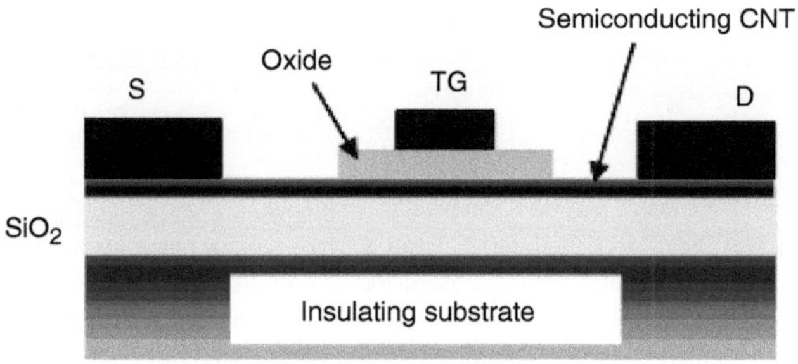

FIGURE 7.11 Schematic configuration of a TG CNT FET.

[31]. This approximation has demonstrated good results in the modelling of CNT FETs with low and high mobilities.

The CNT FET has an analytic model also in the ballistic limit [31,32], which works well for $L < 60$ nm, providing good agreement between theory and experimental results.

7.3 CONCLUSIONS

This chapter demonstrates that graphene and CNTs have impressive transport properties. In the case of graphene, a new chapter of solid-state physics emerges – the Dirac-like electrons produce new unexpected phenomena, such as Klein paradox, minimum conductivity and, above all, ballistic transport of carriers at room temperature with unprecedented high mobility. Graphene FETs and negative differential resistance devices are the beneficiaries of these outstanding physical properties. The graphene FETs are expected to work soon in the sub-millimetre and terahertz region. On the other hand, CNTs can be well separated nowadays in semiconducting and metallic nanotubes with a purity exceeding 99%. CNT FETs with gate lengths of 9 nm are reported, while CNT FETs working in the quantum capacitance limit have cutoff frequencies in the terahertz range.

Presently, the applications of the unique graphene and CNT transport properties are exceeding the area of nanoelectronics and RF [33] and entering the area of bionanoelectronics [34] where CNT FETs and graphene FETs are among the most sensitive nanosensors for detecting biological molecules, such as hybridized DNA and proteins, and can act as forerunners of early signalling of serious diseases. Soon, carbon-based nanomaterials, such as graphene and CNTs, will be present in various aspects of everyday life.

REFERENCES

[1] Geim AK, Novoselov KS. The rise of graphene. Nat Mater 2007;6:183–91.

[2] Castro Neto AH, Guinea F, Peres NMR, Novoselov KS, Geim AK. The electronic properties of graphene. Rev Mod Phys 2009;81:109–62.

[3] Lee C, Wei X, Kysar JW, Hone J. Measurement of the elastic properties and intrinsic strength of monolayer graphene. Science 2008;321:385–8.

[4] Novoselov KS, Geim AK, Morozov SV, Jiang D, Zhang Y, Dubonos SV. Electric field effect in atomically thin carbon films. et al. Science 2004;306:666–9.

[5] Dragoman D. Low-energy conductivity of single-and-double-layer graphene from uncertainty measurements. Phys Scr 2010;81:035702.

[6] Katsnelson MI, Novoselov KS, Geim AK. Chiral tunnelling and the Klein paradox in graphene. Nat Phys 2006;2:620–5.

[7] Dragoman D. Evidence against Klein paradox in graphene. Phys Scr 2009;79:015003.

[8] Liao L, Bai J, Qu Y, Lin Y-C, Huang Y, Duan X. High-κ oxide nanoribbons as gate dielectrics for high mobility top-gated graphene transistors. PNAS 2010;107:6711–5.

[9] Bolotin KI, Sikes KJ, Jiang Z, Klima M, Fundenberg G, Hone J. Ultrahigh electron mobility in suspended graphene. et al. Solid State Commun 2008;146:351–5.

[10] Mayorov AS, Gorbachev RV, Morozov SV, Britnell L, Jail R, Ponomarenko LA. Micrometer-scale ballistic transport in encapsulated graphene at room temperature. et al. Nano Lett 2011;11:2396–9.
[11] Dragoman M, Dragoman D. Nanoelectronics. Principles and devices. Artech House; 2009.
[12] Akinwande D, Nishi Y, Wong H-SP. An analytical derivation of the density of states, effective mass and carrier density of achiral carbon nanotubes. IEEE Trans Electron Dev 2008;55: 289–97.
[13] Dragoman D, Dragoman M. Negative differential resistance of electrons in graphene barrier. Appl Phys Lett 2007;90:143111.
[14] Dragoman D, Dragoman M. Giant thermoelectric effect in graphene. Appl Phys Lett 2007;91:203116.
[15] Dragoman D, Dragoman M. Terahertz Bloch oscillations in periodic graphene structures. Appl Phys Lett 2008;93:103105.
[16] Nanot S, Avriller R, Escoffier W, Broto J-M, Roche S, Raquet B. Propagative Landau states and Fermi level pinning in carbon nanotubes. Phys Rev Lett 2009;103:256801.
[17] Raquet B, Avriller R, Lassagne B, Nanot S, Escoffier W, Broto J-M. Onset of Landau-level formation in carbon-nanotube-based electronic Fabry-Perot resonators. et al. Phys Rev Lett 2008;101:046803.
[18] Zubarev A, Dragoman D. Tunable energy filtering of ballistic electrons in graphene. J Phys D 2012;45:075301.
[19] Schwierz F. Graphene transistors. Nat Nanotechnol 2010;5:487–96.
[20] Kim K, Choi J-Y, Kim T, Cho S-H, Chung H-J. A role for graphene in silicon-based semiconductor devices. Nature 2011;479:338–44.
[21] Liao L, Lin Y-C, Bao M, Cheng R, Bai J, Liu Y. High-speed graphene transistors with a self-aligned nanowire gate. et al. Nature 2010;467:305–8.
[22] Dragoman M, Neculoiu D, Cismaru A, Muller AA, Deligeorgis G, Konstantinidis G. Coplanar waveguide on graphene in the range 40 MHz-110 GHz. et al. Appl Phys Lett 2011;99:033112.
[23] Parrish KN, Akinwande D. An exact solvable model for the graphene transistor in the quantum capacitance limit. Appl Phys Lett 2012;101:053501.
[24] Sabri SS, Levesque PL, Aguirre CM, Guillemette J, Martel R, Szkopek T. Graphene field transistor with parylene gate dielectric. Appl Phys Lett 2009;95:242104.
[25] Reddy D, Register LF, Carpenter GD, Banerjee SK. Graphene field-effect transistors. J Phys D 2011;44:313001.
[26] Huang B-C, Zhang M, Wang Y, Woo J. Contact resistance in top-gated graphene field-effect transistors. Appl Phys Lett 2011;99:032107.
[27] Liae L, Bai J, Cheng R, Liu Y-C, Jiang S, Qu Y. Sub-100 nm channel length graphene transistors. et al. Nano Lett 2010;10:3952–6.
[28] Xu H, Zhang Z, Wang Z, Wang S, Liang X, Peng L-M. Quantum capacitance limited vertical scaling of graphene field-effect transistor. ACS Nano 2011;5:2340–7.
[29] Nougaret L, Happy H, Dambrine G, Derycke V, Bourgoin J-P, Green AA. 80 GHz field-effect transistors produced using high purity semiconducting single-walled carbon-nanotube. et al. Appl Phys Lett 2009;94:243505.
[30] Franklin AD, Luiser M, Han S-H, Tulevski G, Breslin CM, Gignac L. Sub-10 nm carbon nanotube transistor. et al. Nano Lett 2012;12:758–62.
[31] Chek DCY, Tan MLP, Ahmadi MT, Ismail R, Arora VK. Analytical modeling of high performance single-walled carbon field-effect-transistor. Microelectronics J 2010;41:579–84.

[32] Akinwande D, Liang J, Chong S, Nishi Y, Wong H-SP. Analytic ballistic theory of carbon nanotubes transistors: experimental validation, device physics, parameter extraction and performance projection. J Appl Phys 2008;104:124514.

[33] Dragoman M, Neculoiu D, Dragoman D, Deligeorgis G, Konstantinidis G, Cismaru A. Microwave applications on graphene. et al. IEEE Microw Mag 2010;11:81–6.

[34] Dragoman D, Dragoman M. Bionanoelectronics. Springer; 2012.

Chapter 8

Mechanical Properties of Carbon Nanotubes and Graphene

Zhong Hu*, **Xiaoxing Lu**†
**Department of Mechanical Engineering, South Dakota State University, Brookings, South Dakota 57007, USA*
†*Department of Materials Science and Engineering, University of California, Los Angeles, California 90095, USA*

8.1 INTRODUCTION

Diamond, the most well known allotrope of carbon (sp^3 bonded forming four covalent bonds with neighbouring carbon atoms into a diamond cubic atomic structure), and graphite (sp^2 bonded forming three covalent bonds with neighbouring carbon) have very different physical properties. Diamond, which is highly transparent and one of the hardest materials, has very low electrical conductivity. Graphite, on the other hand, is opaque, very soft and an excellent conductor.

In the last two decades, several exotic forms of low-dimensional carbon have been discovered, including fullerenes [1] (zero dimension), carbon nanotubes (CNTs) [2,3] (one dimension) and graphene [4] (two dimension). All of these materials share graphite's sp^2 carbon bonding. The Nobel Prize for Chemistry in 1996 was awarded jointly to Robert F. Curl, Jr., Sir Harold W. Kroto and Richard E. Smalley for their discovery of fullerenes. Graphene, the mother of all graphitic forms, is a monolayer of carbon atoms tightly packed into a two-dimensional hexagonal lattice sheet that can be wrapped up to fullerenes, rolled into CNTs and stacked into graphite. Since the discovery of CNTs by Iijima in 1991 [2], research on their growth, characterization and application development has exploded. Graphene has also been extensively studied in the last several years, although it was only recently isolated for the first time by Novoselov et al. in 2004 [4]. Six years later, Geim and Novoselov won the 2010 Nobel Prize in Physics for their groundbreaking work illustrating the importance of graphene. Graphene has attracted significant interest from the scientific community due to its extraordinary physical properties (mechanical, thermal, electrical, optical, etc.) and is currently being explored for a number of applications including nanoelectromechanical systems and nanoelectronics. Recent mechanical experiments show graphene to be the strongest material ever measured [5] and suggest opportunities for graphene as a nanomechanical material.

Carbon Nanotubes and Graphene. DOI: 10.1016/B978-0-08-098232-8.00008-5

However, rolling up a graphene sheet on a nanometre scale dramatically impacts the electrical properties. The small diameter of a CNT, compared with traditional micron-sized graphitic fibres, also impacts mechanical properties. Perhaps the most striking effect is the ability to associate high flexibility and strength with high stiffness, a property that is absent in graphite fibres [6,7]. These properties of CNTs pave the way for a new generation of high-performance composites [8–11]. These potential usages stimulate the need to explore CNT mechanical properties, study their deformation behaviour under loading and identify any failure mechanisms that may appear.

8.2 BASIC CONCEPTS

Mechanical properties are the responses of a solid to mechanical loads that change its shape and/or volume from its equilibrium configuration. When the atoms of a solid under load are displaced from their equilibrium positions, forces are induced to restore the initial shape as the load is removed. In the elastic regime (particularly for small deformations), the behaviour is wholly reversible. Increasing load leads to the formation and propagation of defects that relieve mechanical stress.

8.2.1 Elastic regime

For anisotropic solids (such as CNT-modified composites at the nanoscale) 21 independent constants are necessary to characterize elastic behaviour. The generalized Hooke's law can be written as:

$$\{\sigma_{ij}\} = \left[C_{ijkl}\right] \cdot \{\varepsilon_{kl}\} \tag{8.1}$$

where $\{\sigma_{ij}\}$ is the stress tensor, $\{\varepsilon_{ij}\}$ the strain tensor and $[C_{ijkl}]$ the elasticity tensor (which is symmetric and composed of 21 elastic moduli or constants).

Graphite, an approximately transversely isotropic material, is symmetric with respect to rotation about an axis of symmetry (e.g. the z axis) perpendicular to the basal plane $x - y$; its perpendicular elastic constants differ greatly from the elastic constants along the basal plane. For such a material, the general Hooke's law is given by:

$$\begin{Bmatrix} \sigma_x \\ \sigma_y \\ \sigma_z \\ \tau_{yz} \\ \tau_{zx} \\ \tau_{xy} \end{Bmatrix} = \begin{bmatrix} C_{11} & C_{12} & C_{13} & 0 & 0 & 0 \\ C_{12} & C_{11} & C_{13} & 0 & 0 & 0 \\ C_{13} & C_{13} & C_{33} & 0 & 0 & 0 \\ 0 & 0 & 0 & C_{44} & 0 & 0 \\ 0 & 0 & 0 & 0 & C_{44} & 0 \\ 0 & 0 & 0 & 0 & 0 & (C_{11} - C_{12}/2) \end{bmatrix} \begin{Bmatrix} \varepsilon_x \\ \varepsilon_y \\ \varepsilon_z \\ \gamma_{yz} \\ \gamma_{zx} \\ \gamma_{xy} \end{Bmatrix} \tag{8.2}$$

Given graphite's symmetry, there are only five independent elastic constants, experimentally determined as [12,13]:

$$\mathrm{C}_{11}=1.06\pm0.02\,\mathrm{TPa}, \mathrm{C}_{12}=180\pm10\,\mathrm{GPa},\ \mathrm{C}_{13}=15\pm5\,\mathrm{GPa},$$
$$\mathrm{C}_{33}=36.5\,\mathrm{GPa}, \mathrm{C}_{44}=0.18\sim0.35\,\mathrm{GPa}$$

More frequently, the inverse Hooke's law for Equation (8.2) can be written as:

$$\begin{Bmatrix} \varepsilon_x \\ \varepsilon_y \\ \varepsilon_z \\ \gamma_{yz} \\ \gamma_{zx} \\ \gamma_{xy} \end{Bmatrix} = \begin{bmatrix} \frac{1}{E_x} & -\frac{\nu_{xy}}{E_x} & -\frac{\nu_{zx}}{E_z} & 0 & 0 & 0 \\ -\frac{\nu_{xy}}{E_x} & \frac{1}{E_x} & -\frac{\nu_{zx}}{E_z} & 0 & 0 & 0 \\ -\frac{\nu_{zx}}{E_z} & -\frac{\nu_{zx}}{E_z} & \frac{1}{E_z} & 0 & 0 & 0 \\ 0 & 0 & 0 & \frac{1}{G_{zx}} & 0 & 0 \\ 0 & 0 & 0 & 0 & \frac{1}{G_{zx}} & 0 \\ 0 & 0 & 0 & 0 & 0 & \frac{1}{G_{xy}} \end{bmatrix} \begin{Bmatrix} \sigma_x \\ \sigma_y \\ \sigma_z \\ \tau_{yz} \\ \tau_{zx} \\ \tau_{xy} \end{Bmatrix} \tag{8.3}$$

where E, G and ν are Young's modulus, shear modulus and Poisson's ratio, respectively and $G_{xy}=\dfrac{E_x}{2(1+\nu_{xy})}$.

As a single layer of graphite, graphene is a two-dimensional material and its Hooke's law can be easily deduced from graphite's:

$$\begin{Bmatrix} \sigma_x \\ \sigma_y \\ \tau_{xy} \end{Bmatrix} = \begin{bmatrix} \mathrm{C}_{11} & \mathrm{C}_{12} & 0 \\ \mathrm{C}_{12} & \mathrm{C}_{11} & 0 \\ 0 & 0 & (\mathrm{C}_{11}-\mathrm{C}_{12})/2 \end{bmatrix} \begin{Bmatrix} \varepsilon_x \\ \varepsilon_y \\ \gamma_{xy} \end{Bmatrix}$$
$$\text{or} \begin{Bmatrix} \varepsilon_x \\ \varepsilon_y \\ \gamma_{xy} \end{Bmatrix} = \begin{bmatrix} \frac{1}{E_x} & -\frac{\nu_{xy}}{E_x} & 0 \\ -\frac{\nu_{xy}}{E_x} & \frac{1}{E_x} & 0 \\ 0 & 0 & \frac{1}{G_{xy}} \end{bmatrix} \begin{Bmatrix} \sigma_x \\ \sigma_y \\ \tau_{xy} \end{Bmatrix} \tag{8.4}$$

where $G_{xy}=\dfrac{E_x}{2(1+\nu_{xy})}$. Therefore, graphene is an elastic isotropic 2D material, and E, G and ν are derived from the above as:

$$E = C_{11}\left[1-\left(\frac{C_{12}}{C_{11}}\right)^2\right] \approx 1.029 \pm 0.024\,\text{TPa},$$

$$\nu = \frac{C_{12}}{C_{11}} \approx 0.170 \pm 0.013, G = \frac{1}{2}(C_{11} - C_{12}) \approx 0.440 \pm 0.015\,\text{TPa}$$

As the thinnest graphite, graphene provides a perfect model to reveal the difference in mechanical properties between low-dimensional and bulk materials.

CNTs are like rolled-up graphene sheets with the slightly curved sp^2 CC bonds that are still very strong. These characteristics, coupled with low defect density, suggested that CNTs would have very high Young's modulus (similar to graphite's) and extremely high breaking strength. However, many mechanical properties of CNTs were found to be very sensitive to their chirality and size. Therefore, a direct measurement of the mechanical properties of individual CNTs with known structures is highly desirable.

Theoretically, simple deformation experiments could determine the corresponding mechanical properties: Young's modulus E can be measured from a tensile test that characterizes solid resistance to uniaxial stretching and is defined as the ratio of stress to strain (σ_x/ε_x), while Poisson's ratio ν is defined as the ratio of transversal (ε_y or ε_z) to longitudinal (ε_x) strain. Shear modulus G is defined as the ratio of shear stress to shear strain, which can be measured from a shear test [14–17].

For CNTs, these moduli can be measured in a cylindrical model based on a cylindrical coordinate system (see Figure 8.1):

$$E = \frac{\sigma_x}{\varepsilon_x} = \frac{F/A}{\delta L/L} = \frac{F/(\pi dt)}{\delta L/L} = \frac{1}{V}\left(\frac{\partial^2 U_s}{\partial \varepsilon_x^2}\right)_{\varepsilon_0} = \frac{1}{\pi dtL}\left(\frac{\partial^2 U_s}{\partial \varepsilon_x^2}\right)_{\varepsilon_0} \tag{8.5}$$

$$G = \frac{\tau_{x\theta}}{\gamma_{x\theta}} = \frac{TL}{J\delta\theta} = \frac{1}{V}\left(\frac{\partial^2 U_s}{\partial \gamma_{x\theta}^2}\right)_{\varepsilon_0} = \frac{1}{\pi dtL}\left(\frac{\partial^2 U_s}{\partial \gamma_{x\theta}^2}\right)_{\varepsilon_0} \tag{8.6}$$

$$\nu_{xr} = -\frac{\varepsilon_r}{\varepsilon_x} \tag{8.7}$$

where F and T are tensile force in tensile test (Figure 8.1(a)) and torque in torsional test (Figure 8.1(b)), corresponding to tensile displacement δL or torsional

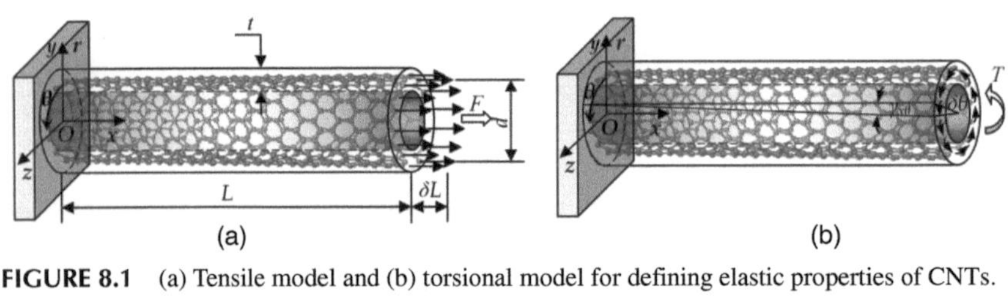

FIGURE 8.1 (a) Tensile model and (b) torsional model for defining elastic properties of CNTs.

angle $\delta\theta$, respectively; J the polar moment of cylinder inertia; A and V equilibrium cross-section area and volume of the cylinder, respectively; d and t the medium diameter and wall thickness, respectively. For single-walled carbon nanotubes (SWCNTs), t is counted by the interlayer spacing in graphite, ca. 0.335 nm. U_s is the strain energy (Table 8.1).

TABLE 8.1 Data summary on CNTs (assuming an effective thickness of 0.335 nm for SWCNTs)

Reference	Method	Mechanical properties
Treacy et al. (1996) [76]	Thermal vibration by TEM	$E = 1.77 - 1.37/{+2.38}$ TPa for $d_0 = 5.6 \sim 24.8$ nm, $d_i = 1.0 \sim 6.6$ nm, $t = 3.3 \sim 18.2$ nm
Iijima et al. (1996) [46]	TEM and atomistic simulations	Single and double kinks in MWCNTs $d_0 = 5 \sim 10$ nm; single kink in SWCNTs $d_0 = 0.8 \sim 1.2$ nm; buckling under bending up to 90°
Yakobson et al. (1996) [18]	Many-body potential in elasticity of shell	$E = 1.08$ TPa, $\nu = 0.19$, compressive $\varepsilon_c \geq 0.05$, for $d = 1$ nm, $L = 6$ nm SWCNT armchair (7, 7)
Wong et al. (1997) [79]	Cantilevered beam by AFM	$E = 1.28 \pm 0.59$ TPa for MWCNTs, $d_0 = 26 \sim 76$ nm
Lu (1997) [102]	Empirical force-constant model	$E = 0.971 \sim 0.975$ TPa, $\nu = 0.277 \sim 0.280$, $G = 0.436 \sim 0.478$ TPa, $B = 0.190 \sim 0.192$ TPa for SWCNTs $d = 0.68 \sim 27.0$ nm; $E = 0.97 \sim 1.11$ TPa, $\nu = 0.269 \sim 0.280$, $G = 0.436 \sim 0.541$ TPa, $B = 0.191 \sim 0.194$ TPa for MWCNTs $(5n, 5n)$ $d = 0.68 \sim 6.78$ nm; $E = 0.430 \sim 0.795$, $B = 0.015 \sim 0.033$ TPa for MWCNTs (n, n) $d = 0.66 \sim 2.02$ nm
Falvo et al. (1997) [124]	Bending by AFM tip	Local ε up to 16% without separating a MWCNT under repeated bending
Krishnan et al. (1998) [77]	Thermal vibration by TEM	$E = 1.25 - 0.35/{+0.65}$ TPa for $d_0 = 1.0 \sim 1.5$ nm SWCNTs
Lourie et al. (1998) [39]	Cooling-induced vibration, compressive buckling and collapse by TEM	$\sigma_{c(min)} = 101.2 \sim 135.0$ GPa, $\sigma_{c(max)} = 118.4 \sim 146.5$ GPa for $d_0 = 7.3 \sim 17.3$, $t/d_0 = 0.30 \sim 0.41$, $L = 47.0 \sim 223.5$ nm; collapse for $d_0 = 19.6 \sim 56.2$ nm, $t/d_0 = 0.035 \sim 0.05$, fragment length $L = 60 \sim 140$ nm

(continued)

TABLE 8.1 Data summary on CNTs (assuming an effective thickness of 0.335 nm for SWCNTs) *(cont.)*

Reference	Method	Mechanical properties
Lourie and Wagner (1998) [83]	Compression monitored by micro-Raman spectroscopy	$E = 2.825 \sim 3.577$ TPa for SWCNTs $d_0 = 1.4$ nm, $E = 1.718 \sim 2.437$ TPa for MWCNTs $d_0 = 10 \sim 20$ nm, $\varepsilon = -(0.55 \sim 0.9)\%$
Hernandez et al. (1998) [103]	Non-orthogonal tight-binding	$E = 1.22 \sim 1.26$ TPa, $\nu = 0.247 \sim 0.275$ for SWCNTs $d = 0.791 \sim 2.034$ nm chirality
Gao et al. (1998) [104]	MM/MD	Bending modulus for SWCNTs $E_{(n, n)} = 963.44$ GPa, $E_{(n, 0)} = 911.64$ GPa, $E_{(2n, n)} = 935.48$ GPa, interlayer spacing $d_{(n, n)} = 0.338$ nm, $d_{(n, 0)} = 0.341$ nm, $d_{(2n, n)} = 0.339$ nm. Tube axial $E_{(10,10)} = 640.30$ GPa, $E_{(17,0)} = 648.43$ GPa, $E_{(12,6)} = 673.94$ GPa
Zhang et al. (1998) [121]	Tight-binding MD	Elastic limit $\sigma_y = 100 \sim 170$ GPa for SWCNTs. A $(n, 0)$ tube nearly twice the elastic limit as a (n, n) tube with the same radius
Salvetat et al. (1999) [6]	Simple-supported beam bending model by AFM tip and TEM	$E = 810 \pm 410$ GPa for arc-grown larger diameter SWCNTs, $E = 1.28 \pm 0.59$ TPa for MWCNTs $d_0 = 26 \sim 76$ nm, $E \approx 1$ TPa and $G \approx 1$ GPa for long bundles with small-diameter SWCNTs
Salvetat et al. (1999) [91]	Simple-supported beam bending model by AFM tip and TEM	$E = 810 - 160/+410$ GPa for arc-discharged MWCNTs $d_0 = 9.6 \sim 20.0$ nm, $L = 240 \sim 420$ nm, $E = 27 - 15/+23$ GPa for catalytic MWCNTs $d_0 = 26.0 \sim 32.0$ nm, $L = 210 \sim 280$ nm,
Salvetat et al. (1999) [92]	Same as above	$E = 810 - 160/+410$ GPa for SWCNTs $d_0 = 3.0 \sim 20.0$ nm, $L = 100 \sim 370$ nm
Yu et al. (2000) [81]	Tensile test by AFM tip and TEM	$E = 1.002 - 0.682/+0.468$ TPa, $\sigma_f = 30 - 17/+22$ GPa at $\varepsilon_f = 1.1 \sim 5.3\%$ for SWCNTs $d_0 = 19 \sim 41$ nm
Yu et al. (2000) [82]	Tensile test in 'nanostressing stage' in SEM and TEM	$E \approx 1$ TPa $G \approx 1$ GPa for outermost layer, $\sigma_{TS} = 11 \sim 63$ GPa at ε_f up to 12% for MWCNTs $d_0 = 13 \sim 36$ nm, $d_i = 4 \sim 10$ nm, $L = 1.80 \sim 10.99$ nm
Tombler et al. (2000) [93]	Three-point bending by AFM	$E \approx 1.2$ TPa for SWCNTs $L \approx 605$ nm, $d_0 = 3.1 \pm 0.2$ nm

TABLE 8.1 Data summary on CNTs (assuming an effective thickness of 0.335 nm for SWCNTs) *(cont.)*

Reference	Method	Mechanical properties
Ru (2000) [108]	CM elastic shell model	Bending stiffness $D = Eh^3/16 \sim Eh^3/12.2$ for MWCNTs $h = 2t \sim 8t$
Ru (2000) [109]	CM elastic shell model	For SWCNTs $d_0 = \sim 1.3$ nm, $p_c = \sim 1.8$ GPa for elastic buckling
Demczyk et al. (2002) [94]	Pulling and bending by TEM	$E \approx 0.9$ TPa, $\sigma_{TS} = 0.15$ TPa for MWCNTs
Li and Chou (2003) [51]	CM elastic beam element FEA	$E = 0.995 \sim 1.033$ TPa for SWCNTs armchair and zigzag
Li and Chou (2003) [114]	CM non-linear truss FEA/ van der Waals force	$E = 1.05 \pm 0.05$ TPa, $G = 0.40 \pm 0.05$ TPa for MWCNTs
Bozovic et al. (2003) [122]	AFM anti-ferromagnetic manipulation	Local $\varepsilon \sim 30\%$ no breaking, $\varepsilon_y = 5\% \sim 10\%$ for SWCNTs and defect formation
Xiao et al. (2005) [115]	CM based on modified Morse potential	$E \approx 1.13$ TPa, SWCNTs $\sigma_{TS} = 126.2$ GPa at $\varepsilon_f = 23.1\%$ for armchair, $\sigma_{TS} = 94.5$ GPa for zigzag at $\varepsilon_f = 15.6\% \sim 17.5\%$
Guhados et al. (2007) [96]	Bending a clamped MWCNT beam by AFM tip and TEM beam mechanics	$E = 350 \pm 110$ GPa, $G = 1.4 \pm 0.3$ GPa for MWCNTs, $d_0 = 20 \sim 50$ nm, $L > 0.8$ μm
Kirtania and Chakraborty (2007) [120]	CM/FEA beam element model	$E = 1.004 \sim 1.04$ TPa, $G = 0.16 \sim 0.34$ TPa for arm-chair $d = 0.5 \sim 2.8$ nm; $E = 1.02$ $0 \sim 1.028$ TPa, $G = 0.31 \sim 0.48$ TPa for zigzag $d = 0.4 \sim 2.8$ nm; $E = 0.5098 \sim 5.3304$ TPa for SWCNT (14, 0); $E = 0.5057 \sim 5.287$ TPa for SWCNT (8, 8), $t = 0.69 \sim 0.066$ nm
Duan et al. (2007) [123]	MM/modified Morse potential/REBO potential	$\sigma_{TS} = 105.38$ GPa, $\varepsilon_f = 0.33$ for zigzag SWCNTs, $\sigma_{TS} = 134.01$ GPa, $\varepsilon_f = 0.43$ for armchair SWCNTs
Khare et al. (2007) [145]	QM/MM based on Tersoff–Brenner potential and DFT	σ_f and ε_f = (158 GPa, 0.13), (95 GPa, 0.122), (131 GPa, 0.186), (103 GPa, 0.121) for CNTs with (10, 0) two-atom vacuum, (10, 0) one-atom vacuum, (5, 5) SW and (5, 5) one-atom vacuum, respectively
Ding et al. (2007) [97]	TEM/AFM tensile loading	$E = 620 \sim 1200$ GPa (mean 940 GPa), $\sigma_{TS} = 10 \sim 66$ GPa for MWCNTs

(continued)

TABLE 8.1 Data summary on CNTs (assuming an effective thickness of 0.335 nm for SWCNTs) *(cont.)*

Reference	Method	Mechanical properties
Wei et al. (2008) [95]	Bending CNTs using AFM with Timoshenko beam model	$E = 126 \sim 937$ GPa, $G = 33 \sim 785$ MPa for MWCNTs $d_0 = 30 \sim 60.9$ nm, $d_i = 7.9 \sim 12.4$ nm, $L = 1.80 \sim 18.4$ μm
Lu and Chen (2008) [116]	CM/FEA	$E \approx 0.881$ TPa d up to 2.4 nm, $G = 0.34 \sim 0.35$ TPa d up to 4.2 nm $\nu \approx 0.27$ d up to 2. 4 nm for armchair and zigzag SWCNTs
Ávila and Lacerda (2008) [119]	MM/FEA beam element	$E = 0.963 \sim 1.025$ TPa SWCNTs $\nu = 0.15 \sim 0.29$ for armchair, zigzag and chiral, $d = 0.4 \sim 1.8$ nm, $L = 1.8130 \sim 2.8986$ nm. $E \approx 1.0$ TPa for graphene
Huang (2009) [84]	Rayleigh scattering spectroscopy/tensile test by AFM cantilever	$E = 0.97 \pm 0.16$ TPa for individual SWCNT or $E = 1.01 - 0.21/{+0.24}$ TPa for SWCNT bundles. $d_0 = 1.18 \sim 2.8$ nm, $L \approx 100$ μm with chirality $\sigma_f \approx 40.6 - 21.6/{+19.4}$ GPa at $\varepsilon_f \approx 3.7 - 2.0/{+2.1}\%$
Rossi and Meo (2009) [113]	MM/FEA	$E = 0.912 \sim 0.920 \pm 0.009$ TPa for SWCNTs chirality; $\sigma_f = 117.1 \sim 117.9$ GPa at $\varepsilon_f = 19.95 \sim 20.00\%$ for (5, 5) ~ (12, 12), $d = 0.678 \sim 1.628$ nm, $L = 7.261 \sim 17.106$ nm, $L/d = 4.5 \sim 21.6$ nm; $\sigma_f = 93.8 \sim 94.7$ GPa at $\varepsilon_f = 17.20 \sim 18.40\%$ for (5, 0) ~ (20, 0) $d = 0.391 \sim 1.566$ nm, $L = 8.313 \sim 16.697$ nm, $L/d = 8.0 \sim 21.4$ nm
Faccio et al. (2009) [148]	QM/DFT for uniaxial tension	$E = 1.01$ TPa for SWCNT (5, 5) unitary for ε within ±0.02%, non-linear for higher ε
Yamamoto et al. (2010) [98]	TEM and AFM tensile loading	$\sigma_{TS} = 2 \sim 48$ GPa (mean 20 GPa) for pristine MWCNTs, $\sigma_{TS} = 1 \sim 18$ GPa (mean 6 GPa) for acid-treated MWCNTs with channellike defects
Mohammad pour et al. (2011) [117]	CM/FEA	$E = 1.195 \sim 1.233$ TPa for zigzag SWCNTs (6, 0) ~ (22, 0) $d = 0.470 \sim 1.722$ nm
Lu and Hu (2012) [60]	MM/FEA elliptical beam element model	$E = 0.989 \sim 1.058$ TPa for SWCNTs, $G = 0.237 \sim 0.469$, $d \approx 0.375 \sim 12.0$ nm, $L = 12.7 \sim 18.1$ nm

Note: B, bulk modulus; σ_c, buckling stress; ε_c, buckling strain; σ_{TS}, ultimate tensile strength; σ_f, failure stress; ε_f, failure strain.

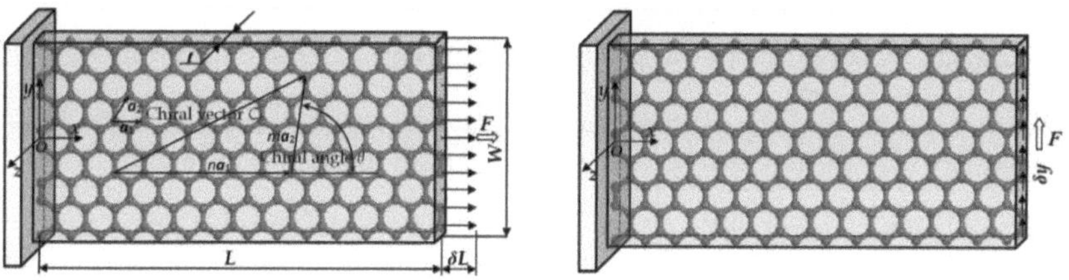

(a) (b)

FIGURE 8.2 (a) Tensile model and (b) shear model for defining elastic properties of graphene.

In the case of graphene sheets, it makes more sense to define the in-plane moduli instead of the classic 3D moduli. Therefore, these moduli can be measured in a thin-plate model based on a Cartesian coordinate system (see Figure 8.2):

$$E = \frac{\sigma_x}{\varepsilon_x} = \frac{F/A}{\delta L/L} = \frac{F/(Wt)}{\delta L/L} = \frac{1}{V}\left(\frac{\partial^2 U_s}{\partial \varepsilon_x^2}\right)_{\varepsilon_0} = \frac{1}{LWt}\left(\frac{\partial^2 U_s}{\partial \varepsilon_x^2}\right)_{\varepsilon_0} \tag{8.8}$$

$$G = \frac{\tau_{x\theta}}{\gamma_{x\theta}} = \frac{F/A}{\delta y/L} = \frac{F/(Wt)}{\delta y/L} = \frac{1}{V}\left(\frac{\partial^2 U_s}{\partial \gamma_{x\theta}^2}\right)_{\varepsilon_0} = \frac{1}{LWt}\left(\frac{\partial^2 U_s}{\partial \gamma_{x\theta}^2}\right)_{\varepsilon_0} \tag{8.9}$$

$$v_{xy} = -\frac{\varepsilon_y}{\varepsilon_x} \tag{8.10}$$

where F is the tensile force in tensile test (Figure 8.2(a)) or shear force in shear test (Figure 8.2(b)), corresponding to tensile displacement δL or shear displacement δy and t graphene sheet thickness (0.335 nm for monolayer) (Table 8.2).

TABLE 8.2 Data summary on graphene (assuming an effective thickness of 0.335 nm)

Reference	Method	Mechanical properties
Blakslee et al. (1970) [13]	Compression by ultrasonic, sonic resonance and static test	Graphite $E_1 = 1.02 \pm 0.03$ TPa, $E_3 = 36.5 \pm 1$ GPa, $G_{12} = 0.23 \pm 0.02$ GPa
Odegard et al. (2002) [110]	CM truss model	Ratio of Young's moduli to SWCNT ($t = 0.34$ nm, $A = 2.14$ nm^2) 52% for MWCNTs $t = 0.57 \sim 0.69$ nm, $A = 3.58 \sim 4.34$ nm^2,

(continued)

TABLE 8.2 Data summary on graphene (assuming an effective thickness of 0.335 nm) *(cont.)*

Reference	Method	Mechanical properties
Meo and Rossi (2006) [111]	MM/FEA	$E = 0.929 \sim 0.988$ TPa for $L = 6.252 \sim 21.173$ nm, $W = 7.137 \sim 19.936$ nm
Bunch et al. (2007) [25]	AFM tapping with Raman signal measuring	$E = 0.5 \sim 2$ TPa for single layer $t = 0.3$ nm, multi-layer $t = 15$ nm, $L = 1.1$ μm, $W = 1.93$ μm
Frank et al. (2007) [143]	AFM/suspended graphene	$E = 0.5$ TPa for layered graphene sheets of $t = 2 \sim 8$ nm
Kirtania and Chakraborty (2007) [120]	CM/FEA beam element model	$E = 0.9657 \sim 1.03114$ TPa for zigzag $H = 0.1368 \sim 14.7784$ nm, $W = 0.4923 \sim 4.9225$ nm; $E = 1.027458 \sim 1.081739$ TPa for armchair $H = 0.4923 \sim 7.1376$ nm, $W = 1.1368 \sim 12.6469$ nm
Liu et al. (2007) [144]	*Ab initio* calculation of tension	$E = 1.050$ TPa, $\nu = 0.186$, instability at $\varepsilon_x = 0.194$, $\sigma_x = 110$ GPa, $\varepsilon_y = -0.016$ or $\varepsilon_y = 0.266$, $\sigma_y = 121$ GPa, $\varepsilon_x = -0.027$
Khare et al. (2007) [145]	QM/MM/Tersoff–Brenner/ DFT	$\sigma_f = 55 \sim 97$ GPa containing a half slit length of $1.9 \sim 0.4$ nm
Bunch et al. (2008) [146]	Vibration by AFM and Raman	$E = \sim 1$ TPa
Lee et al. (2008) [5]	Nano indentation by AFM	$E = 1.0 \pm 0.1$ TPa, $E' = -2.0 \pm 0.4$ TPa, $\sigma_f = 130 \pm 10$ GPa at $\varepsilon_f = 0.25$
Faccio et al. (2009) [148]	QM/DFT for uniaxial stress	$E = 0.96$ TPa ν=0.17 unitary for ε within ± 0.02%, non-linear for higher ε
Shokrieh and Rafiee (2010) [149]	CM/FEA	$E = 1.04$ TPa for analytical, $E_1 = 1.041.14$ TPa and $E_2 = 1.09$ TPa for FEA
Faccio et al. (2011) [15]	First Principles/DFT by SIESTA code	$E = 0.964(9)$ TPa, $G = 0.408$ TPa, $\nu = 0.17$
Min and Aluru (2011) [151]	MD	G_{max} at around 800 K. τ_f and ε_f decreased as the temperature is increased. G of zigzag higher than that of armchair. $\tau_f = 10.5$ GPa for zigzag and 8.7 GPa for armchair. $\sigma_f = 97.54$ GPa zigzag at 300 K

TABLE 8.2 Data summary on graphene (assuming an effective thickness of 0.335 nm) *(cont.)*

Reference	Method	Mechanical properties
Ranjbartoreh et al. (2011) [152]	Tensile, indentation and bending tests	Hardness ≈2.13 GPa of graphene papers is 10 times of graphite and two times of carbon steel. $\sigma_y \approx 6.4$ GPa
Scharfenberg et al. (2011) [153]	AFM/linear elasticity theory	Interlayer $\tau_c \geq 5.6$ MPa for thickness from 1 to 17 atomic layers
Lu and Huang (2011) [154]	Atomistic simulation/ second-generation REBO potential	$E = 707 \sim 780$ GPa for ribbons, $W = 1.2 \sim 8.9$ nm. $\sigma_f \approx 105 \pm 2$ GPa, $\varepsilon_f \approx 0.287 \sim 0.97$ for zigzag, $\sigma_f \approx 82 \pm 7$ GPa, $\varepsilon_f \approx 0.162 \sim 0.172$ for armchair
Lu and Hu (2012) [60]	MM/FEA elliptical beam element	$E = 1.0670 \sim 1.1967$ TPa for $W/L = 0.029 \sim 3.027$
Zheng et al. (2012) [156]	MD LAMMPS package/ AIREBO potential	$\sigma_f = 106$ GPa and $\varepsilon_f = 0.205$ for graphene, $\sigma_f = 2.212 \sim 3.855$ GPa and $\varepsilon_f = 0.159 \sim 0.268$ for nanochains with different random defects

The adoption of thickness can be problematic in nanoscale objects (such as SWCNTs or graphene) because there is no obvious way to define the thickness of a cylinder shell or plate that is one-atom thick, although taking the interlayer spacing in graphite as an alternative to thickness may seem reasonable. Yakobson et al. [18] concluded that the effective thickness of SWCNTs should be taken as $t = 0.066$ nm if the bending rigidity $D = 0.85$ eV and in-plane stiffness $Et = 360$ J m^{-2} are consistent with the classic shell bending theory ($E = 5.45$ TPa). Zhang and Wang [19] led to $t \approx 0.1$ nm and $E \approx 3.5$ TPa for SWCNT(10,10). Discussion of SWCNT and graphene wall thickness can also be found in Refs [20–22]. Here, thickness, and therefore elastic moduli, depends on type of load, as well as nanotube radius and chirality; this explains why the thickness obtained from the atomistic simulations is scattered. To avoid disagreement over the value of the effective thickness of SWCNTs or graphene, in-plane tensile stiffness is defined as Et, which represents the rigidity of the material to the axial loading applied. In the same way, in-plane shear stiffness is defined as Gt, which represents the rigidity of the material to the shear load applied.

The elastic response of CNTs or graphene must be considered non-linear when large strain is involved because the stress–strain response must curve over to a maximum point that defines the intrinsic breaking stress. An elastic response implies the existence of energy potential; this is a function of strain and can be expressed as a Taylor series in powers of strain. The lowest-order (quadratic) term leads to a linear elastic response (Hooke's law). The third-order term gives rise to non-linear elastic behaviour. The resulting isotropic elastic response under uniaxial tension can be expressed as

$$\sigma = E\varepsilon + E'\varepsilon^2 \tag{8.11}$$

where E' is the third-order elastic modulus, which is typically negative and leads to a lessening of stiffness at high tensile strains and an increasingly stiff response at high compressive strains.

According to the classic beam theory [23], when investigating the deflections of a beam with bending stiffness EI (also called flexural stiffness or bending rigidity; I is the moment of inertia), the curvature ($\kappa = 1/\rho$) at any point x along the length of the beam is proportional to the bending moment $M(x)$ at that section (see Figure 8.3):

$$\kappa = \frac{1}{\rho} = -\frac{d^2y/dx^2}{\left[1+(dy/dx)^2\right]^{3/2}} = \frac{M(x)}{EI} \tag{8.12}$$

In classic beam theory, which assumes small deflections, the relationship between bending moment and curvature is linearized by neglecting the square term of the slope $(dy/dx)^2$; therefore, the governing linear equation takes the following form:

$$\frac{d^2y}{dx^2} = -\frac{M(x)}{EI} \tag{8.13}$$

Neglecting $(dy/dx)^2$, however, is not permissible for slender beams when deflections are appreciable compared to length. The second-order term allows for the shortening of the lever arm during deflection when a load is applied at the free end of a cantilever beam. The classic theory is therefore not applicable in the calculation of large deflections. Moreover, classic beam theory offers no solution for deflections in any direction other than the one normal to the original beam shape.

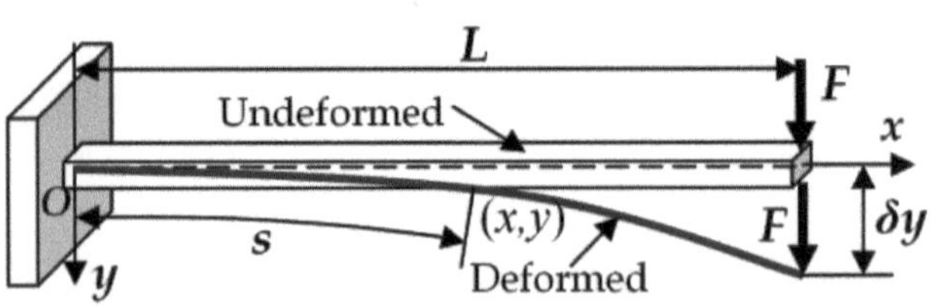

FIGURE 8.3 Cantilever beam with a concentrated load at the free end.

In the case of a thin plate (e.g. graphene) with an applied load, the bending stiffness D of the plate depends on plate thickness, the elastic properties of the plate and the load applied:

$$D = \frac{dM}{d\kappa} = \frac{Et^3}{12\left(1-\nu^2\right)} \tag{8.14}$$

A plate's bending stiffness has one dimension of length less than that of the classic beam, as it refers to the moment per unit length per unit of curvature, not the total moment.

While the in-plan mechanical properties (elastic moduli) of monolayer graphene have been deduced from experiments or easily obtained from computational modelling, direct measurement of monolayer graphene's bending stiffness has not been reported in the literature [5,24–27]. The often-cited experimental value of 1.2 eV (~0.192 nNnm) was derived from the phonon spectrum of graphite [28]. The finite bending stiffness of atomically thin graphene monolayer contrasts with classic theories for plates and shells (Equation (8.14)). The linear relationship between bending stiffness and Young's modulus is a result of the classic Kirchhoff hypothesis, which assumes linear variation of the strain and stress along the thickness of a thin elastic plate. However, the physical thickness of a graphene monolayer cannot be unambiguously defined in the continuum sense; the Kirchhoff hypothesis simply does not apply. Different physical origins must be sought for the bending moment and bending stiffness in graphene monolayers. *Ab initio* energy calculation is a promising method with $D = 3.9$ eVÅ^2/atom, or equivalently, 0.238 nNnm (1.5 eV), or continuum mechanics (CM) based on finite crystal elasticity and the Cauchy–Born rule [29–31]. Analytic formulas based on empirical potential [22,26,32–34] for solid-state carbon atoms can also give a good predicted value of $D = 0.225$ nNnm (1.4 eV) [25].

The bending stiffness of a SWCNT is

$$EI = \frac{1}{L}\frac{d^2U_s}{d\kappa^2} \tag{8.15}$$

From the elastic theory of beam bending, the strain energy of a bent nanotube can be expressed as [35]:

$$U_s = 0.5EtL\oint r^2\kappa^2\,dl \tag{8.16}$$

The integral is taken around the circumference of the nanotube, and r is the distance of atoms from the central line (or the bent axis) of the tube. Therefore, bending stiffness is found to be equal to $Et(\pi r^3)$, and scales as the cube of the tube's radius. Results from molecular dynamics (MD) simulations with Tersoff–Brenner potential indicate that stiffness scales as $R^{2.93}$, which aligns with the scaling predicted by continuum elastic theory [36,37].

8.2.2 Beyond the elastic regime

Although structural applications of CNTs rely on the elastic strength of the sp^2 bonding network, elastic response alone provides a limited picture of mechanical properties. Beyond the elastic limit, materials deform plastically via changes in chemical bonding topology (i.e. yield). A large strain beyond the threshold may cause an abrupt change in strain energy versus deformation (i.e. 'buckling' behaviour), similar to the buckling that occurs in structure mechanics when opposite edges of a thin, slender beam are pressed toward one another. For small loads, the beam is compressed in the axial direction while keeping its linear shape and the strain energy is proportional to the square of axial displacement. Beyond a certain critical load, it suddenly bends archwise and the relationship between strain energy and displacements deviates significantly from the square law. Besides axial compression, bending and torsion give rise to buckling in elastic beams, where the buckle patterns strongly depend on geometric and material parameters.

CNTs will exhibit plasticity, instability and fracture when acted upon by the destabilizing loads of axial tension, compression, torsion, bending, external pressure or large strain. Theoretical studies and experimental investigation indicate that CNTs subject to large deformations reversibly switch into different morphological patterns. Each shape change corresponds to an abrupt release of energy and singularity in the stress–strain curve. These transformations during and after buckling can be simulated by molecular mechanics (MM), MD, *ab initio* electronic structure methods and CM. The wavelike distortion formed in CNTs can be observed by high-resolution transmission electron microscopy (TEM) [18,26,37–45].

CNTs under axial tension will undergo plastic deformation and defect formation, followed by brittle failure. Figure 8.4(a)–(d) depicts the brittle fracture process on the outermost layer of four-walled (5,5), (10,10), (15,15) and (20,20) CNTs under large axial tension with $L/D = 4.5$. Figure 8.4(e) shows the elastoplastic stress–strain response of the multi-walled carbon nanotubes (MWCNTs) by computer modelling [38]. The inner layers of MWCNTs cannot endure tensile loads applied at both ends because the stress transferability between the

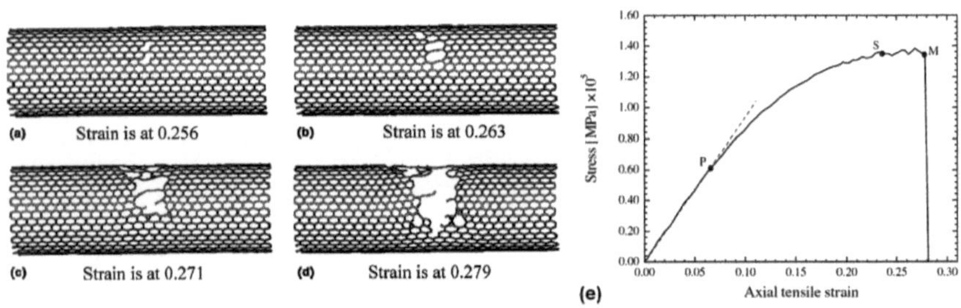

FIGURE 8.4 The failure process of the outermost layer of four-walled CNTs. (Reprinted with permission from Ref. [38], copyright ©2004 Acta Materialia Inc. Published by Elsevier Ltd.)

nanotube layers is very weak and the outermost layer of the nanotubes takes most of the load. Therefore, MWCNT failure could start at the outermost layer by breaking the bonds among carbon atoms. When strain reaches the maximum point of 0.256, two bonds break, creating two holes (see Figure 8.4(a)). As strain increases, more bonds break, the holes grow larger and the CNT eventually fractures (see Figure 8.4(b)–(d)).

CNTs under large compressive strains go through two distinct deformation modes: sideways buckling of thick tubes and plastic collapse of thin tubes (without buckling). Figure 8.5 [18] illustrates the axial compression of a MD-simulated nanotube with length L = 6 nm, diameter d = 1 nm and armchair helicity (7,7). The strain energy (a) displays four singularities corresponding to shape changes. At $\varepsilon_1 = 0.05$ the cylinder buckles into a pattern (b), displaying two identical perpendicular flattenings ('fins'). A further increase in ε gradually enhances this pattern until, at $\varepsilon_2 = 0.076$, the tube switches to a three-fin pattern (c) while maintaining a straight axis. In a sideways buckling, at $\varepsilon_3 = 0.09$, the flattenings serve as hinges and only a plane of symmetry is preserved (d). At $\varepsilon_4 = 0.13$, an entirely squashed asymmetric configuration is formed (e). Another study revealed a mechanism of SWCNT nanoplasticity under uniaxial compression in which the bonding geometry collapsed from a graphitic (sp^2) to a localized diamondlike (sp^3) reconstruction [27]. The computed critical stress (≈153 GPa for 12% compressed tube) and the shape of the resulting plastic deformation are congruent with the experimentally estimated range of values for thin nanotubes [39].

Figure 8.6(a) offers a clear example of the rippled MWCNT structure under high bending [39]. The tube, with a diameter of ~31 nm, is subjected to a bending curvature radius of ~400 nm. The compressed side of the nanotube may undergo yielding and local inward compressive buckling. Enhanced images at the ripple region, though not identical to the specimen in Figure 8.6(a), can be found from Ref. [44]. Figure 8.6(b)–(d) illustrate simulated cross-sectional images of a double-walled CNT (diameters 1.0 and 1.7 nm) under progressively increasing bending: (b) the straight relaxed tube; (c) a single kink forming above a critical curvature and (d) a two-kink complex forming in the middle of

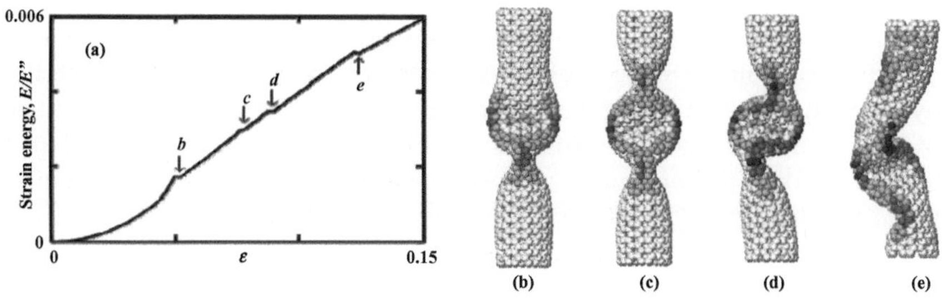

FIGURE 8.5 MD-simulated nanotube under axial compression. (Reprinted with permission from Ref. [18], copyright ©1996 by The American Physical Society.)

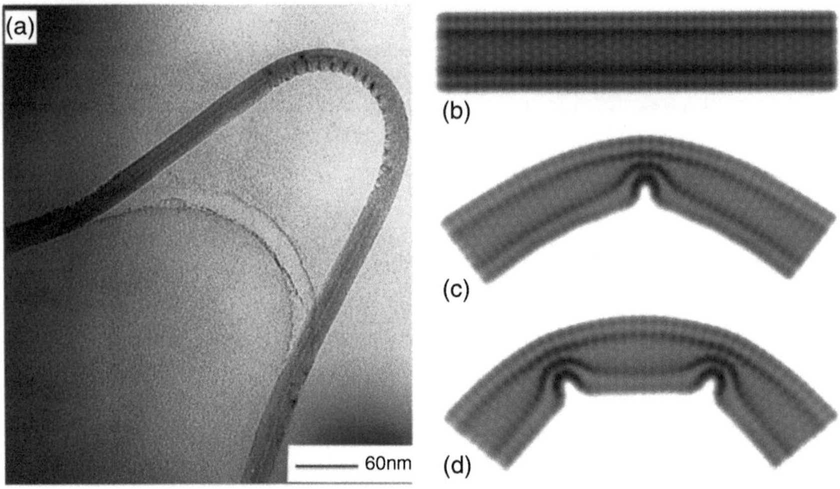

FIGURE 8.6 Under high bending, nanotubes collapse to form kinks on the internal (compression) side of the bend. ((a) Reprinted with permission from Ref. [39], copyright ©1998 by The American Physical Society; (b)–(d) reprinted with permission from Ref. [46], copyright ©1996 by American Institute of Physics.)

a large bend [46]. The distance between the two kinks (~2 nm) was found to be constant over a wide range of bend angles.

The mechanical properties of graphene beyond the elastic regime can also be measured or simulated. The elastic properties and intrinsic breaking strength of free-standing monolayer graphene membranes were measured by nanoindentation in an atomic force microscopy (AFM) [5]. The force–displacement behaviour was interpreted within a framework of non-linear elastic stress–strain response, and yielded second- and third-order elastic stiffness of 340 and -690 N m^{-1}, respectively. The breaking strength of 42 N m^{-1} represents the intrinsic strength of a defect-free sheet. These quantities correspond to a Young's modulus of $E = 1.0$ TPa, third-order elastic stiffness of $D = -2.0$ TPa and intrinsic strength of 130 GPa for bulk graphite. These experiments established graphene as the strongest material ever measured, and showed that atomically perfect nanoscale materials can be mechanically tested to deformations well beyond the linear regime.

Studies of the fracture behaviour of graphene sheets include simulation of the effects of multiple Stone–Wales (SW) defects on their mechanical properties by an atomistic-based finite bond element model [47]. The model used modified Morse potential to predict the Young's modulus and stress–strain relationship of perfect and defective nanotubes and graphene sheets. The variation of ultimate stress and strain at failure has also been investigated. Figure 8.7 shows the crack evolution at different strain levels for a graphene sheet (from left to right). After defect formations in a graphene sheet, one can apply further load (tension) onto the defective sheet until facture failure.

A thin graphene sheet/ribbon also tends to buckle under compression [48].

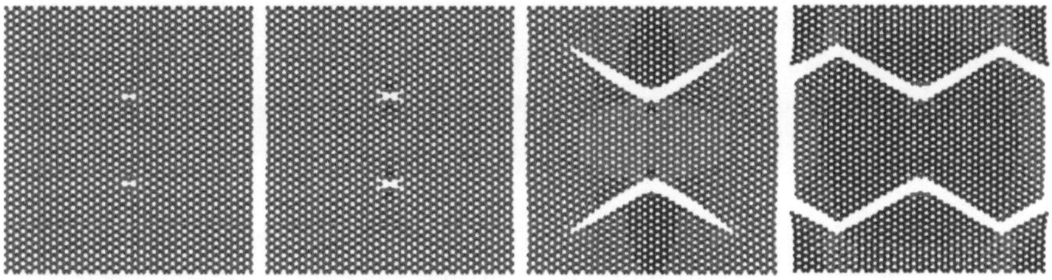

FIGURE 8.7 Fracture pattern of a graphene sheet with two defects along its axial direction under tension. (Reprinted with permission from Ref. [47], copyright ©2009 by Elsevier B.V.)

In addition to the above-mentioned mechanical properties, there are many other property terminologies often used to evaluate mechanical performance, such as hardness, bulk modulus, yield stress and strain corresponding to the yield point, ultimate tensile strength to the maximum stress point, failure (breaking) strength and strain to the failure point, buckling (critical) stress or bending critical angle to instability, etc. The concepts of these properties are similar to the definitions in the solids.

8.3 COMPUTER MODELLING AND EXPERIMENTAL APPROACHES

Extensive experimental and theoretical studies have characterized the mechanical properties of CNTs and graphene. One traditionally relies on a variety of experimental approaches to characterize, measure and manipulate material properties in order to understand, explore and tailor them for new applications. However, in many situations it may be difficult (or even impossible) to perform controlled experiments that mimic the real-process conditions. Because experiments are expensive and difficult to conduct, and there is considerable scatter in the experimental results, development of computer modelling for characterizing and predicting the mechanical behaviour of CNTs and graphene has been pursued with great vigour. Computer modelling is necessary for meaningful and instructive display of mechanical behaviour.

8.3.1 Computational modelling approaches

Computational approaches can be classified into two categories: the 'bottom-up' approach, including *ab initio* quantum mechanics (QM), MM, MD and Monte Carlo (MC) methods; and the 'top-down' approach based on CM and finite element analysis (FEA). The concepts of some of the major approaches are introduced in the following sections.

8.3.1.1 Quantum mechanics

First-principles or *ab initio* QM simulation has become one of the most influential and important techniques for modelling at the atomic level. It considers molecules as a collection of particles (nuclei and electrons), which move according to the laws of quantum mechanics. This is done by solving the electronic Schrödinger equation for the system of interest [49]:

$$\hat{H}\phi = E\phi \tag{8.17}$$

where $\hat{H}$ is the Hamiltonian operator of the system, Φ the wavefunction which describes the positions and motions of all the system's particles (nuclei and electrons) and E energy.

Electronic wavefunctions are very complicated because they depend on the positions of all the electrons. For many-particle systems, such as molecules consisting of n electrons and N nuclei, the wavefunction depends on four variables per particle (i.e. three coordinates and the spin). A solution of the Schrödinger equation for many-particle systems is impossible; certain simplifications and approximations are necessary to make the calculation feasible. The Hartree–Fock (HF) and correlated method, based on molecular orbital theory, enables an approximate solution to the Schrödinger equation based on N-electron wavefunctions. However, wavefunctions grow increasingly complex as the molecular system becomes larger. Density functional theory (DFT) provides another approximation in which the complicated N-electron wavefunction and the associated Schrödinger equation are replaced by much simpler electron density and its associated calculation scheme. These methods can be readily applied to molecules.

Solving the Schrödinger equation using HF or DFT methods requires large computational resources. Moreover, solids (including nanostructured materials) play a very important role in materials science. There are many simplified approaches to extending electronic structure calculations to solids. One possibility is to use empirical data from experiments to adjust quantities, which would have to be computed otherwise. These methods are deemed semi-empirical. Tight-binding theory can be viewed as a semi-empirical theory for periodic solids in which a solid is a collection of individual atoms brought from infinite separation to close proximity. When atoms are far apart, their constituent electrons feel only the potential from the atomic core plus other electrons in that atom. Thus, the electronic wavefunctions are simply the atomic orbitals. As the atoms are brought close together to form a solid, the electrons begin to feel the potential from surrounding atoms, and wavefunction becomes delocalized.

8.3.1.2 Atomistic modelling

In QM, with nuclei and electrons as the basic ingredients, systems are modelled at a deeper level of physics than with atoms and interatomic potential. The exact solution of equations of motion for molecules based on QM is a rather complex problem, and molecules of the size of interest are often too large for QM. If electronic properties are not of direct concern, and structures, conformations, vibrational spectra, etc. are the properties of interest then atomistic modelling can be a useful alternative. Atomistic modelling has proven to be an effective tool for studying CNTs and graphene's mechanical responses. MM and MD modelling are the most widely used atomistic modelling methods for this purpose.

MM uses classical mechanics to model molecular systems. This method involves a classical description of a molecular system through interatomic potentials or force fields. It uses a set of parameterized functions to represent the potential energy hypersurface of a molecule as accurately as possible. The idea behind MM is that character is largely determined by immediate environment; thus, the properties of a bond (e.g. its length and strength) are nearly constant in different molecules if the same atoms are bond together. Each atom is simulated as a single particle, and each particle is assigned a radius, polarizability and a constant net charge (generally derived from QM calculations and/or experiment). Bonded interactions are treated as 'springs' with an equilibrium distance equal to the experimental or calculated bond length. The mathematics of spring deformation can be used to describe the ability of bonds to stretch, bend and twist. Non-bonded atoms (greater than two bonds apart) interact through van der Waals attraction, steric repulsion and electrostatic attraction/repulsion (see Figure 8.8).

In CNTs or graphene, carbon atoms are bonded together with covalent bonds and non-covalent bonds. These bonds constrain the displacement of individual carbon atoms under external forces. The total force on each atomic nuclei or nucleus is the vectorial sum of force generated by the electrons and electrostatic force between the positively charged nuclei themselves. The molecular force

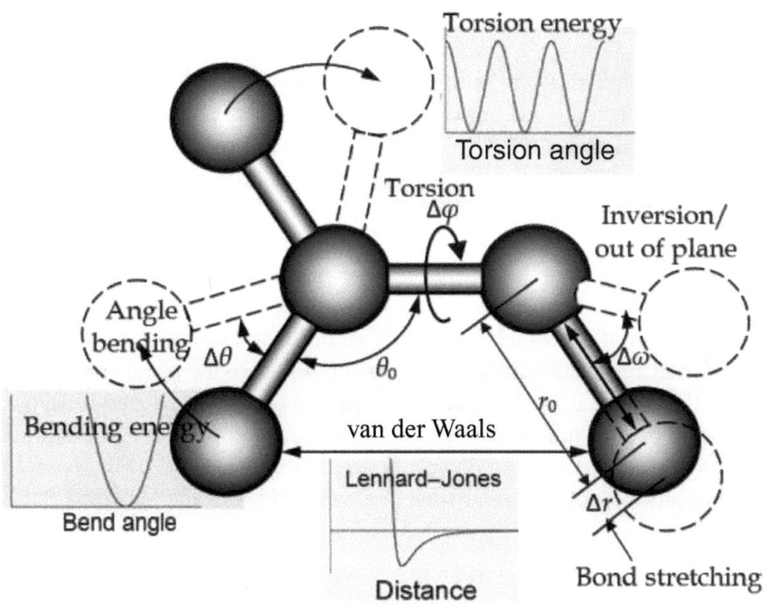

FIGURE 8.8 Interatomic interactions in MM.

field is expressed as steric potential energy, which is the sum of energies due to bonded interaction and non-bonded interactions [50,51]:

$$U = \sum U_r + \sum U_\theta + \sum U_\varphi + \sum U_\omega + \sum U_{vdw} + \sum U_{es} \tag{8.18}$$

where U_r is bond stretching energy, U_θ bond angle bending energy, U_φ dihedral angle torsion energy, U_ω inversion terms energy, U_{vdw} van der Waals interaction energy and U_{es} electrostatic interaction energy. In the case of small strain and covalent bond, the effects of U_ω, U_{vdw} and U_{es} are neglected.

With bond stretching, the accurate shapes of the experimental bond energy curve of diatomic molecules are available from spectroscopy. Beyond simple harmonic functions, there are several different potential functions available to describe CC bonds. The Tersoff–Brenner empirical interatomic potential function [52–55] is widely used to calculate the potential energy of covalent bonds and interatomic force. Though very accurate, this function is complicated. Simpler forms of potential functions are available, such as Morse or modified Morse potential [56,57].

Angle-bending energy is based on Hooke's law and may consider a higher order of the $\Delta\theta$ term to estimate the energy associated with vibration about the equilibrium bond angle.

Torsion energy is primarily used to correct the remaining energy terms, rather than to represent a physical process. Torsion energy represents the amount of energy that must be added to or subtracted from the stretching energy + bending energy + non-bonded interaction energy terms to make the energy total agree with experimental results or rigorous QM calculation of a model dihedral angle. Torsion energy is usually modelled by a simple periodic function.

Non-bonded terms are much more costly to computationally calculate in full because a typical atom, although bonded to only a few of its neighbours, interacts with every other atom in the molecule. Fortunately, the van der Waals term falls off rapidly and is typically modelled using a '6–12 Lennard-Jones potential'. A cutoff radius is generally used to speed up calculation so that atom pairs whose distances are greater than the cutoff have a van der Waals interaction energy of zero.

MD, on the other hand, is a form of computer simulation wherein atoms and molecules are allowed to interact for a period of time under the known laws of physics (such as Newton's second law), thereby giving a view of the motion of the atoms.

Conformational transitions and local vibrations are the usual subjects of MD studies. MD alters the intramolecular degrees of freedom in a step-wise fashion, analogous to energy minimization. The individual steps in energy minimization are merely directed at limiting down-hill direction. The steps in MD represent the changes in atomic position over time (i.e. velocity). For the '*i*' atom of the system, Newton's second law is used in MD formalism to simulate atomic motion:

$$\boldsymbol{F}_i = m_i \frac{d^2 \boldsymbol{r}_i}{dt^2} \tag{8.19}$$

where $\boldsymbol{F}_i$ and $\boldsymbol{r}_i$ are the resultant force vector and position vector on the 'i' atom, respectively, m_i the mass of the 'i' atom. In practice, the atoms are assigned initial velocities that conform to the total kinetic energy of the system, which in turn, is dictated by the desired simulation temperature. This is carried out by slowly 'heating' the system (initially at absolute zero) and then allowing the energy to equilibrate among the constituent atoms; the initial velocity field can also be assigned by the Maxwell–Boltzmann distribution function.

The basic ingredients of MD are the calculation of the force on each atom, and the position of each atom throughout a specified period of time, typically on the order of picoseconds ($=10^{-12}$ seconds). The force on an atom can be calculated from the change in energy between its current position and its position a small distance away. This can be recognized as a derivative of the energy with respect to the change in the atom's position. With proper boundary conditions, appropriate time step and energy conservation, and appropriate algorithms, a variety of physical properties can be calculated.

Given the high frequency of molecular thermal vibration, MD could be used in larger systems (up to millions of atoms) on shorter time scales ($10^{-6} \sim 10^{-9}$ seconds). The simulation of larger systems on longer time scales is currently limited to continuum methods.

8.3.1.3 Continuum modelling

Long MM/MD simulations are mathematically ill-conditioned, generating cumulative errors in numerical integration and time consuming computation procedures. Various phenomena in materials science involve processes over a wide range of length scales, from the atomistic to the continuum and time scale up to the second; processes restricted by current computing power. The coarse-grained computational method was developed as an improvement [58], with the computational complexity of atomistic models reduced by two orders of magnitude without loss of accuracy. Such computation can be carried out for a model with up to several tens of millions of atoms. Meanwhile, CM/FEA provides a powerful tool to fill this gap.

The development of nanoscale continuum theories that integrate CM theories with nanoscale molecular structure has aroused much interest. These modelling techniques are referred to as 'nanoscale mechanics' or 'molecular structural mechanics'; they link the interatomic potential of atomic structure to the continuum level of materials. Yakobson et al. [18] were the first to explicitly acknowledge the analogy between the mechanics of CNTs and the continuum elasticity of shells. Since then, the use of CM concepts in CNTs and graphene has become a common practice. In CM models, CNTs or graphene (composed of covalent bonded carbon atoms) are treated as continuous and homogeneous macrostructures, and their material nanostructures (such as the lattice spacing between individual carbon atoms) are ignored. The stress at a given point is defined as the function of the strain at that point (i.e. a hyperelastic material). In this respect, CM models are also referred to as local continuum models [59]. In such models, molecular physical phenomena define characteristic geometries, and simplifying assumptions are made that may or may not be appropriate. CM rationalizes the mechanics of materials in terms of irreducible sets of parameters, is amenable to analytical calculations, and affords very efficient simulations (e.g. finite elements (FEs)). The classic approach to continuum modelling of atomistic systems adopts a standard continuum theory (e.g. linear elasticity) and adjusts the material parameters to the data available. This phenomenological approach is simple and useful, particularly when the mechanics are well understood *a priori*, and the phenomena are simple enough to be encoded in the model. This approach has enjoyed great popularity for CNTs and graphene, which have been modelled as thin elastic shells and classical elastic beams, among other things.

For covalent systems, the main contributions to the total steric energy contain the first four terms in Equation (8.18). Under the assumption of small deformation, the harmonic approximation is adequate to define energy. By considering dihedral angle torsion and inversion as a single equivalent term, we have the following equations for each energy term:

$$U_r = \frac{1}{2}k_r(\Delta L)^2 = \frac{1}{2}\frac{EA}{L}(\Delta L)^2 \tag{8.20}$$

$$U_\theta = \frac{1}{2}k_\theta(\Delta \theta)^2 = \frac{1}{2}\frac{EI}{L}(\Delta \theta)^2 \tag{8.21}$$

$$U_\tau = U_\varphi + U_\omega = \frac{1}{2}k_\tau(\Delta \varphi)^2 = \frac{1}{2}\frac{GJ}{L}(\Delta \varphi)^2 \tag{8.22}$$

where k_r, k_θ and k_τ are the bond stretching, bending and torsion resistance force constants, respectively; while ΔL, $\Delta\theta$ and $\Delta\varphi$ represent bond or beam element stretching, bending angle and twisting angle variations, respectively. L and A are the beam length and cross-section area, respectively. E, G, I and J are the beam element's elastic modulus, shear modulus, moment of inertia and polar moment of inertia, respectively.

The potential associated with bond inversion can be taken into account indirectly by considering the rolling energy of rolling a graphene sheet into a CNT. During the rolling process, only the 'inclined' and 'transverse' bonds contribute to bending resistance [60]. The coefficients in the above equations refer to Refs [51,57,60–62].

The above fundamentals of CM can be further developed into FEA. The governing differential equations for the stretching, bending and twisting of a beam are written as:

$$EA\frac{d^2u(x)}{dx^2}=f(x),0\leq x\leq L \tag{8.23}$$

$$EI\frac{d^2v(x)}{dx^4}=M(x),\quad 0\leq x\leq L \tag{8.24}$$

$$GJ\frac{d^2\varnothing(x)}{dx^2}=T(x),\quad 0\leq x\leq L \tag{8.25}$$

where $f(x)$, $M(x)$ and $T(x)$ are axial load, bending moment and torque, respectively; $u(x)$, $v(x)$ and $\phi(x)$ are axial displacement, transverse displacement and twist angle, respectively. Each CC bond is treated as a separate beam element with two nodes, i and j. By integrating tension, bending and torsion stiffness matrix, an elemental equilibrium equation in terms of 12 degrees of freedom can be written as:

$$\left[\mathbf{K}^{\mathrm{e}}\right]\left\{\mathbf{u}^{\mathrm{e}}\right\}=\left\{\mathbf{f}^{\mathrm{e}}\right\} \tag{8.26}$$

where $\left\{\mathbf{f}^{\mathrm{e}}\right\}=\left\{f_{ix},f_{iy},f_{iz},M_{ix},M_{iy},M_{iz},f_{jx},f_{jy},f_{jz},M_{jx},M_{jy},M_{jz}\right\}^T$ is the nodal force vector of the beam element, $[\mathbf{K}^{\mathrm{e}}]$ the elemental stiffness matrix and $\left\{\mathbf{u}^{\mathrm{e}}\right\}=\left\{u_{ix},u_{iy},u_{iz},\theta_{ix},\theta_{iy},\theta_{iz},u_{jx},u_{jy},u_{jz},\theta_{jx},\theta_{jy},\theta_{jz}\right\}^T$ the nodal displacement vector of the beam element.

In order to obtain the deformation of a space framelike CNT, the above elemental stiffness equations must be established for every element in local coordinates; all of these equations can then be transformed into a common global reference system. Finally, a system of simultaneous linear equations can be assembled according to the requirements of nodal equilibrium. The nodal displacements, reaction forces and moments can be obtained by substituting the material properties of the CC bond and taking the boundary conditions into account.

With CNTs or graphene under large strain of compression or bending with buckling, the above simplified model is not adequate. From a mechanics viewpoint, the strong crystalline network and hollow cylindrical geometry confer CNTs and graphene with a very rich non-linear elasticity. While such methods have been extended to the finite deformation regime [63], the reduced number of fitted parameters raises doubts about the appropriateness of these models when material non-linearities are important. In other words, material properties at the nanoscale are size dependent; the small length scale effect should thus be considered to improve prediction of the mechanical behaviour of nanostructures. The theory of non-local elasticity, proposed by Eringen [64,65], assumes that the stress at a given point is a function of the strain field at every point in the body. Therefore, the long-range forces between atoms and internal length scales are introduced into the constitutive relations of a homogeneous and isotropic material as a material parameter. Considerable attention has recently been directed at the development and application of non-local continuum models for the analysis of CNTs and graphene [59,66–75].

8.3.2 Experimental approaches

The above-mentioned simple tests for extracting the mechanical properties of CNTs and graphene have been easily realized by computational modelling. However, because most of these models were not experimentally feasible, other setups have been developed to test mechanical properties directly or indirectly. The range of experimental values for different CNTs can vary by up to one order of magnitude. The main reasons for this difference are lack of information on the physical structure of the CNTs and the fact that mechanical properties are very sensitive to chirality and size. The experimental methods commonly used to measure the mechanical properties of CNTs and graphene include TEM observations, indenting/bending/pulling/resonating by AFM tips and Raman spectroscopy.

CNT type (chirality and diameter) can be assigned by visualizing CNT structures through high-resolution TEM, Raman spectroscopy and QM calculation. General observations using TEM have provided useful insight into the stiffness and strength of CNTs and graphene. Bent and buckled MWCNTs are commonly observed. MWCNTs are reasonably stiff because they generally appear fairly straight in TEM images. TEM reveals that, under high bending, nanotubes collapse to form kinks on the internal (compression) side of the bend [39,44]. Mechanical property testing can also be observed by TEM. For example, Treacy et al. [76,77] estimated the Young's modulus of isolated nanotubes by using TEM to measure the amplitude of their intrinsic thermal vibrations. Peng et al. [78] resolved a long-standing discrepancy between theory and experiment by providing the first direct measurements from TEM images of MWCNT tensile fracture strength that approach the ultimate strength estimates of QM calculations.

Scanning probe microscopy (SPM) not only forms images using a physical probe that scans the specimen surface, but also enables individual tubes or graphene to be directly manipulated. One of the most established types of SPM is AFM (including contact AFM, non-contact AFM, tapping AFM and dynamic contact AFM). Figure 8.9 offers schematics of AFM applications with TEM observation. Wong et al. [79] measured the elasticity, strength and toughness of MWCNTs using AFM, based on nanobeam mechanics, in which MWCNTs were pinned at one end to molybdenum disulfide surfaces. The bending force was measured versus displacement along the unpinned length. Cuenot et al. [80] used a resonant contact AFM technique to quantitatively measure the elastic modulus of polymer nanotubes. They applied an oscillating electric field

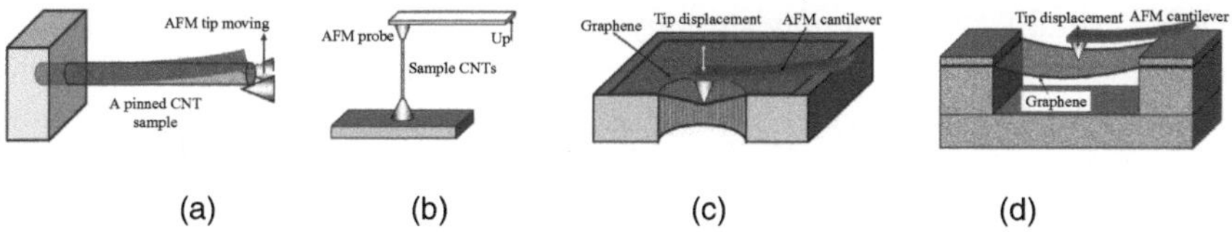

FIGURE 8.9 Schematic of experimental examples conducted by AFM: (a) beam bending, (b) pulling, (c) indenting circle and (d) indenting ribbon.

between the sample holder and the microscope head to excite the oscillation of the cantilever in contact with nanotubes. The nanotubes were suspended over the pores of a membrane. Due to the AFM tip contact, the resonance frequency of the system of the free cantilever shifted to higher values, enabling determination of nanotube stiffness. Yu et al. [81,82] conducted tensile tests by pulling SWCNT or MWCNT rope between an AFM tip and a CNT 'paper' sample with a 'nanostressing stage' inside a scanning electron microscopy to measure elastic modulus, breaking strength and breaking strain. In Lee et al.'s [5] experiment, a large graphene flake spanned an array of circular holes of $1 \sim 1.5\ \mu m$ in diameter. They probed the mechanical properties of the free-standing films by indenting the centre of each film with an AFM tip. The force–displacement behaviour was interpreted within a framework of non-linear elastic stress–strain response. A similar approach was used by Shivaraman et al. [24]. The cantilever tip was pushed against the centre of the suspended sheet in the tapping mode. As soon as the tip made contact with the graphene, the free amplitude of cantilever vibration went to zero. The applied force of the cantilever tip against the sheet caused the tip to deflect and the sheet to bend. Lourie and Wagner [83], Huang [84] and Sfeir et al. [85] used Raman spectroscopy or Rayleigh scattering spectroscopy to monitor the deformation processes of CNTs and graphene.

Although the above discussion primarily focuses on axial properties of CNTs and in-plane graphene, simple geometric considerations suggest that CNTs should be much softer in the radial direction than along the tube axis. Indeed, TEM observation of radial elasticity suggested that even the van der Waals forces can deform two adjacent nanotubes [86]. Nanoindentation experiments performed by several groups on MWCNTs and tapping/contact mode AFM measurement performed on SWCNTs indicated Young's modulus in the order of several GPs, confirming that CNTs are indeed rather soft in the radial direction [87–89].

8.4 MECHANICAL PROPERTY DATA SUMMARY

The many experimental and computational studies undertaken to estimate the mechanical properties of CNTs and graphene allow the analysis of the dependence of their length, diameter and chirality. The differences among the mechanical properties of CNTs and graphene are generally not due solely to gaps in computational and experimental approaches. From a computational point of view, several factors that may strongly influence the results include differences in methods, molecular parameters, reference models and computational conditions. From an experimental point of view, factors that may diffuse the results include differences in sample synthesis and measurement techniques, the presence and type of defect, and difficulties in sample identification.

In 1996, Treacy et al. [76,77] were the first to perform experimental measurement of Young's modulus in CNTs by using TEM to measure thermal vibrations. Wong et al. [79] used a cantilevered beam model in which individual MWCNTs were bent using an AFM tip. Lourie et al. [39,83] used a compression bar model in their experimental measuring of Young's modulus, and buckling and collapse properties monitored by TEM or micro-Raman spectroscopy. Tensile test have been reported for both SWCNTs and MWCNTs with a broad range of stiffness values and breaking strength [81,82,90]. The simple-supported three-point bending model using AFM tip was reported by Salvetat et al. and Tombler et al. [6,91–93] for individual MWCNTs and different-sized SWCNTs (structurally ordered MWCNTs grown by arc discharge or by the catalytic decomposition of hydrocarbons). A pulling and bending test on individual CNTs *in situ* in a TEM measured the intrinsic strength and elastic modulus of a defect-free MWCNT [94]. Wei et al. [95] and Guhados et al. [96] developed a technique using AFM to determine the effective values of both Young's modulus and shear modulus of MWCNTs. The former accomplished it by fitting the experimental *f–L* (natural frequencies–length) data obtained by bending MWCNTs using AFM with the Timoshenko beam model. The latter applied force with the AFM tip to deflect a clamped MWCNT beam, and determined the moduli based on the beam structural mechanics calculation. Huang [84] characterized suspended SWCNTs by Rayleigh scattering spectroscopy before tensile test by AFM cantilever to the clamped SWCNTs. Ding et al. [97] and Yamamoto et al. [98] investigated the fracture strength of MWCNTs under tensile loading by TEM and AFM. These experiments confirmed the exceptional mechanical properties of CNTs [99].

Theoretical studies of the mechanical properties of CNTs and graphene are both more numerous and more advanced than experimental measurements, mainly due to the technological challenges involved in the production of nanotubes and the manipulation of nanometer-sized objects. Dresselhaus et al. [100,101] identified the fundamental relations governing the geometry of CNTs with three main configurations (chiral, zigzag and armchair), enabling computer simulation of CNTs. 'Bottom-up' methods have been extensively applied in the characterization of CNT material properties [18,46,102–107]. Meanwhile, CM/FEA approaches have been established. Yakobson et al. [18], Ru [108,109] and Odegard et al. [110] proposed continuum shell or truss element modelling, and Meo and Rossi [111–113] combined the use of non-linear elastic and torsional elastic springs. Mechanical properties, such as Young's modulus, ultimate strength and strain, the stress–strain curve for CNTs, were calculated. In Li and Chou's work [51,114], a FE model was built with a non-linear truss rod to simulate van der Waals interaction among the layers for MWCNTs. Using the same method, Xiao et al. [115] developed a FE beam model incorporating a modified Morse potential. Young's modulus, Poisson's ratio and stress–strain relationships were evaluated for SWCNTs under axial and torsion load. Lu and Hu [60] adopted an improved FE model, considering bending stiffness of the graphene layer. The out-of-plane deformation of the bonds was distinguished from the in-plane deformation by considering an elliptical cross-section for the beam elements. Additional research uses this approach [116–120].

From structural and mechanical perspectives, the non-linear and instable mechanical response of CNTs (such as plastic deformation and 'buckling' behaviour) is a major topic in the nanotube research community [45]. Zhang et al. [121] studied plastic deformation of CNTs based on tight-binding MD calculations, and found that the onset of plastic deformation depended very strongly on the wrapping index. In addition, the remnant bond rotations remaining after strain release strongly affected the electronic structure of the disordered nanotube. Bozovic et al. [122] used anti-ferromagnetic manipulation to controllably stretch individual metallic SWCNTs, and found that SWCNTs can sustain elongations as great as 30% without breaking. Duan et al. [123] studied the fracture of SWCNTs by non-linear MM modelling, incorporating the modified Morse potential and reactive empirical bond-order (REBO) potential. The tensile strength, fracture strain and fracture angle for perfect SWCNTs under tension were discussed. Given their nanometric scales, similarities and differences in buckled patterns compared with macroscopic counterparts should not be at all surprising. This complexity has stimulated tremendous analysis of CNT buckling under diverse loading conditions.

In 1996, Iijima at al. [46] reported high-resolution TEM observations and atomistic simulations of the bending of SWCNTs and MWCNTs under mechanical duress. Single and multiple kinks were observed at high bending angles and the bending was fully reversible up to very large bending angles, despite the occurrence of kinks and highly strained tube regions. In the same year, Yakobson et al. [18] simulated instabilities of SWCNTs by using a realistic many-body potential method, and found that CNTs were remarkably resilient, sustaining extreme strain with no signs of brittleness, plasticity or atomic rearrangements. In 1997, Falvo et al. [124] and in 1998, Lourie et al. [39,125] showed the similar phenomena in MWCNTs and bundles of SWCNTs by applying load through AFM tips. Since then, research on buckling under axial compression, radial compression, bending, torsion and their combinations has been intensively studied [18,43–46,89]; these peculiar behaviours beyond Hooke's law can now be well described by the continuum model. Shima [45] conducted a detailed review of the dramatic progress in nanotube buckling research during the past few years, in which he cited over 200 research papers. Two facts have been the primary drivers of this research. One is the excellent geometric reversibility of CNTs against mechanical deformation; i.e. their cylindrical shapes are reversible upon unloading without permanent damage to the atomic structure. In addition, because CNTs exhibit high-fatigue resistance, they are a promising medium for mechanical energy storage with extremely high energy density [126]. The other driver is the substantial influence of buckling on CNT physical properties. For example, it was recently shown that CNTs undergoing axial buckling instability have potential utility as a single-electron transistor [127–129] and can play a crucial role in developing nanoelectromechanical systems.

Graphene is a wonder material to which many superlatives have been attributed [130–132]. It is the thinnest known material and the strongest ever measured. Its charge carriers exhibit giant intrinsic mobility, have zero effective mass and can travel for micrometers without scattering at room temperature. Graphene can sustain currents six orders of magnitude higher than copper, shows record thermal conductivity and stiffness, is impermeable to gas and reconciles such conflicting qualities as brittleness and ductility.

Several recent reviews discuss every aspect of graphene [133–141]. In 2007, Bunch et al. [25] used a simple drive and detection system by AFM tapping mode and Raman spectral signal to probe the mechanical properties of single- and multi-layer graphene sheet resonators in order to extract the fundamental resonance frequency and to characterize the quality factor and built-in tension. At least two recent experiments report mechanical tests performed on suspended graphene films; nanoindentation of suspended multi-layer graphene flakes was used to measure bending stiffness [142] and extract Young's modulus [143].

The elastic deformation and failure strength of graphene has been simulated using *ab initio* methods [144,145]. The elastic properties and instability were predicted. The predicted elastic response is highly non-linear for strains above 10%. The intrinsic strength reported serves as a benchmark for structural and mechanical applications, although the strength of macroscopic graphitic materials is still limited by the presence of defects and grain boundaries. In addition, these measurements demonstrate that third-order elastic constants associated with non-linear behaviour can be measured in atomically perfect nanoscale materials. These measurements can be used to validate models of atomic potential far beyond the linear regime [5].

In addition to their use as mechanical resonators, graphene sheets are robust (impermeable) enough to act as a thin membrane between two dissimilar environments [146]. In 2008, Bunch's PhD thesis [147] investigated the properties of resonating drumheads, which consisted of graphene-sealed microchambers containing a small volume of trapped gas. These experiments provided possible ways to probe the membrane properties of single atomic layers of graphene. Lee et al. [5] measured the elastic properties and intrinsic breaking strength of free-standing monolayer graphene membranes by nanoindentation in an AFM. They interpreted force–displacement behaviour within a framework of non-linear elastic stress–strain response. These experiments established graphene as the strongest material ever measured, and showed that atomically perfect nanoscale materials can be mechanically tested to deformations well beyond the linear regime.

Faccio at al. [148] investigated the structural and mechanical properties of zigzag graphene nanoribbons in the presence of stress by applying DFT within a generalized gradient approximation-Perdew–Burke–Ernzerhof approximation. Shokrieh and Rafiee [149] predicted Young's modulus of graphene sheets and CNTs using a CM approach. Ferralis [150] used Raman scattering techniques to review the mechanical properties of graphene. Min and Aluru [151] investigated the temperature- and chirality-dependent shear properties of graphene by MD simulations. Ranjbartoreh et al. [152] investigated mechanical properties of graphene paper by tensile, indentation and bending tests. Graphene paper shows extremely high modulus of elasticity during bending tests. The high strength and stiffness of graphene paper are ascribed to the interlocking-tile microstructure of individual graphene nanosheets. Scharfenberg et al. [153] examined the mechanical properties of graphene samples of thicknesses ranging from 1 to 17 atomic layers, placed on a microscale-corrugated elastic substrate.

Using AFM to deform the substrate through graphene and linear elasticity theory to model the deformation, the graphene's bending rigidity, adhesion, interlayer shear strength have been extracted. Lu and Huang [154] investigated the non-linear elastic properties of monolayer graphene nanoribbons under quasi-static uniaxial tension based on atomistic simulations. Lau et al. [155] investigated graphene's wrinkle behaviour by suspending graphene membranes of 1–50 layers by exfoliating across pre-defined trenches on Si/SiO_2 substrates, and found that most graphene membranes were not flat, but formed periodic ripples. These static ripples can be imaged in titled stages in an SEM, using tapping mode in AFM or even under optical microscopy for thicker films. This interesting phenomenon is more than just a mechanical curiosity because the local electronic properties of graphene are intricately related to its local morphology. Zheng et al. [156] used MD simulation to investigate the mechanical properties of graphene nanochain constructed by sewing up pristine or twisted graphene nanoribbons and interlocking the obtained nanorings. The effects of length, width and twist angle on mechanical performance were analysed. These outstanding mechanical properties of graphene paper could lead to a wide range of engineering applications.

REFERENCES

[1] Guo T, Jin CM, Smalley RE. Doping bucky: formation and properties of boron-doped buckminsterfullerene. J Phys Chem 1991;95(13):4948–50.

[2] Iijima S. Helical microtubules of graphitic carbon. Nature 1991;354:56–8.

[3] Iijima S, Ichihashi T. Single-shell carbon nanotubes of 1-nm diameter. Nature 1993;363(6430):603–5.

[4] Novoselov KS, Geim AK, Morozov SV, Jiang D, Zhang Y, Dubonos SV. Electric field effect in atomically thin carbon films. et al. Science 2004;306(5296):666–9.

[5] Lee C, Wei X, Kysar JW, Hone J. Measurement of the elastic properties and intrinsic strength of monolayer graphene. Science 2008;321(5887):385–8.

[6] Salvetat JP, Bonard JM, Thomson NH, Kulik AJ, Forró L, Benoit W. Mechanical properties of carbon nanotubes. et al. Appl Phys A 1999;69:255–60.
[7] Lau KT, Hui D. The revolutionary creation of new advanced materials—carbon nanotube. Composites B 2002;33(4):263–77.
[8] Lau KT, Gu C, Hui D. A critical review on nanotube and nanotube/nanoclay related polymer composite materials. Composites B 2006;37(6):425–36.
[9] Lau KT, Cheung HY, Lu J, Hui D, Li HL. Carbon nanotubes for space and bio-engineering applications. J Comput Theor Nanosci 2008;5(1):23–35.
[10] Hu Z, Arefin MRH, Yan X, Fan QH. Mechanical property characterization of carbon nanotube modified polymeric nanocomposites by computer modeling. Composites B 2014;56:100–8.
[11] Gibson RF. A review of recent research on mechanics of multifunctional composite materials and structures. Compos Struct 2010;92(12):2793–810.
[12] Kelly BT. Physics of graphite. Applied Science Publishers; 1981.
[13] Blakslee OL, Proctor DG, Seldin EJ, Spence GB, Weng T. Elastic constants of compression-annealed pyrolytic graphite. J Appl Phys 1970;41(8):3373–82.
[14] Salvetat JP, Désarmot G, Gauthier C, Poulin P. Mechanical properties of individual nanotubes and composites. Understanding carbon nanotubes: from basics to applications, lecture notes in physics, vol. 677. Berlin: Springer; 2006. p. 439–94.
[15] Faccio R, Fernández-Werner L, Pardo H, Goyenola C, Denis PA, Mombrú ÁW. Mechanical and electric properties of graphene nanostructures. Physics and applications of graphene – theory. InTech; 2011.
[16] Yamamoto T, Watanabe K, Hernández ER. Mechanical properties, thermal stability and heat transport in carbon nanotubes. In: Carbon nanotubes: advanced topics in the synthesis, structure, properties and applications. Berlin: Springer; 2008. p. 165–197.
[17] Wang Q, Liew KM. Mechanical properties of carbon nanotubes. In: Carbon nanotubes: new research. Nova Science Publisher; 2009. [chapter 6].
[18] Yakobson BI, Brabec CJ, Bernholc J. Nanomechanics of carbon tubes: instabilities beyond linear response. Phys Rev Lett 1996;76:2511–4.
[19] Zhang LC, Wang CY. Effective wall thickness of single-walled carbon nanotubes for multiscale analysis: the problem and a possible solution. IUTAM Symp Model Nanomater Nanosyst 2009;13:53–61.
[20] Wang CY, Zhang LC. A critical assessment of the elastic properties and effective wall thickness of single-walled carbon nanotubes. Nanotechnology 2008;19(7):075705.
[21] Yakobson BI, Avouris P. Mechanical properties of carbon nanotubes. In: Carbon nanotubes, topics in applied physics, vol. 80. Berlin: Springer; 2001. p. 287–327.
[22] Huang Y, Wu J, Hwang KC. Thickness of graphene and single-wall carbon nanotubes. Phys Rev B 2006;74:245413. (9 pages).
[23] Timoshenko S, Woinowsky-Krieger S. Theory of plates and shells. McGraw-Hill; 1987.
[24] Shivaraman S, Barton RA, Yu X, Alden J, Herman L, Chandrashekhar MVS. Free-standing epitaxial graphene. et al. Nano Lett 2009;9(9):3100–5.
[25] Bunch JS, van der Zande AM, Verbridge SS, Frank IW, Tanenbaum DM, Parpia JM. Electromechanical resonators from graphene sheets. et al. Science 2007;315(5811):490–3.
[26] Lu Q, Arroyo M, Huang R. Elastic bending modulus of monolayer graphene. J Phys D Appl Phys 2009;42:102002. (6 pages).
[27] Srivastava D, Menon M, Cho K. Nanoplasticity of single-wall carbon nanotubes under uniaxial compression. Phys Rev Lett 1999;83(15):2973–6.
[28] Nicklow R, Wakabayashi N, Smith HG. Lattice dynamics of pyrolytic graphite. Phys Rev B 1972;5:4951–62.

[29] Kudin KN, Scuseria GE, Yakobson BI. C2F, BN, and C nanoshell elasticity from *ab initio* computations. Phys Rev B 2001;64(23):235406. (10 pages).
[30] Arroyo M, Belytschko T. Finite crystal elasticity of carbon nanotubes based on the exponential Cauchy–Born rule. Phys Rev B 2004;69:115415.
[31] Arroyo M, Belytschko T. Continuum mechanics modeling and simulation of carbon nanotubes. Meccanica 2005;40(4–6):455–69.
[32] Tu ZC, OuYang ZC. Single-walled and multiwalled carbon nanotubes viewed as elastic tubes with the effective Young's modulus dependent on layer number. Phys Rev B 2002;65:233407.
[33] Tersoff J. New empirical approach for the structure and energy of covalent systems. Phys Rev B 1988;37:6991–7000.
[34] Sanchez-Portal D, Artacho E, Soler JM, Rubio A, Ordejon P. *Ab initio* structural, elastic, and vibrational properties of carbon nanotubes. Phys Rev B 1999;59:12678–88.
[35] Landau LD, Lifschitz EM. Theory of elasticity. 3rd ed. Pergamon Press; 1986.
[36] Robertson DH, Brenner DW, Mintmire JW. Energetics of nanoscale graphitic tubules. Phys Rev B 1992;45:12592–5.
[37] Srivastava D, Wei C, Cho K. Nanomechanics of carbon nanotubes and composites. Appl Mech Rev 2003;56(2):215–30.
[38] Liew KM, He XQ, Wong CH. On the study of elastic and plastic properties of multi-walled carbon nanotubes under axial tension using molecular dynamics simulation. Acta Mater 2004;52:2521–7.
[39] Lourie O, Cox DM, Wagner HD. Buckling and collapse of embedded carbon nanotubes. Phys Rev Lett 1998;81(8):1638–41.
[40] Zhang CL, Hui-Shen Shen HS. Buckling and postbuckling analysis of single-walled carbon nanotubes in thermal environments via molecular dynamics simulation. Carbon 2006;44(13):2608–16.
[41] Wang Q, Quek ST, Varadan VK. Torsional buckling of carbon nanotubes. Phys Lett A 2007;367:135–9.
[42] Qian D, Wagner GJ, Liu WK. Mechanics of carbon nanotubes. Appl Mech Rev 2002;55(6):495–533.
[43] Duan XJ, Tang C, Zhang J, Guo WL, Liu ZF. Two distinct buckling modes in carbon nanotube bending. Nano Lett 2007;7:143–8.
[44] Poncharal P, Wang ZL, Ugarte D, de Heer WA. Electrostatic deflections and electromechanical resonances of carbon nanotubes. Science 1999;283:1513–6.
[45] Shima H. Buckling of carbon nanotubes: a state of the art review. Materials 2012;5:47–84.
[46] Iijima S, Brabec C, Maiti A, Bernholc J. Structural flexibility of carbon nanotubes. J Chem Phys 1996;104(5):2089–92.
[47] Xiao JR, Staniszewski J, Gillespie JW. Tensile behaviors of graphene sheets and carbon nanotubes with multiple Stone–Wales defects. Mater Sci Eng A 2010;527(3):715–23.
[48] Lu J, Huang R. Nonlinear mechanical of single-atomic-layer graphene sheets. Int J Appl Mech 2009;1(3):443–67.
[49] Hill J-R, Subramanian L, Maiti A. Molecular modeling techniques in material science. Taylor & Francis; 2005.
[50] Rappe AK, Casewit CJ, Colwell KS III, Goddard WA, Skiff WM. UFF, a full periodic table force field for molecular mechanics and molecular dynamics simulations. J Am Chem Soc 1992;114(25):10024–35.
[51] Li C, Chou TW. A structural mechanics approach for the analysis of carbon nanotubes. Int J Solids Struct 2003;40(10):2487–99.

[52] Brenner DW. Empirical potential for hydrocarbons for use in simulating the chemical vapor deposition of diamond films. Phys Rev B 1990;42:9458–71.

[53] Brenner DW. Erratum: empirical potential for hydrocarbons for use in simulating the chemical vapor deposition of diamond films. Phys Rev B 1992;46:1948. (1 page).

[54] Lindsay L, Broido DA. Optimized Tersoff and Brenner empirical potential parameters for lattice dynamics and phonon thermal transport in carbon nanotubes and graphene. Phys Rev B 2010;81(20):205441. (6 pages).

[55] Lindsay L, Broido DA. Erratum: optimized Tersoff and Brenner empirical potential parameters for lattice dynamics and phonon thermal transport in carbon nanotubes and graphene [Phys. Rev. B 2010;81:205441]. Phys Rev B 2010;82(20):209903(E). (1 page).

[56] Machida K, editor. Principles of molecular mechanics. Kodansha and Wiley co-publication; 1999.

[57] Belytschko T, Xiao SP, Schatz GC, Ruoff RS. Atomistic simulations of nanotube fracture. Phys Rev B 2002;65:235430. (8 pages).

[58] Arias I, Arroyo M. Size-dependent nonlinear elastic scaling of multiwalled carbon nanotubes. Phys Rev Lett 2008;100:085503. (4 pages).

[59] Zhang YY, Wang CM, Duan WH, Xiang Y, Zong Z. Assessment of continuum mechanics models in predicting buckling strains of single-walled carbon nanotubes. Nanotechnology 2009;20:395707. (8 pages).

[60] Lu X, Hu Z. Mechanical property evaluation of single-walled carbon nanotubes by finite element modeling. Composites B 2012;43:1902–13.

[61] Chang TC, Gao HJ. Size-dependent elastic properties of a single-walled carbon nanotube via a molecular mechanics model. J Mech Phys Solids 2003;51:1059–74.

[62] Scarpa F, Adhikari S. A mechanical equivalence for Poisson's ratio and thickness of CC bonds in single wall carbon nanotubes. J Phys D Appl Phys 2008;41:085306. (5 pages).

[63] Pantano A, Parks DM, Boyce MC. Mechanics of deformation of single and multi-wall carbon nanotubes. J Mech Phys Solids 2004;52(4):789–821.

[64] Eringen AC. Linear theory of nonlocal elasticity and dispersion of plane waves. Int J Eng Sci 1972;10:425–35.

[65] Eringen AC. On differential equations of nonlocal elasticity and solutions of screw dislocation and surface waves. J Appl Phys 1983;54:4703. (8 pages).

[66] Pradhan SC. Buckling analysis and small scale effect of biaxially compressed graphene sheets using non-local elasticity theory. Sa¯dhana¯ 2012;37(4):461–80.

[67] Ansari R, Shahabodini A, Alipour A, Rouhi H. Stability of a single-layer graphene sheet with various edge conditions: a non-local plate model including interatomic potentials. J Nanoeng Nanosyst 2012;226(2):51–60.

[68] Sudak LJ. Column buckling of multiwalled carbon nanotubes using nonlocal continuum mechanics. J Appl Phys 2003;94:7281. (7 pages).

[69] Wang Q, Wang CM. The constitutive relation and small scale parameter of nonlocal continuum mechanics for modeling carbon nanotubes. Nanotechnology 2007;18:075702. (4 pages).

[70] Zhang YQ, Liu GR, Wang JS. Small-scale effects on buckling of multiwalled carbon nanotubes under axial compression. Phys Rev B 2004;70(20):205430. (6 pages).

[71] Wang CM, Zhang YY, Ramesh SS, Kitipornchai S. Buckling analysis of micro- and nano-rods/tubes based on nonlocal Timoshenko beam theory. J Phys D Appl Phys 2006;39(17):3904–9.

[72] Wang Q, Han QK, Wen BC. Estimate of material property of carbon nanotubes via nonlocal elasticity. Adv Theor Appl Mech 2008;1(1):1–10.

[73] Wang Y, Wang XX, Ni XG, Wu HA. Simulation of the elastic response and the buckling modes of single-walled carbon nanotubes. Comput Mater Sci 2005;32:141–6.

[74] Zhang YY, Tan VBC, Wang CM. Effect of chirality on buckling behavior of single-walled carbon nanotubes. J Appl Phys 2006;100:074304. (6 pages).
[75] Kutana A, Giapis KP. A Transient deformation regime in bending of single-walled carbon nanotubes. Phys Rev Lett 2007;97:245501.
[76] Treacy MMJ, Ebbesen TW, Gibson JM. Exceptionally high Young's modulus observed for individual carbon nanotubes. Nature 1996;381:678–80.
[77] Krishnan A, Dujardin E, Ebbesen TW, Yianilos PN, Treacy MMJ. Young's modulus of single-walled nanotubes. Phys Rev B 1998;58(20):14013–9.
[78] Peng B, Locascio M, Zapol P, Li S, Mielke SL, Schatz GC. Measurements of near-ultimate strength for multiwalled carbon nanotubes and irradiation-induced crosslinking improvements. et al. Nat Nanotechnol 2008;3(10):626–31.
[79] Wong EW, Sheehan PE, Lieber CM. Nanobeam mechanics: elasticity, strength, and toughness of nanorods and nanotubes. Science 1997;277(5334):1971–5.
[80] Cuenot S, Fretigny C, Demoustier-Champagne S, Nysten B. Measurement of elastic modulus of nanotubes by resonant contact atomic force microscopy. J Appl Phys 2003;93(9):5650–5.
[81] Yu MF, Files BS, Arepalli S, Ruoff RS. Tensile loading of ropes of single wall carbon nanotubes and their mechanical properties. Phys Rev Lett 2000;84(24):5552–5.
[82] Yu MF, Lourie O, Dyer MJ, Kelly TF, Ruoff RS. Strength and breaking mechanism of multiwalled carbon nanotubes under tensile load. Science 2000;287:637–40.
[83] Lourie O, Wagner HD. Evaluation of Young's modulus of carbon nanotubes by micro-Raman spectroscopy. J Mater Res 1998;13(9):2418–22.
[84] Huang M. Studies of mechanically deformed single wall carbon nanotubes and graphene by optical spectroscopy. PhD Thesis, Columbia University, USA; 2009.
[85] Sfeir MY, Wang F, Huang L, Chuang C-C, Jone J, O'Brien SP. Probing electronic transitions in individual carbon nanotubes by Rayleigh scattering. et al. Science 2004;306(5701):1540–3.
[86] Ruoff RS, Tersoff J, Lorents DC, Subramoney S, Chan B. Radial deformation of carbon nanotubes by van der Waals forces. Nature 1993;364(6437):514–6.
[87] Palaci I, Fedrigo S, Brune H, Klinke C, Chen M, Riedo E. Radial elasticity of multiwalled carbon nanotubes. Phys Rev Lett 2005;94(17):175502. (4 pages).
[88] Yu M-F, Kowalewski T, Ruoff R. Investigation of the radial deformability of individual carbon nanotubes under controlled indentation force. Phys Rev Lett 2000;86(7):1456–9.
[89] Yang YH, Li WZ. Radial elasticity of single-walled carbon nanotube measured by atomic force microscopy. Appl Phys Lett 2011;98:041901. (3 pages).
[90] Xie S, Li W, Pan Z, Chang B, Sun L. Mechanical and physical properties on carbon nanotube. J Phys Chem Solids 2000;61:1153–8.
[91] Salvetat JP, Kulik AJ, Bonard JM, Briggs GAD, Stockli T, Metenier K. Elastic modulus of ordered and disordered multiwalled carbon nanotubes. et al. Adv Mater 1999;11(2):161–5.
[92] Salvetat JP, Briggs GAD, Bonard JM, Bacsa RR, Kulik AJ, Stockli T. Elastic and shear moduli of single-walled carbon nanotube ropes. et al. Phys Rev Lett 1999;82(5):944–7.
[93] Tombler TW, Zhou CW, Alexaeyev L, Kong J, Dai HJ, Liu L. Reversible electromechanical characteristics of carbon nanotubes under local-probe manipulation. et al. Nature 2000;405:769–72.
[94] Demczyk BG, Wang YM, Cumings J, Hetman M, Han W, Zettl A. Direct mechanical measurement of the tensile strength and elastic modulus of multiwalled carbon nanotubes. et al. Mater Sci Eng A 2002;334(1–2):173–8.
[95] Wei XL, Liu Y, Chen Q, Wang MS, Peng LM. The very low shear modulus of multi-walled carbon nanotubes determined simultaneously with the axial Young's modulus via in situ experiments. Adv Funct Mater 2008;18(10):1555–62.

[96] Guhados G, Wan WK, Sun XL, Hutter JL. Simultaneous measurement of Young's and shear moduli of multiwalled carbon nanotubes using atomic force microscopy. J Appl Phys 2007;101:033514. (5 pages).

[97] Ding W, Calabri L, Kohlhaas KM, Chen X, Dikin DA, Ruoff RS. Modulus, fracture strength, and brittle vs. plastic response of the outer shell of arc-grown multi-walled carbon nanotubes. Exp Mech 2007;47(1):25–36.

[98] Yamamoto G, Suk JW, An J, Piner RD, Hashida T, Takagi T. The influence of nanoscale defects on the fracture of multi-walled carbon nanotubes under tensile loading. et al. Diamond Relat Mater 2010;19(7–9):748–51.

[99] Shokrieh MM, Rafiee R. A review of the mechanical properties of isolated carbon nanotubes and carbon nanotube composites. Mech Compos Mater 2010;46(2):155–72.

[100] Dresselhaus MS, Dresselhaus G, Eklund PC, editors. Science of fullerenes and carbon nanotubes. Academic Press; 1996.

[101] Dresselhaus MS, Dresselhaus G, Saito R. Physics of carbon nanotubes. Carbon 1995;33(7):883–91.

[102] Lu JP. Elastic properties of carbon nanotubes and nanoropes. Phys Rev Lett 1997;79(7). 1297-1230.

[103] Hernandez E, Goze C, Bernier P, Rubio A. Elastic properties of C and $B_xC_yN_z$ composite nanotubes. Phys Rev Lett 1998;80(20):4502–5.

[104] Gao GH, Cagin T, Goddard WA. Energetics, structure, mechanical and vibrational properties of single-walled carbon nanotubes. Nanotechnology 1998;9(3):184–91.

[105] Lier GV, Alsenoy CV, Doran VV, Geerlings P. *Ab initio* study of the elastic properties of single-walled carbon nanotubes and graphene. Chem Phys Lett 2000;326(1-2):181–5.

[106] Klein ML, Shinoda W. Large-scale molecular dynamics simulations of self-assembling systems. Science 2008;321(5890):798–800.

[107] Vandivort K, Phillips JC, Villa E, Kale LV, Schulten K, Freddolino PL, Gumbart J, Trabuco LG, Chandler DE, Hsin J, Harrison CB. Long time and large size molecular dynamics simulations made feasible through new TeraGrid hardware and software. TeraGrid'08, Arlington, Virginia, June 9–13, 2008. 2008. 9 pages.

[108] Ru CQ. Effective bending stiffness of carbon nanotubes. Phys Rev B 2000;62(15):9973–6.

[109] Ru CQ. Elastic buckling of single walled carbon nanotube ropes under high pressure. Phys Rev B 2000;62(15):10405–8.

[110] Odegard GM, Gates TS, Nicholson LM, Wise KE. Equivalent-continuum modeling of nano-structured materials. Compos Sci Technol 2002;60(14):1869–80.

[111] Meo M, Rossi M. Prediction of Young's modulus of single wall carbon nanotubes by molecular-mechanics-based finite element modelling. Compos Sci Technol 2006;66(11–12):1597–605.

[112] Meo M, Rossi M. A molecular-mechanics based finite element model for strength prediction of single wall carbon nanotubes. Mater Sci Eng A 2007;454–455:170–7.

[113] Rossi M, Meo M. On the estimation of mechanical properties of single-walled carbon nanotubes by using a molecular-mechanics based FE approach. Compos Sci Technol 2009;69(9):1394–8.

[114] Li C, Chou TW. Elastic moduli of multi-walled carbon nanotubes and the effect of van der Waals forces. Compos Sci Technol 2003;63(11):1517–24.

[115] Xiao JR, Gama BA, Gillespie JW Jr. An analytical molecular structural mechanics model for the mechanical properties of carbon nanotubes. Int J Solids Struct 2005;42(11–12):3075–92.

[116] Lu JN, Chen HB. Analysis of single-walled carbon nanotubes using a chemical bond element model. Chin J Chem Phys 2008;21(4):353–60.

[117] Mohammadpour E, Awang M, Abdullah MZ. Predicting the Young's modulus of single-walled carbon nanotubes using finite element modeling. J Appl Sci 2011;11(9):1653–7.
[118] Tserpes KI, Papanikos P. Finite element modeling of single-walled carbon nanotubes. Composites B 2005;36(5):468–77.
[119] Ávila AF, Lacerda GSR. Molecular mechanics applied to single-walled carbon nanotubes. Mater Res 2008;11(3):325–33.
[120] Kirtania S, Chakraborty D. Finite element based characterization of carbon nanotubes. J Reinf Plast Compos 2007;26(15):1557–70.
[121] Zhang P, Lammert PE, Crespi VH. Plastic deformation of carbon nanotubes. Phys Rev Lett 1998;81(24):5346–8.
[122] Bozovic D, Bockrath M, Hafner JH, Lieber CM, Park H, Tinkham M. Plastic deformations in mechanically strained single-walled carbon nanotubes. Phys Rev B 2003;67:033407. (4 pages).
[123] Duan WH, Wang Q, Liew KM, He XQ. Molecular mechanics modeling of carbon nanotube fracture. Carbon 2007;45:1769–76.
[124] Falvo MR, Clary GJ, Taylor RM II, Chi V, Brooks FP Jr, Washburn S. Bending and buckling of carbon nanotubes under large strain. et al. Nature 1997;389:582–4.
[125] Lourie O, Wagner HD. Transmission electron microscopy observations of fracture of single-wall carbon nanotubes under axial tension. Appl Phys Lett 1998;73(24):3527–9.
[126] Zhang R, Wen Q, Qian W, Sheng D, Zhang Q, Wei F. Superstrong ultralong carbon nanotubes for mechanical energy storage. Adv Mater 2011;23:3387–91.
[127] Weick G, Pistolesi F, Mariani E, von Oppen F. Discontinuous Euler instability in nanoelectromechanical systems. Phys Rev B 2010;81:121409.
[128] Naieni AK, Yaghoobi P, Woodsworth DJ, Nojeh A. Structural deformations and current oscillations in armchair-carbon nanotube cross devices: a theoretical study. J Phys D Appl Phys 2011;44:085402.
[129] Weick G, von Oppen F, Pistolesi F. Euler buckling instability and enhanced current blockade in suspended single-electron transistors. Phys Rev B 2011;83:035420.
[130] Geim AK. Graphene: statues and prospects. Science 2009;324:1530–4.
[131] Novoselov KS, Jiang D, Schedin F, Booth TJ, Khotkevich VV, Morozov SV. Two-dimensional atomic crystals. et al. PNAS 2005;102(30):10451–3.
[132] Geim AK, Novoselov KS. The rise of graphene. Nat Mater 2007;6:183–91.
[133] Cooper DR, D'Anjou B, Ghattamaneni N, Harack B, Hilke M, Horth A. Experimental review of graphene. et al. ISRN Condens Matter Phys 2012;2012. 56 pages.
[134] Castro Neto AH, Guinea F, Peres NMR, Novoselov KS, Geim AK. The electronic properties of graphene. Rev Mod Phys 2009;81(1):109–62.
[135] Das Sarma S, Adam S, Hwang EH, Rossi E. Electronic transport in two-dimensional graphene. Rev Mod Phys 2011;83(2):407–70.
[136] Choi W, Lahiri I, Seelaboyina R, Kang YS. Synthesis of graphene and its applications: a review. Crit Rev Solid State Mater Sci 2010;35(1):52–71.
[137] Ni ZH, Wang YY, Yu T, Shen ZX. Raman spectroscopy and imaging of graphene. Nano Res 2008;1(4):273–91.
[138] Avouris P. Graphene: electronic and photonic properties and devices. Nano Lett 2010;10(11):4285–94.
[139] Giannazzo F, Raineri V, Rimini E. Transport properties of graphene with nanoscale lateral resolution. Scanning Probe Microsc Nanosci Nanotechnol 2011;2:247–58.
[140] Schwierz F. Graphene transistors. Nat Nanotechnol 2010;5(7):487–96.
[141] Bonaccorso F, Sun Z, Hasan T, Ferrari AC. Graphene photonics and optoelectronics. Nat Photonics 2010;4(9):611–22.

[142] Poot M, van der Zant HSJ. Nanomechanical properties of few-layer graphene membranes. Appl Phys Lett 2008;92(6):063111. (3 pages).

[143] Frank IW, Tanenbaum DM, van der Zande AM, McEuen PL. Mechanical properties of suspended graphene sheets. J Vacuum Sci Technol B 2007;25(6):2558–61.

[144] Liu F, Ming P, Li J. *Ab initio* calculation of ideal strength and phonon instability of graphene under tension. Phys Rev B 2007;76:064120. (7 pages).

[145] Khare R, Mielke SL, Paci JT, Ballarini R, Schatz GC, Belytschko T. Coupled quantum mechanical/molecular mechanical modeling of the fracture of defective carbon nanotubes and graphene sheets. Phys Rev B 2007;75:075412. (12 pages).

[146] Bunch JS, Verbridge SS, Alden JS, van der Zande AM, Parpia JM, Craighead HG. Impermeable atomic membranes from graphene sheets. et al. Nano Lett 2008;8(8):2458–62.

[147] Bunch JS. Mechanical and electrical properties of graphene sheets. PhD Thesis, Cornell University, USA; 2008.

[148] Faccio R, Denis PA, Pardo H, Goyenola C, Mombru AW. Mechanical properties of graphene nanoribbons. J Phys Condens Matter 2009;21:285304. (7 pages).

[149] Shokrieh M, Rafiee R. Prediction of Young's modulus of graphene sheets and carbon nanotubes using nanoscale continuum mechanics approach. Mater Des 2010;31:790–5.

[150] Ferralis N. Probing mechanical properties of graphene with Raman spectroscopy. J Mater Sci 2010;45:5135–49.

[151] Min K, Aluru NR. Mechanical properties of graphene under shear deformation. Appl Phys Lett 2011;98:013113. (3 pages).

[152] Ranjbartoreh AR, Wang B, Shen X, Wang G. Advanced mechanical properties of graphene paper. J Appl Phys 2011;109:014306. (6 pages).

[153] Scharfenberg S, Rocklin DZ, Chialvo C, Weaver RL, Goldbart PM, Mason N. Probing the mechanical properties of graphene using a corrugated elastic substrate. Appl Phys Lett 2011;98:091908. (3 pages).

[154] Lu J, Huang R. Nonlinear mechanical properties of graphene nanoribbons. Mater Res Soc Symp Proc 2011;1284. 7 pages.

[155] Lau CN, Bao W, Velasco J Jr. Properties of suspended graphene membranes. Mater Today 2012;15(6):238–45.

[156] Zheng Y, Xu L, Fan Z, Wei N, Huang Z. A molecular dynamics investigation of the mechanical properties of graphene nanochains. J Mater Chem 2012;22:9798–805.

Chapter 9

Organometallic Chemistry of Carbon Nanotubes and Graphene

Santanu Sarkar, Elena Bekyarova, Robert C. Haddon
Departments of Chemistry and Chemical & Environmental Engineering, Center for Nanoscale Science and Engineering, University of California, Riverside, CA-92521-0403, USA

9.1 INTRODUCTION

The advent of the fullerenes, carbon nanotubes (CNTs) and graphene represents a new generation of carbon allotropes that have led to a tremendous upsurge in research activity on conjugated carbon nanomaterials. While this has resulted in a great deal of important science with the potential for far-reaching technological developments, there is still a need to develop chemical routes to controllably modify the electronic structure of these materials. Chemical modification of these carbon nanostructures in a rational and predictive manner has established itself as a promising route to producing solution-processable carbon materials and to engineer their structure and properties for a wide range of applications, including printable electronics, energy conversion (fuel cells, solar cells) and energy storage (batteries, supercapacitors) [1–3]. While conventional addition chemistry in which the sp^2-conjugated carbon atoms are rehybridized to sp^3 has been widely explored in the new carbon allotropes [4], the participation of graphene and CNTs in organometallic chemical reactions has received limited attention. The metal complexes of these carbon materials are potential candidates as reusable solid catalysts for organic synthetic applications [5], in organometallic catalysis as electronically conjugated catalyst supports [6], molecular wires [7] and new electronic materials [8,9]. Fullerenes [10–13] and CNTs [14] are curved carbon materials with demonstrated ability to serve as primary ligands, whereas graphene represents a new class of periodic, planar two-dimensional π-ligands which has been recently reported to have a rich organometallic chemistry [6], in analogy with the coordination chemistry of polyaromatic hydrocarbons (PAHs) [15]. The flat two-dimensional π-surface of graphene exhibits a unique chemical reactivity as a result of the electronic structure at the Dirac point and this provides the

Carbon Nanotubes and Graphene. DOI: 10.1016/B978-0-08-098232-8.00009-7

opportunity to perform a wide range of chemical reactions [6,9,16,17]. Functionalization of CNTs with transition metal complexes alters their electronic and chemical properties and single-walled carbon nanotubes (SWCNTs) have been already used as transition metal supports in heterogeneous [18–20] and homogeneous catalysis [5].

In this chapter, we review in parallel the organometallic chemistry of fullerenes [13], CNTs [6,14,21–24] and graphene [6,8]. Our main focus is on the organometallic chemistry of CNTs and graphene, while the fullerene chemistry provides a well-understood point of comparison.

9.2 REACTIONS OF CARBON NANOTUBES AND GRAPHENE

The chemistry of graphene [2,3,16,25–27] has developed in an analogous manner to that of CNTs (SWCNTs) [1,28]. We have distinguished two limiting cases for the interaction between the addend and the graphitic surface, which are discussed below as 9.2.1 and 9.2.2 [9,24].

9.2.1 Destructive hybridization: covalent bond formation involving the creation of sp^3-hybridized carbon atoms in the graphitic surface

Destructive hybridization accompanies the standard addition reactions (Figure 9.1(a)) which typically lead to the formation of covalent C−X (X = H, C, N, O, F, Cl, Br) bonds to the basal plane carbon atoms with concomitant conversion of the carbon atoms at the sites of attachment from sp^2 to sp^3 hybridization and the removal of these atoms from conjugation [9,24]. This type of chemistry has been widely used as a means for bandgap engineering in graphene [29,30], and examples include hydrogenation [31,32], halogenation [29,33], oxidation (ozonolysis) [34], carboxylation [35], hydroxylation and epoxidation [36–39], radical additions [2,4,16,30,40–47], carbene addition [48–50], nitrene addition [51,52], Diels–Alder reactions [17,53] including benzyne addition [54,55] and 1,3-dipolar cycloaddition [56] reactions. The graphene derivatives isolated to date, when incorporated in electronic field effect devices, show low conductivity and significantly reduced carrier mobility.

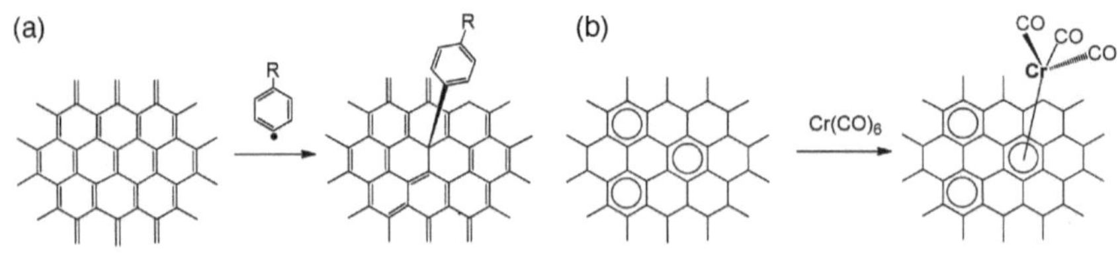

FIGURE 9.1 Schematics illustrating: (a) destructive hybridization: addition of nitrophenyl radicals to graphene and (b) constructive hybridization: organometallic complexation reactions of graphene.

9.2.2 Constructive hybridization: metal complexation

This type of chemistry involves the complexation of the benzenoid rings of CNTs and graphene by a transition metal (Figure 9.1(b)), in which the degree of rehybridization at the site of complexation is insufficient to saturate the conjugated electronic structure. In contrast to destructive hybridization, the organometallic hexahapto (η^6) functionalization of CNTs [22–24], graphene [6,8] and other graphitic materials [57] allow the constructive rehybridization of the π-system of the carbon materials with the vacant d_π-orbitals of appropriate transition metals.

The organometallic hexahapto functionalization preserves the conjugation of these extended periodic π-electron systems and the functionalized carbon atoms remain a part of the electronic band structure. In this chapter, we discuss the organometallic reactivity of fullerenes, CNTs and graphene and point out the differences and similarities in the reactivity pattern based on their molecular geometries and electronic structures.

9.3 ORGANOMETALLIC CHEMISTRY OF CARBON NANOTUBES AND GRAPHENE

Organometallic chemistry of carbon materials has received theoretical attention [58–63], but the synthetic difficulties associated with finding suitable reaction conditions for the formation of metal complexes together with the characterization of the products have been the main impediments to progress in the field [60,64]. In particular, carbon-based periodic π-electron systems, which include CNTs, graphite and graphene, have condensed benzenoid ring systems that are available for exploring the organometallic chemistry of a wide-range of metals.

9.3.1 Reactivity and bonding in organometallic complexes

The interaction between metal atoms (M) and graphitic surfaces can be categorized into four distinct bonding configurations. (1) *Weak physisorption*, which leads to little charge transfer or rehybridization of the graphitic carbon atoms. (2) *Ionic chemisorption*, which leads to charge transfer to the graphitic structure and preservation of the conjugation and band structure. Ionic chemisorption can lead to the formation of salts, in which a large amount of charge is transferred to the graphitic structure and such doped materials are often associated with enhanced conductivities and even metallic and superconducting properties [65–67]. (3) *Covalent chemisorption*, which leads to significant rehybridization of the graphitic band structure and some degree of charge transfer. This type of covalent chemisorption on graphitic structures can lead to a drastic modification of the bonding and a number of metals are associated with the formation of carbides. Although not involving metals, covalent chemisorption is also associated with the destructive hybridization discussed above in relationship to the formation of rehybridized sp^3 carbon atoms in the graphitic lattice.

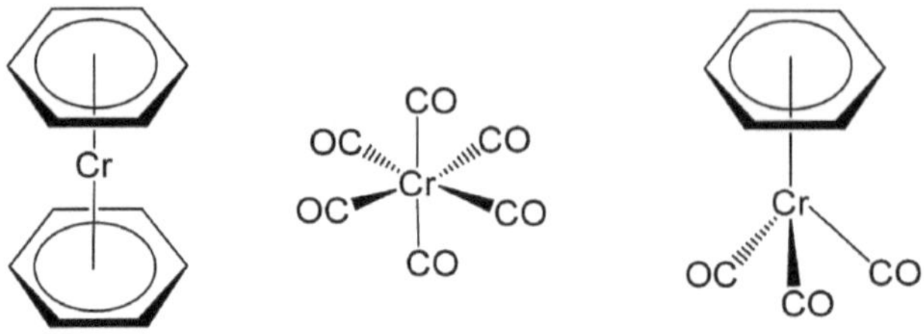

1 2 3

FIGURE 9.2 Structures of (η^6-C_6H_6)$_2$Cr (**1**), $Cr(CO)_6$ (**2**) and (η^6-C_6H_6)$Cr(CO)_3$ (**3**).

(4) *Organometallic hexahapto-complexation*, a new mode of covalent chemisorption on graphitic surfaces, which is based on organometallic chemistry and makes use of the hexahapto-metal bond to electronically conjugate adjacent carbon surfaces, which contain the benzenoid ring system [6,9,22–24,57].

Electronic coupling of graphitic surfaces, comprised of polycyclic benzenoid ring systems, is best exemplified by the well-known molecular complex, bis(benzene)chromium [(η^6-C_6H_6)$_2$Cr] (structure **1**, in Figure 9.2), which is the quintessential case of a bis-hexahapto–metal bonded system. Based on the gas-phase structure of this compound [68], the pyramidalization angle [69] is calculated to be $\theta_P = 1.7°$, in the sense that the hydrogen atoms tilt towards the metal atom, compared to the normal tetrahedral angle of $\theta_P = 19.5°$ for sp^3-hybridized carbon. In the highly condensed ring system of graphene, with a rigid network of benzene rings, the degree of pyramidalization given above represents an upper bound and thus there will be very little geometric distortion on metal complexation of the graphitic benzenoid ring systems. Nevertheless, these bonds are strong [70], allow the metal d-orbitals to couple to the π-systems while preserving the band structure, and are capable of electrically interconnecting graphitic surfaces [6,22–24,57]. In addition to chromium, there are many metals with the ability to form such bonds and are, therefore, candidates for this type of chemistry [71].

9.3.2 General approach to synthesis of organometallic complexes

The organometallic complexes of CNTs and graphene can be prepared by employing techniques used for the synthesis of small-molecule arene–metal complexes, which have been reviewed in detail by Kundig [72]. In subsequent sections, we discuss the synthesis of (arene)metal carbonyls and bis(arene)metal complexes, where chromium metal is taken as the central metal atom and the arene = benzene, higher polycyclic aromatics, CNTs or graphene.

9.3.2.1 *Method A*

(Arene)$Cr(CO)_3$ complexes are synthesized by thermolysis of $Cr(CO)_6$ under an inert atmosphere in the presence of an excess of the arene [6,72]. The reaction requires refluxing the mixture in a high-boiling-point solvent [72]. The solvent can be dibutylether/THF, 1,2-dimethoxyethane (DME), diglyme/THF, heptane/diglyme, α-picoline, decalin, decalin/ethyl formate or decalin/butyl acetate. The polar ether or ester additives (or solvents) promote carbonyl dissociation, stabilize intermediates and the vigorous reflux of lower boiling additives (such as THF) washes the sublimed $Cr(CO)_6$ from the reflux condenser back into the reaction mixture. Prior to heating and mixing, solvents are usually degassed by several freeze/pump/thaw cycles or by bubbling inert gas through the solvent. A wide range of complexes of small aromatic molecules with useful functionalities have been prepared by a combination of dibutyl ether/THF (9:1) in good yield with reaction times typically 1–4 days. Higher temperature shortens the reaction time, but increases the risk of decomposition of the products [72].

9.3.2.2 *Method B*

This complexation method involves a procedure similar to *Method A*, but naphthalene (~0.25 equivalents) is added as an additional ligand in the reaction mixture (consisting of an arene, chromium hexacarbonyl in Bu_2O/THF)[73], which allows the reaction to occur at a lower temperature. This is due to the in-situ formation of (naphthalene)$Cr(CO)_3$ complex, which is labile as a result of haptotropic slippage of the naphthalene ligand (change of hapticity from η^6- to η^4- or η^2-coordination), thus facilitating its dissociation and coordination of the new arene (facile arene exchange reaction) [8,74,75].

Such distinctly different property of naphthalene in facile arene exchange capability has been widely explored for synthesis of (η^6-arene)Cr(η^6-naphthalene), (μ,η^6,η^6-naphthalene)-bis(η^6-benzene)dichromium as well as poly[(μ-η^6,η^6-naphthalene)chromium] compounds [76] under very mild reaction conditions, in which the ligand exchange reaction of bis(η^6-naphthalene)Cr [74] with fluorobenzene, benzene, toluene, mesitylene and hexamethylbenzene in THF has been employed [77].

9.3.2.3 *Method C*

Complexes of condensed aromatics are reported to be unstable towards polar solvents (THF, DMSO and acetone) and their synthesis requires special attention [72], or use of more labile $Cr(CO)_3L_3$ ($L = CH_3CN$, NH_3, pyridine) precursors, which allow the formation of (arene)$Cr(CO)_3$ complexes at a much lower temperature [8,72]. Complexation reaction of graphene and $Cr(CO)_3(CH_3CN)_3$ has afforded (η^6-graphene)$Cr(CO)_3$ complexes at a temperature as low as 50 °C[8]. Room-temperature complexation of arene is accomplished by the reaction of $Cr(CO)_3(NH_3)_3$ with $BF_3.OEt_2$ in the presence of an arene [78,79].

9.3.2.4 *Method D*

Timms first demonstrated the synthesis of organometallic complexes of transition metals using metal vapour synthesis (MVS) in 1969 [80,81], and the process has been used to synthesize a variety of compounds that incorporate metal–ligand bonds [77]. The electron beam evaporation technique was used as a source of metal vapours in 1973 [82], and the technique was shown to yield bis(arene)molybdenum complexes by condensation of molybdenum vapour with benzene, toluene or mesitylene at 77 K [82], whereas the reaction of titanium vapour with benzene afforded extremely air-sensitive $(C_6H_6)_2Ti$ [83]. The MVS technique is generally employed for the synthesis of bis(arene)–metal or related (arene)–metal–(arene) oligomeric complexes, and this method allowed the synthesis of a novel triple-decker sandwich complex: $(\eta^6\text{-mesitylene})_2(\mu\text{-}\eta^6\text{:}\eta^6\text{-mesitylene})Cr_2$[84].

We have employed a derivative of this method in which bis-hexahapto–metal complexes of carbon materials are synthesized by the controlled e-beam evaporation of metal atoms onto thin films of carbon materials at room temperature with in situ measurement of properties, such as the conductivity in a cryogenically pumped high-vacuum chamber [22–24].

In this chapter, we focus on the use of solution chemistry (*Methods A, B,* and *C*) [8] and gas-phase chemistry (*Method D*) [22–24] for the preparation of a series of organometallic derivatives of graphene, graphite and SWCNTs [6,22–24,57].

9.4 ORGANOMETALLIC CHEMISTRY OF CARBON NANOTUBES

The coordination chemistry of SWCNTs, where the nanotube itself can be considered as a primary ligand, is an unexplored area and the reactivity of SWCNTs to metal coordination compounds still remains to be understood. In subsequent sections, we describe various types of bonding of metal atoms to CNT surfaces.

In general, coordination chemistry of SWCNTs (Figure 9.3) can be categorized into the following four subcategories: (i) bihapto (η^2-)–transition metal complexes of pristine SWCNTs, (ii) mono-hexahapto (η^6-)–transition metal complexes of SWCNTs via solution chemistry approach, (iii) bis-hexahapto (η^6)–transition metal complexes of SWCNTs via MVS and (iv) coordination chemistry of oxidized SWCNTs side chains.

9.4.1 Bihapto (η^2-)–transition metal complexes of pristine SWCNTs

CNTs are excellent π-ligands [9], in which the carbon π-orbitals hybridize with the d-orbitals of the transition metal with subsequent modification of the chemical, electronic and magnetic properties of SWCNTs.

SWCNTs undergo functionalization with the transition metal complex, *trans*-$IrCl(CO)(PPh_3)_2$ (Vaska's compound) [14], by η^2-coordination of the metal fragment to the incipient double bonds present in the curved graphene surface of the SWCNTs, which is similar to the behaviour observed in the fullerene

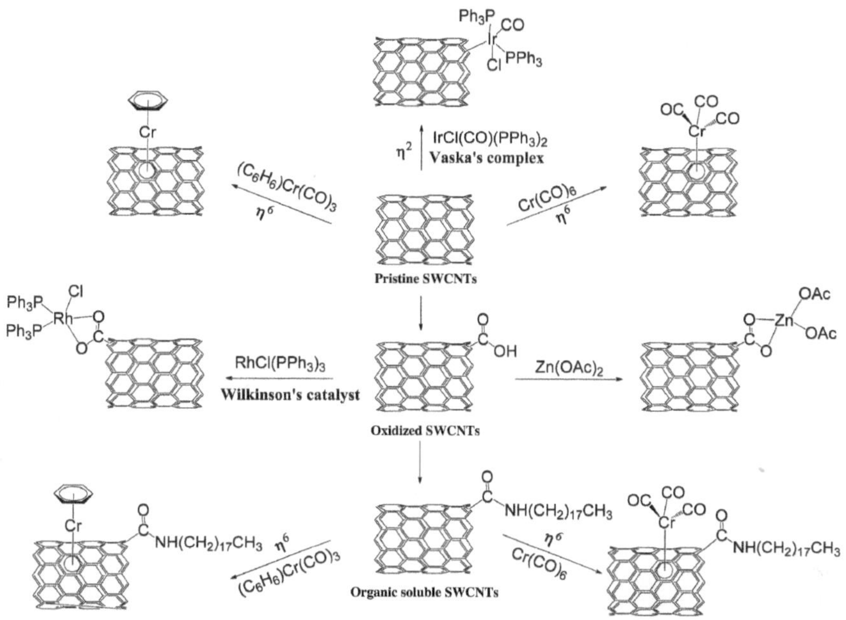

FIGURE 9.3 Organometallic complexation reactions of SWCNTs as a function of the chemical nature of the SWCNTs: pristine (SWCNT), acid-oxidized (SWCNT-COOH) and organic-soluble amide [SWCNT-CONH(CH_2)$_{17}$$CH_3$] forms.

adducts [10]. The metal complexes of CNTs were characterized by ^{31}P-NMR spectroscopy, FT-IR spectroscopy, as well as transmission electron microscopy (TEM) and EDX. The TEM results showed that complexation to Vaska's compound did not damage the CNTs. This study showed that the insertion of metal moieties between bundles of nanotubes inhibits the van der Waals interactions and promotes further intercalation of the nanotube bundles to produce smaller bundles and even individual tubes.

9.4.2 Mono-hexahapto (η^6)–transition metal complexes of SWCNTs via the solution chemistry approach

Transition metal carbonyls were known for many years as important precursors to new organotransition metal chemistry. The occurrence of strong vibrational bands in the IR spectrum, due to the metal-coordinated carbonyl group ($\nu_{C=O}$) in the region ~2050 cm^{-1}, has been a spectral signature of transition-metal carbonyl derivatives [85]. Density-functional calculations have been used to study the interaction of –$Cr(CO)_3$ moieties with a series of semiconducting (SC) nanotubes [58–60]; it was suggested that the chromium fragments coordinate to the walls of the nanotubes and that CNTs with narrower diameters are more reactive than the nanotubes with larger diameters [86].

A SWCNT–chromium complex was studied as a catalyst for the activation of HiPco nanotubes towards Diels–Alder cycloadditions [87]; the Diels–Alder process was found to be enhanced by the electron-donating ability of the diene and by the enhanced electrophilicity of the CNT due to its complexation to chromium [6,8].

The first organometallic complexation chemistry of graphene, graphite and chemically functionalized, soluble CNTs (SWCNT–CONH$(CH_2)_{17}CH_3$) was performed using $Cr(CO)_6$ and (η^6-benzene)$Cr(CO)_3$ [6,8,22]. The compounds were characterized by a combination of analytical techniques – thermogravimetric analysis (TGA); X-ray photoelectron spectroscopy (XPS); scanning electron microscopy (SEM); attenuated total reflectance-infrared (ATR-IR); ultraviolet, visible, near infrared, far infrared (UV-vis-NIR-FIR) and Raman spectroscopy, in order to demonstrate the coordination of the chromium to the sidewalls of the SWCNTs. The chromium complexes of octadecylamine endgroup-functionalized SWCNTs showed high levels of chromium incorporation as a result of the higher dispersibility of these nanotubes in organic solvents, which provides efficient exposure of the exohedral and endohedral SWCNT surface to the chromium reagents. In addition to the simple hexahapto complexation reactions, the process is accompanied by formation of metal clusters on CNT surface (as observed from HR-TEM images), where the graphitic surface of the SWCNT acts as a template for cluster growth.

After complexation of the arene moieties by chromium, the optical absorption spectrum shows the suppression of the interband electronic transitions, which is related to the modification of the π-electron systems of the CNTs. The degree of the suppression of the SWCNT electronic band transition during hexahapto-coordination chemistry is not as strong as in case of the covalent sidewall functionalization chemistry in which carbon–carbon double bonds are converted from sp^2 to sp^3 [49].

9.4.3 Bis-hexahapto (η^6)–transition metal complexes of SWCNTs via metal vapour synthesis

The bis-(η^6)-hexahapto mode of bonding between the graphitic surfaces of adjacent SWCNTs with transition metals has been demonstrated by employing *Method D*, as discussed in Sec. 9.3.2.4. In this situation the transition metal (M) acts as a conduit for electron transport between the SWCNT surfaces in the resulting (η^6-SWCNT)M(η^6-SWCNT) interconnects [23,24]. The metal was deposited on the SWCNT films by controlled electron-beam evaporation of a very small amount of the transition metals on thin films of the SWCNTs (thickness, $t = 8$ nm). The success of this process depends on the high mobility of the transition metals on graphitic surfaces, which are able to diffuse over the benzenoid ring system until they become localized at a SWCNT–SWCNT junction where the geometry of the contact supports the formation of the stable bis-hexahapto bonds [9,22–24].

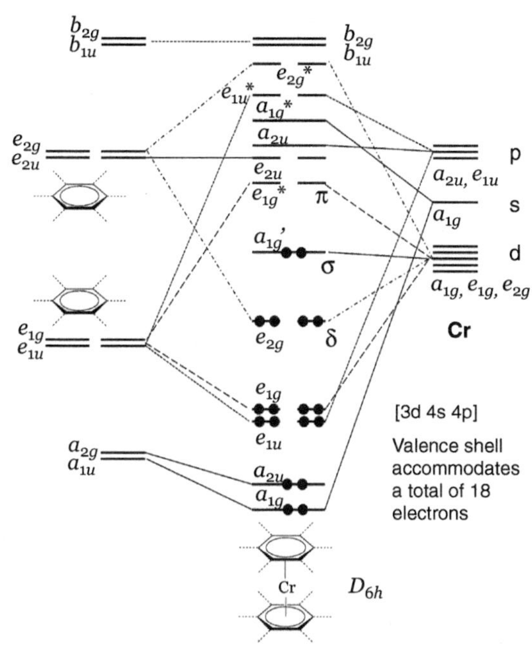

FIGURE 9.4 Orbital interaction diagram for bis(benzene)chromium, [(η^6-C_6H_6)$_2$Cr].

The bis-hexahapto mode of bonding in CNTs and graphene can be understood within the conventional orbital interaction diagram for bis(benzene) chromium [(η^6-C_6H_6)$_2$Cr] (Figure 9.4). All of the extended periodic π-electron graphitic structures are narrow or zero bandgap materials, and thus the electron-donor and electron-acceptor interactions between the HOMOs and LUMOs of the π-systems, and the d-orbitals of the transition metals will be enhanced by the high-lying HOMO and low-lying LUMO of the graphitic surfaces [9,17].

The e_{1g} and e_{2u} benzene π-orbitals (Figure 9.4), which hybridize with the metal d-orbitals, are strongly involved in the construction of the hexahapto-metal-bonds in (η^6-C_6H_6)$_2$Cr, and are available at the Dirac point in graphene [9,17]. Thus, the electronic structure of the graphitic π-electron systems is ideally suited for the realization of organometallic chemistry [6,8,9].

9.4.3.1 Transport Properties of SWCNT Thin Films: Atomic Contacts and Interconnects via the Formation of Bis-Hexahapto–Metal Bonds

The hexahapto–metal bond in bis-SWCNT–metal complexes functions as an electrical interconnect between the side-wall benzene rings of adjacent SWCNTs [22–24].

Deposition of metals on electric arc (EA) produced SWCNT films, both in the form of the usual statistical 1:2 mixture of metallic and semiconducting (SC)

SWCNTs, and as separated SC-SWCNTs, was performed by high-vacuum e-beam evaporation [22–24]. In order to differentiate between covalent organometallic chemistry and the charge transfer effects implicit in ionic chemisorption, we focus on the results obtained from thin films (thickness, $t \approx 8$ nm) of SC-SWCNTs [24].

In order to represent the behaviours introduced in Sec. 9.3.1, we investigated three reactions of the SWCNTs with metals (M): (a) weak physisorption (M = Au), (b) strong physisoption (charge transfer, M = Li) and (c) chemisorption, with hexahapto complexation (M = Cr). The deposition of these three metals brings about disparate effects in the conduction behaviour of SC-SWCNT films (Figure 9.5). The response of the SWCNT film to gold deposition is very small, despite the fact that the bulk electrical conductivity of gold is about six times that of Cr and Li. The conductance increased by a factor of over 100 on Li deposition and the conductivity of the SC-SWCNT film increased monotonically, before reaching a maximum conductivity at about $t \approx 1.2$ nm. The overall response of the SWCNT film to the deposition of Cr was not as pronounced as in the case of Li; however, the variation in the conductivity of the SC-SWCNT film showed a different dependence on the amount of metal deposited. The amount of Cr needed to achieve its maximum effect on the conductivity of the SC-SWCNT film was extremely small, and for a Cr thickness (t), $t < 0.05$ nm, the conductivity of the SC-SWCNT film increased by more than an order of magnitude, whereas for $t = 0.05$ nm Li, there was a relatively small change in the SC-SWCNT film conductivity.

The conductivity increase observed in the case of Au deposition on the SC-SWCNT film can be explained by the parallel conductance of a weakly absorbed, continuous Au film, whereas the Cr and Li depositions involve more complex phenomena. The compositions (MC_x), at the points where the films attained their maximum conductance, were determined to be: $t_M = 1.2$ nm for Li, giving a value of $x \approx 8$ (LiC_8), whereas for Cr the upper limit was reached at $t_M < 0.05$ nm, which gives $x > 180$. The increase in the conductivity of SWCNTs on exposure to alkali metals has been attributed to charge transfer into the conduction band of the SWCNTs to give a composition KC_8, which is reminiscent of the composition of graphite intercalation compounds [65,66].

The SC-SWCNT films consist of a network of individual SC-SWCNTs and SWCNT bundles, and the transport characteristics of the films are dominated by the resistance of the individual SC-SWCNTs, but more importantly, by the highly resistive nature of the inter-CNT junctions [88].

The enhanced film conductivity in case of Cr deposition is ascribed to the reduction of the inter-nanotube resistance due to the formation of a small number of highly conductive bis-hexahapto–metal bonds that serve to bridge individual SC-SWCNTs, and thereby reduce the resistance between individual SWCNTs and bundles. The chromium atoms are highly mobile [89,90]; and thus, they are able to diffuse along the graphitic surfaces until they encounter a SWCNT–SWCNT junction or intra-bundle contact with a geometry that allows

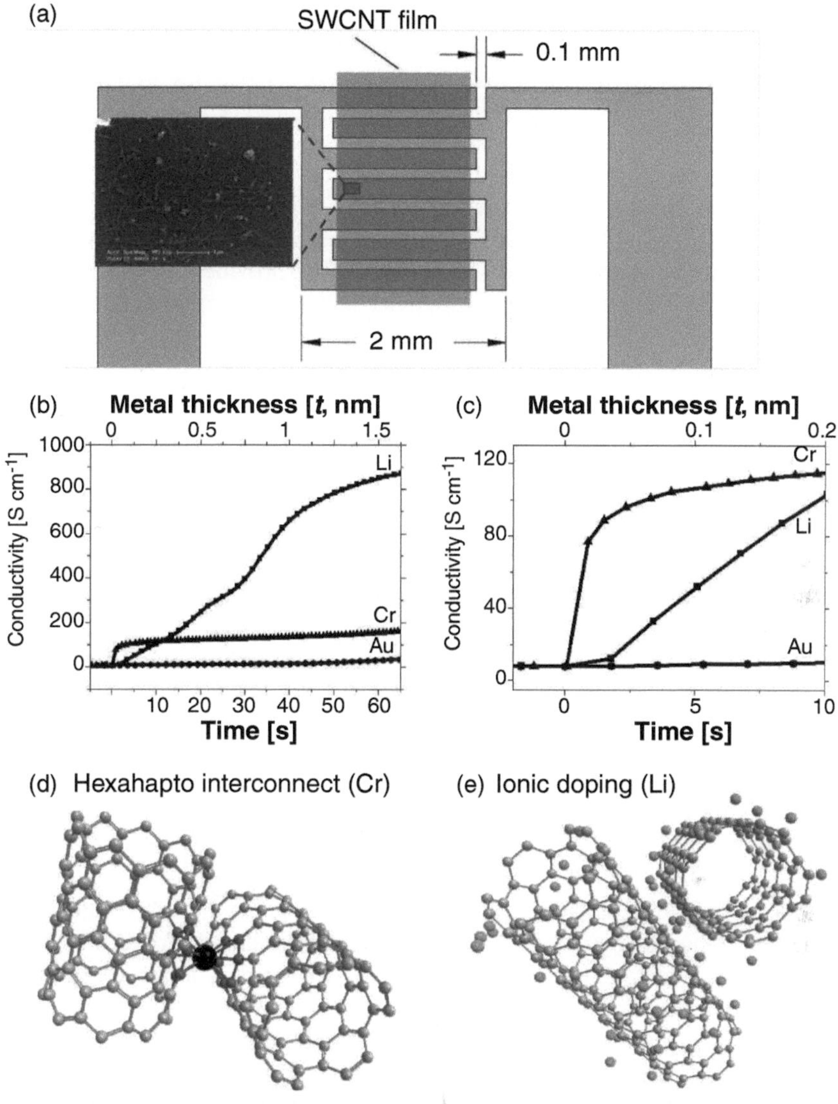

FIGURE 9.5 (a) SC-SWCNT thin film on interdigitated electrodes, together with SEM image. (b, c) Conductivity of SWCNT films as a function of Li, Cr and Au deposition. (d) Chromium atoms interconnecting adjacent nanotubes by formation of a bis-hexahapto [(η^6-SWCNT)Cr(η^6-SWCNT)] linkage, thereby reducing inter-nanotube junction resistance. (e) Electron transfer process (doping), in which lithium atoms donate electrons to the conduction bands of the SWCNTs. Adapted with permission from Ref. [24]. Copyright 2012 American Institute of Physics.

formation of a bis-hexahapto–metal bond (Figure 9.5(d)). The formation of a (η^6-SWCNT)Cr(η^6-SWCNT) bond at a junction is kinetically favourable because the van der Waals gap of 3.15 Å within SWCNT bundles is comparable to the separation of 3.23 Å between the benzene rings in $(\eta^6\text{-}C_6H_6)_2Cr$ (structure **1**) [68]. The chromium atom possesses six valence shell electrons and by covalent coordination of two η^6-benzenoid ligands, shares a total of 18-electrons which fills the 3d 4s 4p first row transition metal valence shell and is known to lead to a stable electronic structure. Due to its filled outer d-orbital, gold is unable to participate in hexahapto complexation and cannot provide a conducting pathway at the CNT junctions; the conductivity increase during gold deposition is ascribed to the parallel conductance of a continuous gold metal film formed over the SWCNTs.

The response of the conductivities of the SC-SWCNT films to the deposition of first row transition metals (M = Ti, V, Cr, Mn and Fe) immediately adjacent to Cr in the periodic table indicates that these metals spontaneously form $(\eta^6\text{-}C_6H_6)_2M$ complexes as in low-temperature MVS experiments [71]; they exhibit the same sharp increase in conductivity that was found in the case of chromium. The effect of metal atom deposition on the conductivity of the SWCNT thin films decreases in the order: Cr > V, Mn > Fe, Ti [24].

We conclude that covalent chemisorption (Ti, V, Cr, Mn and Fe), in which there is constructive rehybridization, preserves the graphitic band structure and leads to the formation of conductive CNT interconnects via bis-hexahapto bond formation and it is apparent that this new mode of bonding to graphitic surfaces may provide a powerful approach for the fabrication of new electronic materials of increased dimensionality.

9.4.4 Coordination chemistry of oxidized SWCNTs side chains

The ligands of an inorganic complex are expected to exfoliate CNT bundles in the same manner as polymers [91–93], and it has been demonstrated that the covalent functionalization of oxidized SWCNTs with transition metal complexes induces solubility in organic solvents [14]. Improved solubility, charge transfer and catalytic activity were noted in SWCNTs functionalized with iridium and rhodium complexes. Direct reaction of carboxylic acid functionalized SWCNTs with *trans*-$IrCl(CO)(PPh_3)_2$ (Vaska's compound) [14] and $RhCl(PPh_3)_3$ (Wilkinson's catalyst) [5] offers functionalized SWCNTs, in which the rhodium and iridium centres are coordinated to the nanotube through the oxygen atoms coating the side walls and form a hexacoordinate structure around the metal atom (Figure 9.3). Oxidative additions to form 18-electron compounds are well known for d^8 complexes. The functionalized SWCNT–Wilkinson's complex adducts were found to be a good catalyst for the hydrogenation reaction of cyclohexene to cyclohexane at room temperature [5,14].

Transition metal complexes based on the coordination of zinc to the surface of SWCNTs have been prepared and characterized by TEM, IR spectroscopy,

XPS, elemental mapping and linear profile analysis [94]. The zinc–SWCNTs complexes are poorly soluble in water and other organic solvents, but they do undergo ligand exchange.

9.5 ORGANOMETALLIC CHEMISTRY OF GRAPHENE

The benzenoid π-surface of graphene together with its unique Fermi level electronic structure suggests that chemical reactions involving bonding configurations, such as the hexahapto (η^6-) metal complexation might be possible [6]. In hexahapto–metal complexation reactions, the vacant d_π-orbitals of a transition metal (M) are expected to constructively overlap with the occupied π-orbitals of graphene, resulting in the coordination of the metal atoms (M) to the individual benzenoid aromatic rings of graphene in (η^6-graphene)−M bonding configuration [6,8]. Such η^6-metal functionalization of graphene systems leads to very little structural reorganization of the π-system and a recent study on exfoliated graphene (XG) has demonstrated that such mono-hexahapto complexation of single-layer graphene (SLG) leads to graphene field-effect transistor (FET) devices which retain high electronic mobility with enhanced on/off ratio [8].

9.5.1 Comparison of the hexahapto complexation ability of fullerenes and graphene

Compared to the fullerene molecules, which are curved in two dimensions, graphene has a flat surface and this makes it a suitable ligand for hexahapto (η^6)-complexation reactions [6,22]. The coordination chemistry of the fullerenes is dominated by mono (η^1)- and bihapto (η^2)- complexation [10]. It has been shown that the curvature in C_{60} significantly inhibits the potential of the molecule to function as a ligand in pentahapto (η^5)- and hexahapto (η^6)-complexation reactions because the fullerene π-orbitals are directed away from the metal as a result of the rehybridization of the ring carbon atoms (Figure 9.6(a)) [95,96]. Nevertheless, the curvature can be ameliorated by functionalization [96–98], and this has allowed the preparation of organometallic fullerene derivatives [97,99]. In C_{60} the π-orbital axis vectors are directed away from the centre of the respective rings and make angles of 16.3° (POAV2) to a normal to the plane of the five-membered rings and 25.8° (POAV2) to a normal to the plane of the six-membered rings; hence, hexahapto (η^6)- coordination is even more strongly disfavoured than pentahapto (η^5)-complexation [96].

9.5.2 Synthesis of organometallic complexes of graphene

Graphene–metal complexes have been synthesized by *Methods A, B* and *C* as described in Sec. 9.3.2. A typical sample of solvent exfoliated graphene XG consists of a mixture of multi-layer-graphene flakes (micrometre dimensions), together with SLG. After reaction of XG with 0.13 equivalents of

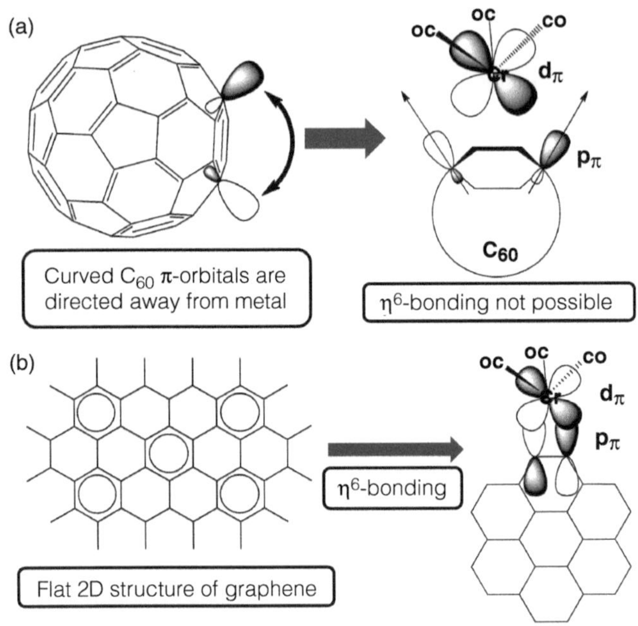

FIGURE 9.6 Comparison of the orientation of (a) fullerene π-orbitals and (b) graphene π-orbitals. The π-orbital axis vectors are shown in the case of C_{60}, and it is apparent that the curvature and rehybridization present in the fullerene structure severely inhibit its ability to function as a hexahapto ligand [96], while the flat two-dimensional structure of graphene makes it an ideal hexahapto ligand [6,8,9].

$Cr(CO)_6$ (Figure 9.7(a)), the IR spectrum showed C–O stretching vibrations at 1939 cm^{-1} (Figure 9.8(c)) and the product was assigned as (η^6-graphene) $Cr(CO)_3$ [6]. After the reaction of XG with 0.02 equivalents of $Cr(CO)_6$ (Figure 9.7(c)), the C–O vibrations could not be observed in the product, but the Cr2p XPS spectrum shows the presence of Cr(0) and we assign the structure of the product as (η^6-graphene)$_2$Cr. The observation of varying product structures as a function of the reagent stoichiometry is consistent with results reported for molecular chromium complexes [84,100]. Such (η^6-graphene)$_2$Cr complexes offer the possibility of three-dimensional interconnection between adjacent graphene sheets, providing the opportunity to extend the electronic structure of the two-dimensional graphene sheets into three dimensions without creating pyramidalized sp^3 carbon centres [2,57].

The ATR-IR spectrum of the reaction product of XG (or HOPG) and (η^6-benzene)$Cr(CO)_3$ showed the aromatic C–H vibration of benzene at a lower frequency compared to that in the starting material [101] and does not show the C–O vibrations, indicating the formation of the (η^6-graphene)Cr(η^6-benzene) product (Figure 9.7(b)).We have also compared the organometallic complexation reactivity of graphene as a function of number of layers (n), and observed that SLG ($n = 1$) is more reactive than few-layer graphene ($n \geq 2$) and HOPG ($n = \infty$) [8].

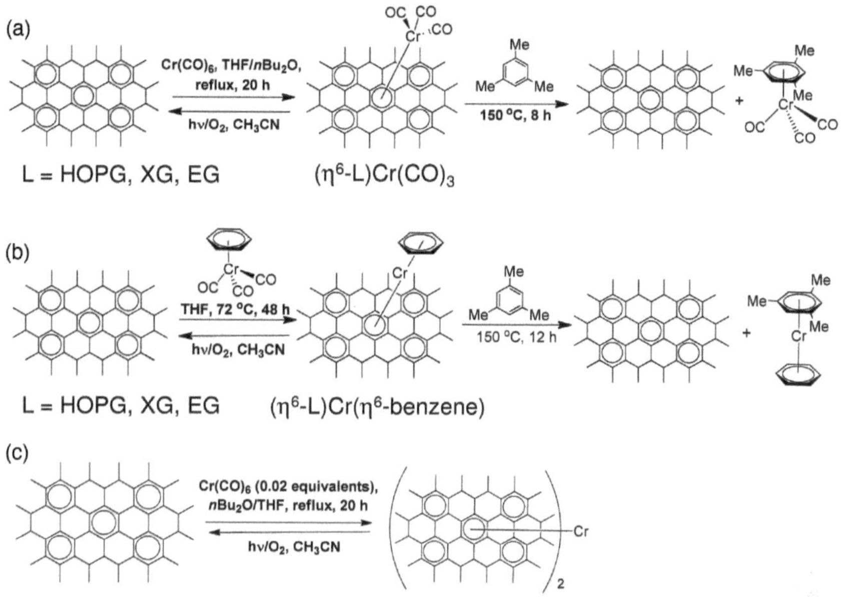

FIGURE 9.7 Metal functionalization reactions of graphene and methods for decomplexation. Reactions of graphene with (a) chromium hexacarbonyl and (b) (η^6-benzene)chromium tricarbonyl. (c) Use of excess XG_{sol} (solvent XG) with $Cr(CO)_6$ allowed the isolation of the fully graphene-coordinated material, (η^6-XG)Cr(η^6-XG), in which two graphene sheets are interconnected by zero-valent chromium metal [6].

9.5.3 Characterization of the organometallic complexes of graphene

Characterization data on the nature of the products from metal complexation reactions with the graphene surface can be obtained from infrared (ATR-IR), Raman and UV-vis-NIR spectroscopy, while quantitative analytical information on the amount of chromium present in these complexes can be obtained using TGA and XPS. Solution UV-vis spectroscopy and the electrochemistry of well-dispersed organometallic complexes of carbon materials can be useful in identifying the nature of the metal–ligand bonding [6,102].

ATR-IR spectroscopy, which relies on the C–O (or C–H) stretching frequencies of the residual ligands, is an extremely useful technique for the characterization of organometallic complexes. Coordination of electron-donating ligands, such as benzene to $Cr(CO)_3$, leads to a decrease in the C–O stretching frequencies of the remaining CO ligands when compared to the $Cr(CO)_6$ starting material. For example, coordination of benzene to –$Cr(CO)_3$ moieties decreases the C–O stretching frequency from 2000 cm^{-1} (Figure 9.8(b) in $Cr(CO)_6$) [103] to 1854 and 1954 cm^{-1} (Figure 9.8(e)). Coordination of HOPG to –$Cr(CO)_3$ moieties leads to a decrease of C–O stretching frequencies of the residual CO ligands to 1948 cm^{-1} (Figure 9.8(d)) [6], while coordination of

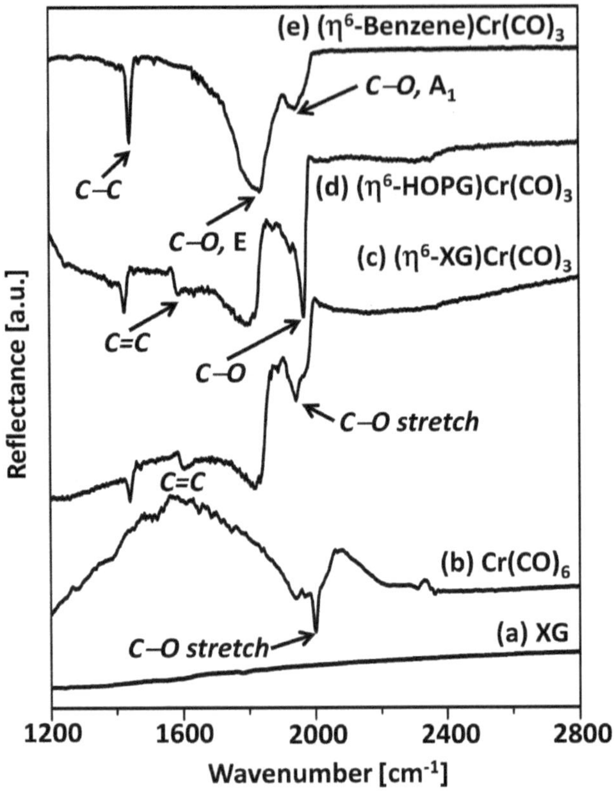

FIGURE 9.8 ATR-IR characterization of (a) XG and the organometallic complexes, (b) chromium hexacarbonyl, (c) (η^6-XG)$Cr(CO)_3$, (d) (η^6-HOPG)$Cr(CO)_3$ and (e) (η^6-benzene)$Cr(CO)_3$. (Adapted with permission from Ref. [6], copyright 2011 Royal Society of Chemistry.)

graphene (XG) and SWCNT-$CONH(CH_2)_{17}CH_3$ to –$Cr(CO)_3$ moieties leads to a decrease of C–O stretching frequencies of the residual CO ligands to 1939 (Figure 9.8(c)) and 1982 cm^{-1} [22], highlighting the trend in donating abilities of the ligands.

On the other hand, coordination of electron-withdrawing –$Cr(CO)_3$ moieties to a benzene ring (where the aromatic C–H stretch vibration occurs at 3062 cm^{-1}) increases the C–H bond strength and therefore the C–H stretching frequency (3088 and 3105 cm^{-1}), whereas the introduction of electron-donating group to benzene–Cr moieties (as in $[(\eta^6\text{-}C_6H_6)]_2Cr$) decreases the C–H bond strength and brings about a decrease in C–H stretching frequency (3044 and 3036 cm^{-1}). From these trends in C–H stretching frequencies, it is clear that electron-donating ligands serve to lower the C–H stretching frequency in the resulting (ligand)$Cr(C_6H_6)$ complex. The coordination of the HOPG ligand to the –$Cr(C_6H_6)$ moieties leads to a decrease in the C–H stretching frequencies (2969 and 3080 cm^{-1}) in (η^6-HOPG)Cr(η^6-benzene), indicating that HOPG is a superior electron donor to benzene [6]. Coordination of graphene or SWCNT ligands to the $Cr(C_6H_6)$ moieties leads to an even more pronounced decrease in C−H stretches (~2851 and 2926 cm^{-1}) indicating the better electron-donating ability of these ligands as compared to HOPG or benzene ligands; this is in agreement with our analysis of the electronic structure of these ligands in terms of the frontier molecular orbitals and energies at the Dirac point of graphene [8,9].

9.5.4 Decomplexation reactions of organometallic complexes of graphene

The ease of metal removal in the graphitic organometallic complexes is similar to that observed previously in small-molecule chemistry and the (arene) metal$(CO)_3$ complexes undergo loss of metal selectively in high yields at the end of a synthetic sequence [72,104,105]. While the arene−metal bond can survive in a large number of reaction environments, the (arene)$Cr(CO)_3$ complexes can be readily cleaved upon oxidation of the metal (with Ce(IV), Fe(III), I_2, hν/O_2). The mildest procedure is the exposure of a solution of the complex in diethylether or acetonitrile to sunlight and air for few hours [6]. In small-molecule chemistry, this method generally allows the isolation of the arene in yields that are usually >80% [72].

The process of decomplexation of graphene−metal bond can be followed by Raman spectroscopy, in which the I_D/I_G ratio quantifies the relative extent of functionalization in any covalent chemistry of graphene [6]. Applying the same chemistry to the organometallic complexes of graphene, we find that the exposure of the (η^6-graphene)Cr(η^6-graphene) complex to light for 3 hours in acetonitrile is sufficient to cleave the complex ($I_D/I_G = 1.59$ and 0.06 for the same sample before and after exposure to sunlight; Figure 9.9(a)). The decomplexation of (η^6-graphene)Cr(η^6-benzene) requires longer exposure to sunlight; after exposure for 6 hours the I_D/I_G ratio decreases from 1.40 to 0.18 (Figure 9.9(b)), suggesting that the (η^6-graphene)Cr(η^6-graphene) complex is more reactive than (η^6-graphene)Cr(η^6-benzene).

Refluxing an (arene)$Cr(CO)_3$ complex in pyridine is reported to cleave the arene−metal bond and allows recycling of the Cr(0) complex in the form of $Cr(CO)_3(Py)_3$[72], while employing the same process to (η^6-graphene) $Cr(CO)_3$ complex led to very small XG flakes, which were well dispersed in pyridine [6].

Another effective way to reverse the metal complexation involves refluxing the organometallic complexes with an electron-rich arene (such as mesitylene, anisole), which is able to replace the original ligand from the starting complex via arene exchange reaction.

For example, decomplexation of (η^6-graphene)$Cr(CO)_3$ complexes with mesitylene or anisole led to the formation of (η^6-mesitylene)$Cr(CO)_3$ (Figure 9.7(a)) [6] or (η^6-anisole)$Cr(CO)_3$ [8] complexes (detected by electrospray ionization (ESI)–mass spectrometry), with regeneration of graphene

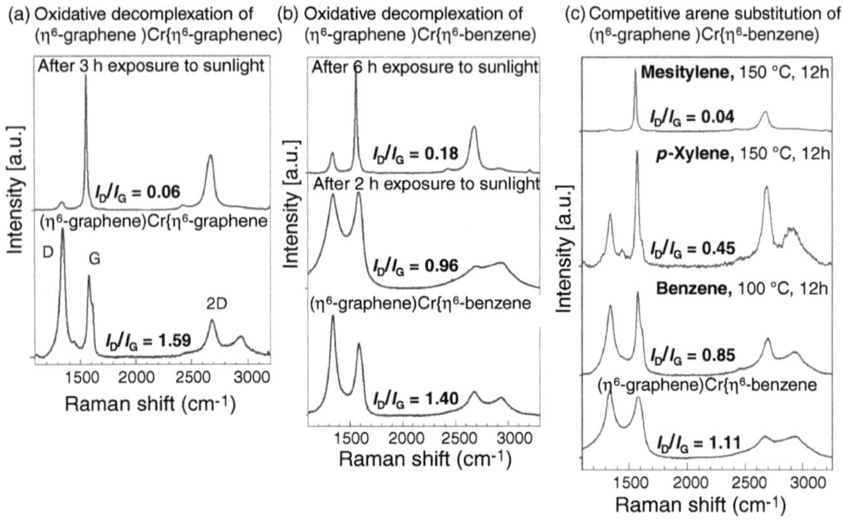

FIGURE 9.9 Raman spectroscopic monitoring of the decomplexation reactions as observed by the effect of sunlight on (a) (η^6-graphene)$_2$Cr and (b) (η^6-graphene)Cr(η^6-benzene) complexes. (c) Decomplexation of (η^6-graphene)Cr(η^6-benzene) with benzene, *p*-xylene and mesitylene via competitive arene exchange reactions. (Adapted with permission from Ref. [6], copyright 2011 Royal Society of Chemistry.)

(as seen by Raman spectroscopy). Similarly, the decomplexation of the (η^6-graphene)Cr(η^6-benzene) sandwich complex with mesitylene led to graphene and a new complex, which was tentatively assigned as (η^6-mesitylene) Cr(η^6-benzene) (Figure 9.7(b)) [6].

The strength of the (η^6-graphene)–Cr bond was investigated in a series of competition reactions with electron-rich arenes [6]. Refluxing the (η^6-graphene) Cr(η^6-benzene) complex (I_D/I_G = 1.1, Figure 9.9(c)) in benzene (oil bath temperature 100 °C; final I_D/I_G = 0.85) or *p*-xylene (oil bath temperature 150 °C; final I_D/I_G = 0.45) was insufficient to fully regenerate XG, whereas heating in mesitylene (oil bath temperature 150 °C) gave a final I_D/I_G = 0.04, indistinguishable from the starting graphene sample. Thus, it is apparent that the (η^6-graphene)–Cr bond is fairly robust [106]; the (η^6-benzene)–Cr bond energy is reported to be 164 kJ mol^{-1} [107].

9.6 CONCLUSIONS AND PERSPECTIVES

The formation of a bis-hexahapto–metal bond (constructive rehybridization) preserves the graphitic band structure, and this bis-η^6-metal functionalization scheme provides a new approach to electrically contacting and interconnecting graphitic surfaces [6]. The bis-η^6-metal functionalization chemistry leads to an enhancement in the conductivity by increasing the dimensionality of the electronic structure (in the case of CNTs) [22–24], and it is also capable of preserving high mobility in mono-hexahapto functionalized SLG devices [8].

In addition, the development of the organometallic chemistry of graphitic materials offers the potential to generate doped structures by use of transition metals which deviate from the traditional 18-electron rule [108, 109], while providing electrical interconnects that are applicable to the bridging of any surface containing the benzenoid ring system, including single-walled CNTs, graphene and other graphitic carbons. In fact the organometallic approach outlined earlier may lead to new materials and phenomena and there are potential applications in a number of fields, such as atomic spintronics and organometallic catalysis.

ACKNOWLEDGEMENTS

The authors acknowledge financial support by NSF under grant DMR-1305724 and the Defense Microelectronics Activity (DMEA) under agreement number H94003-10-2-1003.

REFERENCES

[1] Niyogi S, Hamon MA, Hu H, Zhao B, Bhowmik P, Sen R, et al. Chemistry of single-walled carbon nanotubes. Acc Chem Res 2002;35:1105–13.

[2] Bekyarova E, Sarkar S, Niyogi S, Itkis ME, Haddon RC. Advances in the chemical modification of epitaxial graphene. J Phys D Appl Phys 2012;45:154009.

[3] Sarkar S, Bekyarova E, Haddon RC. Covalent chemistry in graphene electronics. Mater Today 2012;15:276–85.

[4] Sarkar S, Bekyarova E, Haddon RC. Reversible grafting of α-naphthylmethyl radicals to epitaxial graphene. Angew Chem Int 2012;51:4901–4.

[5] Banerjee S, Wong SS. Structural characterization, optical properties, and improved solubility of carbon nanotubes functionalized with Wilkinson's catalyst. J Am Chem Soc 2002;124:8940–8.

[6] Sarkar S, Niyogi S, Bekyarova E, Haddon RC. Organometallic chemistry of extended periodic π-electron systems: hexahapto-chromium complexes of graphene and single-walled carbon nanotubes. Chem Sci 2011;2:1326–33.

[7] Jiang J, Smith JR, Luo Y, Grennberg H, Ottosson H. Multidecker bis(benzene)chromium: opportunities for design of rigid and highly flexible molecular wires. J Phys Chem C 2011;115:785–90.

[8] Sarkar S, Zhang H, Huang J-W, Wang F, Bekyarova E, Lau CN, et al. Organometallic hexahapto functionalization of single layer graphene as a route to high mobility graphene devices. Adv Mater 2013;25:1131–6.

[9] Bekyarova E, Sarkar S, Wang F, Itkis ME, Kalinina I, Tian X, et al. Effect of covalent chemistry on the electronic structure and properties of carbon nanotubes and graphene. Acc Chem Res 2013;46:65–73.

[10] Balch AL, Catalano VJ, Lee JW. Accumulating evidence for the selective reactivity of the 6-6 ring fusion of C_{60}. Preparation and structure of $(\eta^2\text{-}C_{60})Ir(CO)Cl(PPh_3)_2\ 5C_6H_6$. Inorg Chem 1991;30:3980–1.

[11] Balch AL, Catalano VJ, Lee JW, Olmstead MM, Parkin SR. $(\eta^2\text{-}C_{70})Ir(CO)Cl(PPh_3)_2$: The synthesis and structure of an organometallic derivative of a higher fullerene. J Am Chem Soc 1991;113:8953–5.

[12] Balch AL, Lee JW, Olmstead MM. Structure of a product of double addition to C_{70}: [C_{70}{Ir (CO)Cl($PPhMe_2)_2$}$_2$]. $3C_6H_6$. Angew Chem Int Ed Engl 1991;31:1356–8.

[13] Balch AL, Olmstead MM. Reactions of transition metal complexes with fullerenes (C_{60}, C_{70}, etc.) and related materials. Chem Rev 1998;98:2123–65.

[14] Banerjee S, Wong SS. Functionalization of carbon nanotubes with a metal-containing molecular complex. Nano Lett 2002;2:49–53.

[15] Filatov AS, Petrukhina MA. Probing the binding sites and coordination limits of buckybowls in a solvent-free environment: experimental and theoretical assessment. Coord Chem Rev 2010;254:2234–46.

[16] Niyogi S, Bekyarova E, Hong J, Khizroev S, Berger C, de Heer WA, et al. Covalent chemistry for graphene electronics. J Phys Chem Lett 2011;2:2487–98.

[17] Sarkar S, Bekyarova E, Haddon RC. Chemistry at the Dirac point: Diels-Alder reactivity of graphene. Acc Chem Res 2012;45:673–82.

[18] Lordi V, Yao N, Wei J. Method for supporting platinum on single-walled carbon nanotubes for a selective hydrogenation catalyst. Chem Mater 2001;13:733–7.

[19] Planeix JM, Coustel N, Coq B, Brotons V, Kumbhar PS, Dutartre R, et al. Application of carbon nanotubes as supports in heterogenous catalysis. J Am Chem Soc 1994;116:7935–6.

[20] Ramesh P, Itkis ME, Tang JM, Haddon RC. SWNT-MWNT hybrid architecture for proton exchange membrane fuel cell cathodes. J Phys Chem C 2008;112:9089–94.

[21] Banerjee S, Hemraj-Benny T, Wong SS. Covalent surface chemistry of single-walled carbon nanotubes. Adv Mater 2005;17:17–29.

[22] Kalinina I, Bekyarova E, Sarkar S, Wang F, Itkis ME, Tian X, et al. Hexahapto-metal complexes of single-walled carbon nanotubes. Macromol Chem Phys 2012;213:1001–19.

[23] Wang F, Itkis ME, Bekyarova E, Sarkar S, Tian X, Haddon RC. Solid-state bis-hexahapto-metal complexation of single-walled carbon nanotubes. J Phys Org Chem 2012;25:607–10.

[24] Wang F, Itkis ME, Bekyarova E, Tian X, Sarkar S, Pekker A, et al. Effect of first row transition metals on the conductivity of semiconducting single-walled carbon nanotube networks. Appl Phys Lett 2012;100:223111.

[25] Wu JS, Pisula W, Mullen K. Graphenes as potential material for electronics. Chem Rev 2007;107:718–47.

[26] Rao CNR, Sood AK, Subrahmanyam KS, Govindaraj A. Graphene: the new two-dimensional nanomaterial. Angew Chem Int Ed 2009;48:7752–77. Angew Chem 2009;121:7890–7916.

[27] Loh KP, Bao QL, Ang PK, Yang JX. The chemistry of graphene. J Mater Chem 2010;20:2277–89.

[28] Karousis N, Tagmatarchis N, Tasis D. Current progress on the chemical modification of carbon nanotubes. Chem Rev 2010;110:5366–97.

[29] Robinson JT, Burgess JS, Junkermeier CE, Badescu SC, Reinecke TL, Perkins FK, et al. Properties of fluorinated graphene films. Nano Lett 2010;10:3001–5.

[30] Zhang H, Bekyarova E, Huang J-W, Zhao Z, Bao W, Wang F, et al. Aryl functionalization as a route to band gap engineering in single layer graphene devices. Nano Lett 2011;11:4047–51.

[31] Ryu S, Han MY, Maultzsch J, Heinz TF, Kim P, Steigerwald M, et al. Reversible basal plane hydrogenation of graphene. Nano Lett 2008;8:4597–602.

[32] Elias DC, Nair RR, Mohiuddin TMG, Morozov SVBP, Halsall MP, Ferrari AC, et al. Control of graphene's properties by reversible hydrogenation: evidence for graphane. Science 2009;323:610–3.

[33] Nair RR, Ren W, Jalil R, Riaz I, Kravets VG, Britnell L, et al. Fluorographene: a two-dimensional counterpart of teflon. Small 2010;6:2877–84.

[34] Tasis D, Tagmatarchis N, Bianco A, Prato M. Chemistry of carbon nanotubes. Chem Rev 2006;106:1105–36.

[35] Niyogi S, Bekyarova E, Itkis ME, McWilliams JL, Hamon MA, Haddon RC. Solution properties of graphite and graphene. J Am Chem Soc 2006;128:7720–1.
[36] Chattopadhyay J, Mukherjee A, Hamilton CE, Kang J, Chakraborty S, Guo W, et al. Graphite epoxide. J Am Chem Soc 2008;130:5414–5.
[37] Boehm HP. Surface oxides on carbon and their analysis: a critical assessment. Carbon 2002;40:145–9.
[38] Dreyer DR, Park S, Bielawski CW, Ruoff RS. The chemistry of graphene oxide. Chem Soc Rev 2010;39:228–40.
[39] Ramesh P, Itkis ME, Bekyarova E, Wang F, Niyogi S, Chi X, et al. Electro-oxidized epitaxial graphene channel field-effect transistors with single-walled carbon nanotube thin film gate electrode. J Am Chem Soc 2010;132:14429–36.
[40] Bekyarova E, Itkis ME, Ramesh P, Berger C, Sprinkle M, de Heer WA, et al. Chemical modification of epitaxial graphene: spontaneous grafting of aryl groups. J Am Chem Soc 2009;131:1336–7.
[41] Bekyarova E, Itkis ME, Ramesh P, Haddon RC. Chemical approach to the realization of electronic devices in epitaxial graphene. Phys Stat Sol RRL 2009;3:184–6.
[42] Niyogi S, Bekyarova E, Itkis ME, Zhang H, Shepperd K, Hick J, et al. Spectroscopy of covalently functionalized graphene. Nano Lett 2010;10:4061–6.
[43] Sinitskii A, Dimiev A, Corley DA, Fursina AA, Kosynkin DV, Tour JM. Kinetics of diazonium functionalization of chemically converted graphene nanoribbons. ACS Nano 2010;4:1949–54.
[44] Hong J, Niyogi S, Bekyarova E, Itkis ME, Palanisamy R, Amos N, et al. Effect of nitrophenyl functionalization on the magnetic properties of epitaxial graphene. Small 2011;7:1175–80.
[45] Englert JEJM, Dotzer C, Yang GA, Schmid M, Papp C, Gottfried JM, et al. Covalent bulk functionalization of graphene. Nat Chem 2011;3:279–86.
[46] Sharma R, Baik JH, Perera CJ, Strano MS. Anomalously large reactivity of single graphene layers and edges toward electron transfer chemistries. Nano Lett 2010;10:398–405.
[47] Koehler FM, Jacobsen A, Ensslin K, Stampfer C, Stark WJ. Selective chemical modification of graphene surfaces: distinction between single- and bilayer graphene. Small 2010;6:1125–30.
[48] Hu H, Zhao B, Hamon MA, Kamaras K, Itkis ME, Haddon RC. Sidewall functionalization of single-walled carbon nanotubes by addition of dichlorocarbene. J Am Chem Soc 2003;125:14893–900.
[49] Kamaras K, Itkis ME, Hu H, Zhao B, Haddon RC. Covalent bond formation to a carbon nanotube metal. Science 2003;301:1501.
[50] Chua CK, Ambrosi A, Pumera M. Introducing dichlorocarbene in graphene. Chem Commun 2012;48:5376–8.
[51] Strom TA, Dillon EP, Hamilton CE, Barron AR. Nitrene addition to exfoliated graphene: a one-step route to highly functionalized graphene. Chem Commun 2010;46:4097–9.
[52] Liu LH, Lerner MM, Yan MD. Derivitization of pristine graphene with well-defined chemical functionalities. Nano Lett 2010;10:3754–6.
[53] Sarkar S, Bekyarova E, Niyogi S, Haddon RC. Diels-Alder chemistry of graphite and graphene: graphene as diene and dienophile. J Am Chem Soc 2011;133:3324–7.
[54] Zhong X, Jin J, Li S, Niu Z, Hu W, Li R, et al. Aryne cycloaddition: highly efficient chemical modification of graphene. Chem Commun 2010;46:7340–2.
[55] Criado A, Gomez-Escalonilla MJ, Fierro JLG, Urbina A, Pena D, Guitian E, et al. Cycloaddition of benzyne to SWNT: towards CNT-based paddle wheels. Chem Commun 2010;46:1272–4.
[56] Quintana M, Spyrou K, Grzelczak M, Browne WR, Rudolf P, Prato M. Functionalization of graphene via 1,3-dipolar cycloaddition. ACS Nano 2010;4:3527–33.

[57] Tian X, Sarkar S, Moser ML, Wang F, Pekker A, Bekyarova E, et al. Effect of group 6 transition metal coordination on the conductivity of graphite nanoplatelets. Mater Lett 2012;80: 171–4.

[58] Nunzi F, Mercuri F, Sgamelotti A. The coordination chemistry of carbon nanotubes: a density functional study through a cluster model approach. J Phys Chem B 2002;106:10622–33.

[59] Nunzi F, Mercuri F, Sgamelotti A. The interaction of $Cr(CO)_3$ on the (n, 0) nanotube side-walls: a density functional study through a cluster model approach. Mol Phys 2003;101:2047–54.

[60] Nunzi F, Mercuri F, de Angelis F, Sgamelotti A, Re N, Giannozi P. Coordination and haptotropic rearrangement of $Cr(CO)_3$ on (n, 0) nanotube sidewalls: a dynamical density functional study. J Phys Chem B 2004;108:5243–9.

[61] Nunzi F, Sgamellotti A, De Angelis F. $Cr(CO)_3$-activated Diels-Alder reaction on single-wall carbon nanotubes: a DFT investigation. Chem Eur J 2009;15:4182–9.

[62] Avdoshenko SM, Ioffe IN, Cuniberti G, Dunsch L, Popov AA. Organometallic complexes of graphene: toward atomic spintronics using a graphene web. ACS Nano 2011;5: 9939–49.

[63] Li EY, Marzari N. Improving the electrical conductivity of carbon nanotube networks: a first-principles study. ACS Nano 2011;5:9726–36.

[64] Dagani R. Putting the 'nano' into composites. Chem Eng News 1999;77:25.

[65] Dresselhaus MS, Dresselhaus G. Intercalation compounds of graphite. Adv Phys 2002;51: 1–186.

[66] Lee RS, Kim HJ, Fischer JE, Thess A, Smalley RE. Conductivity enhancement in single-walled carbon nanotube bundles doped with K and Br. Nature 1997;388:255–7.

[67] Haddon RC. Electronic structure, conductivity and superconductivity in alkali metal doped C_{60}. Acc Chem Res 1992;25:127.

[68] Haaland A. The molecular structure of gaseous dibenzene chromium, $(C_6H_6)_2Cr$. Acta Chem Scand 1965;19:4146.

[69] Haddon RC. Comment on the relationship of the pyramidalization angle at a conjugated carbon atom to the sigma bond angles. J Phys Chem A 2001;105:4164–5.

[70] King WA, Di Bella S, Lanza G, Khan K, Duncalf DJ, Cloke FGN, et al. Metal-ligand bonding and bonding energetics in zerovalent lanthanide, group 3, group 4, and group 6 bis(arene) sandwich complexes. a combined solution thermochemical and ab initio quantum chemical investigation. J Am Chem Soc 1996;118:627–35.

[71] Elschenbroich C. Organometallics. 3rd ed. Wiley-VCH: Weinheim; 2006.

[72] Kundig EP. Synthesis of transition metal η^6-arene complexes. Top Organomet Chem 2004;7:3–20.

[73] Kondo Y, Green JR, Ho JW. Tartrate-derived aryl aldehyde acetals in the asymmetric directed metalation of chromium tricarbonyl arene complexes. J Org Chem 1993;58:6182–9.

[74] Kundig EP, Timms PL. Metal atom preparation and ligand displacement reactions of bis-naphthalene chromium and related compounds. J Chem Soc Chem Commun 1977;912–3.

[75] Kundig EP, Perret C, Spichiger S, Bernardinelli G. Naphthalene complexes. 5. Arene exchange-reactions in naphthalene-chromium complexes. J Organomet Chem 1985;286: 183–200.

[76] Bush BF, Lagowski JJ. Chemistry of bis(η^6-naphhtalene)chromium ligand exchange reactions: synthesis and characterization of poly[(μ-η^6, η^6-naphthalene)chromium]. Organometallics 1988;7:1945–8.

[77] Bush BF, Lynch VM, Lagowski JJ. Transition-metal organometallic compounds. 8. Arene exchange reactions of bis(naphthalene) chromium. Organometallics 1987;6:1267–75.

[78] Vebrel J, Mercier R, Belleney J. A direct and efficient complexation of some indenes and dihydronaphthalenes with $(NH_3)_3Cr(CO)_3$. J Organomet Chem 1982;235:197–200.

[79] Morley JA, Woolsey NF. Metal arene complexes in organic-synthesis – hydroxylation, trimethylsilylation, and carbethoxylation of some polycyclic aromatic-hydrocarbons utilizing η^6-arene chromium tricarbonyl complexes. J Org Chem 1992;57:6487–95.

[80a] Timms PL. The formation of complexes from transition-metal vapours. J Chem Soc Chem Commun 1969;1033.

[80b] Timms PL. New developments in making compounds and materials by condensing gaseous high-temperature species at atmospheric or low pressure. Chem Soc Rev 1996;25:93–9.

[81] Timms PL. Low temperature condensation of high temperature species as a synthetic method. Adv Inorg Radiochem 1972;14:121 and references therein.

[82] Green MLH, Silverthorn WE. Arene molybdenum chemistry: some π-allyl, dihydride, dinitrogen, and carbonyl derivatives. J Chem Soc Dalton 1973;301.

[83] Benfield FWS, Green MLH, Ogden JS, Young D. Synthesis of bis-π-benzene-titanium and -molybdenum using metal vapours. J Chem Soc Chem Commun 1973;866–7.

[84] Lamanna WM. Metal vapor synthesis of a novel triple-decker sandwich complex: (η^6-mesitylene)$_2$(μ-η^6:η^6-mesitylene)Cr_2. J Am Chem Soc 1986;108:2096–7.

[85] Solladie-Cavallo A. Arene-chromium tricarbonyl complexes: bonding and behavior. Polyhedron 1985;4:901–27.

[86] Mercuri F, Sgamelotti A. Theoretical investigations on the functionalization of carbon nanotubes. Inorg Chim Acta 2007;360:785–93.

[87] Menard-Moyon C, Dumas F, Doris E, Mioskowski C. Functionalization of single-wall carbon nanotubes by tandem high-pressure/$Cr(CO)_6$ activation of Diels-Alder cycloaddition. J Am Chem Soc 2006;128:14764–5.

[88] Hu LB, Hecht DS, Gruner G. Carbon nanotube thin films: fabrication, properties, and applications. Chem Rev 2010;110:5790–844.

[89] Chan KT, Neaton JB, Cohen ML. First-principles study of adatom adsorption on graphene. Phys Rev B 2008;77:235430.

[90] Zan R, Bangert U, Ramasse Q, Novoselov KS. Metal-graphene interaction studied via atomic resolution scanning transmission electron microscopy. Nano Lett 2011;11:1087–92.

[91] Bandyopadhyaya R, Nativ-Roth E, Regev O, Yerushalmi-Rozen R. Stabilization of individual carbon nanotubes in aqueous solutions. Nano Lett 2002;2:25–8.

[92] Chen J, Rao AM, Lyuksyutov S, Itkis ME, Hamon MA, Hu H. Dissolution of full-length single-walled carbon nanotubes. J Phys Chem B 2001;105:2525–8.

[93] O'Connell MJ, Boul P, Ericson LM, Huffman C, Wang Y, Haroz E. Reversible water-solubilization of single-walled carbon nanotubes by polymer wrapping. Chem Phys Lett 2001;342:265–71.

[94] Meng L, Fu C, Fei Z, Lu Q, Dyson PJ. Coordination chemistry on the surface of single-walled carbon nanotubes. Inorg Chim Acta 2010;363:3926–31.

[95] Rogers JR, Marynick DS. Theoretical estimates of the η^6 bonding capability of buckminsterfullerene in transition-metal complexes. Chem Phys Lett 1993;205:197–9.

[96] Haddon RC. Organometallic chemistry of the fullerenes: η^2 and η^5-π complexes. J Comp Chem 1998;19:139–43.

[97] Sawamura M, Iikura H, Nakamura E. First pentahaptofullerene metal complexes. J Am Chem Soc 1996;118:12850–1.

[98] Matsuo Y, Tahara K, Nakamura E. Synthesis and electrochemistry of double-decker buckyferrocenes. J Am Chem Soc 2006;128:7154–5.

[99] Sawamura M, Kuninobu Y, Toganoh M, Matsuo Y, Yamanaka M, Nakamura E. Hybrid of ferrocene and fullerene. J Am Chem Soc 2002;124:9354–5.

[100] Calhorda MJ, Frazao CF, Simoes JAM. Metal-carbon bond strengths in $Cr(CO)_6$, $Cr(\eta^6\text{-}C_6H_6)_2$, and $Cr(CO)_3(\eta^6\text{-}C_6H_6)$. J Organomet Chem 1984;262:305–14.

[101] Stanghellini PL, Diana E, Arrais A, Rossin A, Kettle SFA. Benzene and tropilium metal complexes. Intra- and intermolecular interaction evidenced by vibrational analysis: the blue-shift hydrogen bond. Organometallics 2006;25:5024–30.
[102] Lundquist RT, Cais M. Ultraviolet spectra of organometallic compounds. J Org Chem 1962;27:1167–72.
[103] Cotton FA, Wilkinson G, Murillo CA, Bochmann M. Advanced inorganic chemistry. 6th ed. New York: John Wiley & Sons; 1999.
[104] Pampaloni G. Aromatic hydrocarbons as ligands. Recent advances in the synthesis, the reactivity and the applications of bis(η^6-arene) complexes. Coord Chem Rev 2010;254:402–19.
[105] Pape AR, Kaliappan KP, Kundig EP. Transition-metal-mediated dearomatization reactions. Chem Rev 2000;100:2917–40.
[106] Dai J, Zhao Y, Wu X, Zeng XC, Yang J. Organometallic hexahapto-functionalized graphene: band gap engineering with minute distortion to the planar structure. J Phys Chem C 2013;117:22156–61.
[107] Skinner HA, Connor JA. Metal-ligand bond-energies in organometallic compounds. Pure Appl Chem 1985;57:79–88.
[108] Sarkar S, Moser ML, Tian X, Zhang X, Al-Hadeethi F, Haddon RC. Metals on graphene and carbon nanotube surfaces: from mobile atoms to atomtronics to bulk metals to clusters and catalysts. Chem Mater 2014;26:184–95.
[109] Moser ML, Tian X, Pekker A, Sarkar S, Bekyarova E, Itkis ME, Haddon RC. Hexahapto-lanthanide interconnects between the conjugated surfaces of single-walled carbon nanotubes. Dalton Trans 2014;43:7379–82.

Chapter 10

Preparation and Properties of Carbon Nanopeapods

Toshiya Okazaki
Nanotube Research Center, National Institute of Advanced Industrial Science and Technology (AIST), Tsukuba, 305-8565, Japan

10.1 INTRODUCTION

Filling of the interior space of carbon nanotubes (CNTs) was first achieved with multi-walled carbon nanotubes (MWCNTs) in 1993. When lead particles were deposited on the walls of MWCNTs in order to investigate their surface activity, Ajayan and Iijima accidentally found that many tubes trapped them inside after annealing [1]. Although the encapsulation occurred on nanometer scale, this phenomenon can be explained by macroscopic capillarity [2]. It was found that only relatively low-surface-tension materials (less than 100–200 mN m^{-1}) can be drawn inside MWCNTs.

Incorporation of molecules into the hollow space of single-walled carbon nanotubes (SWCNTs) was also accidentally discovered [3]. In 1998, Luzzi and co-workers found that C_{60} molecules nested inside SWCNT, like many peas in a pod, when they were studying the purified SWCNTs by using a transmission electron microscope (TEM). This hybrid nano-carbon material has, therefore, been called as 'nanopeapods', 'nanotube peapods', 'fullerene peapods' or merely 'peapods'. Since then, nanopeapods have attracted many researchers' attention due to their unique structural properties that may open up a brand new 'nano-world' in both scientific and practical arenas. For instance, the confining geometry on quasi-one-dimensional (1D) crystallization of fullerenes may provide an ideal system to study the physical and chemical properties of the low-dimensional materials. Furthermore, they are also expected as building blocks in future nano-electronics because the mechanical and electronic properties of SWCNTs frequently undergo considerable modification upon encapsulation, which allow us to finely tune these parameters by altering the encapsulated fullerenes.

A few years after the discovery, several groups independently succeeded to synthesize nanopeapods with high efficiency [4,5]. Their synthesis methods are very similar with each other except for details, in which fullerenes are encapsulated into the hollow space of SWCNTs by thermal kinetic energy. It is

Carbon Nanotubes and Graphene. DOI: 10.1016/B978-0-08-098232-8.00010-3

an astonishingly simple method which does not require any special apparatus. Besides fullerenes, other molecules can also be encapsulated by the same procedure. Every researcher in the ordinary physical and chemical laboratory can synthesize nanopeapods by this method.

In this chapter, we review the synthesis methods of nanopeapods and their physical properties that have been revealed so far. Especially, photoluminescence (PL) and resonance Raman studies on the electronic and the phonon properties of fullerene-encapsulated nanopeapods are mainly discussed.

10.2 HIGH-YIELD SYNTHESIS OF CARBON NANOPEAPODS

A bulk production method of nanopeapods was achieved through the gas-phase reaction [4,5]. The method allows us to produce various fullerene nanopeapods in high yield. To obtain high-purity and high-yield nanopeapods, one should prepare isolated fullerenes and high-purity end-opened SWCNTs. Fullerenes are soluble molecules so that one can easily purify them more than 99% with a well-established separation method, such as a high-performance liquid chromatography (HPLC) [6]. On the other hand, SWCNTs can be purified by a combination of several methods.

The purity of SWCNTs is the most important factor for making nanopeapods with high yield. For example, we usually remove the amorphous carbon materials and metal catalyst particles by the following procedure [7]. First, SWCNTs are heated at ~350 °C for 30 min in air to remove most of the amorphous carbon and other carbon materials which coat catalyst metal particles. The obtained SWCNTs are treated in methanol solution of sodium hydroxide (0.2 g/100 ml) for 30 min and filtrated. After the filtration, SWCNTs are dispersed in *i*-propanol under sonication for 20 min. The SWCNT dispersion is centrifuged with ~1600*g* for 10 min and the supernatant is discarded. The washing process with *i*-propanol is repeated until the supernatant becomes transparent. Then the remaining metal particles are washed by hydrochloric acid and heated at 600 °C for 2 hours in vacuum.

The distribution of the diameter of SWCNTs is another important factor. If the diameter of SWCNT is smaller than that of molecules, encapsulation of the molecules will not be possible. The smallest limit of tube diameters was precisely determined for several molecules based on the PL studies [8–11] (see Sec. 10.4). Table 10.1 shows the obtained smallest diameter of SWCNTs that can encapsulate C_{60}, C_{70} fullerenes, azafullerenes $(C_{59}N)_2$ and coronenes. For instance, SWCNTs with the diameters larger than 1.25 nm are required for making C_{60} nanopeapods.

The doping of molecules into inner hollow space of the so-prepared SWCNTs is an easy job. To open the cap of SWCNTs, purified SWCNTs are heated at ~500 °C for 30 min in air. The treated SWCNTs and the molecules are sealed under vacuum ($\sim 1 \times 10^{-4}$ Pa) in quartz tube. Prior to the doping, SWCNTs are degassed at ~200 °C in a vacuum to remove residual gases. One can obtain high-yield nanopeapods after heating the ampoule for 2 days at proper doping temperatures, such as ~600 °C for C_{60} and ~650 °C, for higher fullerenes and metallofullerenes. The obtained nanopeapods were washed with proper solvents to remove the molecules adsorbed on the outside of the walls. After the filtration, we obtained a dark, paperlike sheet, so-called buckypaper (Figure 10.1(a)). A typical high-resolution TEM (HRTEM) image of C_{60} nanopeapods is shown

TABLE 10.1 Smallest limit of tube diameters (*d*) for C_{60}, C_{70}, $(C_{59}N)_2$ and coronene [8–11]

	d/nm
C_{60}	1.25
C_{70}	1.28
$(C_{59}N)_2$	1.28
Coronene	1.32

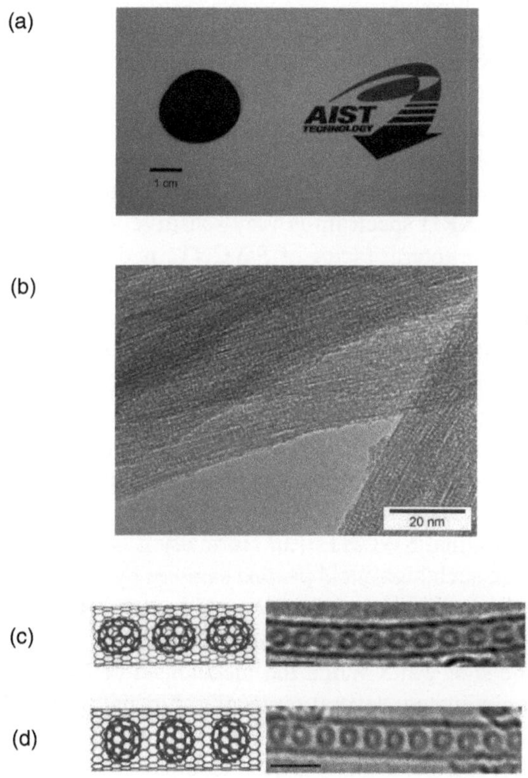

FIGURE 10.1 (a) Picture and (b) HRTEM image of C_{60} nanopeapods buckypaper [7]. HRTEM images of C_{70} nanopeapods with (c) lying and (d) standing alignments. Scale bar = 2 nm [9].

in Figure 10.1(b). It shows a crystalline bundle of close-packed tubes, almost all of which contain C_{60} molecules in a 1D array.

Besides the gas-phase doping, nanopeapods can also be made through liquid-phase reactions [12–15]. The methods involve immersing or refluxing end-opened SWCNTs in a saturated solution of molecules. For example, β-carotene was encapsulated in SWCNTs by refluxing the mixture for 10 hours in a nitrogen atmosphere [14].

The exohedrally functionalized C_{60} was encapsulated in SWCNTs by using super critical carbon dioxide [16,17]. The low-temperature condition of the method allows the insertion of the molecules without damage. Due to the presence of functional groups, interfullerene separations were increased depending on the length of the functional group.

Unfortunately, the filling-yield through the liquid-phase doping has been lower than that of the vapour-phase doping at the present moment probably because considerable amount of solvent molecules were simultaneously encapsulated in SWCNTs [18]. The solution method is, however, especially advantageous in bio-medical use of nanopeapods because it can be applied for thermally unstable molecules, such as large organic and biological molecules.

Even though the HRTEM image clearly shows that fullerenes are closely packed inside the SWCNTs, it is difficult to quantify the filling yield. Several evaluation criteria have so far been proposed based on Raman scattering [5], electron energy-loss spectroscopy (EELS) [19], X-ray diffraction (XRD) [20] and difference in weight before and after the doping [21]. One of the reliable evaluation methods is that by utilizing the XRD spectrum [20]. The (1 0) peak of SWCNTs in the XRD spectrum is very sensitive to the molecular encapsulation because the structural factor of SWCNTs is drastically changed by the additional electron density from the encapsulated molecules. Thus, substantial reduction in the intensity of the (1 0) peak is observed upon encapsulation of molecules, which allows us to quantitatively estimate the filling yield.

PL spectroscopy is also effective for estimating the filling yield because the encapsulated molecules usually affect the electronic structure of SWCNTs, which cause the spectral shifts of the PL peaks [8]. For instance, the filling yield of C_{60} in SWCNTs that were synthesized by the arc-discharging method was estimated as more than 95% from the relative intensity of the PL peaks of (13, 5) nanotube peapods and the original SWCNTs [9]. Normally it is very difficult to find out empty SWCNTs in such high-yield peapod samples by the TEM observations.

Here a question arises as to how SWCNTs catch fullerenes. Do fullerenes enter the interior space from the open end of the tube or from defects that might be present on the side wall? While the mechanism of fullerene encapsulation is still not fully elucidated, several molecular dynamics simulations have provided important clues to it [22,23]. In general, the number and the thickness of bundles of SWCNTs increase, as the yield and purity of SWCNTs increase. In most cases, researchers have used such high-purity SWCNTs for synthesizing nanopeapods most of which formed bundles rather than the isolated. The end of the bundles may not be flat, i.e. some SWCNTs are sticking out of the others and some standing back from the neighbours. When sublimating fullerenes approach to the end of such bundles, they may first adsorb in the sticking neighbourhoods and then enter the inner hollow space of the standing-back SWCNT through the open end by 'head-on' collisions [23]. The probability for encapsulation via the 'head-on' collision was found to be approximately 100 times larger than those through the defects on the side wall and an isolated open-end.

Once nanopeapods are formed, their structure is very stable. Theoretical studies predict that the encapsulation process of the C_{60} molecules is exothermic for SWCNTs with proper diameter [24,25]. A decrease of the total entropy is compensated by the gain in the potential energy. For example, each C_{60} molecules is stabilized by 0.51 eV upon the encapsulation into (10, 10) SWCNT, indicating that it is very difficult to extract C_{60} from the SWCNTs by high-temperature annealing, for example. Indeed, it has experimentally been observed that C_{60} molecules react with each other within SWCNTs before extraction [26].

However, it is possible to take C_{60} molecules out from SWCNTs by sonication process in solution [27,28]. It is likely that C_{60} molecules can escape from SWCNTs if the solvents form very effective interaction with C_{60}. The prominent tube diameter and solvent dependences suggest that the efficiency of the release is resultant from the balance of the interaction between SWCNTs and the encapsulated molecules, and that between the solvents and the molecules [27,28].

10.3 PACKING ALIGNMENT OF THE MOLECULES IN SWCNTS

As the interior space of SWCNTs is similar in size to the diameter of fullerenes, the encapsulated fullerenes form quasi-1D crystals inside SWCNTs. Intermolecular distances between the adjacent fullerenes have been obtained by the TEM observations, and the electron and the XRD patterns (Table 10.2) [4,29–33]. In the case of C_{60} nanopeapods, intermolecular distance of the encapsulated C_{60} molecules is shorter than that of the three-dimensional (3D) bulk crystal of C_{60}. This means that the interaction between the adjacent C_{60} molecules in nanopeapods is stronger than that in the bulk crystal. However, the C_{60} molecules do not form covalent bonds because the obtained distance of 0.97 nm is longer than that of the C_{60} polymers. The CNT walls should affect the electronic structure of the encapsulated C_{60}, which results in the subtle structural change.

The elongated C_{70} and C_{80} (D_{5d}) exhibit different intermolecular distances depending on their orientations inside SWCNTs (Table 10.2). Figure 10.1(c) and (d) shows HRTEM images of $(C_{70})_n$@SWCNTs with different diameters. It is clearly seen that C_{70} have to lie inside SWCNTs when the diameter of SWCNTs is smaller than the longer axis of the C_{70} molecules (Figure 10.1(c)). In contrast, C_{70} fullerenes are 'standing' to the tube axis when they are encapsulated in SWCNTs with larger diameter (Figure 10.1(d)). PL studies reveal that the transition from the 'lying' alignment to the 'standing' alignment occurs at $d_t \sim 1.41$ nm [9]. This transition diameter value is in good agreement with the previously reported values of 1.42 and 1.435 nm, which were estimated by XRD experiments and theoretical calculations, respectively [34,35].

Generally endohedral metallofullerenes undergo the charge transfer in the ground states. Two or three valence electrons transfer from the encaged atom to fullerenes. For instance, the electronic structure of $Sm@C_{82}$ is well represented as $Sm^{2+}@C_{82}{}^{2-}$ as a result of the transfer of the two electrons [36,37]. On the other hand, three electrons are transferring in $Gd@C_{82}$ and $La@C_{82}$ leading to $Gd^{3+}@C_{82}{}^{3-}$ and $La^{3+}@C_{82}{}^{3-}$ configurations [38,39]. The obtained intermolecular distance of $Sm@C_{82}$ inside SWCNTs is 1.10 nm, which is identical to those of C_{82} in $(C_{82})_n$@SWCNTs and $Gd@C_{82}$ in $(Gd@C_{82})_n$@SWCNTs (Table 10.2) [4,29]. It is likely that the intermolecular distance inside SWCNT is not sensitive to the electronic structure and is rather determined by the size of the fullerene cage.

Except for fullerenes, it has been reported that planner π-conjugated molecules, coronenes, form 1D regular structure inside SWCNTs [11]. Figure 10.2(a) shows HRTEM images of self-assembled coronene columns inside SWCNTs. It is clearly seen that coronene molecules are accommodated along the tube axis with regular intervals. The formation of the columnar structure is due to two dominant factors: π–π stacking between coronenes and the interaction between the encapsulated

TABLE 10.2 Intermolecular distances of various fullerenes in their nanopeapods structures, together with those in the bulk crystal and C_{60} polymer for references

Fullerene (symmetry)	In nanopeapods (nm)*	In nanopeapods (nm)†	In 3D bulk crystal (nm)	In polymerized crystal (nm)
C_{60}	0.97	0.97	1.002	0.91
C_{70}	1.01, 1.10	1.00, 1.10	1.044	
C_{78} (C_{2v}(3))	1.00			
C_{80} (D_{5d})	1.01, 1.15			
C_{82} (C_2)	1.10		1.14	
$Sm@C_{82}$ (I)	1.10			
$Gd@C_{82}$	1.10			
$La@C_{82}$ (I)	1.11			
C_{84} (D_2/D_{2d})	1.10		1.12	
$Sc_2@C_{84}$	1.10		1.12	

All measurements were carried out at room temperature.
**Based on the electron diffraction and TEM data [30,31].*
†Based on the XRD data [33].

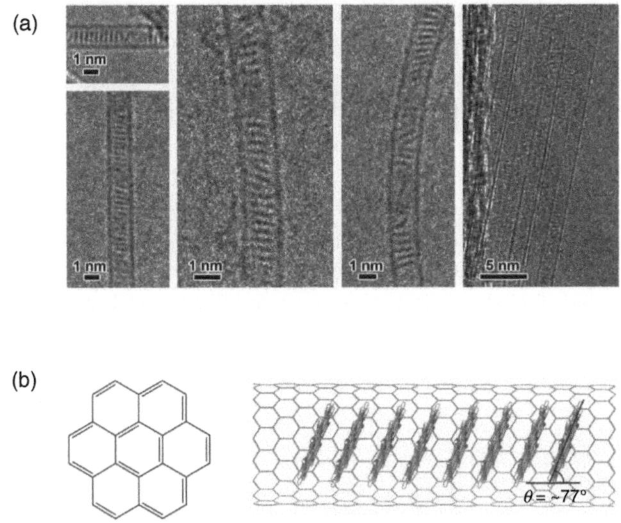

FIGURE 10.2 (a) HRTEM images of coronenes@SWCNTs. (b) Molecular structure of coronene and schematic illustration of coronenes@SWCNTs [11].

coronene and the host SWCNT. In the monoclinic crystal, coronenes have an interplanar distance of 0.340 nm with a tilt angle of about 46° relative to the stacking axis [40]. On the other hand, within SWCNTs, statistical analysis of the HRTEM images shows that the distance between the molecular planes is 0.35 ± 0.03 nm and the angle between the molecular plane and the tube axis is ~77° in average (Figure 10.2(b)). Although the interplanar distances are identical between 1D and solid coronenes within experimental error, a substantial difference is observed in the tilt angle. The larger angle of ~77° is resultant from the balance between the coronene–coronene interaction and the coronene–SWCNT interaction, which is supported by the density functional theory (DFT) calculations [11].

The observed results indicate that the 1D crystal structures of the encapsulated molecules in SWCNTs are sharply distinct from those of the molecule in 3D solids. This implies that SWCNTs can possess fairly unique nanospaces that can be used for molecular nanostructures.

10.4 ELECTRONIC PROPERTIES OF NANOPEAPODS

10.4.1 C_{60} encapsulation effects on electronic structures of semiconducting SWCNTs

The electronic structures of SWCNTs are effectively affected by the encapsulated molecules. PL spectroscopy allows us to investigate the molecular-encapsulated effects on the electronic states of each (*n, m*) SWCNTs because the emission and excitation spectra show characteristic peaks depending on the molecular structure of SWCNTs.

Figure 10.3(a) and (d) shows two-dimensional (2D) PL contour plots of the unfilled SWCNTs in SDBS-D_2O as a function of emission (λ_{11}) and excitation (λ_{22}) wavelengths [7,8]. Here two SWCNTs having different diameter distributions are used: one is synthesized by pulsed-laser vapourization (PLV) method (Figure 10.3(a)–(c)) [8] and the other by arc-discharging method (Figure 10.3(d)–(f)) [7]. The PL maxima (spots) on the map are clearly seen in the second interband (E_{22}) excitation region (λ_{22} = 800–1050 nm) and the first interbands (E_{11}) emission region (λ_{11} = 1400–1800 nm) of SWCNTs with 10–1.6 nm in diameter, which can be assigned to the specific (*n*, *m*) SWCNTs [41]. It is noted that all peak positions were slightly and uniformly red-shifted after the tube opening (~9 meV in E_{11} and ~15 meV in E_{22}). This can be explained by the presence of the solvents in the interior space of the open-ended

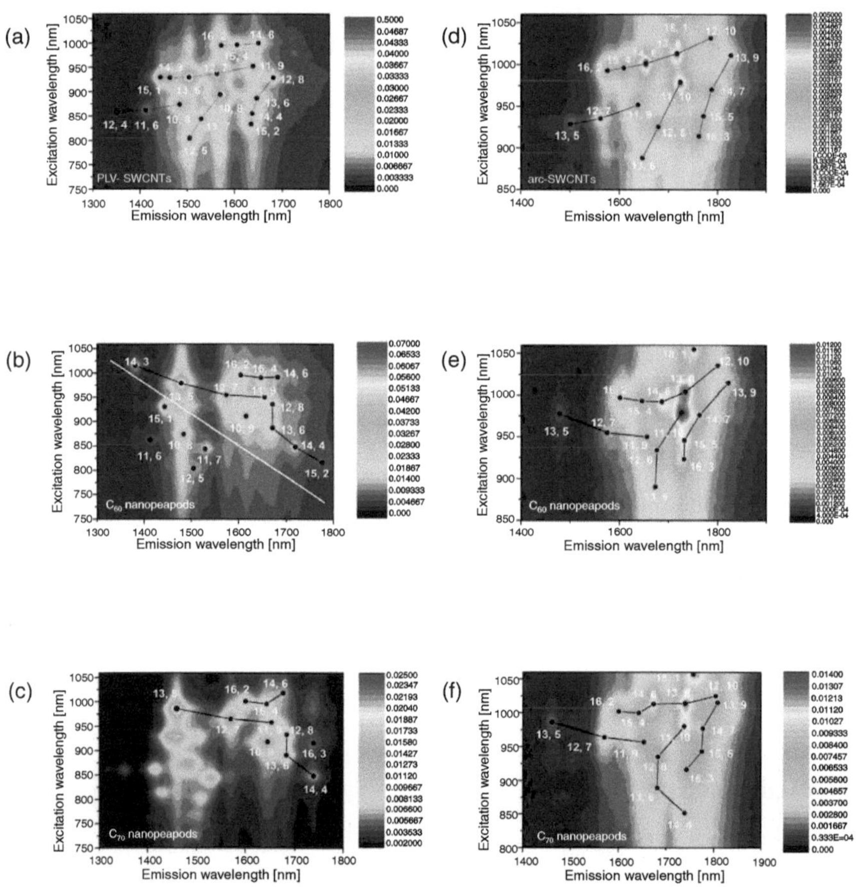

FIGURE 10.3 PL maps obtained from micelle aqueous solutions of (a, d) SWCNTs, (b, e) C_{60} nanopeapods and (c, f) C_{70} nanopeapods, respectively. SWCNTs were synthesized by (a–c) the laser-vapourization method (d_t = 10–1.4 nm) [8,9] and (d–f) the arc-discharging method (d_t = 1.3–1.55 nm) [7,9].

SWCNTs. Actually, it has been known that water molecules are encapsulated inside SWCNTs [42]. The filling water solvents should cause a reduction of the electron–electron repulsion and the exciton-binding energy by dielectric screen effects [43], which results in the observed red shifts.

Overall features of PL behaviour of SWCNTs drastically change upon C_{60} encapsulations. Figure 10.3(b) and (e) shows the 2D PL contour maps of SWCNTs after C_{60} encapsulations. The PL peaks in Figure 10.3(b) can be divided into two groups (the white line in Figure 10.3(b)). One is the peaks whose positions are the same as those of unfilled SWCNTs. The origin of these peaks are thus easily assigned to the (11, 6), (15, 1), (10, 8), (10, 5) and (11, 7) tubes, respectively (Figure 10.3(b)). The characteristic feature of this group is that the tube diameters are less than ~1.25 nm. On the other hand, the original PL peaks of larger-diameter tubes with $d_t > 1.25$ nm such as (13, 5), (10, 7) and (10, 9) tubes almost completely disappears in Figure 10.3(b) and (d). Alternatively, new PL peaks appear at different positions to those of the unfilled SWCNTs. The threshold diameter of ~1.25 nm closely matches the theoretically predicted value of ~1.28 nm for unstrained filling of C_{60} [24]. This agreement strongly suggests that C_{60} is preferentially encapsulated into SWCNTs with $d_t > 1.25$ nm and that the filling causes spectral shifts for those SWCNTs.

The observed new PL peaks were assigned to each (n, m) nanopeapods by examining many samples that has a different chirality distribution and a different filling yield [7,8]. The results are shown in Figure 10.3(b) and (d).

In order to investigate the detail mechanisms of bandgap modification, the energy differences in E_{11} and E_{22} between C_{60} nanopeapods and SWCNTs ($\Delta E_{ii} = E_{ii}^{NPDs} - E_{ii}^{SWCNTs}$, $i = 1.2$) are plotted as a function of tube diameter (Figure 10.4(a) and (b)). Apparently, the diameter dependencies of ΔE_{11} and ΔE_{22} are different between the '$2n + m$' family types (type I (mod$(2n + m, 3) = 1$) and type II (mod$(2n + m, 3) = 2$), respectively). For example, ΔE_{11} for type I tubes shows positive values in a smaller-diameter regime (Figure 10.4(a)). With increasing tube diameter, it exponentially decreases towards -0.02 eV and approaches to zero. On the contrary, the ΔE_{11} for type II tubes increases towards 0.02 eV as d_t increases, and then gradually decreases again (Figure 10.4(a)). As is the case of ΔE_{11}, ΔE_{22} also shows clear family-type dependence (Figure 10.4(b)). Namely, ΔE_{22} for type I tubes increases with increasing tube diameter and then decreases at $d_t \sim 1.4$ nm, whereas ΔE_{22} for type II tubes exhibits a totally opposite trend (Figure 10.4(b)).

Such a strong dependence on '$2n + m$' family type is a characteristic feature of the strain-induced spectral shift [44]. Theoretical calculations predicted that the bandgaps of E_{11} and E_{22} change towards opposite directions to each other upon the structural deformation of SWCNTs caused by the mechanical strain [44]. Furthermore, the strain-induced bandgap shifts for type I tubes are entirely contrary to those of type II tubes. For example, E_{11} of type I tubes increases under a radial expansion and/or a compressive strain along the tube, whereas that of type II tubes decrease [44]. The observed ΔE_{11} and ΔE_{22} for smaller-diameter

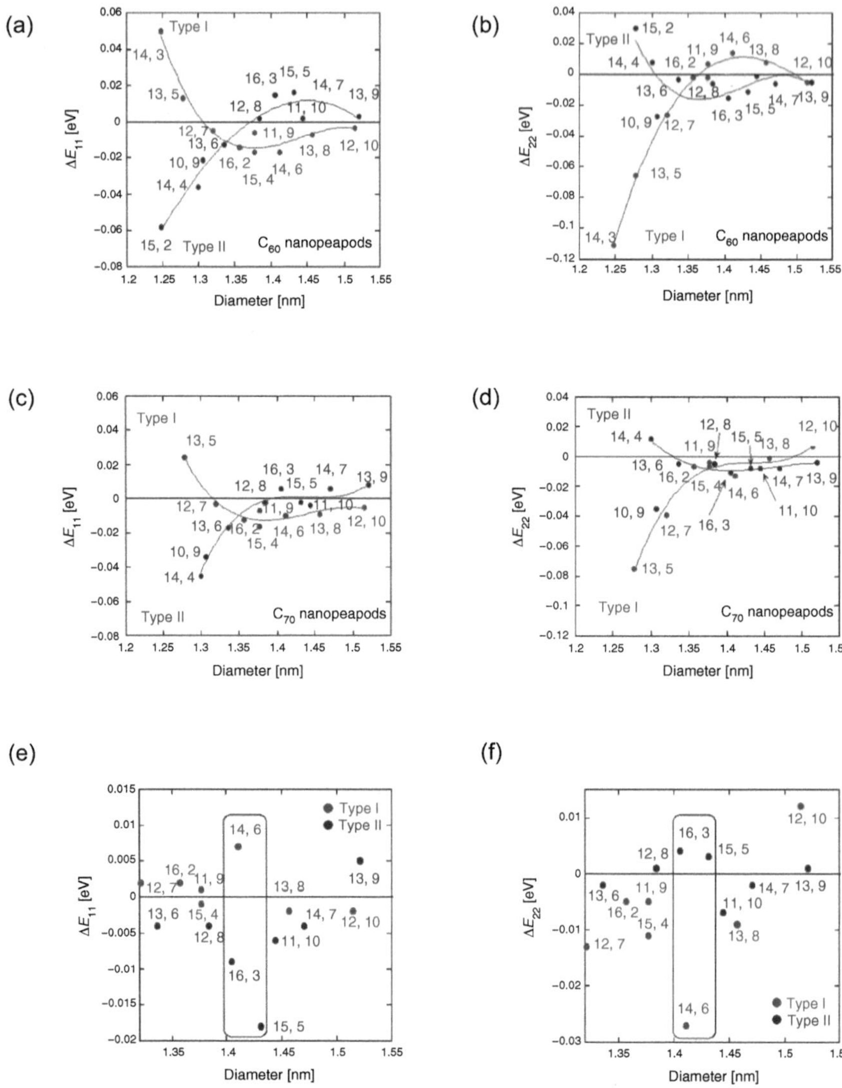

FIGURE 10.4 Differences in optical transition energies E_{11} and E_{22} (ΔE_{11} and ΔE_{22}, respectively) for (a, b) C_{60} nanopeapods and SWCNTs, and those for (c, d) C_{70} nanopeapods and SWCNTs, as a function of tube diameter [7–9]. Difference in (e) ΔE_{11} and (f) ΔE_{22} between C_{70} nanopeapods and C_{60} nanopeapods ($\Delta E_{11}'$ and $\Delta E_{22}'$, respectively) [9].

tubes ($d_t < \sim 1.32$ nm) correspond to this situation (Figure 10.4(a) and (b)). The tube diameter expansion due to the fullerene encapsulation is responsible for the observed PL peak shifts.

On increasing the tube diameter, the ΔE_{11} and ΔE_{22} approach to zero line and change their signs at around $1.30 \sim 1.35$ nm (Figure 10.4(a) and (b)).

Surprisingly, this value matches well with the nearest neighbouring distance between C_{60} and SWCNTs obtained by a simple calculation. The intermolecular distance between C_{60} molecules inside SWCNTs was found to be 0.97 nm by electron-diffraction methods (Table 10.2). It is well known that the interlayer distance between two graphenes is 0.34 nm in graphite [45]. Hence, the nearest distance from the centre of C_{60} to the tube wall can be calculated to be 0.66 (=(0.97 + 0.34)/2) nm, which suggests that C_{60} requires the tube diameter of larger than 1.31 (=0.97 + 0.34) nm to enter the interior space without friction. Excellent agreement about the threshold diameters strongly suggests the van der Waals nature of the interaction between C_{60} and SWCNTs. The attractive intermolecular interaction between SWCNTs and fullerenes causes the PL shifts in larger-diameter SWCNTs ($d_t > \sim 1.32$ nm).

It is generally accepted that the PL behaviours of SWCNTs are dominated by the Coulomb interaction between the produced electron-hole pairs (excitons). For example, although the one-electron theory predicts that the energy ratio between E_{22} and E_{11} (E_{22}/E_{11}) must approach two as tube diameter increases, experimental results show that it approaches a value smaller than two. This 'ratio problem' can be explained by the strong exciton effects in SWCNTs [46]. Interestingly, the obtained E_{22}/E_{11} values approach almost same values (~1.75) in this case [7]. This strongly suggests the exciton effects are very similar between SWCNTs and nanopeapods.

10.4.2 C_{60} encapsulation effects on electronic structures of metallic SWCNTs

One of the remarkable properties of SWCNTs is that they show metallic or semiconducting properties depending on the tube diameter and the wrapping angle. For monitoring the interaction between metallic SWCNTs and the encapsulated molecules, Raman spectroscopy is one of the most suitable and efficient techniques because metallic SWCNTs do not exhibit PL. The radial breathing mode (RBM) phonon of SWCNTs is particularly useful, since its frequency is directly related to the inverse tube diameter ($1/d_t$), and this frequency is sensitive to molecular encapsulation. Because of the resonance effect, the Raman intensity of SWCNTs drastically changes with the molecular structure and the excitation wavelength. Although the shift in the RBM frequencies due to chemical and electrochemical doping has been frequently reported, the observed shifts appear to reflect the change in resonance enhancement in a mixture of SWCNTs rather than an actual shift in the RBM mode of a particular (*n*, *m*) tube. Therefore, for comprehensive characterizations, it is necessary to measure the Raman spectra by using several laser excitation lines.

Figure 10.5 shows the resonance Raman spectra of the original SWCNTs and C_{60} nanopeapods in the RBM frequency region under different excitation energies [47]. In order to assign these RBM phonon peaks to the respective (*n*, *m*) tubes, we constructed a 2D contour map from the resonance Raman spectra of the RBM phonon region from 1.50 to 1.76 eV. In the spectra of the original SWCNTs (Figure 10.5(a)), prominent RBM phonon peaks are observed at around 160 and 170 cm^{-1} under excitations of ~1.60 and ~1.70 eV, respectively. These excitation peaks correspond to the absorption peaks of metallic SWCNTs with diameters of 1.3–1.5 nm [47]. The solid circles in Figure 10.5(b) indicate the RBM phonon peak positions and the optical transition energies (E^{M}_{11}) of the individual (*n*, *m*) tubes obtained by line-shape fitting [47]. It is noteworthy that all of the RBM peaks can be reasonably assigned to each (*n*, *m*) SWCNT. It was unnecessary to assume that the additional RBM peaks originated from another species.

On the other hand, the 2D RBM intensity map of C_{60} nanopeapods is shown in Figure 10.5(c). The general features are very similar to those of the SWCNT control sample (Figure 10.5(a)). The curve fitting analyses show the RBM-peak positions and E^{M}_{11} in the map as depicted in Figure 10.5(d). Although the peak positions of some RBMs are shifted, these peaks still show the typical 'hooked' family pattern. Therefore, it is highly reasonable to assign the observed RBM peaks of C_{60} nanopeapods to each (*n*, *m*) tube after accounting for the 'empty' SWCNTs, as denoted in Figure 10.5(d).

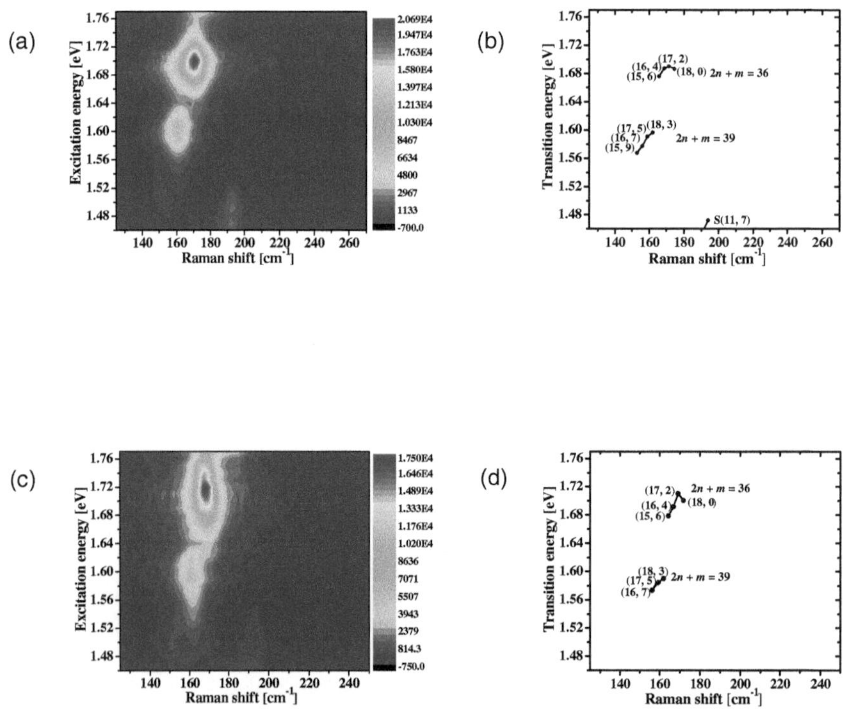

FIGURE 10.5 2D Raman intensity maps of RBM phonon regions of (a) original SWCNTs and (c) C_{60} nanopeapods. The RBM peak positions of (b) SWCNTs and (d) C_{60} nanopeapods obtained by curve-fitting analyses [47].

In order to study the C_{60} encapsulation effect on the electronic properties, we investigated the differences in the optical transition energies before and after insertion. Figure 10.6(a) shows the optical transition energy shifts ($\Delta E = E^{M}_{11}$(NPDs) − E^{M}_{11}(SWCNTs)) as a function of tube diameter (grey circles (red circles in the web version)), together with reference data of type II semiconducting nanopeapods ($\Delta E = E^{S}_{11}$(NPDs) − E^{S}_{11}(SWCNTs)) that are obtained

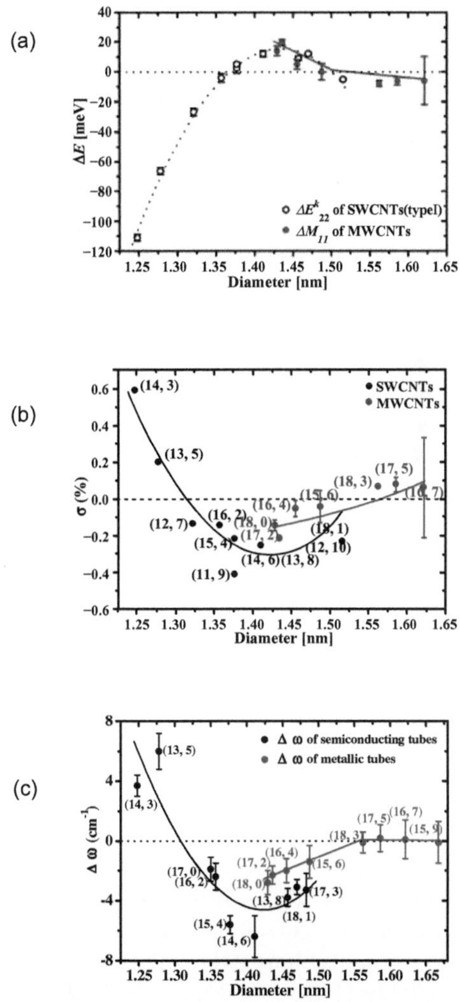

FIGURE 10.6 (a) Optical transition energy shifts ($\Delta E = E^{M}_{11}$(NPDs) − E^{M}_{11}(SWCNTs)) of metallic C_{60} nanopeapods as a function of tube diameter (grey circles (red circles in the web version)), together with reference data of type II semiconducting nanopeapods ($\Delta E = E^{S}_{11}$(NPDs) − E^{S}_{11}(SW CNTs)) that are obtained by PL measurements (Figure 10.4(a)). (b) The values of σ calculated from ΔE. (c) Experimentally obtained $\Delta\omega$ (solid circles) for metallic C_{60} nanopeapods (grey circles (red circles in the web version)) and semiconducting C_{60} nanopeapods (black circles) [47].

by the PL measurements (open circles) (Figure 10.4(a)) [7]. It appears that ΔE for metallic nanopeapods almost overlaps with that for type II semiconducting nanopeapods.

For type II semiconducting nanopeapods, it has been found that the red shifts ($\Delta E < 0$) for $d_t < \sim 1.32$ nm and the blue shifts ($\Delta E > 0$) for $d_t > \sim 1.32$ nm can also be explained by the van der Waals-type intermolecular interaction between C_{60} and SWCNTs (see Sec. 10.4.1). A quantitative similarity of the blue shifts of metallic nanopeapods suggests that the observed shifts are induced by the attractive forces between metallic SWCNTs and C_{60}. In particular, an increase in attractive interaction can be interpreted as an effect similar to that caused by reduction in the effective tube diameter. The degree of these effects is estimated by the following equation:

$$\Delta E = -\mathrm{sgn}(2p+1)3t_0(1+\nu)\sigma\cos 3\theta \qquad (10.1)$$

Here, $p = \mathrm{mod}(n - m, 3)$, t_0 (=2.66 eV) denotes the hopping parameter, ν(=0.2) represents the Poisson's ratio and σ denotes the radial compression [44].

According to this equation, the calculated σ of the metallic nanopeapods is plotted as a function of tube diameter (grey circles (red circles in the web version)), together with that for the type II semiconducting nanopeapods (black circles) (Figure 10.6(b)). It is clearly seen that both metallic and semiconducting SWCNTs show similar diameter dependence of σ; however, the absolute values of σ for the metallic SWCNTs are slightly smaller than those for the semiconducting SWCNTs with similar diameters. These results suggest that the interaction between metallic SWCNTs and the encapsulated C_{60} is weaker than that of semiconducting SWCNTs. The smaller interaction between metallic SWCNTs and the encapsulated C_{60} is also supported by the RBM frequency shifts upon C_{60} insertions (see Sec. 10.5).

10.4.3 C_{70} encapsulation effects on electronic structures of semiconducting SWCNTs

While C_{60} can be regarded as a spherical 'soccer ball' molecule, C_{70} has an ellipsoidal 'rugby ball' shape (Figure 10.1(c) and (d)). Such ellipsoidal molecules are of particular interest because distinct orientations are possible for C_{70} within the SWCNTs, which should lead to a unique modification of the electronic properties of the host SWCNTs. It is known that C_{70} molecules can take two conformations inside SWCNTs, called 'lying' and 'standing' alignments, corresponding to the molecular long axis being parallel and perpendicular to the tube axis, respectively (Figure 10.1(c) and (d)). Theoretical calculations predict that the electronic states of SWCNTs are sensitive to the orientation of the encapsulated C_{70} molecules [48]. The difference in the interwall spacing of C_{70} and SWCNTs results in the dependence of the electronic structures on the orientation. This implies that the electronic properties of SWCNTs can be tuned by controlling the orientation of the encapsulated C_{70}.

Figure 10.3(c) shows the PL map of C_{70} nanopeapods, where PLV–SWCNTs were used as templates (d_t = 1.2–1.4 nm). It is apparent that the PL pattern of the C_{70} nanopeapods is completely different from that of the SWCNTs (Figure 10.3(a)), but very similar to that of the C_{60} nanopeapods (Figure 10.3(b)). The similarity in the maps of C_{70} nanopeapods and C_{60} nanopeapods suggests that the PL peaks in the map of the C_{70} nanopeapods can be assigned to each (n, m) tubes as shown in Figure 10.3(c), where the PL peaks with identical $2n + m$ values are connected.

Even though the PL patterns of the C_{70} nanopeapods and C_{60} nanopeapods are similar, several differences can be observed between the two PL maps. For instance, while the PL peaks of (14, 3) C_{60} nanopeapods (d_t = 1.248 nm) are observed in Figure 10.3(b), the corresponding peak is absent in the case of the C_{70} nanopeapods (Figure 10.3(c)). A diameter of 1.25 nm is the lower limit for C_{60} encapsulation (Table 10.1). The absence of the (14, 3) peak for the C_{70} nanopeapods suggests that C_{70} require larger interior space than C_{60}. In fact, as shown in Figure 10.3(c), the PL peak positions change for SWCNTs with $d_t \geq 1.28$ nm ((13, 5) tubes), whereas those of SWCNTs with $d_t \leq 1.25$ nm were invariant. The observed values of the smallest diameters, i.e. 1.28 nm for C_{70} encapsulations, are in good agreement with the theoretically derived values, which are approximately 1.26 nm [32,48].

We also carried out PL characterization of large-diameter C_{70} nanopeapods (d_t = 1.3–1.55 nm), where arc-SWCNTs were used as templates. Figure 10.3(f) shows the PL map of arc-SWCNTs after C_{70} encapsulation along with reference spectra of C_{60} nanopeapods (Figure 10.3(e)) and the original SWCNTs (Figure 10.3(d)). Again, the peak pattern of the C_{70} nanopeapods shows close similarity to that of the C_{60} nanopeapods. Hence, all the peaks of the C_{70} nanopeapods in the map are assignable to each (n, m) tube by analogy with the corresponding PL pattern of the C_{60} nanopeapods. However, while the '$2n + m$' family pattern shows a characteristic 'hook' shape, the patterns are not smooth near the (16, 3), (14, 6) and (15, 5) tubes.

In Sec. 10.4.1, it was shown that the optical transition energy of SWCNTs is substantially modified by C_{60} encapsulation, depending on the tube diameter. Because the length of the short axis of C_{70} (=0.712 nm) [49] is similar to the diameter of C_{60} (=0.710 nm) [50], C_{70} are expected to exhibit lying alignment in relatively small-diameter SWCNTs (Figure 10.2(c)), resulting in the enlargement of the tube diameter. Figure 10.4(c) and (d) shows the dependence of ΔE_{11} and ΔE_{22} on the diameter of the tube for C_{70} nanopeapods. As observed in the case of C_{60} nanopeapods, strong family-type dependence behaviours are also observed. For instance, ΔE_{11} of the C_{70} nanopeapods (type I) exhibits a positive value for (13, 5) tubes, becomes negative at around $d_t \sim 1.32$ nm, and then approaches zero again (Figure 10.4(c)). It should be noted that, for relatively small diameters ($d_t < \sim 1.32$ nm), ΔE_{11} and ΔE_{22} of the (13, 5) tube (type I) have opposite signs as opposed to those of the (14, 4) tube (type II) (Figure 10.4(c) and (d)). While this family-type dependence behaviour can be explained by the radial expansion of SWCNTs upon C_{70} encapsulation, the observed shifts are somewhat larger than those of (13, 5) and (14, 4) C_{60} nanopeapods (Figure 10.4(a) and (b)).

The degree of radial expansion can be estimated by Equation (10.1) (see Sec. 10.4.2). For the (13, 5) tube (d_t = 1.248 nm), we observed shifts of 24 and 13 meV in the value of E_{11} for C_{70} and C_{60} nanopeapods, respectively (Figure 10.4(c) and (a)); these values correspond to the radial expansion in the tube of 0.37% and 0.20%, respectively, according to Equation (10.1). Hence, the difference in the tube diameter expansion (Δd_t) for C_{70} and C_{60} insertions can be calculated to be 0.0021 nm. The same analysis for the (14, 4) tube (d_t = 1.300 nm) results in a Δd_t = 0.0010 nm. Surprisingly, these values are almost identical with the difference between the length of the short axis of C_{70} and the diameter of C_{60} (=0.002 nm) [49,50]. SWCNTs can be used to detect such a minute difference in molecular size with a sensitivity of ~0.001 nm.

ΔE_{11} and ΔE_{22} of the C_{70} nanopeapods exhibit analogous diameter dependence in the larger-diameter region (d_t > ~1.32 nm) (Figure 10.4(c) and (d)); however, the magnitude of the shifts becomes smaller than those of C_{60} nanopeapods. Figure 10.4(e) and (f) shows the differences in the values of E_{11} and E_{22} for the C_{70} and C_{60} nanopeapods ($\Delta E'_{ii} = E_{ii}^{C70NPDs} - E_{ii}^{C60NPDs}$, i = 1.2) as a function of the tube diameter (d_t > ~1.32 nm). From the figure, it is clear that $\Delta E'_{11}$ and $\Delta E'_{22}$ of the (14, 6) (type I), and the (16, 3) and (15, 5) (type II) tubes (d_t = 1.405, 1.411 and 1.431 nm, respectively) are distinguishable from the other type I and type II tubes. (The characteristic behaviours of (12, 10) tubes will be discussed later.)

The difference between the length of the long axis of C_{70} (=0.796 nm) [49] and the diameter of C_{60} (=0.710 nm) [50] is 0.086 nm. Figure 10.4(a) and (b) suggests that the smallest tube diameter for encapsulating C_{60} without friction is ~1.32 nm. Thus, if the tube diameter is more than ~1.406 nm, C_{70} molecules can find enough space to 'stand up' inside SWCNTs. Surprisingly, the diameter of 1.406 nm is almost identical to that of the (16, 3) tubes (=1.405 nm). This remarkable agreement strongly suggests that C_{70} molecules start to change their alignments in the (16, 3) tubes, which causes the optical transition energy changes observed here. Note that this transition diameter value is in good agreement with the reported values of 1.42 and 1.435 nm, which were estimated by XRD experiments and theoretical calculations, respectively [33,34].

The standing alignment of C_{70} results in an interspacing between C_{70} and the tube wall, which is substantially smaller than that of the C_{60} nanopeapods. As a result of the decrease in the interspacing, the interaction between C_{70} and SWCNTs becomes less attractive. The lower interaction between C_{70} and the SWCNTs should result in smaller spectral shifts. Notably, the values of ΔE_{11} and ΔE_{22} of the (16, 3), (14, 6) and (15, 5) C_{70} nanopeapods shift towards the zero line as compared to those of C_{60} nanopeapods (Figure 10.4(a)–(d)).

Beyond the transition diameter, the standing C_{70} are energetically favourable [48]. Consequently, the attractive interaction between C_{70} and the SWCNTs occurs at $d_t \sim 1.5$ nm because the difference between the length of the long axis of C_{70} and the diameter of C_{60} is 0.086 nm [49,50]. This type of attractive interaction should cause a large shift in E_{11} and E_{22}. The ΔE_{11} and ΔE_{22} values of the (12, 10) and (13, 9) C_{70} nanopeapods are consistent with this supposition (Figure 10.4(c) and (d)). For instance, the ΔE_{11} and ΔE_{22} values of the (12, 10) C_{70} nanopeapods are smaller and larger than those of the (12, 10) C_{60} nanopeapods, respectively, presumably because of the more attractive interaction between C_{70} and the SWCNTs than that between C_{60} and the SWCNTs (Figure 10.4(a)–(d)). The large $\Delta E'_{22}$ of (12, 10) tubes can also be explained by this mechanism (Figure 10.4(f)).

In addition to C_{60} and C_{70} fullerenes, the effect of azafullrerene $(C_{59}N)_2$ and coronene encapsulations on the electronic states of semiconducting SWCNTs have also been investigated by PL spectroscopy [10,11]. The interaction between SWCNTs and the encapsulated molecules (azafullerenes and coronenes) can be explained by van der Waals-type intermolecular force as is the case of C_{60} and C_{70} fullerenes.

10.4.4 Bandgap modulation in fullerene nanopeapods at nanometer scale

Low-temperature scanning tunnelling microscope (STM) and scanning tunnelling spectroscopy (STS) studies performed on Gd@C_{82} nanopeapods and C_{60} nanopeapods revealed significant bandgap modulations at a molecular scale [51,52]. For instance, the bandgap of a semiconducting (11, 9) SWCNT is altered by the insertion of Gd@C_{82} [51]. The original bandgap of 0.43 eV is narrowed down to 0.17 eV where the fullerene is expected to be located.

A down shift in the first van Hove bandgap of ~60 meV was observed for semiconducting SWCNTs with $d_t \sim 1.3$ nm upon C_{60} encapsulation [52]. Interestingly, the value of 60 meV is in remarkable agreement with a decrease of 36 meV in ΔE_{11} for C_{60}@(14, 4) nanopeapods having $d_t = 1.30$ nm, which was obtained by PL spectroscopy (see Sec. 10.4.1).

10.4.5 Valence states of encapsulated atoms in metallofullerene nanopeapods

The electronic structures of doubly (i.e. first by a fullerene and then by a SWCNT) encapsulated atoms in nanopeapods have been investigated by EELS. Especially, the peak positions of the *M*-edge of the lanthanoid metal atoms have served as guides for determining the valence state and consequently the amount of charge transfer [37, 38]. For example, the highest peak positions for the M_5 and M_4 edges for $(Gd@C_{82})_n$@SWCNTs are experimentally measured as 1184 and 1014 eV, respectively [4]. The peak positions of Gd M_{45} edges for $(Gd@C_{82})_n$@SWCNTs are very similar to those for the intact Gd@C_{82}, indicating that the valence state for the encaged Gd atoms of Gd@C_{82} remains unchanged upon encapsulation by SWCNTs.

Similar experiments have been carried out for $(Sc@C_{82}(I))_n$@SWCNTs, $(Ti_2C_2@C_{78})_n$@SWCNTs, $(La@C_{82})_n$@SWCNTs, $(La_2@C_{80})_n$@SWCNTs, $(Ce_2@C_{80})_n$@SWCNTs, $(Sm@C_{82})_n$@SWCNTs and $(Gd_2@C_{92})_n$@SWCNTs [30,53–55]. Even in these nanopeapods, the EELS spectra of the encapsulated atoms in the nanopeapods were same as those in the corresponding intact metallofullerenes and insensitive to the chirality of SWCNTs. The encapsulated atoms of Sc, Ti, La, Sm and Gd still remain Sc^{2+}, Ti^{2+}, La^{3+}, Sm^{2+} and Gd^{3+} states, respectively, in their nanopeapods structures.

10.5 PHONON PROPERTIES OF NANOPEAPODS

10.5.1 Radial breathing modes (RBMs)

As described in Sec. 10.4.2, Raman spectroscopy is one of the most suitable and efficient techniques for characterizing nanopeapods. Figure 10.5 shows the resonance Raman intensity map of metallic C_{60} nanopeapods in RBM frequency region. On the other hand, the 2D contour map of RBM intensity of semiconducting C_{60} nanopeapods (d_t = 1.2–1.4 nm) is shown in Figure 10.7, together with a reference map of original SWCNTs [56–58]. The white circles in Figure 10.7 indicate the RBM phonon peak positions in the individual (n, m) tubes obtained by the line-shape fitting. Typical $2n + m$ family patterns are clearly observed in the maps. On the basis of the PL and Raman results reported in a previous study [8,41], the observed $2n + m$ branches can be assigned to the $2n + m = 28$, 29, 31, 32 and 34 families, as shown in Figure 10.7.

To investigate the C_{60} fullerene encapsulation effects on the RBM frequencies of semiconducting and metallic SWCNTs in further detail, the frequency difference before and after insertion ($\Delta\omega = \omega^{\text{nanopeapods}} - \omega^{\text{SWCNTs}}$) is plotted as a function of the tube diameter (Figure 10.6 (c)) [47,56–58]. Apparently, the frequency shifts show strong tube diameter dependence. In the case of semiconducting nanopeapods with smaller diameters, such as (14, 3) and (13, 5) tubes, high-frequency shifts are observed. With increasing tube diameter, the $\Delta\omega$ value monotonically decreases towards about -6 cm^{-1} at around 1.41 nm and then increases again approaching to zero.

The upward shifts observed in the case of smaller diameter tubes can be explained by the fact that in these tubes, there is steric hindrance provided by the encapsulated C_{60} for RBM motion of SWCNTs [59]. The insertion of an oversized C_{60} molecule directly suppresses the radial vibration motion of SWCNTs. The local strain on the SWCNTs after C_{60} encapsulation may disappear at ~1.32 nm because the frequency shift at this diameter changes the sign. Interestingly, this value matches well with the threshold diameter for C_{60} encapsulation without friction estimated based on the PL measurements (Figure 10.4(a) and (b)). Excellent agreement about the threshold diameter strongly suggests that the RBM phonon shifts observed in the case of (14, 3) and (13, 5) tubes are induced by the repulsive interaction between SWCNTs and C_{60} arising from the limited spacing between them.

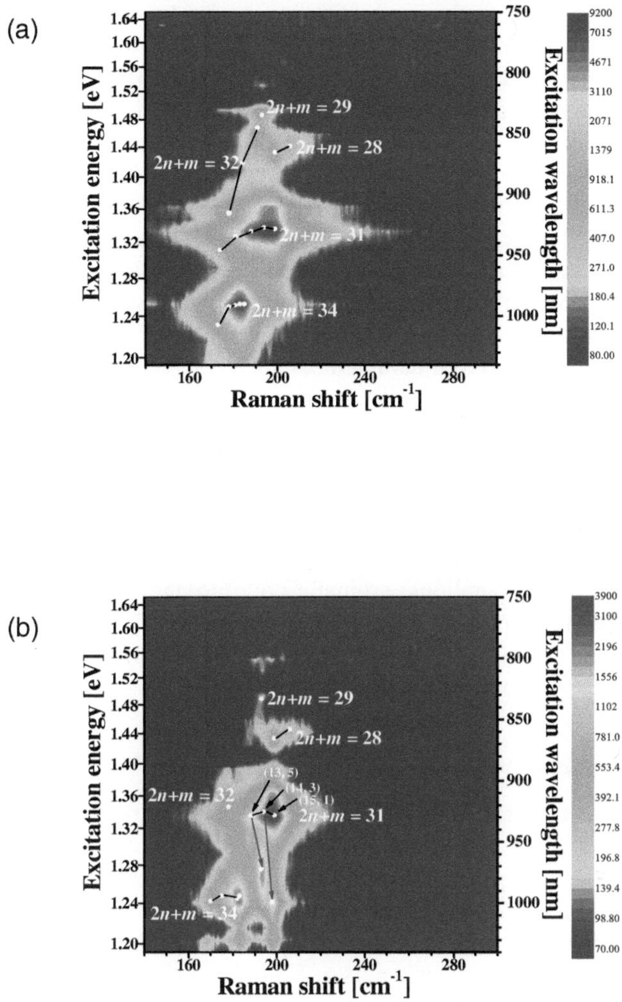

FIGURE 10.7 2D Raman intensity maps of RBM-phonon regions of (a) original SWCNTs and (b) C_{60} nanopeapods [58].

With increasing diameter ($d_t > \sim 1.3$ nm), the attractive intermolecular interaction becomes dominant, and this results in the downward shift of RBM phonon frequencies [47,56–58]. Interestingly, the smallest value of $\Delta\omega (=-6\ \text{cm}^{-1})$ was observed at around an interspacing of 0.35 nm ($d_t = 1.41$ nm); this distance is similar to the interlayer distance for graphite (0.335 nm) [45]. Naturally, the interaction between the SWCNTs and the encapsulated C_{60} becomes weaker as the distance between the tube wall and C_{60} increases. Consequently, the RBM frequency difference approaches zero ($\Delta\omega = 0$).

Interestingly, the diameter dependence of the RBM phonon frequency shifts is in good agreement with the PL results (Sec. 10.4.1). These results strongly suggest that the basic mechanism of C_{60} encapsulation on bandgap modification is the same as that on RBM frequency shifts.

On the other hand, the observed downshifts ($\Delta\omega < 0$) for metallic nanopeapods in the larger-diameter region can be explained by the same mechanism (Figure 10.6(c)). In fact, the effective interaction between the π states of the metallic SWCNTs and C_{60} has been predicted using theoretical calculations [59].

Although the diameter dependence of $\Delta\omega$ of metallic nanopeapods is consistent with that of semiconducting nanopeapods, the amount of the frequency shifts is about half of that for the semiconducting nanopeapods (Figure 10.6(c)). A simple explanation for the smaller-frequency shifts is that the interaction between metallic SWCNTs and C_{60} is weaker than that between semiconducting SWCNTs and C_{60}. Even though theoretical calculations usually predict that aromatic molecules more strongly adsorb onto the surface of metallic SWCNTs than that of semiconducting SWCNTs [60], a close similarity between the diameter dependence of $\Delta\omega$ (Figure 10.6(c)) and σ (Figure 10.6(b)) strongly suggests that the intermolecular interaction between metallic SWCNTs and the encapsulated C_{60} is weaker than that of semiconducting SWCNTs and C_{60}. This seems to be consistent with recent experimental results for metallic-semiconducting SWCNTs separations; aromatic polyfluorene was used to selectively extract semiconducting SWCNTs from a mixture of both metallic and semiconducting SWCNTs [61,62], and the fast eluent behaviours of metallic SWCNTs observed in the agarose gel method [63].

10.5.2 G-band

It is well known that the charge transfer between encapsulated molecules and SWCNTs causes the peak shift of G-band [64]. Figure 10.8(a) shows G-band spectra of C_{60} nanopeapods and SWCNTs at different excitation energies; 1.33 eV (930 nm), 1.29 eV (960 nm), 1.28 eV (970 nm) and 1.25 eV (990 nm) [57]. These excitation energies correspond to the absorption peaks of the semiconducting SWCNTs with diameters of 1.3–1.5 nm [57]. Apparently, there is no peaks shift in these spectra, which indicates absence of apparent charge transfer between semiconducting SWCNTs and the encapsulated C_{60}.

Under the excitation energies of 1.60 and 1.70 eV, we observed typical G-mode signals corresponding to metallic C_{60} nanopeapods (Figure 10.8(b) and (c)) [47]. The spectral shape of the G-mode does not change after C_{60} encapsulation. In particular, the peak positions and the line width of G^--bands at around 1560–1570 cm^{-1} do not show significant change within an experimental resolution. In the case of metallic SWCNTs, the tangential G-mode is very sensitive to the *e*-ph interaction as well as charge transfer [65]. In particular, the peak positions and the line width of the longitudinal optical (LO) G^--band severely decrease upon doping. Hence, the strength of the *e*-ph coupling should be similar for both SWCNTs and nanopeapods, and it may be reasonable to neglect the *e*-ph effects on the RBM shifts observed in C_{60} nanopeapods (Sec. 10.5.1).

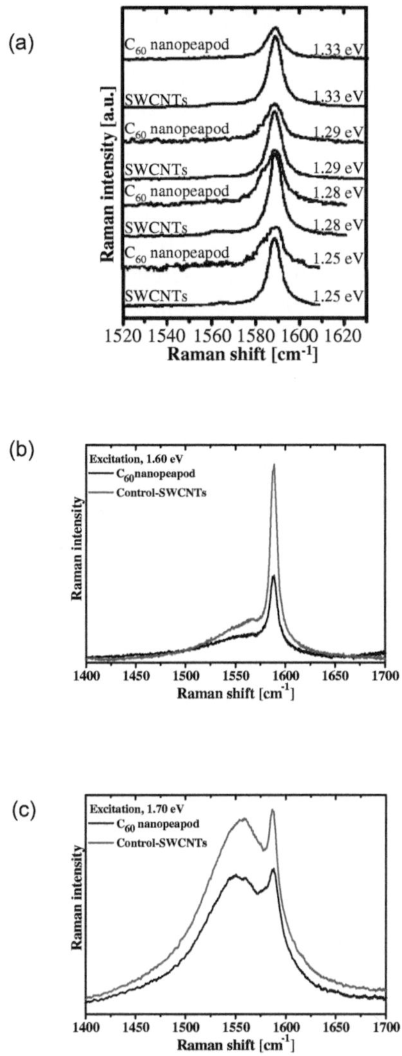

FIGURE 10.8 Resonance Raman spectra in the G-band region of C_{60} nanopeapods and SWCNTs at excitation energies of (a) 105 ~ 1.33 eV, (b) 1.60 eV and (c) 1.70 eV [49,59].

10.6 TRANSPORT PROPERTIES OF NANOPEAPODS

Controlled electronic states of SWCNTs have been achieved by encapsulating fullerenes as described in Sec. 10.4. Such bandgap modification leads to the substantial change in transport properties of SWCNTs. The electric transport properties of several nanopeapods are investigated by using these as channels in the field-effect transistor (FET), which has a back-gate structure [66,67].

Normally, SWCNTs exhibit *p*-type FET properties because oxygen easily converts SWCNTs into *p*-type semiconductors. On the other hand, most nanopeapods such as C_{60} nanopeapods and Gd@C_{82} nanopeapods show ambipolar *p*- and *n*-type behaviour [66]. The bandgap of some SWCNTs becomes smaller by fullerene encapsulations. The small bandgap can lead to significant carrier transport through both conduction and valence bands.

Azafullerenes nanopeapods have been reported to change the transport properties of SWCNTs from *p*-type to *n*-type in FET structures [68,69]. However, spectroscopic study suggests that azafullerene nanopeapods may show *p*-type FET behaviour similar to that of C_{60} nanopeapods [10]. Recent first-principles electronic-structure calculations reveal that the unique transport behaviour of azafullerene nanopeapods originates from the monomer form of $C_{59}N$ encapsulated in SWCNTs [69]. The singly occupied (SO) state of $C_{59}N$ lies in the energy gap of the SWCNT and the energy of this state increases substantially when electrons are injected. Because of this shift to higher energy, the SO state acts as a shallow donor state for the conduction band of the nanotube, which leads to *n*-type behaviour in FET measurements.

10.7 NANOPEAPOD AS A SAMPLE CELL AT NANOMETER SCALE

Since SWCNTs are extremely thin and almost transparent to high-energy electron beams, TEM can allow us to directly observe the structures and motion of encapsulated molecules and their chemical reactions [17,26,53,55,70–77].

The initial stage of electron-induced coalescence reaction of metallofullerenes was investigated by means of in-situ TEM and EELS [71]. Figure 10.9(a) shows HRTEM images of an individual Sm@C_{82} encapsulating SWCNT at the beginning of the observation (before the reaction). The Sm@C_{82} molecules are perfectly aligned in a 1D way with a constant intermolecular spacing.

Following the irradiation for a period of 4 min, a molecular rearrangement in the nanotubes starts to occur (Figure 10.9(b)). Some Sm@C_{82} molecules are already so close with each other as to form dimers or clusters as seen in the HRTEM image. The appearance of dark spots by Sm atoms in Figure 10.9(b) suggests that the rotational motion of Sm-metallofullerenes was frozen due to the formation of intermolecular covalent bondings. Further exposure for ~10 min resulted in coalescence or fusion, which generates the dimers and clusters of the Sm-metallofullerenes (Figure 10.9(c)). These were finally converted to larger nano-capsule after ~20 min irradiation (Figure 10.9(d)). The dark spots observed near the wall of the newly produced capsule correspond to the Sm atoms.

Further information on such reaction dynamics can be obtained by simultaneous in-situ EELS measurements (Figure 10.10). The EELS spectrum observed from the bundles of the intact Sm@C_{82} encapsulating SWCNTs is shown in Figure 10.10(a) together with the reference spectra of

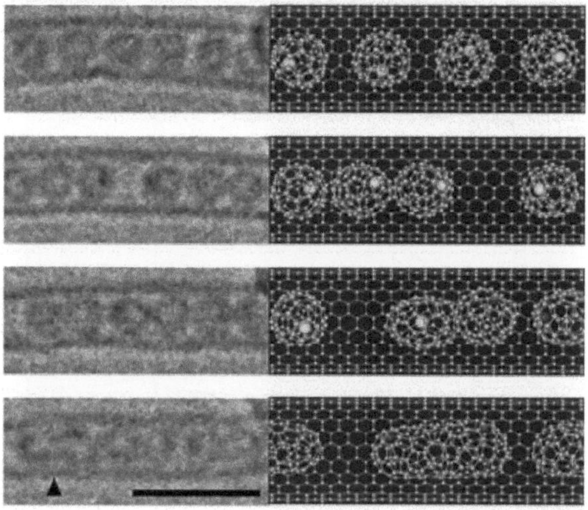

FIGURE 10.9 Time evolution of the HRTEM images of $Sm@C_{82}$ encapsulating SWCNTs (left) and the schematic illustrations (right). The HRTEM images were observed after (a) ~0 min, (b) ~4 min, (c) ~10 min and (d) ~20 min irradiations, respectively. The produced nanocapsules can be clearly seen inside the SWCNT (arrows). The light grey (yellow in the web version) and the dark grey (red in the web version) balls in the illustrations denote the divalent Sm^{2+} and the trivalent Sm^{3+} atoms, respectively. Scale bar = 3 nm [71].

Sm^{2+} in $Sm@C_{82}$ and Sm^{3+} in Sm_2O_3. The peak positions of Sm M_{45} edges for $Sm@C_{82}$ nanopeapods before the reaction are the same as those for the intact $Sm@C_{82}$. After electron beam irradiation for a period of ~10 min, the peak widths of M_{45} edges became broader, showing additional intense peaks overlapping with the main peaks (Figure 10.10(b)). At the same time, the intensity of the divalent Sm^{2+} peaks decreased with the irradiation, and finally new M_{45} peaks appeared (Figure 10.10(c)). In particular, the M_5 edge at ~1080 eV can be clearly seen in the spectrum; whose peak position for this M_{45} edge corresponds to that of the trivalent Sm^{3+}.

At this point the valency of the Sm atoms has transformed from +2 to +3 with a time scale of ~10 min and the conversion was completed within ~20 min irradiation. The time-dependent HRTEM and EELS measurements indicate that the oxidation state of Sm has transformed from +2 to +3 as a result of a new chemical bond formation due to Sm carbides upon coalescence of the fullerene cages (Figure 10.10(c)). Sm atoms may not be encaged in the nanocapsule but should at least be incorporated into the cage network, since the oxidation state of the Sm carbides (SmC_2) is +3 and those of various Sm-metallofullerenes are known to be +2 irrespective of the fullerene size. The Sm atoms tend to be incorporated into the carbon networks to form much stable carbide bonds and do not form Sm clusters.

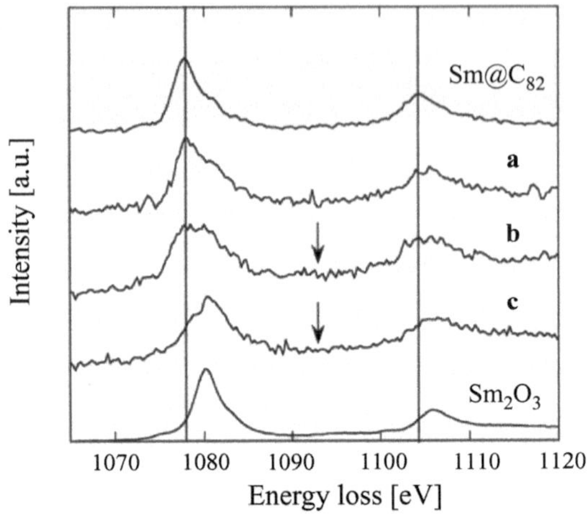

FIGURE 10.10 Time evolution of the EELS spectrum for Sm@C_{82} encapsulating SWCNTs, together with reference spectra for the divalent Sm^{2+} in the intact Sm@C_{82} molecule (upper) and the trivalent Sm^{3+} in Sm_2O_3 (bottom). The EELS spectra were taken at ~10 min intervals [71].

10.8 CONCLUSION

The interior spaces of SWCNTs have almost same size for many organic molecules so that the effective interaction between the guest molecules and the host materials is expected. In fact, as seen in this chapter, the ability to modify the electronic properties of nanopeapods can be realized by encapsulating fullerenes. Such hybrid structures are also useful for direct observations of shape and motion of molecules, and their chemical reactions. One of the great challenges for nanotechnology is the creation of novel functional materials through molecular-level control of material composition. We believe that SWCNTs will lead to a new class of multifunctional materials by utilizing their 'empty' spaces.

ACKNOWLEDGEMENT

We thank Dr. Y. Iizumi (AIST) for her assistance in preparing the illustrations.

REFERENCES

[1] Ajayan PM, Iijima S. Capillarity-induced filling of carbon nanotubes. Nature 1993;361:333–4.
[2] Dujardin E, Ebbesen TW, Hiura H, Tanigaki K. Capillarity and wetting of carbon nanotubes. Science 1994;265:1850–2.
[3] Smith BW, Monthioux M, Luzzi DE. Encapsulated C_{60} in carbon nanotubes. Nature 1998;396:323–4.
[4] Hirahara K, Suenaga K, Bandow S, Kato H, Okazaki T, Shinohara H, et al. One dimensional metallofullerenes crystal generated inside single-walled carbon nanotube. Phys Rev Lett 2000;85:5384–7.

[5] Kataura H, Maniwa Y, Kodama T, Kikuchi K, Hirahara K, Suenaga K, et al. High-yield fullerene encapsulation in single-wall carbon nanotubes. Synth Metals 2001;101:1195–6.

[6] Shinohara H. Endohedral metallofullerenes. Rep Prog Phys 2000;63:843–92.

[7] Okubo S, Okazaki T, Kishi N, Joung S-K, Nakanishi T, Okada S, et al. Diameter-selective interaction between single-walled carbon nanotubes and encapsulated fullerenes. J Phys Chem C 2009;113:571–5.

[8] Okazaki T, Okubo S, Nakanishi T, Joung S-K, Saito T, Otani M, et al. Optical band gap modification of single-walled carbon nanotubes by encapsulated fullerenes. J Am Chem Soc 2008;130:4102–8.

[9] Okubo S, Okazaki T, Hirose-Takai K, Suenaga K, Okada S, Bandow S, et al. Electronic structures of single-wall carbon nanotubes encapsulating ellipsoidal C_{70}. J Am Chem Soc 2010;132:15252–8.

[10] Iizumi Y, Okazaki T, Liu Z, Suenaga K, Nakanishi T, Iijima S, et al. Host-guest interactions in azafullerene ($C_{59}N$)-single-wall carbon nanotube (SWCNT) peapod hybrid structures. Chem Commun 2010;46:1093–5.

[11] Okazaki T, Iizumi Y, Okubo S, Kataura H, Liu Z, Suenaga K, et al. Coaxially stacked coronene column inside single-wall carbon nanotube. Angew Chem Int Ed 2011;50:4853–7.

[12] Yudasaka M, Ajima K, Suenaga K, Ichihashi T, Hashimoto A, Iijima S. Nano-extraction and nano-condensation for C_{60} incorporation into single-wall carbon nanotubes in liquid phases. Chem Phys Lett 2003;380:42–6.

[13] Simon F, Kuzmany H, Rauf H, Pichler T, Bernardi J, Peterlik H, et al. Low temperature fullerene encapsulation in single wall carbon nanotubes: synthesis of N@C_{60}@SWCNT. Chem Phys Lett 2004;383:362–7.

[14] Yanagi K, Miyata Y, Kataura H. Highly stabilized β-carotene in carbon nanotubes. Adv Mater 2006;18:437–41.

[15] Yanagi K, Iakoubovskii K, Mitsui H, Matsuzaki H, Okamoto H, Miyata Y, Maniwa Y, Kazaoui S, Minami N, Kataura H. Photosensitive Function of Encapsulated Dye in Carbon Nanotubes. J Am Chem Soc 2007;109:4992–7.

[16] Britz A, Khlobystov AN, Wang J, O'Neil AS, Poliakoff M, Ardavan A, et al. Selective host-guest interaction of single-walled carbon nanotubes with functionalized fullerenes. Chem Commun 2004;176–7.

[17] Chamberlain TW, Camenisch A, Champness NR, Briggs GAD, Benjamin SC, Ardavan A, et al. Toward controlled spacing in one-dimensional molecular chains: alkyl-chain-functionalized fullerenes in carbon nanotubes. J Am Chem Soc 2007;109:8609–14.

[18] Yanagi K, Iakoubovskii K, Kazaoui S, Minami N, Maniwa Y, Miyata Y, et al. Light-harvesting function of β-carotene inside carbon nanotubes. Phys Rev B 2006;74:155420.

[19] Liu X, Pichler T, Knupfer M, Golden MS, Fink J, Kataura H, et al. Filling factors, structural, and electronic properties of C_{60} molecules in single-wall carbon nanotubes. Phys Rev B 2002;65:045419.

[20] Kataura H, Maniwa Y, Abe M, Fujiwara A, Kodama T, Kikuchi K, et al. Optical properties of fullerene and non-fullerene peapods. Appl Phys A 2002;74:349–54.

[21] Smith BW, Russo RM, Chikkannanavar SB, Luzzi DE. High-yield synthesis and one-dimensional structure of C_{60} encapsulated in single-wall carbon nanotubes. J Appl Phys 2002;91:9333–40.

[22] Berber S, Kwon Y-K, Tománek D. Microscopic formation mechanism of nanotube peapods. Phys Rev Lett 2002;88:185502.

[23] Ulbricht H, Moos G, Hertel T. Interaction of C_{60} with carbon nanotubes and graphite. Phys Rev Lett 2003;90:095501.

[24] Okada S, Saito S, Oshiyama A. Energetics and electronic structures of encapsulated C_{60} in a carbon nanotube. Phys Rev Lett 2001;86:3835–8.

[25] Okada S, Saito S, Oshiyama A. Electron-state control of carbon nanotubes by space and encapsulated fullerenes. Phys Rev B 2003;67:205411.

[26] Bandow S, Takizawa M, Hirahara K, Yudasaka M, Iijima S. Raman scattering study of double-wall carbon nanotubes derived from the chains of fullerenes in single-wall carbon nanotubes. Chem Phys Lett 2001;337:48–54.

[27] Fan J, Yudasaka M, Yuge R, Futaba DN, Hata K, Iijima S. Efficiency of C_{60} incorporation in and release from single-wall carbon nanotubes depending on their diameters. Carbon 2007;45:722–6.

[28] Simon F, Peterlik H, Pfeiffer R, Bernardi J, Kuzmany H. Fullerene release from the inside of carbon nanotubes: a possible route toward drug delivery. Chem Phys Lett 2007;445:288–92.

[29] Hirahara K, Suenaga K, Bandow S, Kato H, Okazaki T, Shinohara H, et al. Electron diffraction study of one-dimensional crystals of fullerenes. Phys Rev B 2001;64:115420.

[30] Okazaki T, Suenaga K, Hirahara K, Bandow S, Iijima S, Shinohara H. Electronic and geometric structures of metallofullerene peapods. Phys B 2002;323:97–9.

[31] Abe M, Kataura H, Kira H, Kodama T, Suzuki S, Achiba Y, et al. Structural transformation from single-wall to double-wall carbon nanotube bundles. Phys Rev B 2003;68. 041405(R).

[32] Maniwa Y, Kataura H, Abe M, Fujiwara A, Fujiwara R, Kira H, et al. C_{70} molecular stumbling inside single-walled carbon nanotubes. J Phys Soc Jpn 2003;72:45–8.

[33] Chorro M, Cambedouzou J, Iwaseiewicz-Wabnig A, Noé L, Rols S, Monthioux M, et al. Discriminated structural behavior of C_{60} and C_{70} peapods under extreme conditions. Europhys Lett 2007;79:56003.

[34] Chorro M, Delhey A, Noe L, Monthioux M, Launois P. Orientation of C_{70} molecules in peapods as a function of the nanotube diameter. Phys Rev B 2007;75:035416.

[35] Verberck B, Michael KH. Nanotube field and orientational properties of C_{70} molecules in carbon nanotubes. Phys Rev B 2007;75:045419.

[36] Okazaki T, Lian Y, Gu Z, Suenaga K, Shinohara H. Isolation and spectroscopic characterization of Sm-containing metallofullerenes. Chem Phys Lett 2000;320:435–40.

[37] Okazaki T, Suenaga K, Lian Y, Gu Z, Shinohara H. Direct EELS observation of the oxidation states of Sm atoms in $Sm@C_{2n}$ metalfullerenes ($74 \leq 2n \leq 84$). J Chem Phys 2000;113:9593–7.

[38] Suenaga K, Kato H, Shinohara H, Iijima S. Fine-structure analysis of Gd M_{45} near-edge EELS on the valence state of $Gd@C_{82}$ microcrystals. Phys Rev B 2000;62:1627–30.

[39] Nishibori E, Takata M, Sakata M, Tanaka H, Hasegawa M, Shinohara H. Giant motion of La atom inside C_{82} cage. Chem Phys Lett 2000;330:497–502.

[40] Robertson JM, White JG. The crystal structure of coronene: a quantitative X-ray investigation. J Chem Soc 1945;607–17.

[41] Weisman BR, Bachilo SM. Dependence of optical transition energies on structure for single-walled carbon nanotubes in aqueous suspension: an empirical Kataura plot. Nano Lett 2003;3:1035–8.

[42] Maniwa Y, Kataura H, Abe M, Suzuki S, Achiba Y, Kira H, et al. Phase transition in confined water inside carbon nanotubes. J Phys Soc Jpn 2002;71:2863–6.

[43] Ando T. Excitons in carbon nanotubes revisited: dependence on diameter, Aharonov-Bohm flux, and strain. J Phys Soc Jpn 2004;73:3351–63.

[44] Yang L, Han J. Electronic structure of deformed carbon nanotubes. Phys Rev Lett 2000;85:154–7.

[45] Palser HAR. Interlayer interactions in graphite and carbon nanotubes. Phys Chem Chem Phys 1999;1:4459–64.

[46] Kane CL, Mele EJ. Ratio problem in single carbon nanotube fluorescence spectroscopy. Phys Rev Lett 2003;90:207401.

[47] Joung S-K, Okazaki T, Okada S, Iijima S. Weak response of metallic single-walled carbon nanotubes to C_{60} encapsulation studied by resonance Raman spectroscopy. J Phys Chem C 2010;116:23844–50.

[48] Otani M, Okada S, Oshiyama A. Energetics and electronic structures of one-dimensional fullerene chains encapsulated in zigzag nanotubes. Phys Rev B 2003;68:105424.

[49] Nikolaev AV, Dennis TJS, Prassides K, Soper AK. Molecular structure of the C_{70} fullerene. Chem Phys Lett 1994;223:143–8.

[50] Johnson RD, Bethune DS, Yannoni CS. Fullerene structure and dynamics: a magnetic resonance potpourri. Acc Chem Res 1992;25:169–75.

[51] Lee J, Kim H, Kahng S-J, Kim G, Son Y-W, Ihm J, et al. Band gap modulation of carbon nanotubes by encaptured metallofullerenes. Nature 2002;415:1005–8.

[52] Hornbaker DJ, Kahng S-J, Misra S, Smith BW, Johnson AT, Mele EJ, et al. Mapping the one-dimensional electronic states of nanotube peapod structures. Science 2002;295:828–31.

[53] Suenaga K, Okazaki T, Wang C-R, Bandow S, Shinohara H, Iijima S. Direct imaging of Sc_2@C_{84} molecules encapsulated inside single-wall carbon nanotubes by high resolution electron microscopy with atomic sensitivity. Phys Rev Lett 2003;90:055506.

[54] Suenaga K, Hirahara K, Bandow S, Iijima S, Okazaki T, Kato H, et al. Core-level spectroscopy on the valence state of encaged metal in metallofullerene-peapods. In: Kuzmany H, Fink J, Mehring M, Roth S, editors. Proceedings of the 15th International Winterschool, AIP Conference Proceedings, American Institute of Physics, New York; 2001;591056-260.

[55] Suenaga K, Taniguchi R, Shimada T, Okazaki T, Shinohara H, Iijima S. Evidence for the intermolecular motion of Gd atoms in a Gd_2@C_{92} nanopeapod. Nano Lett 2003;3:1395–8.

[56] Joung S-K, Okazaki T, Kishi N, Okada S, Bandow S, Iijima S. Effect of fullerene encapsulation on radial breathing mode frequencies of single-wall carbon nanotubes. Phys Rev Lett 2009;103:027403.

[57] Joung S-K, Okazaki T, Okada S, Iijima S. Interaction between single-wall carbon nanotubes and encapsulated C_{60} probed by resonance Raman spectroscopy. Phys Chem Chem Phys 2010;10:8118–22.

[58] Joung S-K, Okazaki T, Okada S, Iijima S. Host-guest interaction between single-wall carbon nanotubes and encapsulated C_{60} probed by resonance Raman spectroscopy. Phys Stat Sol B 2010;247:2700–2.

[59] Okada S. Radial-breathing mode frequencies for nanotubes encapsulating fullerenes. Chem Phys Lett 2007;438:59–62.

[60] Lu J, Nagase S, Zhang X, Wang D, Ni M, Maeda Y, et al. Selective interaction of larger or charge-transfer aromatic molecules with metallic single-wall carbon nanotubes: critical role of the molecular size and orientation. J Am Chem Soc 2006;108:5114–8.

[61] Nish A, Hwang J-Y, Doig J, Nicholas RJ. Highly selective dispersion of single-walled carbon nanotubes using aromatic polymers. Nat Nanotechnol 2007;2:640–6.

[62] Chen F, Wang B, Chen Y, Li L-J. Toward the extraction of single species of single-walled carbon nanotubes using fluorene-based polymers. Nano Lett 2007;7:3013–7.

[63] Tanaka T, Jin H, Miyata Y, Kataura H. High-yield separation of metallic and semiconducting single-wall carbon nanotubes by agarose gel electrophoresis. Appl Phys Exp 2008;1:114001.

[64] Kavan L, Dunsch L. Spectroelectrochemistry of carbon nanostructures. ChemPhysChem 2007;8:974–98.

[65] Farhat H, Son H, Samsonidze Ge G, Reich S, Dresselhaus MS, Kong J. Phonon softening in individual metallic carbon nanotubes due to the Kohn anomaly. Phys Rev Lett 2007;99:145506.

[66] Shimada T, Okazaki T, Taniguchi R, Sugai T, Shinohara H, Suenaga K, et al. Ambipolar field-effect transistor behavior of Gd@C_{82} metallofullerene peapods. Appl Phys Lett 2002;81:4067–9.

[67] Shimada T, Ohno Y, Okazaki T, Sugai T, Suenaga K, Kishimoto S, et al. Transport properties of C_{78}, C_{90} and Dy@C_{82} fullerenes-nanopeapods by field effect transistors. Phys E 2004;21:1089–92.

[68] Kaneko T, Li Y, Nishigaki S, Hatakeyama R. Photoswitching in azafullerene encapsulated single-walled carbon nanotube FET devices. J Am Chem Soc 2008;130:2714–5.

[69] Cuong NT, Otani M, Iizumi Y, Okazaki T, Rotas G, Tagmatarchis N, et al. Origin of the n-type transport behavior of azafullerenes encapsulated in single-walled carbon nanotubes. Appl Phys Lett 2011;99:053105.

[70] Luzzi DE, Smith BW. Carbon cage structures in single wall carbon nanotubes: a new class of materials. Carbon 2000;38:1751–6.

[71] Okazaki T, Suenaga K, Hirahara K, Bandow S, Iijima S, Shinohara H. Real time reaction dynamics in carbon nanotubes. J Am Chem Soc 2001;103:9673–4.

[72] Liu Z, Yanagi K, Suenaga K, Kataura H, Iijima S. Imaging the dynamic behaviour of individual retinal chromophores confined inside carbon nanotubes. Nat Nanotechnol 2007;2:422–5.

[73] Liu Z, Koshino M, Suenaga K, Mrzel A, Kataura H, Iijima S. Transmission electron microscopy imaging of individual functional groups of fullerene derivatives. Phys Rev Lett 2006;96:088304–8.

[74] Guan L, Suenaga K, Shi Z, Gu Z, Iijima S. Polymorphic structures of iodine and their phase transition in confined nanospace. Nano Lett 2007;7:1532–5.

[75] Koshino M, Tanaka T, Solin N, Suenaga K, Isobe H, Nakamura E. Imaging of single organic molecules in motion. Science 2007;316:853.

[76] Liu Z, Suenaga K, Iijima S. Imaging the structure of an individual C_{60} fullerene molecule and its deformation process using HRTEM with atomic sensitivity. J Am Chem Soc 2007;109:6666–7.

[77] Koshino M, Niimi Y, Nakamura E, Kataura H, Okazaki T, Suenaga K, et al. Towards imaging atomic resolved molecular reactions: study of fullerene dimerization. Nat Chem 2010;2:117–24.

Chapter 11

Applications of Carbon Nanotubes and Graphene in Spin Electronics

Akihiro Ito, Kazuyoshi Tanaka
Department of Molecular Engineering, Kyoto University, Nishikyo-ku, Kyoto 615-8510, Japan

11.1 SPINTRONICS

Electrons carry an intrinsic angular momentum called spin, as well as their electric charge in electronic devices. In conventional charge-based CMOS (complementary metal-oxide-semiconductor) devices composed of metal-oxide-semiconductor field-effect transistors (MOSFETs), however, logic operations have been achieved only by controlling the height of a barrier to the electric charge current with the gate voltage. In conjunction with the recent so-called 'beyond-CMOS' debate [1], the importance of discarded spin degree-of-freedom is therefore being more and more appreciated. Unfortunately, the spin angular momentum of electrons conducting in electronic devices is not conserved by rapid spin relaxation (or spin-flip scattering) due to spin–orbit coupling and/or hyperfine interaction from the surroundings in solids, and therefore, the spin information completely disappears within a range of few hundred nanometres. Recent progress on microfabrication technology originating from the microelectronics industry, however, enabled the spin-dependent measurements in various mesoscopic devices. The research field of electronics based on the active injection, manipulation and detection of spin degree of freedom in the solid-state devices is collectively named *spintronics* [2], and spintronic devices are regarded as promising beyond-CMOS devices, simply because the high-speed and energy-saving devices are believed to be realized in the spin-based devices as compared to the conventional charge-based devices [3].

Among a number of strategies pursuing a detection of spin-dependent response to an external stimulus, the hitherto most simple and successful method is to measure the external magnetic field dependency on the electrical resistance of spin devices, and this phenomenon is generally referred to as *magnetoresistance* (MR). In particular, the change of the electrical resistance of metallic ferromagnets by the direction of external magnetic field is known as anisotropic magnetoresistance (AMR) [4]. However, the change of AMR amounts to only

Carbon Nanotubes and Graphene. DOI: 10.1016/B978-0-08-098232-8.00011-5

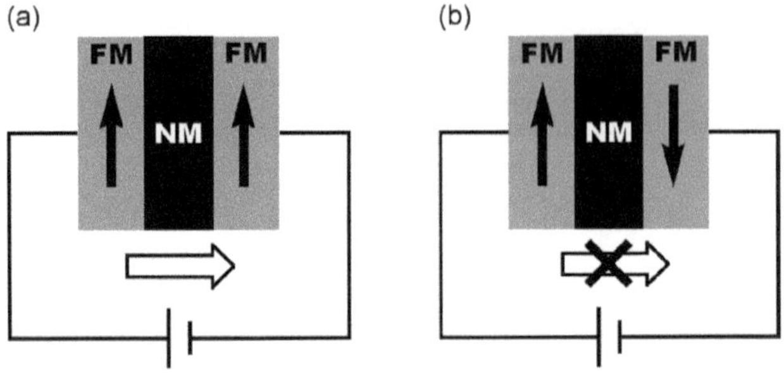

FIGURE 11.1 Schematic representation of a SV, where NM spacer is sandwiched by two FM electrodes. (a) Minimum resistance in the case of P orientation between the magnetizations of the two FM electrodes. (b) Maximum resistance in the case of AP orientation between them.

a few per cent at the most. In 1975, the first breakthrough was brought by the experiment about a multi-layered device comprised of the thin-film ferromagnetic (FM) electrodes and a very thin insulating film sandwiched between the electrodes [5], and the interest in spintronics has been accelerated since then.

It is certain that the prototypical and the most successful spin-based device is the spin valve (SV) [2], in which a change in the MR is effectively utilized as electrical detection of spin-polarized current. Schematic representation of a SV is shown in Figure 11.1. Two FM materials with different coercive fields are applied as spin-injector and spin-detector electrodes, respectively, and they are separated by a non-magnetic (NM) material as a decoupler of direct magnetic communication between them. An external magnetic field allows to tune the relative orientation of the magnetization of the two FM electrodes, by making use of the difference in the coercive fields of the two electrodes. The electrical resistance of the entire device will depend on the relative orientation between the magnetizations of the injector and the detector electrodes. The two extremal configurations, the parallel (P) and anti-parallel (AP) orientations between the magnetizations of the two electrodes, exhibit the minimum and the maximum electrical resistances, respectively, as described in the subsequent sections.

11.1.1 Tunnelling magnetoresistance (TMR)

By using ferromagnet/insulator/superconductor (FM/I/S) junctions, Tedrow and Meservey reported on the first experimental determination of the spin polarization for the conduction band in a FM material [5–7]. Their findings bridged the discovery of tunnelling magnetoresistance (TMR) in FM/I/FM junctions by Jullière [8]. He measured a change in tunnelling conductance between the P and AP orientations between the magnetizations for the two FM layers in FM/I/FM junctions, and the TMR ratio in an FM/I/FM magnetic tunnel junction (MJT) is defined as

$$\mathrm{TMR} \equiv \frac{R_{\mathrm{AP}} - R_{\mathrm{P}}}{R_{\mathrm{P}}} = \frac{G_{\mathrm{P}} - G_{\mathrm{AP}}}{G_{\mathrm{AP}}} = \frac{2P_1P_2}{1 - P_1P_2} \tag{11.1}$$

where $R_{P(AP)}$ is the resistance in the P (AP) configuration, $G_{P(AP)} = 1/R_{P(AP)}$ the conductance in the P (AP) configuration and $P_{1(2)}$ the polarization of the first (second) FM layer with

$$P_i = \frac{N_{i\uparrow}(E_F) - N_{i\downarrow}(E_F)}{N_{i\uparrow}(E_F) + N_{i\downarrow}(E_F)} \quad (i = 1,2) \tag{11.2}$$

where $N_{\uparrow(\downarrow)}(E_F)$ is the density of state (DOS) at Fermi energy (E_F) for spin-up (spin-down) band. The last expression of the TMR ratio in Equation (11.1) is known as Jullière's model, in which the constant tunnelling matrix elements and the electron tunnelling without spin flip are assumed. TMR effect is explainable by the difference in the spin-resolved DOS $N_\uparrow(E_F)$ and $N_\downarrow(E_F)$ at E_F. Electrons can only tunnel from a given spin sub-band in the first FM layer to the same spin sub-band in the second FM layer, if spin-flip tunnelling is not allowed. The tunnelling rate is proportional to the product of the corresponding DOS of the spin sub-band and, thus, is dependent on the relative orientation of the magnetization between the two FM layers. As a consequence, it can be deduced that the conductance in the P configuration is higher than in the AP configuration. It should be therefore noted that *TMR effect does not necessitate the spin transport in the insulator layer as tunnelling barrier.*

In 1982, Maekawa and Gäfvert observed a few per cent of the TMR ratio at 4.2 K in the Ni/NiO/(Fe, Co, Ni) junctions [9]. However, such a low MR ratio at low temperatures is not adequate for commercial applications in magnetic recording. The MR ratio exceeding 10% was accomplished by the multi-layered Fe (FM layer) – Cr (NM layer) structures in 1988 [10,11], and therefore, such kind of MR effect was referred to as 'giant' magnetoresistance (GMR), as described in later sections. The discovery of GMR effect triggered an enormous research activity towards spintronic devices.

11.1.2 Giant magnetoresistance (GMR)

As described in Sec. 11.1.1, the TMR effect stems from the DOS asymmetry in both spin sub-bands between the P and AP configurations for the two FM layers bearing the tunnel barrier. The GMR effect is also related to such a DOS asymmetry between the P and AP configurations for the two FM electrodes, but the underlying mechanism of GMR effect differs fundamentally from that of the TMR effect. While the electrons tunnel between the two FM layers through an insulating thin layer in TMR, the spin-polarized electric current flows through a NM spacer between the two FM electrodes in GMR. The discovery of GMR effect, in particular, current-perpendicular-to-plane GMR, as described later, opened the new way for spintronics, i.e. spin injection or spin accumulation into NM materials is applicable to realization of spin manipulation, which will lead to the spin transistors and quantum computers [3].

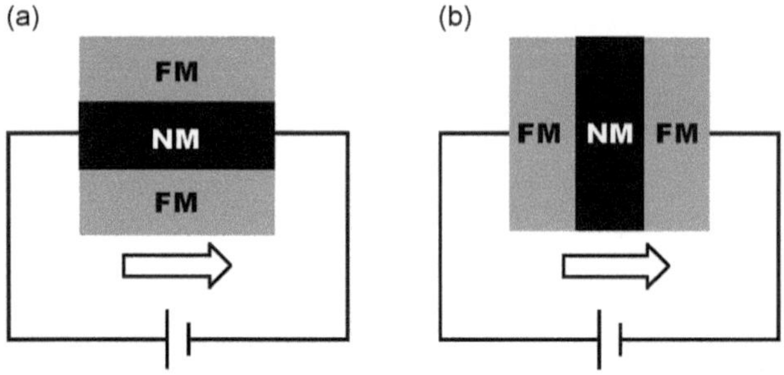

FIGURE 11.2 Schematic representation of (a) the CIP and (b) the CPP GMR geometries.

A current drives a non-equilibrium spin population from a FM electrode as an injector to a NM material, due to the spin polarization in the FM electrode. If the spin polarization reaches another FM electrode as a detector, the resistance in the entire device depends on the mutual orientation between the magnetizations of the injector and the detector. To get a phenomenological insight into the GMR effect, we often make use of so-called two-channel model introduced by Mott [12–14]. When we assume that spin-flip scattering, i.e. the change of spin-up to spin-down, or vice versa, is negligible while a current flows through the NM spacer, we can treat the transport process as the sum of two independent transport channels originating from spin-up (for majority) and spin-down (for minority) electrons. As depicted in Figure 11.2, the GMR effect can be classified into two types due to the device geometry: (1) the GMR observed by flowing the current parallel to the interfaces between the FM/NM/FM layers is referred to as CIP (current in plane) GMR, which was found independently by Grünberg's group in Germany [10] and Fert's group in France [11]; (2) the GMR observed by flowing the current perpendicular to the interfaces between the FM/NM/FM layers as current perpendicular to plane (CPP) GMR, which was first realized by Pratt *et al.* in 1991, and moreover, was found to be several times larger than the CIP GMR [15]. The difference between the two device geometries is described in Sections 11.1.2.1 and 11.1.2.2. In the following, it is assumed that the electron mean free path is much shorter than the dimensions of the conductors (so-called diffusive limit).

11.1.2.1 Current in Plane (CIP) GMR

In a FM metal, the spin sub-bands are split by the exchange interaction, thus resulting in a different DOS and Fermi velocity for spin-up and spin-down electrons. This brings a finite magnetization to the FM metal. In the picture of the two-channel model by Mott, two channels for spin-up and spin-down electrons have different bulk conductivities, $\sigma_\uparrow$ and $\sigma_\downarrow$, respectively. The electric current in a FM metal is mostly carried by spin-up electrons as the majority, and is thus spin polarized. The spin polarization α of the bulk current polarization in a FM metal is defined as

$$\alpha = \frac{\sigma_\uparrow - \sigma_\downarrow}{\sigma_\uparrow + \sigma_\downarrow} \tag{11.3}$$

Spin-up electrons parallel to the magnetization of the FM layer do not suffer from scattering, and thus the resistivity of spin-up channel is low. On the other hand, the resistivity of spin-down channel is high due to enhanced scattering. As a result, in a CIP GMR device, the P configuration for both FM layers gives an overall low resistance, while electrons in both spin channels undergo considerable scattering, thus leading to an overall high resistance. It should be noted here that the critical length scale in a CIP GMR device is the *electron mean free path*, l_e. Thus, the NM spacer and the FM layers must be thinner than the electron mean free path of carrier electrons.

11.1.2.2 Current Perpendicular to Plane (CPP) GMR

The spin-polarized transport was realized by the discovery of the CIP GMR effect. The critical length scale for the CIP GMR device is however provided by the electron mean free path. This indicates that the CIP GMR effect is irrelevant to the spin dynamics in the NM spacer. In contrast, the GMR effect found in the CPP geometry made it possible to generate a non-equilibrium spin population into NM materials, which is now called *spin accumulation* [16]. When the electric current is driven through the interface between a FM material and a NM material, the spin-up electrons in the FM material injects into the spin-up channel in the NM material, while the spin-down electrons in the FM material into the spin-down channel in the NM material. The ratio of spin-up and spin-down electrons injected from the FM material is determined by the bulk current polarization in the FM metal, α, and therefore, a non-equilibrium spin-accumulated state is induced in the vicinity of the contact interface. The spin accumulation, μ_s, is generally defined as the difference $\mu_\uparrow - \mu_\downarrow$ between the electrochemical potentials of spin-up electrons, $\mu_\uparrow$, and spin-down electrons, $\mu_\downarrow$. By interacting with the surroundings, such a spin-accumulated state in the NM material continuously relaxes to reach the equilibrium, where there exists no spin imbalance so that the injected electrons lose their initial spin polarization.

The magnitude of the spin accumulation depends on the spin injection rate into the NM material and the *spin relaxation time*, τ_s, which is defined as,

$$\frac{1}{\tau_s} = \frac{1}{\tau_{\uparrow\downarrow}} + \frac{1}{\tau_{\downarrow\uparrow}} \tag{11.4}$$

where the spin-flip time $\tau_{\uparrow\downarrow}$ (or $\tau_{\downarrow\uparrow}$) corresponds to the average time for spin-up (or spin-down) state of a carrier electron to be scattered into spin-down (or spin-up) state. Moreover, the spin relaxation time is closely related to the length scale for a carrier electron to lose its spin polarization, i.e. the spin relaxation length, l_s. Apparently, τ_s and l_s are related by $l_s = v\tau_s$, where v is the average carrier electron velocity. For instance, in the case of NM metals and/or degenerate Fermi gas conductors [17,18], l_s is represented below by τ_s,

$$l_s = \sqrt{\frac{\tau_s}{4e^2 N(E_F)\rho_{NM}}}, \tag{11.5}$$

where $N(E_F)$ is the DOS at the Fermi level; E_F and ρ_N the resistivity of a NM material. Away from the interface, the decay of the spin accumulation in NM materials is characterized by the corresponding spin relaxation length,

$$\mu_s(l) \equiv \mu_\uparrow - \mu_\downarrow = \mu_s(0)\exp\left(\frac{l}{l_s}\right) \tag{11.6}$$

where l is the distance from the interface, and a finite value of $\mu_s(0)$ indicates the existence of non-equilibrium magnetization in the NM material as a result of electrical spin injection from the FM electrode.

A finite spin accumulation at the contact interface with another FM electrode as detector implies different density of spin-up and spin-down carrier electrons, thus resulting in the appearance of the GMR effect. Thus, the length of l_s to the thickness of the NM spacer, L, determines the ability for the SV device utilizing the GMR effect to work. If $l_s > L$, the spin-polarized current can reach the interface with the FM electrode as detector, and hence, the total resistance differs between the P and AP configurations of the magnetization of the two FM electrodes. As a consequence, it should be noted that the critical length scale for the SV devices utilizing the CPP GMR is the spin relaxation length, l_s, whereas the electron mean free path determines and regulates the critical length scale for the SV devices utilizing the CIP GMR.

11.1.3 Tunnelling or spin injection?

As described in the preceding sections, we have seen that the TMR and GMR effects, in short, play the same role, as far as the operation of the SV devices is concerned. Then which is the better for commercial applications?

Since the discovery of the GMR effect in multi-layer structures, the non-volatile magnetic memory applications of the GMR were actively engineered [19–21]. For example, IBM already produced the first GMR-based read head for hard disk drives in 1997 [22]. Furthermore, realization of large room-temperature GMR [23,24] spurred a change from basic physical research to commercial applications towards non-volatile magnetic random access memory (MRAM) [25].

Although earlier spintronic applications are based on the GMR effects, the discovery of large room-temperature TMR [26,27] has renewed interest in the study of MTJ, which are now the basis for the several MRAM prototypes [28,29]. In addition, room-temperature TMR ratios of several hundred per cent have been recently realized [30–32], sufficiently large to make TMR-based read heads for hard disk drives and/or MRAM. Hence, future generations of magnetic read heads are expected to utilize TMR effect instead of CIP GMR effect.

On the other hand, one of the long-standing goals in the field of spintronics is considered as the realization of spin transistors, in which the non-equilibrium spin-polarized current can be controlled or manipulated by the external stimuli, in particular, by the gate electrode. To achieve the goal, the progress on the both theoretical and experimental understanding of the SV devices based on the CPP GMR effect is highly desirable, simply because it necessitates fulfilment of (1) facile spin injection or spin accumulation, (2) comprehensive understanding of spin relaxation processes and (3) secure spin detection.

Although various spin transistors have been proposed so far, the most promising two are the Datta–Das spin field-effect transistor (FET) in 1990 [33] and the MOSFET by Sugahara and Tanaka in 2004 [34], as sketched in Figure 11.3. In the following, it is assumed that a half-metal-ferromagnet (HMF) such as $LaSrMnO_3$ (LSMO) [35,36], in which only one spin sub-band is occupied at the

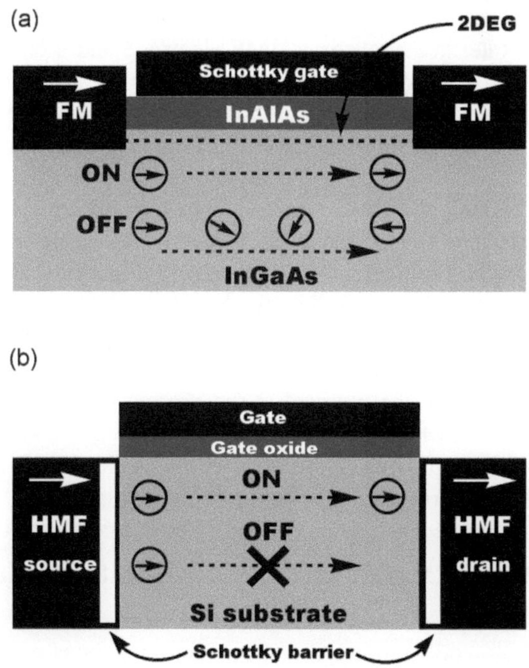

FIGURE 11.3 Schematic structures of (a) Datta–Das spin FET and (b) Sugahara–Tanaka spin MOSFET. Spin-polarized current is regulated by gate-controllable spin–orbit interaction in (a) and by gate-controllable Schottky barrier in (b).

Fermi level, while the other spin sub-band has semiconductive character, and thus the spin polarization, α, approaches 100% at low temperatures, is used as source and drain electrodes.

In the Datta–Das spin FET, the source (spin injector) and drain (spin detector) electrodes are HMF with parallel magnetization configuration, and the injected spin-polarized electrons are transported through a strictly 1D-ballistic channel [37], formed in, for instance, a 2D interface of an InGaAs/InAlAs semiconductor heterojunction. Unlike the SV devices, the spin-dependent device operation is executed not by the external magnetic fields, but by the gate voltage at the top of the channel. The spin precession induced by gate-controllable spin–orbit interaction, which is known as Rashba spin–orbit interaction [38,39], changes the direction of electron spin. In a simplified picture, when the electron spin at the drain points in the initial direction, the electron can enter the drain electrode (ON). Otherwise it is scattered at the interface with the drain electrode (OFF).

On the other hand, the device structure of the spin MOSFET by Sugahara and Tanaka is fundamentally similar to that of Schottky MOSFET [40], except that the spin-polarized current flows through a NM channel. For example, when the HMF source/drain electrodes and Si channel layer are used for the spin MOSFET, it was demonstrated that the high (low) current drive capability in the P (AP) configuration and the large magnetocurrent ratios can be obtained on the basis of 2D numerical analysis [34]. The width of Schottky barrier, which is formed in the interface between a HMF electrode and a NM channel layer, can be modulated by applying a gate-source bias, V_{GS}. In the P configuration of magnetizations for the HMF source/drain electrodes, only spin-up electrons can be injected into the channel by application of V_{GS}, thus resulting in a spin-polarized drain current. By flipping the magnetization of the HMF drain electrode (the AP configuration), the HMF barrier height for spin-up electrons becomes larger at the drain, and thus the spin-up electrons can hardly pass through the HMF drain electrode. This indicates that the HMF drain plays the role of a spin filter. It should be, however, noted that the operation of the spin MOSFET is not guaranteed unless the injected spin-polarized current survives at the HMF drain.

Although both propositions described in earlier sections are intriguing, the spin transistors are yet to be realized at the present stage. We are confronted with several obstacles, which are described in the subsequent sections, to realize the spin transistors employing spin-polarized electrons travelling through the NM layer. Further experimental efforts are being required.

11.1.4 Fundamental obstacles for spin injection, spin detection and spin manipulation

To achieve the spin injection/manipulation/detection, it is considered as one of the promising approaches to combine semiconductor physics and spintronic concepts. This is simply because semiconductors give a long spin relaxation time, which allows the electrical control of the spin dynamics. For instance, the spin relaxation time up to 1 μs was reported for GaAs [41]. However, spin injection/detection in semiconductor devices was very hard to deal with. Only very recently, an all-electrical spin injection/detection was demonstrated by using the inorganic semiconductor GaAs as NM spacer and Fe as FM electrodes [42]. Let us here look at fundamental problems towards the realization of spintronic devices.

11.1.4.1 Conductivity Mismatch

In the CPP-GMR-based SV devices, it is essential to monitor the resistance change due to the mutual orientation between the magnetizations of the FM injector and detector, i.e. the P and AP configurations. It can be interpreted that the individual resistances of the two spin channels can be treated separately, according to the two-channel model by Mott (a parallel resistor model). Hence, different equivalent spin-resistance circuits are depicted for the P and AP configurations of the FM injector and detector, as shown in Figure 11.4. A more elaborate theoretical consideration of CPP GMR was carried out by Valet and Fert on the basis of the Boltzmann equation [43].

The conductivity of a semiconductor as a NM spacer is usually much lower than that of a metal. As is apparent from the simplistic parallel resistor model

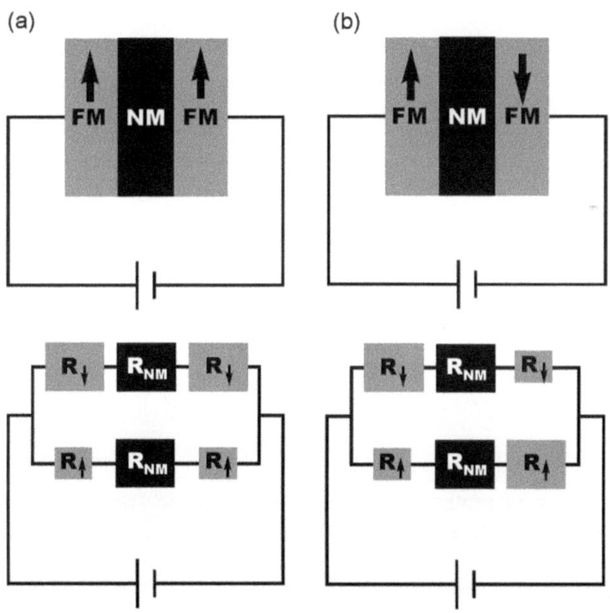

FIGURE 11.4 Schematic representation of the parallel resistor model for a CPP GMR device in (a) the P and (b) the AP configurations. The magnetization of the FM electrodes is depicted by the black arrow. The colour corresponds to each layer. Bigger resistor represents a larger resistance for the designated spin channel.

(Figure 11.4), when the resistance of the NM material, R_{NM}, is much larger than the other resistances, R_{NM} determines the entire resistance of a SV device, and thus no resistance change between the P and AP configurations is observed. This is the so-called conductivity mismatch problem [44]. In fact, the spin injection from FM electrodes (Ni, Co and permalloy) into a semiconductor (based on high-mobility InAs/AlSb 2D electron gas) was reported [45].

The first remedy for this problem is to use a fully spin-polarized FM material as FM electrodes. In the classical FM materials such as Fe, Ni, Co and so on, a spin polarization at the Fermi level is far below 100%, and Co is considered as the best element (45% [7]). Instead, HMF such as LSMO [35,36] are often used to inject fully spin-polarized electrons into a semiconductor material. However, clear-cut solution has not been obtained so far.

The second remedy for the conductivity mismatch problem is to improve the device structure, and such an improvement is off to the races. Three research groups proposed virtually the same approach that the conductivity mismatch problem can be solved by introducing a spin-dependent interface resistance (tunnel junction preferably) at the interfaces in the FM/NM/FM layered CPP-GMR-SV devices [17,18,46,47]. The resistivities of the majority spin (spin-up) and minority spin (spin-down) channels in the FM material, $\rho_{\uparrow(\downarrow)}^{NM}$, are written by,

$$\rho_{\uparrow(\downarrow)}^{FM} = 2[1-(+)\beta]\,\rho_{FM}^{*} \tag{11.7}$$

where β is the bulk asymmetry coefficient, and thus the resistivity of the bulk FM material is represented as $\left(1-\beta^2\right)\rho_{FM}^{*}$. On the one hand, the resistivity of the NM material, $\rho_{\uparrow(\downarrow)}^{NM}$, is denoted as,

$$\rho_{\uparrow(\downarrow)}^{NM} = 2\rho_{NM}^{*} \tag{11.8}$$

When l_s^{FM} and l_s^{NM} are the spin relaxation lengths for the FM and NM materials, the important parameters in the conductivity mismatch problem are the products of the resistivity by the spin relaxation length,

$$r_{FM} = \rho_{FM}^{*} \times l_s^{FM} \tag{11.9}$$

$$r_{NM} = \rho_{NM}^{*} \times l_s^{NM} \tag{11.10}$$

for the FM and NM materials, respectively, and the spin-dependent interface resistance, r_{int}^{*}, which is written by a spin-dependent drop of the chemical potentials at the interface,

$$\mu_{\uparrow(\downarrow)}(z=0^{+}) - \mu_{\uparrow(\downarrow)}(z=0^{-}) = r_{\uparrow(\downarrow)}\,j_{\uparrow(\downarrow)}(z=0) \tag{11.11}$$

where $\mu_{\uparrow(\downarrow)}(z=0^{+})$ and $\mu_{\uparrow(\downarrow)}(z=0^{-})$ are the chemical potentials on the right and left sides of the interface, and $j_{\uparrow(\downarrow)}(z=0)$ is the current density at the interface.

The interface resistivity for the spin-up (spin-down) channel, $r_{\uparrow(\downarrow)}$, is related to r^*_{int} by the equation,

$$r_{\uparrow(\downarrow)} = 2r^*_{\text{int}}\,[1-(+)\gamma] \tag{11.12}$$

where γ is the interface spin asymmetry coefficient. According to the analyses by Fert and Jaffrès [17], the polarization of the current at the interface is deduced below in the case of no spin-dependent interface resistance,

$$\frac{j_\uparrow - j_\downarrow}{j_\uparrow + j_\downarrow} = \frac{\beta}{1+\dfrac{r_{\text{NM}}}{r_{\text{FM}}}} \tag{11.13}$$

This formula clearly indicates that the spin-polarized current is almost completely diminished when it enters the NM material when $r_{\text{NM}} >> r_{\text{FM}}$. This case corresponds to the conductivity mismatch. On the other hand, with the spin-dependent interface resistance originating from a tunnel junction or a tunnelling regime of a Schottky junction, the spin polarization of the current at the interface becomes

$$\frac{j_\uparrow - j_\downarrow}{j_\uparrow + j_\downarrow} = \frac{\beta r_{\text{FM}} + \gamma r^*_{\text{int}}}{r_{\text{FM}} + r_{\text{NM}} + r^*_{\text{int}}} \tag{11.14}$$

under appropriate conditions [17], thus indicating that the conductivity mismatch problem can be obviated by introduction of the spin-dependent interface resistance, r^*_{int}. More importantly, Fert and Jaffrès clarified that r^*_{int} must be chosen in a well-defined window, which is derived from the conditions for spin injection and spin conservation [17]. These theoretical considerations lead to the success of all-electrical spin injection/detection using a semiconductor as a NM material [42].

11.1.4.2 Spurious Effects

Whatever the conductivity mismatch problem was removed, there still exists various obstacles in the measurements of the GMR effects based on the spin injection or spin accumulation. In fact, a number of phenomena gave spurious signals misleading into the wrong observation of the GMR effect (*spurious effects*). To interpret correctly the results on spintronic experiments, it is crucial to discern an intrinsic GMR signal from such spurious signals.

The *Lorentz magnetoresistance* (LMR), which originates from the Lorentz force, is significant for systems with a relatively large mean free path, l_e, and has a MR effect on the order of $(l_e/l_B)^4$, where $l_B = \sqrt{h/(eB)}$ is the magnetic length [48]. In the coherent, diffusive transport regime, electron interference phenomena such as *weak localization* (WL) and *universal conductance fluctuations* (UCF) affect the magnetic field dependence on the entire resistance of the device, which is related to the coherence length, l_ϕ [49]. In addition, stray fields originating from the FM contacts can induce local Hall voltages, and therefore, the change from the P (or AP) to the AP (or P) configurations in the magnetizations of the FM contacts causes the change of the Hall voltages, thus leading to another spurious MR effect, i.e. *local Hall effect* [50]. More seriously, the so-called Coulomb charging effect becomes non-trivial as the system size decreases. In the nano-carbon materials, such as carbon nanotubes (CNTs), the magneto-Coulomb effect (MCE) easily gave a discontinuous change of conductance in parallel with the switch of the direction of magnetization of the FM contact [51].

11.1.4.3 Spin Relaxation

As described in Sec. 11.1.2.2, the spin relaxation length in the NM layer, l_s, is required to be longer so that the spin-polarized current injected from a FM contact as an injector reaches another FM contact as a detector. The injected electric current continuously loses the initial spin polarization by interacting with the environment (*spin relaxation*). Although the underlying mechanism for spin relaxation in solids is too complicated to explain quite simply, the origins are roughly classified into two interactions: (i) *spin–orbit coupling* and (ii) *hyperfine interaction.*

Spin–orbit coupling is one of the relativistic effects, and it occurs whenever a particle with non-zero spin moves around a region with a finite electric field. Hence, it should be noted that the spin degree of freedom is indirectly affected by an electric field via the spin–orbit coupling. The spin–orbit coupling Hamiltonian, H_{SO}, is represented by,

$$H_{\mathrm{SO}} = \zeta \sum_i \boldsymbol{l}_i \cdot \boldsymbol{S}_i \tag{11.15}$$

where $\boldsymbol{l}_i$ and $\boldsymbol{S}_i$ are the orbital and spin angular momentum operators for the electron i, respectively. The spin–orbit coupling constant, ξ, which gives the strength of interaction, is proportional to fourth powers of the atomic number Z in the case of hydrogenlike atom [52], and therefore the spin–orbit coupling, which is indispensable to inorganic semiconductors, is rather weak in organic materials because most organic materials are made of light elements, such as C, H, N, O and so forth. For example, the spin–orbit coupling splits the valence band of GaAs into two bands with a bandgap of 340 meV, while this is only 13 meV for diamond [52].

In 2D III–V semiconductors such as GaAs, there exist two distinct Hamiltonian terms contributing to spin–orbit coupling: the bulk inversion asymmetry term and the structure inversion asymmetry term, which appears only in asymmetric system [2]. The first one comes from the Dresselhaus contribution [53], whereas the second one arises from the Bychkov–Rashba contribution [54]. In addition, three different spin–orbit-coupling-related spin relaxation mechanisms were found for NM materials: the Elliot–Yafet [55,56], D'yakonov–Perel' [57] and Bir–Aronov–Pikus [58] mechanisms.

On the one hand, hyperfine interaction is derived from the interaction between the electron spin and the nuclear spins in the materials, and also causes the spin relaxation. The electron–nuclear coupling Hamiltonian is given by,

$$H_{\text{hyp}} = \sum_i A_i \, \boldsymbol{I}_i \cdot \boldsymbol{S} \tag{11.16}$$

where $\boldsymbol{I}_i$ and $\boldsymbol{S}$ are the nuclear spin angular momentum operator for nucleus i and the electron spin angular momentum operator, respectively [59]. The hyperfine coupling constant, A_i, gives the strength of interaction between them. The hyperfine interaction is not ignorable for spin relaxation in 2D III–V semiconductors such as GaAs because all the lattice nuclei in GaAs carry the magnetic moment of 3/2 spin.

11.2 NANO-CARBON AS NON-MAGNETIC MATERIALS FOR SPINTRONICS

In the preceding sections, several issues to be overcome to realize the spin injection/detection from the FM contacts into the NM layer have been reviewed. Although the utilization of inorganic semiconductors has been assumed so far, organic (or molecular) materials are also applicable as a NM medium to transport and control spin-polarized signals in spintronics devices. Recently, organic spintronics is being considered as an emerging and promising research field [60,61]. For organic materials, long spin relaxation time, τ_s, is a great advantage to spintronics application. Such a characteristic stems from the fact that all the possible mechanisms for spin relaxation are weak. Organic materials consist mainly of light elements, such as C, H, N, O and so forth, and thus the spin–orbit coupling is usually negligible in organic materials. In addition, hyperfine interaction is also weak, since there are only a few light elements possessing a nuclear spin except for hydrogen atom ($I = 1/2$), and the natural abundances of the isotopes ^{13}C ($I = 1/2$) and ^{14}N ($I = 2$) are quite low. Moreover, the molecular orbitals (MOs) responsible for electrical transport mainly consist of p_z atomic orbitals extended over 1D and/or 2D connected carbon networks in π-conjugated organic materials, and therefore there is virtually no orbital overlap with nuclei of C and H atoms, due to the coincidence of the nodal plane of MOs with the molecular plane.

On the other hand, short spin relaxation length, l_s, is a significant shortcoming for spintronics application, simply because the mobility in organic materials is usually rather poor. However, since the discovery of CNTs in 1991 [62] and, more astonishingly, the first deposition on a SiO_2 substrate by mechanical exfoliation of single or a few layers of free-standing graphene in 2004 [63], nano-carbon materials have rapidly emerged as promising candidates for spin information transport. This is simply because both CNTs and graphene are considered to combine a potentially high carrier mobility with a potentially very long spin relaxation time. In this section, recent progress on characterization of nano-carbon materials towards spintronics devices is reviewed.

11.2.1 Carbon nanotube (CNT) devices

Tsukagoshi et al. led off by making the first CNT-based SV device, which consisted of a single multi-walled CNT (MWCNT) contacted by polycrystalline Co FM contacts, as shown schematically in Figure 11.5(a) [64]. The simple two-terminal MR measurements at 4.2 K exhibited the hysteretic resistance increase of ca. 50-mT width around the applied external magnetic field $B = 0$. A maximum +9% MR ratio, which can be calculated by Equation (11.1) in Sec. 11.1.1, was observed at 4.2 K, and the observed MR decreased approximately exponen-

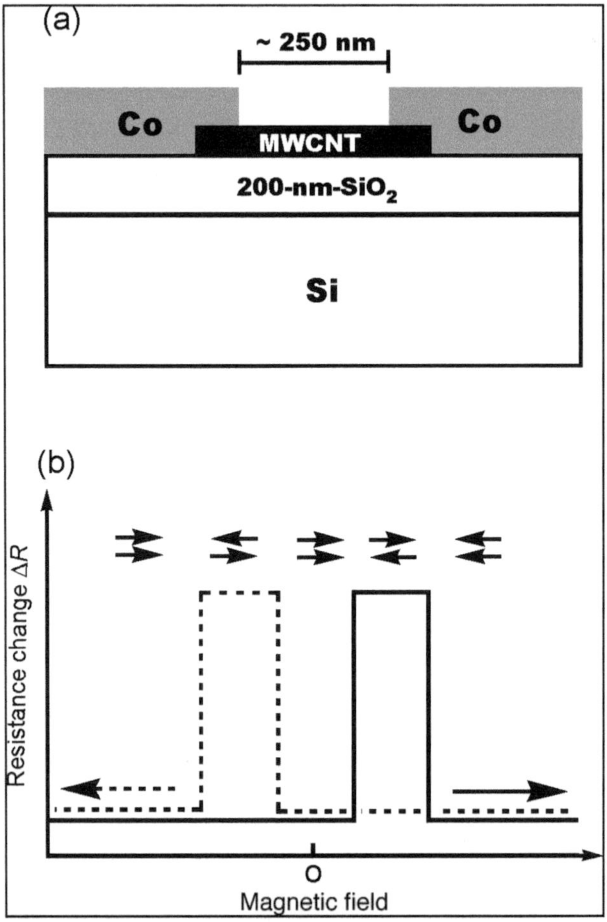

FIGURE 11.5 (a) Schematic structure of a SV device, consisting of a MWCNT connected by 65-nm thick polycrystalline Co contacts, fabricated by electron-beam lithography and thermal evaporation [64]. (b) The resistance change in a SV device as a function of the external magnetic field. The sweep directions of the applied magnetic field are indicated by the horizontal arrows, and the relative orientation between the magnetizations for the FM contacts is illustrated by the pairs of horizontal small arrows.

tially with temperature increase and disappeared at 20 K [64]. In addition, the spin relaxation length, l_s, was estimated to be round 130 nm, on the basis of a simple Jullière's model [8].

Here the outline of the observed MR signal (so-called spin signal) based on SV devices is worth a mention in passing. To realize the P and AP configurations of the magnetization orientations between two FM contacts by applying the external magnetic field, long and narrow FM contacts with the same easy-axis of magnetization along the length of FM contacts but with different coercive forces, which can be actually realized by changing the width of two FM contacts, are usually prepared. As shown schematically in Figure 11.5(b), when the external field is applied parallel to a pair of FM contacts, the magnetization of the FM contact with the weaker coercive force is reversed and the AP configuration is realized to cause the resistance change. Further application of the external magnetic field finally induces a magnetization flip of another FM contact with the stronger coercive force and again returns to the P configuration, thus resulting in the initial device resistance. When the polarization of the applied external magnetic field is changed, the same pattern of resistance change appears symmetrically with respect to $B = 0$ line (resistance hysteresis).

11.2.1.1 Difficulties in CNT-Based SV Devices

Since the report by Tsukagoshi et al., similar devices and the others have been checked by several groups [60]. In those experiments, MWCNTs and single-walled CNTs (SWCNTs) have been used as NM spin transporters, and metals such as Co, NiFe, PdNi and so on, HMF such as LSMO and semiconductors such as GaMnAs have been selected as FM contacts. As a consequence, while the MR effect was repeatedly confirmed, there exist a lack of reproducibility and inconsistency between the reported results both qualitatively and quantitatively [60]. This situation then posed the question of whether or not spin injection was really occurring.

There are three main causes for the large variation in reported MR ratios: For one thing, the quality in the CNT used in the device fabrications affects the results. Impurities, dopants or incommensurate between the outer and the inner tubes in MWCNTs can be considered as a cause of irreproducibility [65]. What is more, the poor quality of the CNT/FM contact interface leads to a wide range of contact resistance. Although a low-contact resistance is not required for spintronic devices, as was pointed out in Sec. 11.1.4.1, low-contact resistance can alleviate charging effects such as MCE, as discussed in Sec. 11.1.4.2. Sahoo *et al.* demonstrated that an improvement in realizing reliable low-ohmic contact resistance was achieved by the introduction of $Pd_{0.3}Ni_{0.7}$ contact [66]. A PdNi/MWCNT/PdNi SV device with a back-gate electrode exhibited a strong gate voltage dependency, and a maximum MR ratio reached about 2% at most at 1.8 K and 2 V back-gate voltage [66]. In 2007, Hueso *et al.* attempted to use single-crystal $La_{2/3}Sr_{1/3}MnO_3$ (LSMO), which is believed to behave like a half-metallic at low temperatures, and successfully achieved a maximum MR ratio of 61% in a single MWCNT SV device with two LSMO contact electrodes at

5 K [67]. The MWCNT/LSMO interface can be regarded as tunnel barriers, which is favourable for avoiding conductivity mismatch and also permits large-bias measurements (~25 mV) to remove the influence of Coulomb blockade and level quantization effects as a CNT quantum dot (QD). Furthermore, the MR ratio did not decrease to zero until 120 K. In short, whether CNT behaves like a QD [68] or quantum interferometer such as a Fabry–Pérot interferometer [69] depends on the strength of coupling between CNT and the FM contacts; thus, resulting in a variety of MR ratios. One final point is not the least but the most serious one. The two-terminal measurements of the MR effect based on CNT–SV devices described so far have been contaminated by spurious signals as described in Sec. 11.1.4.2. To measure only spin accumulation, a smart and rigorous four-terminal method to exclude the spurious effects has been devised by Jedema et al. [16,70]. In the next section, the results observed in this measurement, as well as the details of the measurement, are described.

11.2.1.2 Pure Spin Current and Its Detection Using Non-Local Geometry

For clarity, the ohmic contact at a FM/NM interface is assumed. As shown in Figure 11.6(a), when a spin-polarized current is injected from the FM contact

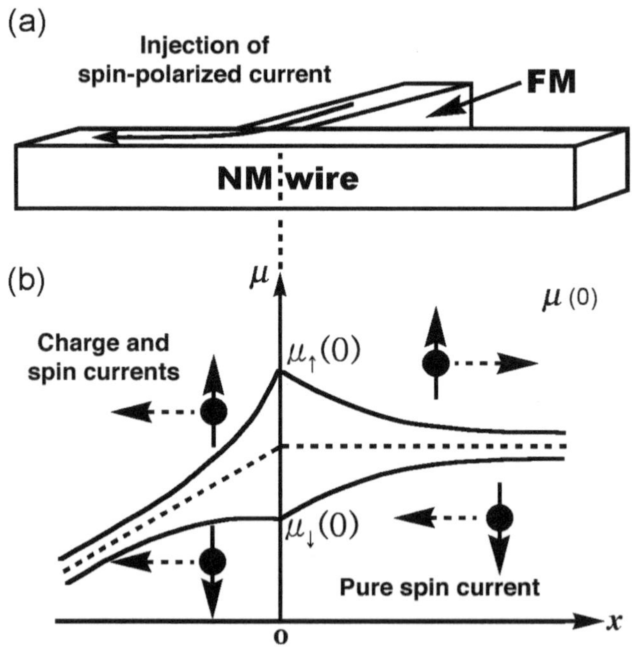

FIGURE 11.6 (a) Schematic view of a NM/FM junction and (b) the change of chemical potentials for spin-up and spin-down electrons in the NM wire for the NM/FM junction as a function of distance from the injection point.

into the NM wire, the spin accumulation, $\mu_s(0)$, occurs at the interface, and then, the accumulated spin begins to diffuse from the injected point into the right and left directions along the NM wire. According to the Valet and Fert model [17], the electrochemical potentials of the two spin channels can be represented, as schematically illustrated in Figure 11.6(b). On the left side of the NM wire, both charge and spin currents flow through along the electrochemical potential gradient. On the other hand, the spin-up and the spin-down currents flow in the opposite directions; thus, indicating that a spin current, $j_s = j\uparrow - j\uparrow$, exists, whereas a charge current does not on the right side of the NM wire. The spin current containing no charge current is called *pure spin current*. Hence, if only the pure spin current is detectable, the spurious effects are removable in the MR measurements.

Turning now to the SV geometry as depicted in Figure 11.7(a), when the spin-polarized current flows from the left FM contact to the left side of the NM wire, only the pure spin current flows to the right side of the NM wire. In the case where the right FM contact is within the spin relaxation length, the pure spin current can thus be probed by the right FM contact. Therefore, a voltage difference, ΔV_s, is observed between the P and AP configurations of the two FM contacts (Figure 11.7(b)). Finally, the spin signal, ΔR_s, is obtained by dividing a voltage difference by an injected current, I_{inj},

$$\Delta R_s \equiv \frac{\Delta V_s}{I_{inj}}$$

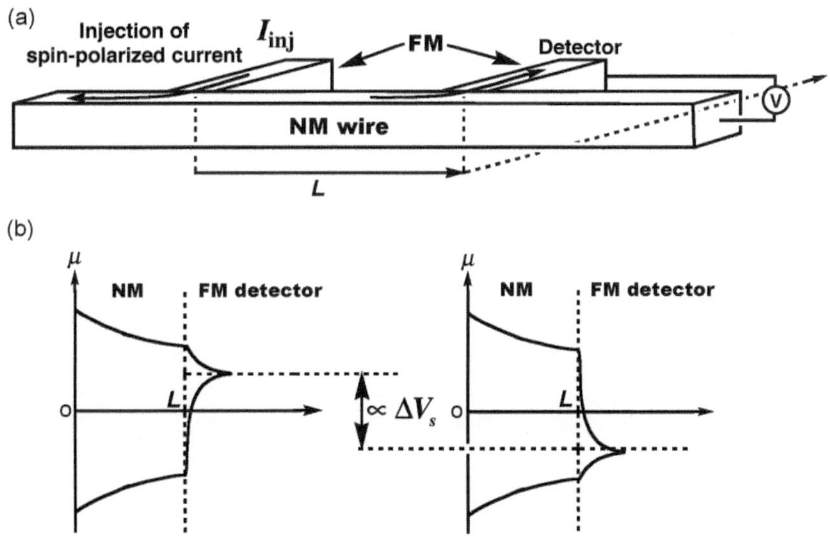

FIGURE 11.7 (a) Schematic view of a SV device and (b) the change of chemical potentials for spin-up and spin-down electrons in the NM wire and in the FM detector by the use of a non-local geometry.

In this way, the measured signal is only due to spin accumulation, and this kind of method is referred to as a (four-terminal) *non-local measurement*.

Tombros *et al.* for the first time applied the above-mentioned four-terminal non-local geometry to measure spin accumulation in SWCNTs [71]. As a consequence, they concluded that the MR signals by using conventional two-terminal local measurements are obscured by the spurious effects. For example, it was demonstrated that 90% of the MR signal in the two-terminal local measurement cannot be attributed to spin accumulation.

11.2.1.3 Summary for CNT-Based Spintronic Devices

A considerable number of works on the CNT-based devices have been carried out so far. As expected, relatively long spin relaxation time (up to 30 ns) and lengths (1.4–50 mm) have been obtained. However, the observed MR ratios varied from device to device, due to both the contact problems and the spurious effects, and more seriously, the SV effect was observable only at very low temperatures. In addition, more experimental work including gate-voltage dependence of the MR effect in CNT-based SV devices is needed on the basis of the non-local geometry. At the present stage, the observed device performance is unsatisfactory from the viewpoint of application to future spin transistors.

11.2.2 Graphene devices

Similar to the application of CNT to spintronic devices, graphene has attracted considerable attention not only due to the expected weak spin–orbit and hyperfine interaction, which guarantee a long spin relaxation time, but also due to extremely high room-temperature mobility higher than 10^4 cm^2 V^{-1} s^{-1} [72,73]. Hence, the realization of room-temperature SV devices was anticipated by using graphene as a NM spin transporter.

So far the formations of single-layer graphenes (SLGs) and multi-layer graphenes (MLGs) including bilayer graphenes (BLG) are classified into two methods: (i) micromechanical exfoliation of crystals of natural graphite and (ii) epitaxial growth on a SiC substrate. In the former method, the desired samples are selected from the exfoliated and transferred graphitic flakes on a substrate by using a combination of optical microscopy and atomic force microscopy [63,74]. In the latter method, ultrathin epitaxial graphite films (from SLG to MLG) were grown by thermal decomposition on the carbon-terminated (0001) surface of 6H-SiC [75–77]. In this section, applications of SLG and/or MLG to spintronic devices are overviewed based on their high anisotropic transport property.

11.2.2.1 Graphene as a Spin Transporter

The first demonstration of MR effects in SLG-based SV devices was provided by adopting a lateral device structure from the Manchester group [78]. Although they observed a 10% MR ratio at room temperature, spurious effects due to the shortcomings by the two-terminal local measurements (soft ferromagnet NiFe was used as FM contacts) probably drowned out the intrinsic spin signal. Reliable experimental evidence of room-temperature spin injection into SLG and MLG were reported independently by Tombros *et al.* [79] and Shiraishi and co-workers [80], respectively, by using the four-terminal non-local measurements. Figure 11.8 shows schematic representations of the four-terminal SV devices that were employed by the two groups. Significant difference between the two devices is found in the interface between the graphene layer and the FM contact. Tombros *et al.* inserted a thin Al_2O_3 layer as a tunnel barrier between them because they failed to observe spin transport in SLG-based devices with transparent low ohmic FM contacts. This closely related to the conductivity mismatch problem (see Sec. 11.1.4.1). On the other hand, no contact problem due to the conductivity mismatch problem was seen in the MLG-based devices.

In addition to the non-local measurements, Tombros *et al.* performed Hanle-type spin precession experiments originating from the Hanle effect [79,81,82]. In the Hanle precession measurements, the external magnetic field, $B_\perp$, is applied perpendicular to the graphene thin-film plane in the P configuration of the magnetization orientations of FM contacts, as shown in Figure 11.9. In this situation, the injected spin begins to precess by the magnetic field, $B_\perp$, and thus, the spin signal exhibits an oscillatory behaviour as a function of $B_\perp$. Consequently, the spin relaxation time and length, τ_s and l_s, can be estimated from the curve fitting to the theoretical formula for the Hanle precession [79]. From such a measurement, τ_s and l_s for SLG-based SV devices were extracted to be a few micrometres and a few hundred picoseconds, respectively. Since then, Shiraishi

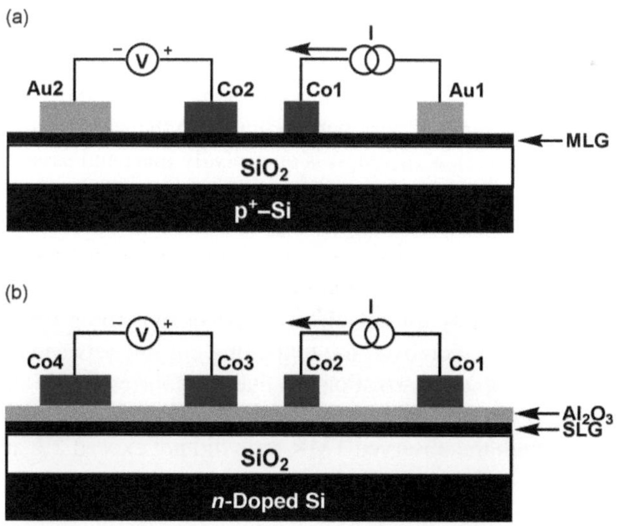

FIGURE 11.8 Schematic structures of the graphene-based SV devices in four-terminal non-local geometry fabricated by (a) Tombros *et al.* [79] and (b) Shiraishi and co-workers [80].

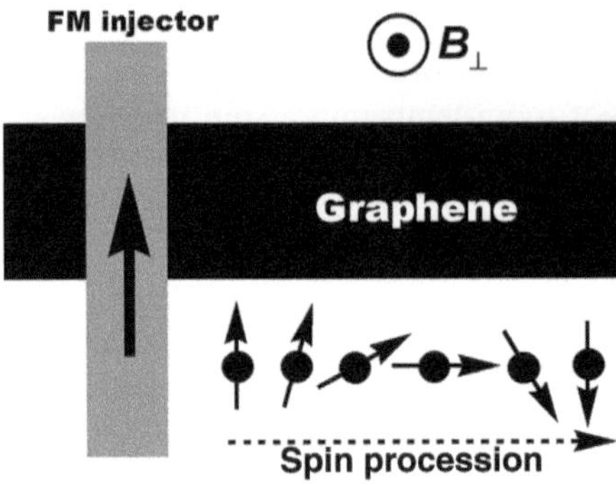

FIGURE 11.9 Schematic view of a Hanle-type spin precession measurements. In association with the rotation of the injected spin, the spin signal observed in the FM detector exhibits an oscillatory behaviour as a function of applied magnetic field.

et al. also carried out the similar experiments for their MLG-based SV devices, and 1.6 μm and 120 ps were estimated for τ_s and l_s, respectively, which are in good accordance with those for SLG-based SV devices [83].

Since the first reports on room-temperature spin injection into SLG and/or MLG, several publications have been issued on graphene SV devices including the gating effects [84–89]. Recently, in the SV devices based on the SLG grown epitaxially on SiC substrate, the long spin relaxation time up to 2.3 ns at 4.2 K [90] and the comparatively large MR ratio and long spin relaxation length up to 246 μm at 2 K [91] were reported. In conjunction with the possibility of fabrication of large-scale spin-based logic devices and networks, epitaxially grown graphene may open the possibility for spin information processing.

11.2.2.2 Graphene as a Tunnel Barrier

As is also pointed out in other chapters, graphene as atomically thin 2D sheet material exhibits poor conductivity perpendicular to the sheet, while it is highly conductive in the sheet. In addition, it is chemically inert and essentially impervious to migration of ions from one side of the sheet to the other [92], simply due to the strong in-plane covalent bonding between carbon atoms in the hexagonal lattice. Such an ultimate property is applicable to a tunnel barrier for TMR-based SV devices. In fact, Jonker and co-workers demonstrated that graphene serves effectively as a very thin insulator layer for tunnelling perpendicular to the graphene sheet, and moreover, the tunnelling current between two FM contacts preserves a net spin polarization, so that the fabricated SV devices (Figure 11.10(a)) show the measurable TMR effect with good signal-to-noise ratio up to 425 K, although the observed TMR ratio did not exceed 2% at 4 K [93].

More interestingly, Jonker and co-workers attempted furthermore to utilize a SLG as a spin-polarized tunnel barrier for spin injection into silicon from a FM metal, which successfully circumvents the conductivity mismatch problem for electrical spin injection into a NM semiconductor from a FM metal [94]. As is

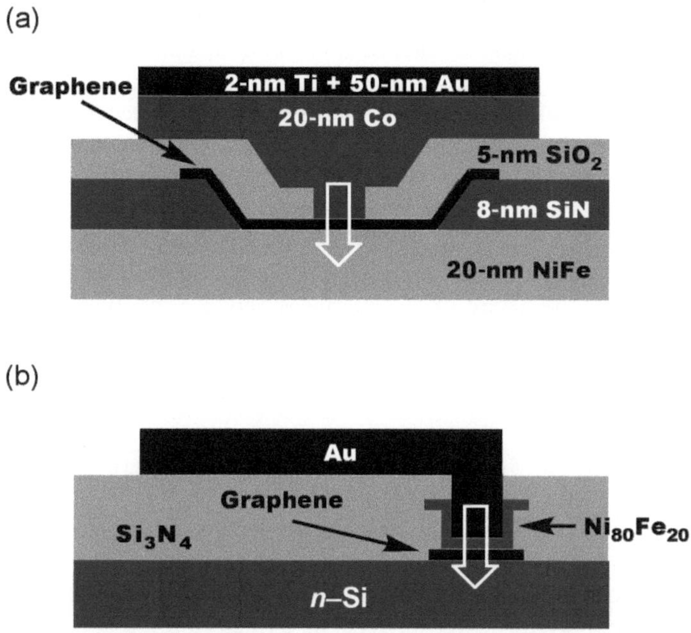

FIGURE 11.10 (a) Graphene-based TMR SV device [93] and (b) the device fabricated for spin injection experiments through a graphene tunnel barrier by Jonker and co-workers [94].

already discussed in Sec. 11.1.4.1, the large difference in conductivity between metal and semiconductor causes no resistance change between the P and AP configurations in a CPP-GMR-based SV device with FM/NM/FM layers. A possible solution for such a conductivity mismatch problem is the introduction of a large spin-dependent resistance (tunnel junction preferably) between the FM and NM layers (see Sec. 11.1.4.1). In this context, extensive effort has therefore been made towards developing appropriate tunnel barriers. Although most work has focused on metal Schottky barriers and insulating oxide layers such as Al_2O_3, MgO and so forth [95–99]; these materials are susceptible to inter-diffusion, interface defects and trapped charge and/or spin, which can, in turn, compromise spin injection/transport/detection. In a device structure shown in Figure 11.10(b), effective spin accumulation in Si was confirmed above room temperature, and moreover the contact resistance-area product, which is considered as an important parameter in determining the practical applications, was found to be three orders of magnitude lower than those achieved with oxide tunnel barriers.

11.3 SUMMARY

Since the first reported organic spintronic device with MWCNTs, a variety of organic NM spacer materials have been experimentally tested so far. As a consequence, spintronic devices with CNTs and graphene have revealed the advantage of both long spin relaxation time and length, which are anticipated for nano-carbon materials. In particular, the spin injection into SLG and MLG samples at room temperature has been successfully accomplished by using the reliable non-local four-probe measurement [79,80], and hence the field of nano-carbon-based spintronics shifted from the work on carbon nanotubes to that on graphene. However, the reported spin relaxation length of around 1 μm for graphene-based devices is still in the same order as that for NM metallic materials, such as Al [70], Cu [82] and so forth. As for the spin coherence, the inherent advantage of graphene has not been fully extracted. Controlling the quality and understanding of the nature of the interfaces between the 2D form of carbon and the inorganic elements of the device becomes one of the critical issues, and considerable efforts are needed in this line. All-organic spintronic devices with organic FM materials might evade several issues occurring on the interface in the near future.

As pointed out in Sec. 11.1, besides their electric charge, spin degree of freedom is utilized in spintronics. In addition, when it comes to a graphene sheet and/or a carbon nanotube, their electronic band structure comprises two equivalent corners of the Brillouin zone, usually denoted by K and K', originating from two carbon atoms per unit cell. Thus, both CNTs and graphenes offer a unique twofold orbital degree of freedom, termed *valley*, that can be regarded as a pseudo-spin and therefore used for quantum manipulation. Together, the twofold degeneracy and twofold orbital degeneracy yield a fourfold degenerate electronic energy spectrum in clean nanotubes and/or graphenes. The valley degree of freedom might be utilized to control an electronic device, in the same way as the electron spin is used in spintronics. In fact, Beenakker and co-workers proposed *valleytronics* [100], in which a valley polarization is caused by occupying a single valley in graphene, and furthermore, an electrostatically controlled valley valve can be considered as a zero-magnetic-field counterpart to the SV described in this chapter. Such a development might afford a promising basis for more advanced quantum computing.

In any case, further progress in spintronics will be largely driven by the materials advances, and therefore chemistry has to play an important role to tow this emerging field. In this sense, carbon-based spintronics will increasingly become an interdisciplinary field of research, where physicists, chemists and electrical engineers integrally cooperate.

REFERENCES

[1] Zhirnov VV, Cavin RK III, Hutchby JA, Bourianoff GI. Limits to binary logic switch scaling–a Gedanken model. Proc IEEE 2003;69:1934–9.

[2] Žutić I, Fabian J, Sarma SD. Spintronics: fundamentals and applications. Rev Mod Phys 2004;76:323–410.

[3] Awschlom DD, Flatté ME. Challenges for semiconductor spintronics. Nat Phys 2007;3:153–9.

[4] Thomson W. On the electro-dynamic qualities of metals: effect of magnetization on the electric conductivity of nickel and of iron. Proc R Soc London 1857;8:546–50.

[5] Tedrow PM, Meservey R. Spin-dependent tunneling into ferromagnetic nickel. Phys Rev Lett 1971;26:192–5.
[6] Tedrow PM, Meservey R. Spin polarization of electrons tunneling from films of Fe, Co, Ni. Gd Phys Rev 1973;B7:318–26.
[7] Tedrow PM, Meservey R. Spin-polarized electron tunneling. Phys Rep 1994;238:173–243.
[8] Jullière M. Tunneling between ferromagnetic films. Phys Lett 1975;54A:225–6.
[9] Maekawa S, Gäfvert U. Electron tunneling between ferromagnetic films. IEEE Trans Magn 1982;18:707–8.
[10] Binasch G, Grünberg P, Saurenbach F, Zinn W. Enhanced magnetoresistance in layered magnetic structures with antiferromagnetic interlayer exchange. Phys Rev 1989;B39:4828–30.
[11] Baibich MN, Broto JM, Fert A, Nguyen van Dau F, Petroff F, Etienne P, et al. Giant magnetoresistance of (001)Fe/(001)Cr magnetic superlattices. Phys Rev Lett 1988;61:2472–5.
[12] Mott NF. The electrical conductivity of transition metals. Proc R Soc London 1936;A153:699–717.
[13] Mott NF. The resistance and thermoelectric properties of the transition metals. Proc R Soc London 1936;A156:368–82.
[14] Mott NF. Electrons in transition metals. Adv Phys 1964;13:325–422.
[15] Pratt WP Jr, Lee S-F, Slaughter JM, Loloee R, Schroeder PA. Perpendicular giant magnetoresistance of Ag/Co multilayers. Phys Rev Lett 1991;66:3060–3.
[16] Johnson M, Silsbee RH. Interfacial charge-spin coupling: injection and detection of spin magnetization in metals. Phys Rev Lett 1985;55:1790–3.
[17] Fert A, Jaffrès H. Conditions for efficient spin injection from a ferromagnetic metal into a semiconductor. Phys Rev 2001;B64:1–9. 184420.
[18] Fert A, George J-M, Jaffrès H, Mattana R. Semiconductors between spin-polarized sources and drains. IEEE Trans Elec Dev 2007;54:921–32.
[19] Hartman U, editor. Magnetic multilayers and giant magnetoresistance. Springer: Berlin; 2000.
[20] Hirota E, Sakakima H, Inomata K. Giant magneto-resistance devices. Springer: Berlin; 2002.
[21] Parkin SSP. Applications of magnetic nanostructures. In: Maekawa S, Shinjo T, editors. Spin dependent transport in magnetic nanostructures. New York: Taylor and Francis; 2002.
[22] Narisetti R. IBM unveils powerful PC disk drive, confirms plans to join two divisions. Wall Street J 1997; November 10, B8.
[23] Parkin SSR, Bharda R, Roche KP. Oscillatory magnetic exchange coupling through thin copper layers. Phys Rev Lett 1991;66:2152–5.
[24] Parkin SSR, Li ZG, Smith DJ. Giant magnetoresistance in antiferromagnetic Co/Cu multilayers. Appl Phys Lett 1991;58:2710–2.
[25] Parkin SSR, Jiang X, Kaiser C, Panchula A, Roche K, Samant M. Magnetically engineered spintronic sensors and memory. Proc IEEE 2003;91:661–80.
[26] Miyazaki T, Tezuka N. Giant magnetic tunneling effect in Fe/Al_2O_3/Fe junction. J Magn Magn Mater 1995;139:L231–4.
[27] Moodera JS, Kinder LR, Wong TM, Meservey R. Large magnetoresistance at room temperature in ferromagnetic thin film tunnel junctions. Phys Rev Lett 1995;74:3273–6.
[28] Parkin SSP, Roche KP, Samant MG, Rice PM, Beyers RB, Scheuerlein RE, et al. Exchange-based magnetic tunnel junctions and application to nonvolatile magnetic random access memory. J Appl Phys 1999;85:5828–33.
[29] Tehrani S, Engel B, Slaughter JM, Chen E, DeHerrera M, Durlam M, et al. Recent developments in magnetic tunnel junction MRAM. IEEE Trans Magn 2000;36:2752–7.
[30] Parkin SSP, Kaiser C, Panchula A, Rice PM, Hughes B, Samant M, et al. Giant tunnelling magnetoresistance at room temperature with MgO(100) tunnel barriers. Nat Mater 2004;3:862–7.

[31] Yuasa S, Nagahama T, Fukushima A, Suzuki Y, Ando K. Giant room-temperature magnetoresistance in single-crystal Fe/MgO/Fe magnetic tunnel junctions. Nat Mater 2004;3:868–71.

[32] Ikeda S, Hayakawa J, Lee YM, Tanakawa T, Matsukura F, Ohno H. Tunnel magnetoresistance in MgO-barrier magnetic tunnel junctions with bcc-CoFe(B) and fcc-CoFe free layers. J Appl Phys 2006;99:1–3. 08A907.

[33] Datta S, Das B. Electronic analog of the electrooptic modulator. Appl Phys Lett 1990;56:665–7.

[34] Sugahara S, Tanaka M. A spin metal-oxide-semiconductor field-effect transistor using half-metallic-ferromagnet contacts for the source and drain. Appl Phys Lett 2004;84:2307–9.

[35] De Groot RA, Mueller FM, Van Engen PG, Buschow KHJ. New class of materials: half-metallic ferromagnets. Phys Rev Lett 1983;50:2024–7.

[36] Park J-H, Vescovo E, Kim H-J, Kwon C, Ramesh R, Venkatesan T. Direct evidence for a half-metallic ferromagnet. Nature 1998;392:794–6.

[37] Tang HX, Monzon FG, Lifshitz R, Cross MC, Roukes ML. Ballistic spin transport in a two-dimensional electron gas. Phys Rev 2000;B61:4437–40.

[38] Levitov LS, Rashba EI. Dynamical spin-electric coupling in a quantum dot. Phys Rev 2003;B67:1–5. 115324.

[39] Rashba EI, Efros AL. Orbital mechanisms of electron-spin manipulation by an electric field. Phys Rev Lett 2003;91:1–4. 126405.

[40] Hattori R, Shirafuji J. Numerical simulation of tunnel effect transistors employing internal field emission of Schottky barrier junction. Jpn J Appl Phys 1994;33:612–8.

[41] Malajovich I, Kikkawa JM, Awschalom DD, Berry JJ, Samarth N. Coherent transfer of spin through a semiconductor heterointerface. Phys Rev Lett 2000;84:1015–8.

[42] Lou X, Adelmann C, Crooker SA, Garlid ES, Zhang J, Reddy KSM, et al. Electrical detection of spin transport in lateral ferromagnet-semiconductor devices. Nat Phys 2007;3:197–202.

[43] Valet T, Fert A. Theory of the perpendicular magnetoresistance in magnetic multilayers. Phys Rev 1993;B48:7099–113.

[44] Schmidt G, Ferrand D, Molenkamp LW, Filip AT, Van Wees BJ. Fundamental obstacle for electrical spin injection from a ferromagnetic metal into a diffusive semiconductor. Phys Rev 2000;B62:R4790–3.

[45] Filip AT, Hoving BH, Jedema FJ, Van Wees BJ, Dutta B, Borghs S. Experimental search for the electrical spin injection in a semiconductor. Phys Rev 2000;B62:9996–9.

[46] Rashba EI. Theory of electrical spin injection: tunnel contacts as a solution of the conductivity mismatch problem. Phys Rev 2000;B62:R16267–70.

[47] Yu ZG, Flattè ME. Electric-field dependent spin diffusion and spin injection into semiconductors. Phys Rev 2002;B66:1–4. R201202.

[48] Prigodin VN, Bergeson JD, Lincoln DM, Epstein AJ. Anomalous room temperature magnetoresistance in organic semiconductors. Synth Met 2006;156:757–61.

[49] Beenacker CWJ, Van Houten H. Quantum transport in semiconductor nanostructures. Solid State Phys 1991;44:1–228.

[50] Monzon FG, Roukes ML. Spin infection and the local Hall effect in InAs quantum wells. J Magn Magn Mater 1999;198–199:632–5.

[51] Van der Molen SJ, Tombros N, Van Bees BJ. Magneto-Coulomb effect in spin-valve devices. Phys Rev 2006;B73:1–4. R220406.

[52] Sanvito S, Rocha AR. Molecular spintronics: the art of driving spin through molecules. J Comput Theor Nanosci 2006;3:624–42.

[53] Dresselhaus G. Spin-orbit coupling effects in zinc blend structures. Phys Rev 1955;100:580–6.

[54] Bychkov YA, Rashba EI. Properties of a 2D electron gas with lifted spectral degeneracy. JETP Lett 1984;39:78–81.

[55] Elliot RJ. Theory of the effect of spin-orbit coupling on magnetic resonance in some semiconductors. Phys Rev 1954;96:266–79.
[56] Yafet Y. *g* Factors and spin-lattice relaxation of conduction electrons. Solid State Phys 1963;14:1–98.
[57] D'yakonov MI, Perel' VI. Spin orientation of electrons associated with the interband absorption of light in semiconductors. Sov Phys JETP 1971;33:1053–9.
[58] Bir GL, Aronov AG, Pikus GE. Spin relaxation of electrons due to scattering by holes. Sov Phys JETP 1976;42:705–12.
[59] Slichter CP. Principles of magnetic resonance. 3rd ed. Berlin: Springer; 1989.
[60] Naber WJM, Faez S, Van der Wiel WG. Organic spintronics. J Phys D Appl Phys 2007;40:R205–28.
[61] Sanvito S. Molecular spintronics. Chem Soc Rev 2011;40:3336–55.
[62] Iijima S. Helical microtubules of graphitic carbon. Nature 1991;354:56–8.
[63] Novoselov KS, Geim AK, Morozov SV, Jiang D, Zhang Y, Dubonos SV, et al. Electric field effect in atomically thin carbon films. Science 2004;306:666–9.
[64] Tsukagoshi K, Alphenaar BW, Ago H. Coherent transport of electron spin in a ferromagnetically contacted carbon nanotube. Nature 1999;401:572–4.
[65] Krompiewski S, Nemec N, Cuniberti G. Spin transport in disordered single-wall carbon nanotubes contacted to ferromagnetic leads. Phys Stat Sol 2006;B243:179–82.
[66] Sahoo S, Kontos T, Schönenberger C, Sürgers C. Electrical spin injection in multiwall carbon nanotubes with transparent ferromagnetic contacts. Appl Phys Lett 2005;86:1–3. 112109.
[67] Hueso LE, Pruneda JM, Ferrari V, Burnell G, Valdés-Herrera JP, Simons BD, et al. Transformation of spin information into large electrical signals using carbon nanotubes. Nature 2007;445:410–3.
[68] Sahoo S, Kontos T, Furer J, Hoffmann C, Graber M, Cotter A, et al. Electric field control of spin transport. Nat Phys 2005;1:99–102.
[69] Man HT, Wever IJW, Morpurgo AF. Spin-dependent quantum interference in single-wall carbon nanotubes with ferromagnetic contacts. Phys Rev 2006;B73:1–4. R241401.
[70] Jedema FJ, Filip AT, Van Wees BJ. Electrical spin injection and accumulation at room temperature in an all-metal mesoscopic spin valve. Nature 2001;410:345–8.
[71] Tombros N, Van der Molen SJ, Van Wees BJ. Separating spin and charge transport in single-wall carbon nanotubes. Phys Rev 2006;B73:1–4. 233403.
[72] Novoselov KS, Geim AK, Morozov SV, Jiang D, Katsnelson MI, Grigorieva IV, et al. Two-dimensional gas of massless Dirac fermions in graphene. Nature 2005;438:197–200.
[73] Zhang Y, Tan Y-W, Stormer HL, Kim P. Experimental observation of the quantum Hall effect and Berry's phase in graphene. Nature 2005;438:201–4.
[74] Novoselov KS, Jiang D, Schedin F, Booth TJ, Khotkevich VV, Morozov SV, et al. Two-dimensional atomic crystals. Proc Natl Acad Sci USA 2005;102:10451–3.
[75] Berger C, Song Z, Li T, Li X, Ogbazghi AY, Feng R, et al. Ultrathin epitaxial graphite: 2D electron gas properties and a route toward graphene-based nanoelectronics. J Phys Chem 2004;B108:19912–6.
[76] Berger C, Song Z, Li X, Wu X, Brown N, Naud C, et al. Electronic confinement and coherence in patterned epitaxial graphene. Science 2006;312:1191–6.
[77] De Heer WA, Berger C, Ruan M, Sprinkle M, Li X, Hu Y, et al. Large area and structured epitaxial graphene produced by confinement controlled sublimation of silicon carbide. Proc Natl Acad Sci USA 2011;108:16900–5.
[78] Hill EW, Geim AK, Novoselov K, Schedin F, Blake P. Graphene spin valve devices. IEEE Trans Magn 2006;42:2694–6.

[79] Tombros N, Jozsa C, Popinciuc M, Jonkman HT, Van Wees BJ. Electrical spin transport and spin precession in single graphene layers at room temperature. Nature 2007;448:571–4.

[80] Ohishi M, Shiraishi M, Nouchi R, Nozaki T, Shinjo T, Suzuki Y. Spin injection into a graphene thin film at room temperature. Jpn J Appl Phys 2007;46:L605–7.

[81] Hanle W. Über magnetische beeinflussung der polarisation der resonanzfluoreszenz. Z Phys 1924;30:93–105.

[82] Jedema FJ, Heersche HB, Filip AT, Baselmans JJA, Van Wees BJ. Electrical detection of spin precession in a metallic mesoscopic spin valve. Nature 2002;416:713–6.

[83] Shiraishi M, Ohishi M, Nouchi R, Mitoma N, Nozaki T, Shinjo T, et al. Robustness of spin polarization in graphene-based spin valves. Adv Funct Mater 2009;19:3711–6.

[84] Wang WH, Pi K, Li Y, Chiang YF, Wei P, Shi J, et al. Magnetotransport properties of mesoscopic graphite spin valves. Phys Rev 2008;B77:1–4. R020402.

[85] Cho S, Chen Y-F, Fuher MS. Gate-tunable graphene spin valve. Appl Phys Lett 2007;91:1–3. 123105.

[86] Józsa C, Popinciuc M, Tombros N, Jonkman HT, Van Wees BJ. Electrical spin drift in graphene field-effect transistors. Phys Rev Lett 2008;100:1–4. 236603.

[87] Tombros N, Tanabe S, Veligura A, Jozsa C, Popinciuc M, Jonkman HT, et al. Anisotropic spin relaxation in graphene. Phys Rev Lett 2008;101:1–4. 046601.

[88] Goto H, Kanda A, Sato T, Tanaka S, Ootuka Y, Odaka S, et al. Gate control of spin transport in multilayer graphene. Appl Phys Lett 2008;92:1–3. 212110.

[89] Han W, Pi K, McCreary KM, Li Y, Wong JJI, Swartz AG, et al. Tunneling spin injection into single layer graphene. Phys Rev Lett 2010;105:1–4. 167202.

[90] Maassen T, Van den Berg JJ, Ijbema N, Fromm F, Seyller T, Yakimova R, et al. Long spin relaxation times in wafer scale epitaxial graphene on SiC(0001). Nano Lett 2012;12:1498–502.

[91] Dlubak B, Martin M-B, Deranlot C, Servet B, Xavier S, Mattana R, et al. Highly efficient spin transport in epitaxial graphene on SiC. Nat Phys 2012;8:557–61.

[92] Chen S, Brown L, Levendorf M, Cai W, Ju S-Y, Edgeworth J, et al. Oxidation resistance of graphene-coated Cu and Cu/Ni alloy. ACS Nano 2011;5:1321–7.

[93] Cobas E, Friedman AL, Van't Erve OMJ, Robinson JT, Jonker BT. Graphene as a tunnel barrier: graphene-based magnetic tunnel junctions. Nano Lett 2012;12:3000–4.

[94] Van't Erve OMJ, Friedman AL, Cobas E, Li CH, Robinson JT, Jonker BT. Low-resistance spin injection into silicon using graphene tunnel barriers. Nat Nanotech 2012;7:737–42.

[95] Hanbicki AT, Van't Erve OMJ, Magno R, Kioseoglou G, Li CH, Jonker BT, et al. Analysis of the transport process providing spin injection through an Fe/AlGaAs Schottky barrier. Appl Phys. Lett. 2003;82:4092–4.

[96] Motsnyi VF, De Boeck J, Das J, Van Roy W, Borghs G, Goovaerts E, et al. Electrical spin injection in a ferromagnet/tunnel barrier/semiconductor heterostructure. Appl Phys Lett 2002;81:265–7.

[97] Jonker BT, Kioseoglou G, Hanbicki AT, Li CH, Thompson PE. Electrical spin-injection into silicone from a ferromagnetic metal/ tunnel barrier contact. Nat Phys 2007;3:542–6.

[98] Dash SP, Sharma S, Patel RS, De Jong MP, Jansen R. Electrical creation of spin polarization in silicon at room temperature. Nature 2009;462:491–4.

[99] Sasaki T. Electrical spin injection into silicon using MgO tunnel barrier. Appl Phys Exp 2009;2:1–3. 053003.

[100] Rycerz A, Tworzydło J, Beenakker CWJ. Valley filter and valley valve in graphene. Nat Phys 2007;3:172–5.

Chapter 12

Biological Application of Carbon Nanotubes and Graphene

Zhao Jun Han,* Amanda E. Rider,* Caitlin Fisher,*,† Timothy van der Laan,*,† Shailesh Kumar,* Igor Levchenko,*,† Kostya (Ken) Ostrikov*,†,‡,§,¶

**CSIRO Materials Science and Engineering, Bradfield Road, Lindfield, New South Wales 2070, Australia, †School of Physics, The University of Sydney, Sydney, New South Wales 2006, Australia, ‡Australian Institute for Innovative Materials, The University of Wollongong, New South Wales 2522, Australia, §Faculty of Science, University of Technology, Sydney, New South Wales 2007, Australia, ¶School of Chemistry, Physics, and Mechanical Engineering, Queensland University of Technology, Brisbane, QLD 4000, Australia*

12.1 INTRODUCTION

Carbon nanotubes (CNTs) and graphene are two representative nanomaterials comprised of purely element carbon [1,2]. Graphene is the two-dimensional, hexagonal sp^2-carbon ring networks with one atomic layer thickness, while CNTs can be envisaged as one or several graphene sheets concentrically rolled up into a one-dimensional cylindrical structure, so-called single-walled (SW) or multi-walled (MW) CNTs, respectively. Figure 12.1 shows the schematic diagram of structures of graphene, SWCNT and MWCNT. Owing to their exceptional mechanical, electrical, optical and thermal properties, CNTs and graphene have been widely considered as a new type of materials with great potentials to revolutionalize many of the biological and medical fields [3–5].

In this chapter, we will first elaborate the recent progress of CNTs in the biological applications, such as biosensors, diagnostic and imaging tools, therapeutic drug delivery systems, anti-cancer, tissue engineering scaffolds and neuron prosthesis. We will then discuss the emerging biological applications for graphene. In particular, we focus on the mechanism of how CNTs and graphene interact with different biological hosts at both the cellular and molecular levels. Some of the most stimulating in vitro and in vivo results are also demonstrated. We aim to show the advantages and challenges in the current development of CNT- and graphene-based biomedical devices, which may lead to better design of these functional nanomaterials in the future.

Carbon Nanotubes and Graphene. DOI: 10.1016/B978-0-08-098232-8.00012-7

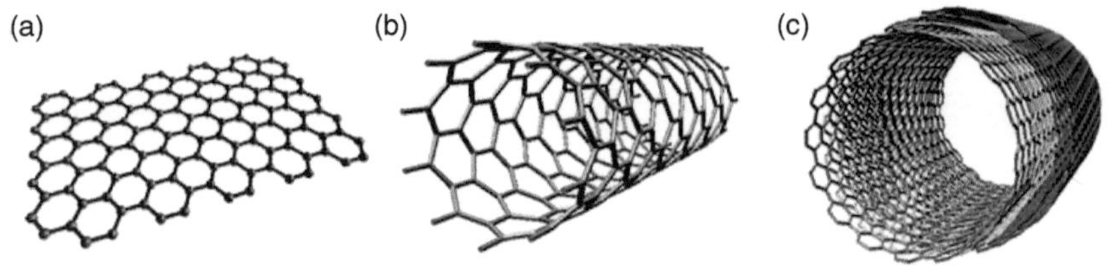

FIGURE 12.1 Schematic diagram of (a) graphene, (b) SWCNT and (c) MWCNT.

12.2 BIOLOGICAL APPLICATION OF CNTS

12.2.1 Synthesis of CNTs

Over the past two decades, intense research efforts have been directed at the synthesis of various CNT-based structures [6–9]. Techniques such as arc discharge, laser ablation and chemical vapour deposition (CVD) are often used to synthesize CNTs [7,10]. Owing to its low-temperature processing and the ability to precisely control the position of CNTs, CVD is currently the most popular method.

CVD refers to a process that a solid material is deposited onto a heated substrate using a gaseous precursor. A catalyst material, commonly the transition metal Fe, Ni and Co nanoparticles, is also needed in the growth of CNTs. The gaseous carbon-containing precursors (e.g. CH_4, C_2H_4, C_2H_2, CO) firstly dissociate into atomic or molecular carbon species on the surface of catalyst nanoparticles. Nucleation then occurs as these carbon species diffuse into the catalyst nanoparticles, reaching a supersaturated state, and then segregating from the surface of nanoparticles to form a nanotube cap [11,12]. The growth of nanotubes is then sustained by the continuous incorporation of carbon atoms via bulk and/or surface diffusions and is terminated by stopping the gas supplies.

Recently, plasma-enhanced CVD, or PECVD, is also used to increase the growth efficiency and reduce the growth temperature of CNTs. PECVD refers to a process where plasma (the fourth state of matter, consisting of a complex mixture of ions, electrons, neutrals, free radicals, etc.) assists the growth in addition to the conventional thermal process [13]. Both experiments and modelling of PECVD indicated that the plasma effects could better control the diameter, length and even chirality of SWCNTs [14–18], leading to optimal structure for desirable biological applications.

In general, two morphologies of CNTs can be grown in CVD and PECVD processes, the random-CNT network and the aligned arrays. Figure 12.2 shows some typical examples of CNTs synthesized in our laboratory, including the random-CNT networks [12,19,20], the free-standing CNTs, the vertically aligned CNTs [21,22] and the three-dimensional CNT scaffolds [23]. These sophisticated structures greatly enable the fabrication and integration of next-generation devices with versatility, reproducibility and robustness in the biologically relevant environments [23,24].

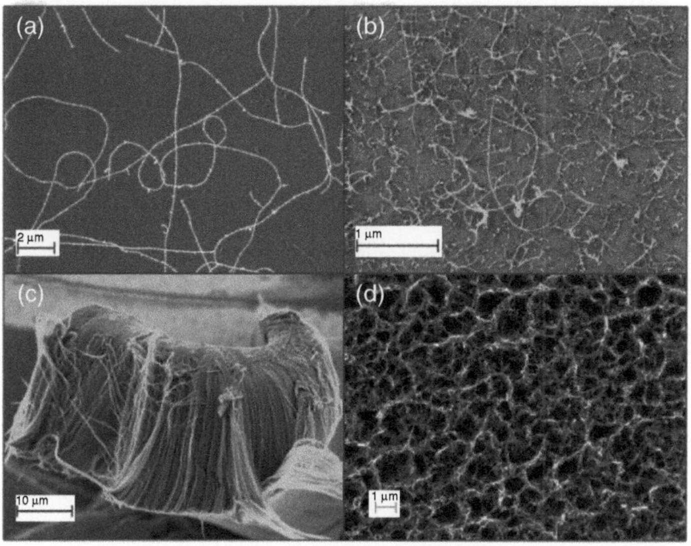

FIGURE 12.2 Some typical CNTs synthesized in our laboratory at CSIRO: (a) random CNT networks, (b) free-standing CNTs, (c) vertically aligned CNTs and (d) three-dimensional CNT scaffolds.

12.2.2 Functionalization of CNTs

One of the major advantageous features of CNTs in biological applications is the ability to link a diversity of species such as sugars, cells, proteins and DNA. These biological or non-biological species can serve to enhance electrical properties, solubility and biocompatibility, to prevent non-specific binding and/or to interact with other biological species in the micro-environment [25–27]. A prerequisite for the successful linkage between CNTs and these species is by proper functionalization. As illustrated in Figure 12.3, there are generally two CNT functionalization methods: covalent bonding and non-covalent wrapping [28].

Covalent bonding involves chemical attachment of the desirable species to the CNT, often at the tube ends or at defect sites [26,27,29]. Oxidation processes, such as annealing in air or exposing the CNTs to H_2SO_4 or HNO_3 acids, are commonly used as a preparation step to create chemically active sites for covalent bonding [30]. However, this may alter the intrinsic properties of the CNTs and/or the bioactivity of the attached biomolecules, which is an undesirable outcome [31].

Non-covalent bonding, or physical adsorption, is the process of wrapping a species around the CNT walls via van der Waal's interactions [32]. This method is preferable to covalent bonding in many cases as it causes less structural damage to the CNT and the wrapped species, and the chemical environment required during fabrication is less harsh [33]. Various species have been reported to successfully wrap around CNTs, such as polysaccharides, DNA, proteins, polypeptides and polymers [34–37].

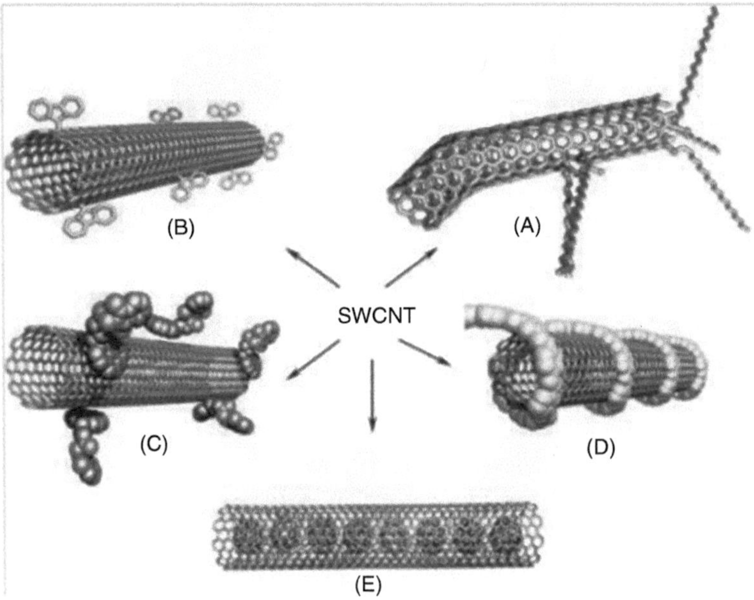

FIGURE 12.3 Typical functionalization methods for SWCNTs: (A) defect-group functionalization, (B) covalent sidewall functionalization, non-covalent functionalization with (C) surfactants and (D) polymers and (E) endohedral functionalization with C_{60}; reproduced with permission from Ref. [28].

12.2.3 CNT-based biosensors

A biosensor consists of a receptor that interacts with the biological analyte, and this interaction is detected by a transducer that transforms the signal (electrical, optical, etc.) into a form that can be easily measured [38]. Biosensors modified with CNTs have the advantages of high signal-to-volume ratio, miniaturization from the CNTs' size being comparable to many biological species, function at ambient temperatures, low surface fouling and high specificity [39,40]. Two common types of CNT-based biosensors are electrical sensors using CNT field-effect transistors (FETs) [41,42] and electrochemical sensors [31,43]. In the following section, we will focus on these two types of biosensors and provide details on the role of CNTs in these devices, giving notable examples.

12.2.3.1 Electrical Biosensors

CNT-FET biosensors consist of a current-carrying channel connecting a source and a drain, where the current can be regulated by a gate voltage [39,44]. The current-carrying channel is typically a semiconducting SWCNT or a randomly orientated net of CNTs, as shown in Figure 12.4(a) and (b) [39]. The sensor consists of monitoring changes to the current, which result from electrical exchanges within biological interactions in the surrounding environment or the binding of biomolecules onto designated sites on the CNT.

CNT-FETs have been reported to perform better than traditional metal-oxide-semiconductor FETs (MOSFETs) and are highly sensitive to electrical and chemical changes in the environment [45]. The one-dimensional nature of the CNTs confines the current to make them highly sensitive to stray charge carriers [42]. The current is transported on the surface, coming into direct contact with the environment. The small size of CNTs also allows for circuit miniaturization, which minimizes the signal-to-noise ratio and increases the sensitivity.

CNTs are usually functionalized with species that interact with the analyte, making the biosensor highly selective. Such species include aptamers, antibodies, sugars, DNA and proteins [46–50]. The wide variety of species allows the tailoring of biosensors to detect many different analytes, such as DNA at concentrations of 100 fM [51]. CNTs are chemically inert and thus resistant to surface fouling or non-specific binding [39]. Further resistance to non-specific binding of proteins can be provided by polymer coatings, such as poly(ethylene glycol) (PEG) and poly(ethylene oxide) (PEO) [38].

The electrical and chemical properties of CNT-FETs can be further improved by the deposition of metallic nanoparticles, such as Au, Pt and Ag. Rajesh et al. demonstrated that the sensitivity of DNA detection was 2.5 times greater after the addition of ZnS nanoparticles into CNT-FETs [52]. One difficulty with using CNT-FETs is to control the Schottky barrier effect, which arises from differences between the work functions of the CNT and the metal contacts and is often hard to predict [42,53]. Methods such as coating the metal contacts with polymers have been used to minimize this effect and stabilize the response [53].

CNT-FET biosensors have also been used to make real-time measurements of biological interactions, such as those involving DNA, proteins, immunoglobulin, morphine and phagocytosis [50,51,54,55]. Glucose detection in real time is very important for health monitoring purposes especially for patients with diabetes. The first CNT-based real-time glucose sensor at room temperature was reported by Besteman et al. in 2003 [40]. Again, decoration by metal nanoparticles can be used to improve the performance of CNT-FETs in real-time measurements, as has been demonstrated in the cases of glucose and heroin [51,55].

12.2.3.2 *Electrochemical biosensors*

Electrochemical CNT-based biosensors are very well documented and perhaps the most common type of CNT-based biosensor [31]. They consist of functionalized CNTs attached to metal or glassy carbon electrodes (GCEs) which, when analytes bind to the CNTs, detect the charge exchange from chemical redox reactions [31,41]. The benefits of incorporating CNTs include their excellent charge-carrying abilities and their tiny size, which could be the same order of magnitude as the analytes. Typical SWCNTs have diameters of around ~1 nm, which is comparable to DNA, active sites in proteins and many other cellular structures. CNT electrochemical biosensors have been used to detect DNA, enzyme, proteins, RNA, dopamine, cholesterol and many other biomolecules [48,56–58].

The detection of DNA is extremely important for research in cellular mutations and understanding human development. Many electrochemical biosensors have been targeted at detecting DNA. For instance, Cai et al. described a very typical configuration where MWCNTs functionalized with –COOH groups on GCEs were used to detect 5'-amino group oligonucleotides at concentrations as low as 10^{-10} mol L^{-1} [59]. Another very common analyte is glucose, which is used to monitor many health issues, such as diabetes. CNTs have been functionalized with species such as glucose oxidase (GO_x) or flavin adenine dinucleotide (FAD) to detect glucose concentrations. Liu et al. found that when biosensors were prepared with a network of CNTs and chitosan (CS), the GO_x provided twice the amount of electron transfer to the electrodes than the FAD due to the structural accessibility of the GO_x bioactive sites [60].

The CNTs can be grown on the electrodes themselves either by self-assembly or by lithography. The 1D structure of CNTs allows them to detect single biomolecules, particularly when they are vertically aligned (Figure 12.4(c)) [61,62]. Single-molecule detection is particularly important when detecting biologically relevant gases, e.g. O_2, NO_2 and NH_3. The sensitivity of biosensors in this respect is dependent on the size, alignment and spatial distribution of the CNT arrays [38,49].

A very simple way of improving the contact between the electrode and the CNTs is by rubbing a layer of randomly orientated CNTs on the surface [31,63]. The properties of these thin films of CNTs are the average of all the CNTs in the film. This method of biosensor fabrication is very simple and readily replicable as opposed to those requiring attention to individual CNTs. A recent CNT-based electrochemical biosensor has been made using spun CNT fibres, which was found to be more sensitive in detecting glucose than the traditional Pt–Ir sensors [64,65].

The decoration of metallic nanoparticles onto CNT-biosensors has been commonly used to increase the conductivity of the sensor. An example of this

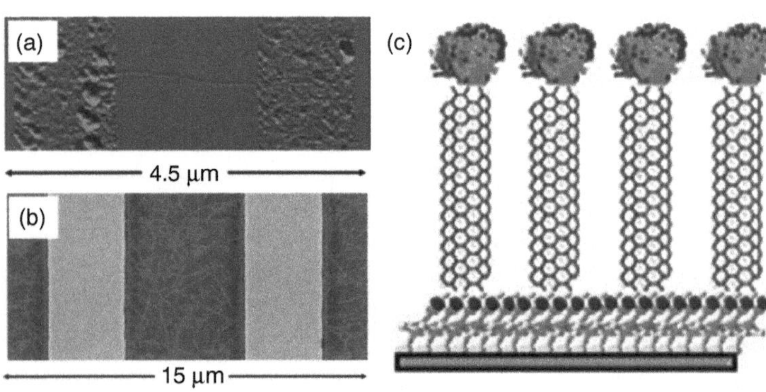

FIGURE 12.4 CNT–FET biosensing devices based on (a) a single SWCNT and (b) a randomly orientated net of SWCNTs; reprinted with permission from Ref. [39]. (c) Illustration of electrochemical sensors based on vertically aligned SWCNTs; reprinted with permission from Ref. [61].

is given by Hrapovic et al. [66], where both nafion-coated SWCNTs and Pt nanoparticles were used to enhance the detection of hydrogen peroxide (H_2O_2). This combination of Pt and CNTs detected lower concentrations than using only the SWCNTs or nanoparticles alone. Other methods to improve sensitivity include gas treatment and plasma etching which can add extra functional groups (e.g. -COOH, -NH_2) to the CNTs [67,68]. Lastly, optimizing the ratio of semiconducting and metallic CNTs can also improve the sensitivity and response speed of CNT electrochemical sensors [49].

Similar to electrical sensors, non-specific binding of proteins can reduce the sensitivity of the electrochemical biosensors, and thus has to be minimized. This can be mitigated by layering polymer coatings to resist non-specific binding. Typical polymers used in electrochemical biosensing are PEG and PEO [69]. The chosen polymer coatings must have a minimal impact on the properties of CNT and the biological receptor [41].

12.2.4 CNTs as diagnostic and imaging tools

Tools for imaging and tracking the location and movement of biological objects such as proteins, cells and tissues are critical in understanding their functionalities and activities in the host biological system [70,71]. Such tools also provide us the ability to diagnose and possibly cure various diseases. The current techniques use organic fluorophores or quantum dots (QDs, i.e. luminescent semiconducting nanoparticles). However, these organic fluorophores and QDs show serious photobleaching effect. They also have a limited lifetime in aqueous solutions (approximately 1–2 weeks) [72]. CNTs make excellent optical sensors compared to these fluorescent probes due to photobleaching-resistance, non-quenching, long life-time and high robustness in a biological environment [54,73,74]; thus, they can be exploited as non-invasive imaging tools for both in vitro and in vivo applications at cellular or sub-cellular levels [75].

12.2.4.1 Raman Imaging

SWCNTs have a high Raman cross-section and can produce strong, resonance-enhanced Raman bands at $\sim$1600 cm^{-1} away from the excitation wavelength (Figure 12.5(a)) [72,76]. The Raman signal is easily detectable, does not blink or quench and does not diminish under prolonged excitation. Heller et al. firstly demonstrated the Raman imaging of SWCNTs in fibroblast and myoblast in 2005 [72]. The SWCNTs used were encapsulated by single-strand DNA (ssDNA). Complementary transmission electron microscopy (TEM) and light microscopy images confirmed that such ssDNA–SWCNT complexes were successfully translocated into the perinuclear endosome. Furthermore, Dai and co-workers used arginine–glycine–asparatic acid (RGD) peptide-linked SWCNTs (RGD–SWCNTs) to selectively bind to $\alpha_v\beta_3$ integrin-linked tumour U87MG cells [77,78]. RGD has a high affinity to $\alpha_v\beta_3$ integrin, which is

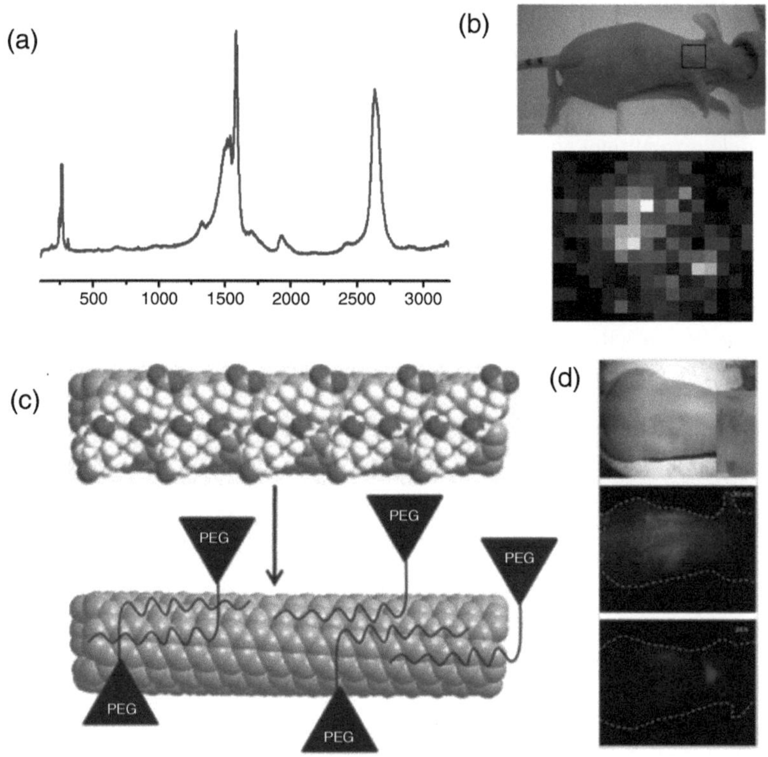

FIGURE 12.5 (a) Resonant Raman spectra of SWCNTs. (b) Raman map of a tumour sites from mouse receiving RGD–SWCNTs at 72 hours; reprinted with permission from Ref. [78]. (c) Fluorescent CNT labels functionalized with PL–PEG to increase biocompatibility. (d) NIR photoluminescence images of mice injected with functionalized SWCNTs at 30 min and 24 hours; reprinted with permission from Ref. [86].

over-expressed in tumour neovasculature. The in vivo Raman image clearly revealed that the injected RGD–SWCNTs preferentially accumulated on the tumour site (Figure 12.5(b)) [78].

The capability of CNT Raman imaging can be enhanced by blending with other functional components. Choi et al. developed SWCNT/iron oxide nanoparticle complexes as multi-modal biomedical imaging agents, combining the high-contrast imaging power of magnetic resonance imaging (MRI) and the sensitivity of Raman [79]. Liu et al. used phospholipid (PL)–PEG functionalized SWCNTs to investigate the circulation and long-term fate of SWCNTs in mice [80]. They found that PL–PEG–SWCNTs have enabled the longest SWCNT blood circulation and a near-complete clearance from the main organs in ~2 months. Raman spectroscopy detected SWCNTs in the intestine, faeces, kidney and bladder of mice, suggesting excretion and clearance of SWCNTs from mice via the biliary and renal pathways. No side effect of SWCNTs to mice was observed [80].

A further improvement in CNT Raman imaging is by utilizing the surface-enhanced Raman scattering (SERS) mechanism. It is known that when CNTs adsorbed onto metal surfaces, their Raman signals can be significantly increased. Chu et al. decorated well-separated SWCNTs with uniform gold nanoparticles. The coupled surface plasmon resonance (SPR) greatly enhanced the Raman fingerprints of SWCNTs, including those non-resonant with laser energy and those with originally undetectable radial-breathing mode (RBM) [81]. Kumar et al. synthesized Ag-core Au-shell (Ag@Au) nanoparticles and carefully tuned their plasmon resonance peak at 541 and 642 nm, which closely matched to the commonly used Raman excitation sources of 532 and 633 nm [82]. A greater enhancement of Raman signal was obtained when compared to nanotubes adsorbed with conventional Ag, Au or other 'off resonant' Ag@Au nanoparticles.

12.2.4.2 Photoluminescence

The near-infrared (NIR) photoluminescence of SWCNTs was discovered in 2002 [83], which enabled us to study the dynamics at the single-molecule level. The NIR emission spectrum of CNTs is in the range of 900–1600 nm [74], particularly useful for skin and other biological membrane transparency [84]. CNTs used as molecular fluorophores show no apparent photobleaching and blinking effect. The fluorescence quantum yield of SWCNTs can be as high as 20% [85].

Because of the deep tissue penetration and low autofluorescence background, fluorescence CNT labels allow high-resolution intravital microscopy imaging of tumour vessels beneath thick skin. Welsher et al. prepared PL–PEG-coated SWCNTs and found that the in vivo imaging agents were both bright and biocompatible (Figure 12.5(c)) [86]. They further linked the RGD peptides to PL–PEG-coated SWCNTs, which could selectively bind to $\alpha_v\beta_3$-integrin positive cells. The whole-animal in vivo imaging was demonstrated in the 1–1.7 μm spectra range by detecting the NIR fluorescence of PL–PEG-coated SWCNTs at a low dose of 17 mg L^{-1} (Figure 12.5(d)) [86]. In addition, real-time video imaging was obtained to reveal how SWCNTs circulated through organs, such as lungs, kidneys, spleen and liver [87]. Very recently, chirality-selected SWCNTs, i.e. enriched semiconducting (12, 1) and (11, 3) SWCNTs which have an identical resonance absorption at ~808 nm and an emission near ~1200 nm, have also been used to obtain the real-time in vivo video imaging of whole mouse body and tumour vessels at a much lower injected dose [88].

Gas detection in relevant biological processes, which requires very high sensitivity, can also be realized using fluorescence CNT labels. For example, Kim et al. fabricated a reusable nitric oxide (NO) detector using a NIR fluorescence sensor modified with SWCNTs, non-covalently wrapped with 3,4-diaminophenyl functionalized dextran (DAP-dex) [89]. Penza et al. showed that the addition of a thin film of SWCNTs onto a quartz crystal microbalance sensor and a silica optical fibre sensor increased the sensitivity of the sensors by an order of magnitude when sensing both gaseous alcohol and volatile organic compounds at room temperature [90,91].

12.2.4.3 *Other Imaging Tools*

CNTs have also been utilized in other novel imaging applications, such as photoacoustic and endoscopic devices. de la Zerda et al. reported a new concept of CNT imaging based on the photoacoustic effect (i.e. ultrasound) [92]. In the design, the energy of non-ionizing laser pulses were absorbed by SWCNTs and converted into heat, leading to the ultrasonic emission. The generated ultrasonic waves were then detected by the transducers to form images. The nanotubes were also linked to RGD peptides through grafting PL–PEG and the photoacoustic images clearly showed the 3D distribution of nanotubes in the tumour, with a higher spatial resolution and deeper depth of penetration than the fluorescence imaging using QDs [92].

CNTs also resemble the endoscope because of the 1D shape. Singhal et al. firstly reported a CNT-based endoscope for interrogating cells, transporting fluids and performing optical and electrochemical diagnostics at the single-organelle level [93]. The endoscope was made by placing a MWCNT (diameter of 50–60 nm) at the tip of a glass pipette, which could probe the intracellular environment with a spatial resolution of ~100 nm (Figure 12.6(a) and (b)). The sub-micrometre size and high mechanical strength of CNT make it possible to penetrate the cell membrane without disrupting the cells [93]. Niu et al. further used this endoscope for SERS measurements [94]. They decorated CNTs with gold nanoparticles and assembled them onto the tips of pulled glass capillaries to form a SERS-enabled endoscope (Figure 12.6(c)). The SERS-enabled endoscopes exhibited high selectivity and sensitivity for detecting trace amounts of analytes (≈1 pM) in biofluid environments, highlighting their capabilities as label-free, biological sensors for real-time in situ cellular diagnostics, biological detection and pharmaceutical research (Figure 12.6(d)).

12.2.5 CNTs as therapeutic drug delivery systems

Owing to their large surface areas and the ability to link with different molecules, CNTs serve as the ideal carrier for drugs. For therapeutic purposes, CNTs can provide precise intracellular delivery of bioactive macromolecules. Furthermore, functionalized nanotubes are able to target specific cells, become ingested and release the conjugated contents in response to a chemical trigger [95,96].

The first case of CNT-mediated intracellular delivery of plasmid DNA was demonstrated by Pantarotto et al. [97]. Plasmid DNA was loaded on ammonium-functionalized SWCNTs through electrostatic interactions. These functionalized CNTs were then taken up into the HeLa cells through penetrating the cell membranes. Subsequently, the gene expression levels were observed to increase by up to 10-fold than that of naked DNA [97]. Richard et al. functionalized CNTs by two cationic amphiphiles and also observed that the efficiency of plasmid DNA transfection using these complexes was several orders of magnitude higher than that of naked DNA [98]. On the other hand, Nunes et al. grafted

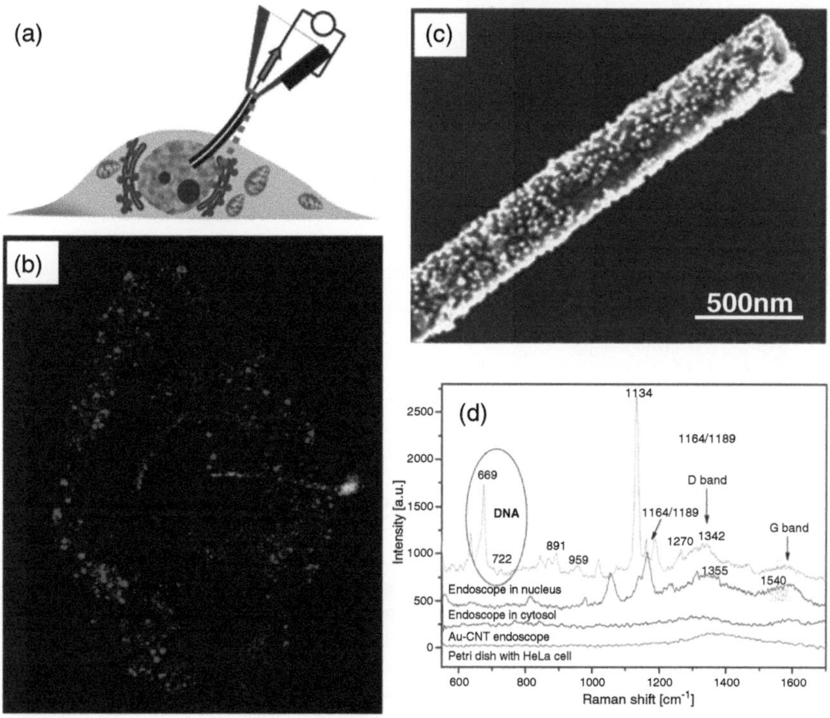

FIGURE 12.6 (a) Schematic of CNT-based endoscope for a single cell. (b) Fluorescence micrograph showing a cell being interrogated with an endoscope filled with FITC-labelled (light grey (green in the web version)) fluorescent polymer nanoparticles; reprinted with permission from Ref. [93]. (c) SEM micrograph of the gold-decorated CNT tip for SEMR-enabled endoscope. (d) SERS spectra obtained from SERS-active endoscope inside different compartments of a living HeLa cell in comparison to the spectra of the petri-dish surface and the endoscope tip in extracellular media; reprinted with permission from Ref. [94].

different cationic polymer on CNTs and effectively transferred plasmid DNA [99]. Gene expression in human lung epithelial (A549) cells was subsequently improved as compared to the equivalent amounts of PEI polymer alone.

CNTs could also be used in RNA interference (RNAi) technology, which utilizes short-interfering RNA sequences (siRNA). Ladeira et al. described a novel approach for siRNA cellular delivery using siRNA coiling into carboxyl-functionalized SWCNTs [100]. The CNT–siRNA delivery system successfully demonstrated transfection efficiency greater than 95%. Efficient reduction of InsP3R-II expression in neonatal cardiomyocytes after CNT–siRNA–InsP3R-II complex transfection was observed (Figure 12.7(a)). These studies confirmed the possibility of delivering DNA efficiently to cells using different surface-charged CNTs for the treatment of both genetic and acquired diseases.

Therapeutic proteins were delivered by CNTs. Kam et al. loaded *cytochrome c*, a small haeme protein which may be released from mitochondria during cells

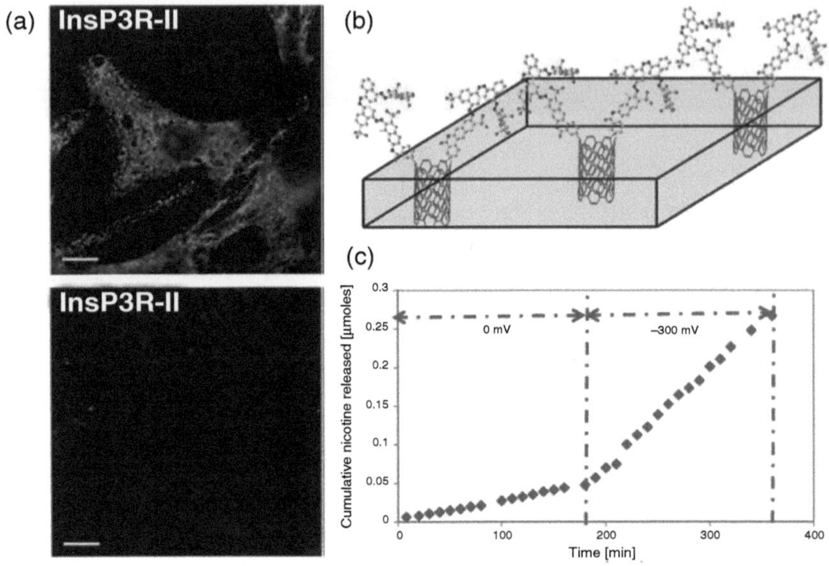

FIGURE 12.7 (a) Top: immunofluorescence of control sample of neonatal cardiomyocytes double-labelled with anti-InsP3R-II (light grey (green in the web version)). Bottom: immunofluorescence of neonatal cardiomyocytes following transfection with CNT–siRNA–InsP3R-II complex for 48 hours. Scale bars are 10 μm; reprinted with permission from Ref. [100]. (b) Schematic of the membrane with vertically aligned CNTs. (c) Flux of nicotine through CNT membrane with/without applying a −300 mV bias; reprinted with permission from Ref. [105].

apoptosis (programmed cell death), on the sidewalls of acid-oxidized SWCNTs [101]. The proteins were readily transported inside NIH-3T3 cells via the endocytosis pathway with nanotubes acting as the transporter. Once released from the endosomes, the internalized protein–nanotube conjugates could enter into the cytoplasm of cells and induce apoptosis [101]. Joshi et al. designed alcohol dehydrogenase (ADH)–nanotube conjugates to selectively destroy the anthrax toxin by NIR irradiation [102]. Through generating reactive oxygen species (ROS), the nanotube conjugates could remotely control the activity of proteins, e.g. protein deactivation [102]. This photochemical approach for the selective degradation of toxin may be broadly applicable for the targeted destruction of a variety of pathogens and cells.

Another application of CNTs as therapeutic drug delivery is on anti-inflammation. Nakamura et al. loaded prednisolone on both the outside and the inside of the oxidized single-walled carbon nanohorns (SWNHs) to treat chronic inflammatory diseases, such as rheumatoid arthritis [103]. After receiving the injection of prednisolone–SWNHs, rats with collagen-induced arthritis (CIA) showed retarded progression of the arthritis compared with controls [103]. Indomethacin, a non-steroidal anti-inflammatory drug commonly used to reduce fever, pain, stiffness and swelling, was also prepared with different CNT membranes for controlled release [104]. It was shown that the pristine MWCNTs undesirably decreased the release rate of indomethacin, whereas MWCNTs functionalized with cellulose acetate increased the release rate, due to the increased hydrophilicity and the changed structure and porosity of the CNT membranes [104].

Lastly, vertically aligned CNTs have also been implemented to in-drug delivery. Wu et al. prepared membranes composed of vertically aligned CNTs and integrated with a nicotine formulation [105]. Due to the fast flow through CNT cores, high charge density and small pore dimensions, highly efficient electrophoretic pumping through CNT membrane was achieved. Such device could be programmed with high efficiency to switch transdermal nicotine delivery rates for nicotine cessation treatment, without severe skin irritation and skin barrier disruption (Figure 12.7(b) and (c)) [105].

12.2.6 CNTs for anti-cancer applications

CNTs for cancer therapy have been a major research subject in the last few years. The advantages of using CNTs include: (1) CNTs can selectively diagnose and image the tumours with high resolution and stability [106]; (2) CNTs can destroy cancer cells based on the photothermal effect due to the significant amounts of heat generated upon NIR excitation [107,108]; (3) CNTs can be internalized into cancer cells and act as the effective carriers for therapeutic anti-cancer drugs [70,109].

The first in vitro photothermal destruction of cancer cells was performed by Kam et al. [107]. Selective internalization of SWCNTs inside cancer cells was realized by functionalization of SWCNT with a folate moiety. Continuous NIR radiation was used to cause cell death because of excessive local heating of SWCNTs in vitro (Figure 12.8(a)) [107]. This selective binding of CNTs to cancer cells was also demonstrated by Wang et al. [106], who prepared SWCNTs–Au–PEG–FA nanocomposite and utilized the strong SPR absorption contributed by the Au shell for targeted photothermal ablation. Moon et al. demonstrated in vivo experiment of mice with complete destruction of the tumours without harmful side effects or recurrence of tumours over 6 months (Figure 12.8(b)) [110]. Furthermore, Ghosh et al. found that DNA-encased CNTs could be more efficient at converting the NIR irradiation into heat compared to pristine CNTs, hence DNA-encased MWCNTs can be used safely and effectively for the selective thermal ablation of malignant tissue in vivo [111].

Similar to the photothermal effect, a photoacoustic 'bomb' effect of CNTs was reported by Kang et al. [112]. SWCNTs were firstly non-covalently functionalized with CS, and then the CS–SWCNTs were conjugated with folate acid, which selectively bound to cancer cells over-expressing folate receptor. Under the treatment of 1064-nm Q-switched millisecond-pulsed laser, 85% of Hep G2 cancer cells with SWCNT uptake died within 20 s, while 90% of the normal cells remained alive due to the lack of SWCNTs inside cells [112].

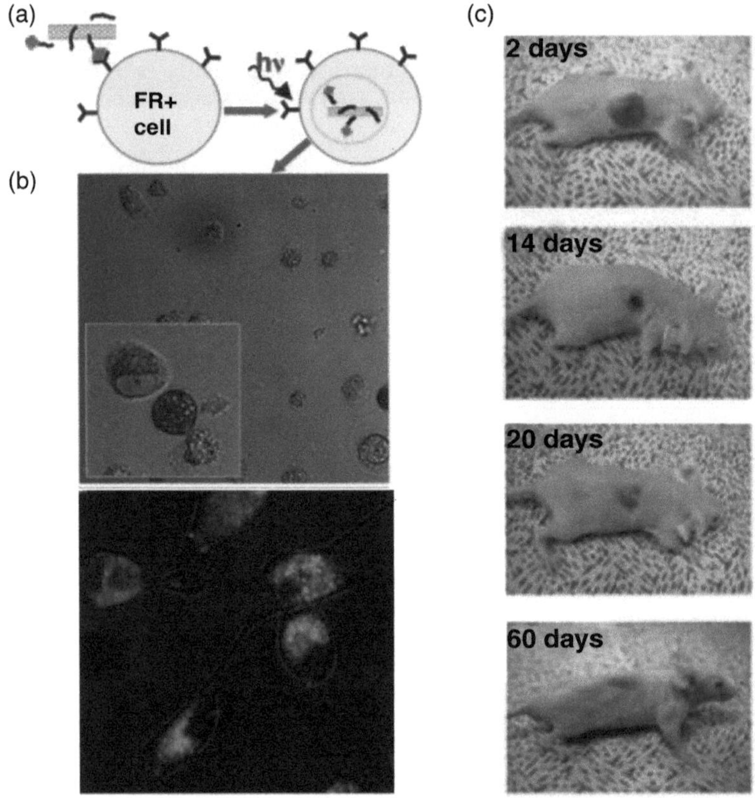

FIGURE 12.8 (a) Schematic of the selective internalization of PL–PEG–FA–SWCNTs into folate-overexpressing (FR+) cells via receptor binding and then NIR laser irradiation. (b) Top: image showing death of FR+ cells after NIR laser irradiation. Inset is the higher-magnification image showing details of the killed cells. Bottom: confocal image of the FR+ cells confirmed the SWCNT uptake with FA and FITC cargoes. Magnifications: ×20; reprinted with permission from Ref. [107]. (c) Representative photos of the mice treated with PEG–SWCNTs followed by NIR irradiation at various time points, demonstrating the in vivo photothermal effects for tumour destruction; reprinted with permission from Ref. [110].

Temperature changes during laser treatment were monitored and no temperature increases of more than ±3 °C were observed. Using this photoacoustic approach, the laser power used for cancer killing was reduced 150–1500 times and the therapy efficiency is improved.

For anti-cancer using chemotherapeutic drugs, Liu et al. prepared water-soluble SWCNTs with PEG functionalization and loaded a high degree of doxorubicin (DOX) [113]. Controllable release of such anti-cancer drugs was realized by using nanotube materials with suitable diameter as well as varying the pH. RGD–integrin recognition was also used to enhance DOX delivery to cancer cells labelled with corresponding biomarkers [113]. Wu et al. covalently bound the anti-tumour agent 10-hydroxycamptothecin (HCPT) to avoid the early release [114]. The pharmacological activity of MWCNTs–HCPT was confirmed by the significantly reduced tumour size. Bhirde et al. combined both cisplatin and epidermal growth factor (EGF; as a targeting agent to squamous cancer) with SWCNTs via the coordination and amidation of nanotube-bound carboxylic groups, respectively, and found that the system had good in vitro and in vivo anti-tumour (HN13 cells) activity with a low drug loading [115].

For in vivo imaging, Liu et al. firstly conjugated paclitaxel (PTX), a widely used cancer chemotherapy drug, to branched PEG chains on SWCNTs via a cleavable ester bond to obtain a water-soluble SWCNT–PTX conjugate [116]. SWCNT–PTX afforded higher efficacy in suppressing tumour growth than clinical Taxol in a murine 4T1 breast cancer model, owing to prolonged blood circulation and 10-fold higher tumour PTX uptake by SWCNT delivery. The strong therapeutic efficacy of SWCNT–PTX is shown even at a low drug dose (5 mg kg^{-1} PTX), desired for higher treatment efficacy and lower side effect [116].

Lastly, Wu et al. demonstrated the delivery of a combination of chemotherapeutic agents such as PTX (Taxol) and C6-ceramide to L3.6 pancreatic cancer cells using CNTs, adopting the 'Trojan-Horse' concept [109]. The compounds they developed could be released inside cells from within the nanotube 'on-command' by inductive heating with an external alternating current or pulsed magnetic field.

12.2.7 CNTs for hard tissue engineering

CNTs also have a great promise in hard tissue engineering, such as bone and teeth formation and regeneration [117]. Both 2D thin-film and 3D nanoporous CNT-based scaffolds were utilized and it was demonstrated that there are at least two benefits by using CNTs in bone formation: (i) accelerating bone formation from osteoblasts and (ii) directing differentiation of human mesenchymal stem cells (hMSCs) into the osteogenic lineage [118–120]. The exact effects of CNTs depend on the type, size, length and surface properties of CNTs, as well as the conditions in cell culture. For example, cells were more active on smaller-diameter CNTs and expressed higher levels of alkaline phosphatase (ALP; a marker in early cell differentiation) and calcium (a marker in bone mineralization) [118].

Tutak et al. prepared SWCNT films on multi-cellulose ester (MCE) membranes by vacuum filtration and incubated these with MC3T3-E1 preosteoblastic cells. They found that surface energy was important in promoting the initial attachment of the cell to the substrate, whereas surface morphology was crucial in supporting the subsequent cell proliferation [121]. Similarly, Nayak et al. demonstrated that surface roughness of CNT thin films may affect the proliferation and differentiation of hMSCs through: (1) the specific adsorption of proteins at the surface and the subsequent cell reactions and (2) the adhesion of cells and their further differentiation into bone lineage [122].

Nayak et al. also prepared CNT scaffolds with a number of surface functional groups, such as carboxylic (–COOH), hydroxylic (–OH) and PEG [122]. By subsequently culturing hMSCs onto these films, the PEG–MWCNTs accelerated the differentiation of hMSC to a higher extent than the carboxylated nanotubes and simultaneously induced bone matrix mineralization. Additionally, hMSC grown on the PEG–MWCNT presented an elongated, fibroblast-like morphology; while in the case of carboxylated MWCNT substrates, cells showed slightly irregular shapes, implying a poor adherence to the substrate, and thus lower cell growth rate [122].

A more effective structure of CNT-based scaffolds is perhaps the 3D porous structure, which can mimic the native microenvironment of the natural tissue scaffolds [123,124]. The 3D structure provides not only the mechanical support but also the nanotextured surfaces for enhanced cell adhesion and signalling. As compared to the 2D scaffolds, additional parameters such as porosity, pore size, pore configurations, dimension and conformation of the 3D scaffolds can be tailored to control the cellular activities. Correa-Duarte et al. prepared such a network of interconnected nanotubes with a regular 3D sieve structure [125]. After incubating with L929 fibroblasts, they found that it took only 7 days for the cells to grow a confluent layer covering the surface of the network [125].

Hybrid 3D scaffolds composed of CNTs and polymers are more commonly reported in the recent literature. For example, Shi et al. fabricated poly(propylene fumarate) (PFF)–CNT 3D nanocomposites by a thermal-cross-linking particulate-leaching technique [126]. Investigations of their mechanical properties revealed that strong CNT–PFF interactions were formed and cross-linking densities were increased at a relatively low CNT concentration, resulting in the effective load transfer in the nanocomposites. Li and co-workers also prepared biodegradable poly(lactide-*co*-glycolide) (PLGA)/carboxyl-functionalized MWCNT nanocomposites via a solvent-casting technique and measured the improved mechanical properties upon the addition of CNTs [127]. In vitro testing of rat mesenchymal stem cells (MSCs) on these nanocomposites showed that the cells had better adhesion and viability than on the pure PLGA. da Silva et al. prepared the CNT/collagen 3D composite and observed no contraction of the composite when incubated with osteoblasts, as usually occurs on pure collagen [128]. Hirata et al. cultured osteoblasts on a CNT/collagen sponge in a dynamic flow cell culture system. They found that CNTs helped the initial adhesion of osteoblasts on the CNT/collagen sponge and the cells spread during the early stages [129]. Significantly higher levels of cell differentiation markers, such as ALP, calcium and osteopontin contents, were also measured on the CNT/collagen composites compared to that on the pure collagen sponge [129].

12.2.8 CNTs for neuron prosthesis

The nervous system is the most complex cellular communication network in mammals, which contains over 10^{11} neurons. Each neuron has an elaborated morphology with extended neurites (axons and dendrites) over long distances. The damage of neurons could lead to severe diseases such as Alzheimer's and Parkinson's diseases. To rebuild these damaged nerves, one has to precisely control the direction and structural polarization of individual neurons. Unfortunately, currently there is only limited success (e.g. by using the synthetic 'bridge') to promote the regeneration of nerves in the brain and spinal cord. CNTs are strong, flexible and electrically conductive. They have shown a great promise as the substrate (known as neuron prosthesis) for regulating the outgrowth of neurites and the differentiation of neural stem cells (NSCs).

The first example of CNTs used in neuroscience is given by Mattson et al. in 2000 [130]. They showed that surface functionalization played a critical role in promoting the neurites and branching of neurons. For example, when embryonic rat-brain neurons were incubated on thin films of MWCNTs, neurons extended only one or two neurites on pristine nanotubes, whereas extensive branching could be observed on nanotubes coated with 4-hydroxynonenal (4-HNE) molecules [130]. 4-HNE can induce intracellular Ca^{2+} levels and modify cytoskeletal proteins, the signalling mechanisms that regulate neurite outgrowth in many types of neurons.

The importance of chemical functionalization on neurite outgrowth is further demonstrated by grafting the surface of nanotubes with neutral, positive and negative charges [131]. It was found that more growth cones, longer average neurite length and longer neurite branching could be obtained when the neurons were incubated onto positively charged MWCNT-ethylenediamine (EN; at physiological pH = 7.35) substrate [131]. The ability to functionalize nanotubes with these molecules warrants clinical applications of this technology.

Very recently, Liedtke et al. investigated the signalling mechanism of neurons exposed to CNTs and the possibility of using CNTs to enhance neural engineering devices for curing neural injury conditions [132]. By exposing neurons to gum arabic (GA)-dispersed CNTs, they found the up-regulation of chloride transporter, potassium chloride cotransport 2 (KCC2). This up-regulation of KCC2 suppressed the intraneuronal chloride (Cl^-) concentration, which then accelerated the neuronal maturization. Neurons cultured on GA-dispersed CNT substrate thus showed significantly enlarged soma size and increased expression of neuronal differentiation marker (β3-tubulin) [132]. These findings proved the ability of CNTs in altering neurons' gene expression and shed light on curing neural injury conditions associated with elevated intraneuronal Cl^- concentration, such as pain, epilepsy, traumatic neural injury and ischaemia.

Selective growth and control of the neuronal orientation and the differentiation of human NSCs (hNSCs) can also be realized on patterned-CNT networks. Park et al. firstly prepared a self-assembled monolayer of methyl-terminated 1-octadecanethil (ODT) pattern on thin Au films on cover glass substrates [133]. In a square pattern of CNTs as shown in Figure 12.9 (a), hNSCs was bound strongly on the CNT pattern regions and outgrew and extended over the ODT regions. After differentiation (Figure 12.9 (b) and (c)), neuronal outgrowth

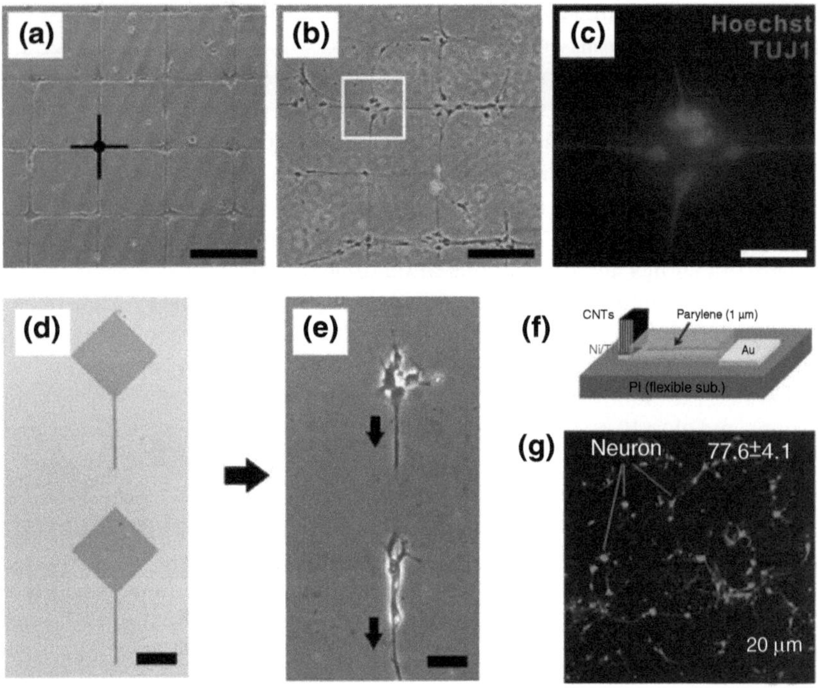

FIGURE 12.9 (a)–(c) Neural network formed on narrow line shape CNT patterns. (a) Phase contrast images of hNSC. (b) Phase contrast image of the differentiated cells. (c) Immunofluorescence image of the differentiated cells as squared in (b). (d) SEM image of a CNT pattern with a single narrow strip. (e) Phase contrast images of the differentiated cell on the CNT patterns. Note that the hNSC grew along the CNT single narrow strip regions during the differentiation. Scale bars are 200 μm in (a); 50 μm in (b), (d) and (e) and 20 μm in (c); reprinted with permission from Ref. [133]. (f) A schematic side view of the flexible CNT electrode device. (f) Fluorescent image of neuron cells cultured on UV-ozone treated CNTs for 40 min; reprinted with permission from Ref. [136].

extended on the nearby CNT patterns, forming the neural networks where the cells grown on different CNT square patterns were connected. Furthermore, the differentiation of individual hNSCs was controlled by CNTs composed of one square and one line shape (Figure 12.9 (d) and (e)): selective hNSC adhesion inside the square regions; then the hNSCs outgrew along the narrow line shape regions during the growth and differentiation stage.

Another important application in neuron prosthesis is to use CNT microelectrode arrays. Neurons use electrical potential difference between the cell membranes to transmit electrical signals. This enables us to use sharpened metal electrodes to electrically stimulate neuronal activities and investigate the functions of the brain. Currently metals and conductive polymers are used to fabricate into electrodes for neural stimulation [134]. CNT arrays show promise in stimulating neurons based on the capacitive charge-injection mechanism. CNTs microelectrodes have a wide operational potential window (-1.5 to 1.0 V), a large charge-injection limit (1–1.6 mC cm^{-2}), and are more electrochemically stable [134,135]. These properties outperformed the conventional metal and metal alloy electrodes commonly used for neural stimulation, such as Pt and activated IrO_x [135]. Hsu et al. recorded the extracellular neuronal activities of crayfish by the flexible-CNT electrodes where CNTs were directly grown on the polyimide (PI) substrate at <400 °C (Figure 12.9(f) and (g)) [136]. The interfacial capacitance per unit area of these CNT electrodes could approach 10^6 F mm^{-2}. The low interfacial impedance and high capacitance greatly facilitated charge transfer capability and neuronal recording.

12.3 GRAPHENE-BASED BIOLOGICAL APPLICATIONS

Graphene and related materials (GRMs, including few layer graphene, functionalized graphene, graphene oxide or GO, nanocomposites containing graphene, etc.) are of interest for a range of applications, from ultracapacitors to touch screens, due to their promising electrical and optical responses [137,138]. In this section, we will first discuss various synthesis methods for horizontal and vertical graphenes; this will be followed by Sec. 12.3.2 which will describe the use of GRM in biosensing devices. Sec. 12.3.3 will cover other biological/biomedical applications of GRM, e.g. imaging, therapeutics, drug/gene delivery and tissue engineering.

12.3.1 Synthesis of graphene and related materials

Horizontal, planar graphene has been synthesized via three main routes: mechanical exfoliation, chemical synthesis and CVD (Figure 12.10 (a)). Currently CVD shows the most promise for eventual upscaling and industrial synthesis. Vertical graphene was also synthesized and is currently being investigated with great interest due to the much higher reactivity exhibited by the larger density of open edges (Figure 12.10 (b)). The main fabrication method used to synthesize vertical graphenes is PECVD [139].

12.3.1.1 Mechanical Exfoliation

Mechanical exfoliation is a technique where layers of graphite are mechanically separated until a single monolayer is left. Using Scotch tape and constant lifting and sticking a graphene sheet can eventually be isolated from a graphite sheet [140]. To achieve this, the weak van der Waals force between graphene sheets in graphite must be broken. This force has been shown to be 300 nN μm^{-2} [141]. Whilst not a large force, it is still enough to present problems. Another problem with this technique is that the tape leaves residues on the surface and heating is generally used to remove these residues. On the other hand, this method allows the placement of graphene directly onto a targeted material and in general requires less post-synthesis transferring. Whilst this sticking and lifting process is the most common exfoliation method, another method instead

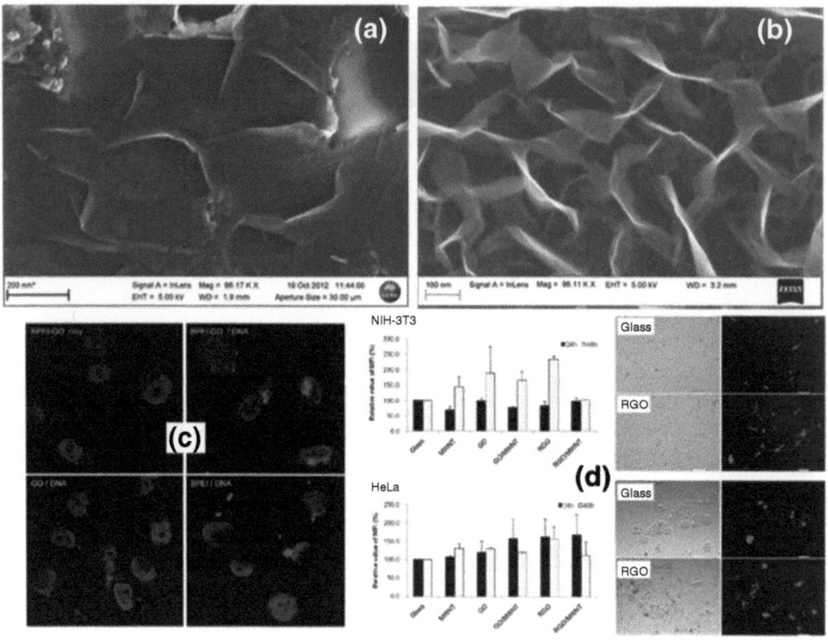

FIGURE 12.10 (a) Horizontal and (b) vertical graphene sheets, synthesized in our laboratory at CSIRO. (c) Confocal fluorescence microscopic images demonstrating the use of BPEI-GO (branched PEI-conjugated to GO) as a bioimaging reagent; reproduced with permission from Ref. [174]. (d) The impact of surface-type on the gene transfection efficiency of NIH-3T3 (fibroblast) and HeLa cells; reproduced with permission from Ref. [179].

utilized high voltages to stick the graphene to a substrate [142]. The graphite can be lifted and leaves the graphene stuck to the surface. This is a marked improvement over the scotch tape method; however, again it does not produce large amounts of graphene and the separation is not perfect.

The major drawback of this technique is the time and effort required. This is because when the graphene is isolated, it is often mixed with multi-layered graphene and can be difficult to find [143]. Considering the simple nature of this process it may be somewhat surprising that very high-quality graphene can be produced. The reason is that the crystalline quality of the graphene is determined by the quality of the graphite it comes from. High-quality graphite is easily purchased, and thus the quality of graphene produced from this method is very good when compared to that of other synthesis methods.

12.3.1.2 Chemical Synthesis

Chemical synthesis of graphene is achieved through the reduction of GO. Brodie first synthesized GO in 1859 using volatile reagents; however, Hummers in 1958 presented another synthetic route and GO is now easily produced with a safe and simple chemical reaction [144]. Hummers process heats a graphite dispersion with concentrated sulphuric acid, sodium nitrate and potassium permanganate to 45 °C for a couple of hours. GO is the product of oxidation of graphite. The GO in this process is in general not homogenous and contains varying amounts of carbon, oxygen and hydrogen. The GO sheets are then isolated so they can be reduced to produce graphene.

Complete conversion of graphene films from GO is yet to be achieved despite a plethora of different methods being used. The methods commonly used vary from complicated multi-staged reduction processes to very simple treatment in alkaline solution to achieve deoxygenation [145,146]. The difficulty with this technique is that a lot of partially converted GO is left [147]; however, it is noted that many applications of graphene (particularly in the context of biological applications) are improved or indeed are reliant on a GO presence. Despite this technique not being the most efficient at producing graphene, it is easily upscaled and GO is much cheaper than high-quality graphite. Furthermore, when pure graphene is not required or not desired, the graphene produced via chemical synthesis is generally sufficient.

12.3.1.3 Chemical Vapour Deposition

Growth of graphene on catalysts using CVD is by far the most researched synthesis avenue. This is because CVD affords a greater degree of control over conditions during synthesis and is scalable. Common carbon sources include hydrocarbons and fluorocarbons injected in the gas phase. There is a wide variety of different catalysts that have been used to grow graphene including Pt(111) [148], Ru(001) [149], Ni(111) [150], Ir(111) [151], Rh(111) [152] and Co(111) [153]. Most growth occurs on a single-crystalline face of a metal substrate. Reports of polycrystalline growth are generally at higher temperatures or at temperatures past that for crystallization [154–157].

Growth mechanisms differ between metals and there is no one mechanism to describe the growth on all metals. Primarily this is a consequence of the lattice mismatch between graphene and the metal, carbon solubility of carbon in the metal and the melting point of the metal. For Cu, graphene is grown in a surface-catalysed process whereby the carbon interacts on the surface [157]. This process is self-limiting as once graphene coverage is achieved then no more carbon can interact with the surface and form new graphene, one of the reasons copper is favoured for future growth. When Ni is used, the dissolution of carbon into the metal above 800 °C and subsequent segregation on cooling below 800 °C determines graphene growth [158].

12.3.1.4 Vertical Graphene Nanosheets

The most popular method for the production of vertical graphenes is PECVD. This is because the plasma is pivotal in producing strong electric fields that direct the growth of the graphene sheets. Vertical alignment evolves over time;

initially an interface layer of carbon is formed on the substrate. These islands provide nucleation sites for the vertical graphene nanosheets. Due to this initial coverage vertically aligned graphene sheets can be synthesized on a plethora of different substrates. These sheets are commonly grown at low pressures in a plasma consisting of hydrogen, argon and a carbon-containing precursor. Efforts continue to produce vertically aligned graphene consisting of a single sheet but these are yet to bear fruit. Most importantly, when the graphene sheets are aligned vertically, open edges are exposed to the environment, making them interesting for sensing and functionalizing.

12.3.2 Graphene-based biosensors

GRM has been employed as components in a range of biosensing devices, acting either as a biomolecular recognition element (i.e. interacting with the analyte to be sensed) or as a transducer (processing the sensed information) [24]. The unique properties of GRM lead them to be important components in biosensors working via electrical or optical routes. In this chapter, we will first discuss electrical-based GRM sensors, i.e. Förster resonance energy transfer (FRET), FET and electrochemical sensors, followed by optical-based GRM sensors, i.e. SPR spectroscopy, SERS, UV-Vis spectrometry, colorimetric, etc.

Cao et al. developed a microfluidic chip with an integrated GO-based FRET aptamer sensor for the in situ detection of a range of cancer cells [159]. GO, combined with an aptamer (FAM-Sgc8), was the FRET probe – initially GO served to quench the fluorescence of FAM-Sgc8 (hence, there was a very low background signal as desired), until cancer cells (CCRF-CEM, T-cell acute lymphoblastic leukaemia cells) were added, whereupon the FAM-Sgc8 attached to them and detached from the GO, with the fluorescence recovered, signalling the presence of the target cells. Using this method a very low detection limit of 25 cells ml^{-1} was obtained [159]. GO has also been demonstrated to work well as an important component in electrochemical sensors where it may act as the electrode. For example, Liu et al. demonstrated that free-standing Au nanoparticle–reduced GO and SWCNT-reduced GO composite films exhibited enhanced performance as an electrochemical sensor in the cyclic voltammogram [160]. Aptamer-modified graphene-FETs as label-free immunosensors were also examined by Ohno et al. [161].

Wu et al. designed a surface plasmon resonance sensor based on Au-graphene structure [162]. They took advantage of the biocompatibility of the graphene surface, specifically its affinity for aromatic ring containing biomolecules and used it as the biomolecular recognition element of the sensor. Rider et al. presented Au-nanoparticle-decorated vertical graphenes as a promising SERS substrate, with an increase in the effective sensor sites at 224% afforded by the bookshelf-style morphology [163]. Seo et al. presented optical measurements of vertical graphenes of differing nitrogen contents, demonstrating their promise as an optical biosensing platform through the detection of bovine serum albumin via both Raman spectroscopy and UV-Vis transmission measurements [164]. Getting very close to cell imaging is a droplet-based colorimetric biosensor based on a GO nanoprobe, presented by Xiang et al. [165]. The droplet changes from a dark colour (due to fluorescence quenching by the GO) to a bright colour when DNA single strands combine to form a double helix.

12.3.3 Graphene-based other biological applications

12.3.3.1 Imaging

Graphene QDs as bioimaging probes for human osteosarcoma (MG-63) cells were reviewed by Shen et al. [166]. Gollavelli and Ling used multi-functional graphene (modified to be magnetic, fluorescent and water dispersible) as a probe for in vitro imaging of HeLa cells and in vivo imaging of zebrafish embryos [167].

In many cases, it is more advantageous to use functionalized graphene or GO as an imaging probe, due to the greater bio/cytocompatibility. Liu et al. demonstrated cell imaging using GO decorated with gold nanoparticles (to boost the G peak from GO which was used to image the HeLa 229 cells via Raman spatial mapping) as a flexible SERS probe [168]. Another GO composite was studied by Wate et al. [169], who employed a composite material consisting of magnetite nanoparticles, a dendrimer, a cyanine dye (the NIR fluorescent probe for the imaging) and GO (the surface to which the aforementioned components covalently attached). They showed that the composite system was successfully taken up by MCF-7 breast cancer cells. Qian et al. used GO nanoparticles conjugated with PEG as a fluorescent probe to image HeLa cells and for in vivo imaging of mice brains [170]. Sreejith et al. used porous silica nanoparticles, inject with squaraine dyes, then wrapped up with ultrathin GO sheets to image HeLa cells [171]. The GO sheet served to stop the dye leaking out into the cell and to protect it from 'nucleophilic attack' – the composite material itself was shown to be biocompatible.

12.3.3.2 Therapeutic Drug Delivery

Whilst GRMs have been demonstrated as good cellular probes for imaging as they are taken up by the cell, for the same reason they are good candidates for both therapy and drug or gene delivery. Hu et al. demonstrated the use of GO tagged with core/shell CdSe/ZnS QDs as both an imaging probe in the visible range and a 'photothermal therapy agent' to kill cancerous cells (MCF-7) in the NIR range [172].

GO has also been shown to be an effective drug delivery agent for the anticancer drug DOX [173]. However, Zhang et al. note that whilst GO is a good platform for the non-covalent physisorption of drugs, ssDNA and RNA, it must be modified (e.g. by grafting PEI) in order to deliver genes and siRNA [173]. Hence, by grafting PEI onto GO, they demonstrated the successful sequential delivery of siRNA followed by DOX – the first delivery of siRNA was intended to overcome the common multi-drug resistance problem and hence lead to greater chemotherapy efficacy. Kim et al. showed that GO-polyethylenimine (PEI) was useful as both a cellular imaging probe with high cell viability and an efficient non-viral gene delivery vector, which was demonstrated in HeLa and PC-3 cells (Figure 12.10(c)) [174]. Whilst the majority of studies seem to focus on GO as a drug/gene delivery platform, graphene nanosheets have also been used for the controlled release of an antibiotic (gentamicin sulphate) by Pandey et al. [175].

12.3.3.3 Tissue Engineering

The three main cellular functions that a scaffold must facilitate are growth, proliferation and adhesion [176]. In the case of stem cells (e.g. hMSC), there is an additional desired function of differentiation [177]. Graphene and its derivatives largely satisfy these requirements as they are typically biocompatible and biofunctional, exhibit high elasticity and flexibility, and do not interact with the surrounding bio-environment (i.e. bio-inert) [176,178]. GRMs have been used as scaffolds for the cell growth, proliferation, adhesion and (where applicable) differentiation of mammalian NIH-3T3 fibroblast cells [179] (Figure 12.10 (d); also for the effect of carbon nanostructured surfaces on the gene transfection efficiency of NIH-3T3), mammalian colorectal adenocarcinoma HT-29 cells [176], human osteoblasts [177,178], hMSCs [177], mesenchymal stromal cells [180] and hNSCs [181].

12.4. SUMMARY AND OUTLOOK

In summary, we have discussed the promising future of incorporating CNTs and graphene into the field of biology and medicine, specifically their current roles in biosensing, diagnostic and imaging tools, anti-cancer, drug delivery, tissue engineering and neuron prosthesis. We attempt to demonstrate the advantages and challenges in the current development of CNT- and graphene-based biomedical devices, as well as to shed light on better utilizing these materials by possibly controlling their structure and functionalities. For future work, extensive research effort is required to reach the long-term goals associated with incorporating CNTs and graphene into biomedical technology with sufficient performance standards. The main challenges required, in our opinion, include the following:

- Controllable syntheses of functional CNT- and graphene-based structures.
- Proper functionalization of these materials to further increase biocompatibility.
- New strategies to improve the sensitivity, selectivity and robustness of biosensors based on CNT and graphene.
- Protocol and extensive evaluation to determine the exact nature of the interactions between CNT- and graphene-based materials and biological objects.

- Innovative ideas and further experiments for advanced and preclinical assessment of these nanomaterials in biomedical applications.
- Innovative solutions are required to reduce fabrication and running costs of CNT and graphene biomedical devices to make them economically viable.

ACKNOWLEDGEMENTS

This work was partially supported by CSIRO's OCE Science Leadership Program; CSIRO's Office of the Chief Executive Postdoctoral Fellowship Scheme; Transformational Biology, Advanced Materials and Sensors and Sensor Networks Transformational Capability Platforms; as well as the Australian Research Council (ARC).

REFERENCES

[1] Iijima S. Helical microtubules of graphitic carbon. Nature 1991;354:56.

[2] Novoselov KS, Geim AK, Morozov SV, Jiang D, Zhang Y, Dubonos SV, et al. Electric Field Effect in Atomically Thin Carbon Films. Science 2004;306:666.

[3] Yang W, Ratinac KR, Ringer SP, Thordarson P, Gooding JJ, Braet F. Carbon Nanomaterials in Biosensors: Should You Use Nanotubes or Graphene? Angew Chem Int Ed 2010;49:2114.

[4] Yang W, Thordarson P, Gooding JJ, Ringer SP, Braet F. Carbon nanotubes for biological and biomedical applications. Nanotechnology 2007;18:412001.

[5] Liu Z, Tabakman S, Welsher K, Dai H. Carbon Nanotubes in Biology and Medicine: *In vitro* and *in vivo* Detection, Imaging and Drug Delivery. Nano Res 2009;2:85.

[6] Hata K, Futaba DN, Mizuno K, Namai T, Yumura M, Iijima S. Water-Assisted Highly Efficient Synthesis of Impurity-Free Single-Walled Carbon Nanotubes. Science 2004;306:1362.

[7] Robertson J. Growth of nanotubes for electronics. Mater Today 2007;10:36.

[8] Zhang M, Fang S, Zakhidov AA, Lee SB, Aliev AE, Williams CD, et al. Strong, Transparent, Multifunctional, Carbon Nanotube Sheets. Science 2005;309:1215.

[9] Harutyunyan AR, Chen G, Paronyan TM, Pigos EM, Kuznetsov OA, Hewaparakrama K, et al. Preferential Growth of Single-Walled Carbon Nanotubes with Metallic Conductivity. Science 2009;326:116.

[10] Nessim GD. Properties, synthesis, and growth mechanisms of carbon nanotubes with special focus on thermal chemical vapor deposition. Nanoscale 2010;2:1306.

[11] Raty J-Y, Gygi F, Galli G. Growth of Carbon Nanotubes on Metal Nanoparticles: A Microscopic Mechanism from *Ab Initio* Molecular Dynamics Simulations. Phys Rev Lett 2005;95:096103.

[12] Han ZJ, Levchenko I, Yick S, Ostrikov K. 3-Orders-of-Magnitude Density Control of Single-Walled Carbon Nanotube Networks by Maximizing Catalyst Activation and Dosing Carbon Supply. Nanoscale 2011;3:4848.

[13] Ostrikov K. Reactive plasmas as a versatile nanofabrication tool. Rev Mod Phys 2005;77: 489–511.

[14] Ostrikov K, Mehdipour H. Thin Single-Walled Carbon Nanotubes with Narrow Chirality Distribution: Constructive Interplay of Plasma and Gibbs-Thomson Effects. ACS Nano 2011;5:8372.

[15] Ostrikov K, Mehdipour H. Nanoscale plasma chemistry enables fast, size-selective nanotube nucleation. J Am Chem Soc 2012;134:4303.

[16] Neyts EC, Ostrikov K, Han ZJ, Kumar S, van Duin ACT, Bogaerts A. Defect healing and enhanced nucleation of carbon nanotubes by low-energy ion bombardment. Phys Rev Lett 2013;110:065501.
[17] Neyts EC, Shibuta Y, van Duin ACT, Bogaerts A. Catalyzed Growth of Carbon Nanotube with Definable Chirality by Hybrid Molecular Dynamics-Force Biased Monte Carlo Simulations. ACS Nano 2010;4:6665–72.
[18] Neyts EC, van Duin ACT, Bogaerts A. Changing Chirality during Single-Walled Carbon Nanotube Growth: A Reactive Molecular Dynamics/Monte Carlo Study. J Am Chem Soc 2011;133:17225.
[19] Han ZJ, Mehdipour H, Li X, Shen J, Randeniya L, Yang HY, et al. SWCNT Networks on Nanoporous Silica Catalyst Support: Morphological and Connectivity Control for Nanoelectronic, Gas-Sensing, and Biosensing Devices. ACS Nano 2012;6:5809.
[20] Han ZJ, Yick S, Levchenko I, Tam E, Yajadda MMA, Kumar S, et al. Controlled Synthesis of a Large Fraction of Metallic Single-Walled Carbon Nanotube and Semiconducting Carbon Nanowire Networks. Nanoscale 2011;3:3214.
[21] Fisher C, Han ZJ, Levchenko I, Ostrikov K. Control of Dense Carbon Nanotube Arrays via Hierarchical Multilayer Catalyst. Appl Phys Lett 2011;99:143104.
[22] Han ZJ, Ostrikov K. Uniform, Dense Arrays of Vertically Aligned, Large-Diameter Single-Walled Carbon Nanotubes. J Am Chem Soc 2012;134:6018.
[23] Han C, Doepke A, Cho W, Likodimos V, de la Cruz AA, Back T, et al. A Multiwalled-Carbon-Nanotube-Based Biosensor for Monitoring Microcystin-LR in Sources of Drinking Water Supplies. Adv Funct Mater 2013;.
[24] Fisher C, Rider AE, Han ZJ, Kumar S, Levchenko I, Ostrikov K. Applications and Nanotoxicity of Carbon Nanotubes and Graphene in Biomedicine. J Nanomater 2012;2012:315185.
[25] Lin Y, Taylor S, Li H, Fernando KAS, Qu L, Wang W, et al. Advances toward bioapplications of carbon nanotubes. J Mater Chem 2004;14:527.
[26] Bianco A, Kostarelos K, Partidos CD, Prato M. Biomedical applications of functionalised carbon nanotubes. Chem Commun 2005;571.
[27] Katz E, Willner I. Biomolecule-functionalized carbon nanotubes: applications in nanobioelectronics. ChemPhysChem a European journal of chemical physics and physical chemistry 2004;5:1084–104.
[28] Hirsch A. Functionalization of Single-Walled Carbon Nanotubes. Angew Chem Int Ed 2002;41:1853.
[29] Peng X, Wong SS. Functional Covalent Chemistry of Carbon Nanotube Surfaces. Adv Mater 2009;21:625–42.
[30] Karousis N, Tagmatarchis N, Tasis D. Current Progress on the Chemical Modification of Carbon Nanotubes. Chem Rev 2010;110:5366.
[31] Agui L, Yanez-Sedeno P, Pingarron JM. Role of carbon nanotubes in electroanalytical chemistry: a review. Anal Chim Acta 2008;622:11–47.
[32] Rivas GA, Rubianes MD, Rodriguez MC, Ferreyra NF, Luque GL, Pedano ML, et al. Carbon nanotubes for electrochemical biosensing. Talanta 2007;74:291s.
[33] Sinha N, Ma J, Yeow JTW. Carbon Nanotube-Based Sensors. J Nanosci Nanotechnol 2006;6:573–90.
[34] Zhang X, Meng L, Lu Q. Cell Behaviors on Polysaccharide-Wrapped Single-Wall Carbon Nanotubes: A Quantitative Study of the Surface Properties of Biomimetic Nanofibrous Scaffolds. ACS Nano 2009;3:3200.
[35] Staii C, Johnson AT. DNA-Decorated Carbon Nanotubes for Chemical Sensing. Nano Lett 2005;5:1774.

[36] Weizmann Y, Chenoweth DM, Swager TM. Addressable Terminally Linked DNA-CNT Nanowires. J Am Chem Soc 2010;132:14009.
[37] Williams KA, Veenhuizen PTM, de la Torre BG, Eritja R, Dekker C. Carbon nanotubes with DNA recognition. Nature 2002;420:761.
[38] Dai L, Soundarrajan P, Kim T. Sensors and sensor arrays based on conjugated polymers and carbon nanotubes. Pure Appl Chem 2002;74:1753.
[39] Allen BL, Kichambare PD, Star A. Carbon Nanotube Field-Effect-Transistor-Based Biosensors. Adv Mater 2007;19:1439–51.
[40] Besteman K, Lee J-O, Wiertz FGM, Heering HA, Dekker C. Enzyme-Coated Carbon Nanotubes as Single-Molecule Biosensors. Nano Lett 2003;3:727.
[41] Balasubramanian g, Burghard M. Biosensors based on carbon nanotubes. Anal Bioanal Chem 2006;385:452.
[42] Gruner G. Carbon nanotube transistors for biosensing applications. Anal Bioanal Chem 2006;384:322.
[43] Yu X, Chattopadhyay D, Galeska I, Papadimitrakopoulos F, Rusling JF. Peroxidase activity of enzymes bound to the ends of single-wall carbon nanotube forest electrodes. Electrochem Commun 2003;5:408.
[44] Kim SN, Rusling JF, Papadimitrakopoulos F. Carbon Nanotubes for Electronic and Electrochemical Detection of Biomolecules. Adv Mater 2007;19:3214.
[45] Roy S, Gao Z. Nanostructure-based electrical biosensors. Nano Today 2009;4:318.
[46] Bradley K, Briman M, Star A, Gruner G. Charge Transfer from Adsorbed Proteins. Nano Lett 2004;4:253.
[47] Cella LN, Chen W, Myung NV, Mulchandani A. Single-Walled Carbon Nanotube-Based Chemiresistive Affinity Biosensors for Small Molecules: Ultrasensitive Glucose Detection. J Am Chem Soc 2010;132:5024.
[48] Chen RJ, Bangsaruntip S, Drouvalakis KA, Kam NWS, Shim M, Li Y, et al. Noncovalent functionalization of carbon nanotubes for highly specific electronic biosensors. Proc Natl Acad Sci USA 2003;100:4984.
[49] Fu D, Okimoto H, Lee CW, Takenobu T, Iwasa Y, Kataura H, et al. Ultrasensitive Detection of DNA Molecules with High On/Off Single-Walled Carbon Nanotube Network. Adv Mater 2010;22:4867.
[50] Maehashi K, Katsura T, Kerman K, Takamura Y, Matsumoto K, Tamiya E. Label-Free Protein Biosensor Based on Aptamer-Modified Carbon Nanotube Field-Effect Transistors. Anal Chem 2007;79:782.
[51] Ko JW, Woo J-M, Ahn J, Cheon JH, Lim JH, Kim SH, et al. Multi-Order Dynamic Range DNA Sensor Using a Gold Decorated SWCNT Random Network. ACS Nano 2011;5:4365.
[52] Rajesh, Das BK, Srinives S, Mulchandani A. ZnS nanocrystals decorated single-walled carbon nanotube based chemiresistive label-free DNA sensor. Appl Phys Lett 2011;98:013701.
[53] Heller I, Janssens AM, Mannik J, Minot ED, Lemay SG, Dekker C. Identifying the Mechanism of Biosensing with Carbon Nanotube Transistors. Nano Lett 2008;8:591.
[54] Heller I, Smaal WTT, Lemay SG, Dekker C. Probing Macrophage Activity with Carbon-Nanotube Sensors. Small 2009;5:2528.
[55] Tey JN, Gandhi S, Wijaya IPM, Palaniappan A, Wei J, Rodriguez I, et al. Direct Detection of Heroin Metabolites Using a Competitive Immunoassay Based on a Carbon-Nanotube Liquid-Gated Field-Effect Transistor. Small 2010;6:993.
[56] Cui R, Huang H, Yin Z, Gao D, Zhu J-J. Horseradish peroxidase-functionalized gold nanoparticle label for amplified immunoanalysis based on gold nanoparticles/carbon nanotubes hybrids modified biosensor. Biosens Bioelectron 2008;23:1666.

[57] So H-M, Park D-W, Jeon E-K, Kim Y-H, Kim BS, Lee C-K, et al. Detection and Titer Estimation of *Escherichia coli* Using Aptamer-Functionalized Single-Walled Carbon-Nanotube Field-Effect Transistors. Small 2008;4:197.
[58] Wang H, Yang R, Yang L, Tan W. Nucleic Acid Conjugated Nanomaterials for Enhanced Molecular Recognition. ACS Nano 2009;3:2451.
[59] Cai H, Cao X, Jiang Y, He P, Fang Y. Carbon nanotube-enhanced electrochemical DNA biosensor for DNA hybridization detection. Anal Bioanal Chem 2003;375:287.
[60] Liu Y, Wang M, Zhao F, Xu Z, Dong S. The direct electron transfer of glucose oxidase and glucose biosensor based on carbon nanotubes/chitosan matrix. Biosen Bioelectron 2005;21:984–8.
[61] Gooding JJ, Wibowo R, Liu J, Yang W, Losic D, Orbons S, et al. Protein Electrochemistry Using Aligned Carbon Nanotube Arrays. J Am Chem Soc 2003;125:9006.
[62] Lin Y, Lu F, Tu Y, Ren Z. Glucose Biosensors Based on Carbon Nanotube Nanoelectrode Ensembles. Nano Lett 2004;4:191.
[63] Cao Q, Rogers JA. Ultrathin Films of Single-Walled Carbon Nanotubes for Electronics and Sensors: A Review of Fundamental and Applied Aspects. Adv Mater 2009;21:29–53.
[64] Wang J. Electrochemical Glucose Biosensors. Chem Rev 2008;108:814.
[65] Zhu Z, Song W, Burugapalli K, FrancisMoussy, Li Y-L, Zhong X-H. Nano-yarn carbon nanotube fiber based enzymatic glucose biosensor. Nanotechnology 2010;21:165501.
[66] Hrapovic S, Liu Y, Male KB, Luong JHT. Electrochemical Biosensing Platforms Using Platinum Nanoparticles and Carbon Nanotubes. Anal Chem 2004;76:1083.
[67] Fernandez-Sanchez C, Pellicer E, Orozco J, Jimenez-Jorquera C, Lechuga LM, Mendoza E. Plasma-activated multi-walled carbon nanotube-polystyrene composite substrates for biosensing. Nanotechnology 2009;20:335501.
[68] Xu X, Jiang S, Hu Z, Liu S. Nitrogen-Doped Carbon Nanotubes: High Electrocatalytic Activity toward the Oxidation of Hydrogen Peroxide and Its Application for Biosensing. ACS Nano 2010;4:4292.
[69] Star A, Gabriel J-CP, Bradley K, Gruner G. Electronic Detection of Specific Protein Binding Using Nanotube FET Devices. Nano Lett 2003;3:459.
[70] Kostarelos K, Bianco A, Prato M. Promises, facts and challenges for carbon nanotubes in imaging and therapeutics. Nat Nanotechnol 2009;4:627.
[71] Strano MS, Jin H. Where is it Heading? Single-Particle Tracking of Single-Walled Carbon Nanotubes. ACS Nano 2008;2:1749.
[72] Heller DA, Baik S, Eurell TE, Strano MS. Single-Walled Carbon Nanotube Spectroscopy in Live Cells: Towards Long-Term Labels and Optical Sensors. Adv Mater 2005;17:2793.
[73] Bachilo SM, Strano MS, Kittrell C, Hauge RH, Smalley RE, Weisman RB. Structure-assigned optical spectra of single-walled carbon nanotubes. Science 2002;298:2361–6.
[74] Barone PW, Baik S, Heller DA, Strano MS. Near-infrared optical sensors based on single-walled carbon nanotubes. Nat Mater 2005;4:86.
[75] Avouris P, Freitag M, Perebeinos V. Carbon-nanotube photonics and optoelectronics. Nat Photonics 2008;2:341.
[76] Bohn JE, Etchegoin PG, Ru ECL, Xiang R, Chiashi S, Maruyama S. Estimating the Raman Cross Sections of Single Carbon Nanotubes. ACS Nano 2010;4:3466.
[77] Liu Z, Cai W, He L, Nakayama N, Chen K, Sun X, et al. In vivo biodistribution and highly efficient tumour targeting of carbon nanotubes in mice. Nat Nanotechnol 2007;2:47–52.
[78] Zavaleta C, de la Zerda A, Liu Z, Keren S, Cheng Z, Schipper M, et al. Noninvasive Raman Spectroscopy in Living Mice for Evaluation of Tumor Targeting with Carbon Nanotubes. Nano Lett 2008;8:2800.

[79] Choi JH, Nguyen FT, Barone PW, Heller DA, Moll AE, Patel D, et al. Multimodal Biomedical Imaging with Asymmetric Single-Walled Carbon Nanotube/Iron Oxide Nanoparticle Complexes. Nano Lett 2007;7:861.

[80] Liu Z, Davis C, Cai W, He L, Chen X, Dai H. Circulation and long-term fate of functionalized, biocompatible single-walled carbon nanotubes in mice probed by Raman spectroscopy. Proc Natl Acad Sci USA 2008;105:1410.

[81] Chu H, Wang J, Ding L, Yuan D, Zhang Y, Liu J, et al. Decoration of Gold Nanoparticles on Surface-Grown Single-Walled Carbon Nanotubes for Detection of Every Nanotube by Surface-Enhanced Raman Spectroscopy. J Am Chem Soc 2009;131:14310.

[82] Kumar GVP. Surface-enhanced Raman scattering studies of carbon nanotubes using Ag-core Au-shell nanoparticles. J Raman Spectrosc 2009;40:2069.

[83] O'Connell MJ, Bachilo SM, Huffman CB, Moore VC, Strano MS, Haroz EH, et al. Band Gap Fluorescence from Individual Single-Walled Carbon Nanotubes. Science 2002;297:593.

[84] Heller DA, Jin H, Martinez BM, Patel D, Miller BM, Yeung TK, et al. Multimodal optical sensing and analyte specificity using single-walled carbon nanotubes. Nat Nanotechnol 2009;4:114–20.

[85] Ju S-Y, Kopcha WP, Papadimitrakopoulos F. Brightly Fluorescent Single-Walled Carbon Nanotubes via an Oxygen-Excluding Surfactant Organization. Science 2009;323:1319.

[86] Welsher K, Liu Z, Sherlock SP, Robinson JT, Chen Z, Daranciang D, et al. A route to brightly fluorescent carbon nanotubes for near-infrared imaging in mice. Nat Nanotechnol 2009;4:773.

[87] Welsher K, Sherlock SP, Dai H. Deep-tissue anatomical imaging of mice using carbon nanotube fluorophores in the second near-infrared window. Proc Natl Acad Sci USA 2011;108:8943.

[88] Diao S, Hong G, Robinson JT, Jiao L, Antaris AL, Wu JZ, et al. Chirality Enriched (12,1) and (11,3) Single-Walled Carbon Nanotubes for Biological Imaging. J Am Chem Soc 2012;134:16971.

[89] Kim J-H, Heller DA, Hong Jin, Barone PW, Song C, Zhang J, et al. The rational design of nitric oxide selectivity in single-walled carbon nanotube near-infrared fluorescence sensors for biological detection. Nat Chem 2009;1:473.

[90] Penza M, Cassano G, Aversa P, Antolini F, Cusano A, Cutolo A, et al. Alcohol detection using carbon nanotubes acoustic and optical sensors. Appl Phys Lett 2004;85:2379.

[91] Penza M, Cassano G, Aversa P, Cusano A, Cutolo A, Giordano M, et al. Carbon nanotube acoustic and optical sensors for volatile organic compound detection. Nanotechnology 2005;16:2536–47.

[92] de la Zerda A, Zavaleta C, Keren S, Vaithilingam S, Bodapati S, Liu Z, et al. Carbon nanotubes as photoacoustic molecular imaging agents in living mice. Nat Nanotechnol 2008;3:557.

[93] Singhal R, Orynbayeva Z, Sundaram RVK, Niu JJ, Bhattacharyya S, Vitol EA, et al. Multifunctional carbon-nanotube cellular endoscopes. Nat Nanotechnol 2011;6:57.

[94] Niu JJ, Schrlau MG, Friedman G, Gogotsi Y. Carbon Nanotube-Tipped Endoscope for In Situ Intracellular Surface-Enhanced Raman Spectroscopy. Small 2011;7:540.

[95] Farokhzad OC, Langer R. Impact of Nanotechnology on Drug Delivery. ACS Nano 2009;3:16.

[96] Hilder TA, Hill JM. Modeling the Loading and Unloading of Drugs into Nanotubes. Small 2008;5:300.

[97] Pantarotto D, Singh R, McCarthy D, Erhardt M, Briand JP, Prato M, et al. Functionalized carbon nanotubes for plasmid DNA gene delivery. Angew Chem Int Ed 2004;43:5242–6.

[98] Richard C, Mignet N, Largeau C, Escriou V, Bessodes M, Scherman D. Functionalization of Single- and Multi-Walled Carbon Nanotubes with Cationic Amphiphiles for Plasmid DNA Complexation and Transfection. Nano Res 2009;2:638.

[99] Nunes A, Amsharov N, Guo C, Van den Bossche J, Santhosh P, Karachalios TK, et al. Hybrid Polymer-Grafted Multiwalled Carbon Nanotubes for In vitro Gene Delivery. Small 2010;6:2281.

[100] Ladeira MS, Andrade VA, Gomes ERM, Aguiar CJ, Moraes ER, Soares JS, et al. Highly efficient siRNA delivery system into human and murine cells using single-wall carbon nanotubes. Nanotechnology 2010;21:385101.

[101] Kam NWS, Dai H. Carbon Nanotubes as Intracellular Protein Transporters: Generality and Biological Functionality. J Am Chem Soc 2005;127:6021.

[102] Joshi A, Punyani S, Bale SS, Yang H, Borca-Tasciuc T, Kane RS. Nanotube-assisted protein deactivation. Nat Nanotechnol 2008;3:41.

[103] Nakamura M, Tahara Y, Ikehara Y, Murakami T, Tsuchida K, Iijima S, et al. Single-walled carbon nanohorns as drug carriers: adsorption of prednisolone and anti-inflammatory effects on arthritis. Nanotechnology 2011;22:465102.

[104] Madaeni SS, Derakhshandeh K, Ahmadi S, Vatanpour V, Zinadini S. Effect of modified multi-walled carbon nanotubes on release characteristics of indomethacin from symmetric membrane coated tablets. J Membr Sci 2012;389:110.

[105] Wu J, Paudel KS, Strasinger C, Hammell D, Stinchcomb AL, Hinds BJ. Programmable transdermal drug delivery of nicotine using carbon nanotube membranes. Proc Natl Acad Sci USA 2010;107:11698.

[106] Wang X, Wang C, Cheng L, Lee S-T, Liu Z. Noble Metal Coated Single-Walled Carbon Nanotubes for Applications in Surface Enhanced Raman Scattering Imaging and Photothermal Therapy. J Am Chem Soc 2012;134:7414.

[107] Kam NWS, O'Connell M, Wisdom JA, Dai H. Carbon nanotubes as multifunctional biological transporters and near-infrared agents for selective cancer cell destruction. Proc Natl Acad Sci USA 2005;102:11600.

[108] Marches R, Mikoryak C, Ru-HungWang, Pantano P, Draper RK, Vitetta ES. The importance of cellular internalization of antibody-targeted carbon nanotubes in the photothermal ablation of breast cancer cells. Nanotechnology 2011;22:095101.

[109] Wu C-H, Cao C, Kim JH, Hsu C-H, Wanebo HJ, Bowen WD, et al. Trojan-Horse Nanotube On-Command Intracellular Drug Delivery. Nano Lett 2012;12:5475.

[110] Moon HK, Lee SH, Choi HC. *In Vivo* Near-Infrared Mediated Tumor Destruction by Photothermal Effect of Carbon Nanotubes. ACS Nano 2009;3:3707.

[111] Ghosh S, Dutta S, Gomes E, Carroll D, Ralph D'Agostino J, Olson J, et al. Increased Heating Efficiency and Selective Thermal Ablation of Malignant Tissue with DNA-Encased Multiwalled Carbon Nanotubes. ACS Nano 2009;3:2667.

[112] Kang B, Yu D, Dai Y, Chang S, Chen D, Ding Y. Cancer-Cell Targeting and Photoacoustic Therapy Using Carbon Nanotubes as "Bomb" Agents. Small 2009;5:1292.

[113] Liu Z, Sun X, Nakayama-Ratchford N, Dai H. Supramolecular Chemistry on Water-Soluble Carbon Nanotubes for Drug Loading and Delivery. ACS Nano 2007;1:50.

[114] Wu W, Li R, Bian X, Zhu Z, Ding D, Li X, et al. Covalently Combining Carbon Nanotubes with Anticancer Agent: Preparation and Antitumor Activity. ACS Nano 2009;3:2740.

[115] Bhirde AA, Patel V, Gavard J, Zhang G, Sousa AA, Masedunskas A, et al. Targeted Killing of Cancer Cells *in Vivo* and *in Vitro* with EGF-Directed Carbon Nanotube-Based Drug Delivery. ACS Nano 2009;3:307.

[116] Liu Z, Chen K, Davis C, Sherlock S, Cao Q, Chen X, et al. Drug Delivery with Carbon Nanotubes for *In vivo* Cancer Treatment. Cancer Res 2008;68:6652.

[117] Han ZJ, Rider AE, Ishaq M, Kumar S, Kondyurin A, Bilek MMM, et al. Carbon nanostructures for hard tissue engineering. RSC Adv 2013;.

[118] Elias KL, Price RL, Webster TJ. Enhanced functions of osteoblasts on nanometer diameter carbon fibers. Biomaterials 2002;23:3279.

[119] Khang D, Sato M, Price RL, Ribbe AE, Webster TJ. Selective adhesion and mineral deposition by osteoblasts on carbon nanofiber patterns. Int J Nanomed 2006;1:65.

[120] Stout DA, Webster TJ. Carbon nanotubes for stem cell control. Mater Today 2012;15:312.

[121] Tutak W, Chhowalla M, Sesti F. The chemical and physical characteristics of single-walled carbon nanotube film impact on osteoblastic cell response. Nanotechnology 2010;21:315102.

[122] Nayak TR, Jian L, Phua LC, Ho HK, Ren Y, Pastorin G. Thin Films of Functionalized Multiwalled Carbon Nanotubes as Suitable Scaffold Materials for Stem Cells Proliferation and Bone Formation. ACS Nano 2010;4:7717.

[123] Ng R, Zang R, Yang KK, Liu N, Yang S-T. Three-dimensional fibrous scaffolds with microstructures and nanotextures for tissue engineering. RSC Adv 2012;2:10110.

[124] Saffar KP, JamilPour N, Rouhi G. Carbon nanotubes in bone tissue engineering. In: de Mello CAB, editor. Biomedical engineering. InTech; 2009.

[125] Correa-Duarte MA, Wagner N, Rojas-Chapana J, Morsczeck C, Thie M, Giersig M. Fabrication and Biocompatibility of Carbon Nanotube-Based 3D Networks as Scaffolds for Cell Seeding and Growth. Nano Lett 2004;4:2233.

[126] Shi X, Hudson JL, Spicer PP, Tour JM, Krishnamoorti R, Mikos AG. Injectable Nanocomposites of Single-Walled Carbon Nanotubes and Biodegradable Polymers for Bone Tissue Engineering. Biomacromolecules 2006;7:2237.

[127] Yan K, Peng H, Zhou Y, Li H, Liu Z. Formation of Bilayer Bernal Graphene: Layer-by-Layer Epitaxy via Chemical Vapor Deposition. Nano Lett 2011;11:1106.

[128] da Silva EE, Della Colleta HHM, Ferlauto AS, Moreira RL, Resende RR, Oliveira S, et al. Nanostructured 3-D collagen/nanotube biocomposites for future bone regeneration scaffolds. Nano Res 2009;2:462.

[129] Hirata E, Uo M, Nodasaka Y, Takita H, Ushijima N, Akasaka T, et al. 3D collagen scaffolds coated with multiwalled carbon nanotubes: Initial cell attachment to internal surface. J Biomed Mater Res B 2010;93:544.

[130] Mattson MP, Haddon RC, Rao AM. Molecular Functionalization of Carbon Nanotubes and Use as Substrates for Neuronal Growth. J Mol Neurosci 2000;14:175.

[131] Hu H, Ni Y, Montana V, Haddon RC, Parpura V. Chemically Functionalized Carbon Nanotubes as Substrates for Neuronal Growth. Nano Lett 2004;4:507.

[132] Liedtke W, Yeo M, Zhang H, Wang Y, Gignac M, Miller S, et al. Highly Conductive Carbon Nanotube Matrix Accelerates Developmental Chloride Extrusion in Central Nervous System Neurons by Increased Expression of Chloride Transporter KCC2. Small 2013;.

[133] Park SY, Choi DS, Jin HJ, Park J, Byun K-E, Lee K-B, et al. Polarization-Controlled Differentiation of Human Neural Stem Cells Using Synergistic Cues from the Patterns of Carbon Nanotube Monolayer Coating. ACS Nano 2011;5:4704.

[134] Jan E, Hendricks JL, Husaini V, Richardson-Burns SM, Sereno A, Martin DC, et al. Layered Carbon Nanotube-Polyelectrolyte Electrodes Outperform Traditional Neural Interface Materials. Nano Lett 2009;9:4012.

[135] Wang K, Fishman HA, Dai H, Harris JS. Neural Stimulation with a Carbon Nanotube Microelectrode Array. Nano Lett 2006;6:2043.

[136] Hsu H-L, Teng I-J, Chen Y-C, Hsu W-L, Lee Y-T, Yen S-J, et al. Flexible UV-Ozone-Modified Carbon Nanotube Electrodes for Neuronal Recording. Adv Mater 2010;22:2177.

[137] Bae S, Kim H, Lee Y, Xu X, Park J-S, Zheng Y, et al. Roll-to-roll production of 30-inch graphene films for transparent electrodes. Nat Nanotechnol 2010;5:574.
[138] Stoller MD, Park S, Zhu Y, An J, Ruoff RS. Graphene-Based Ultracapacitors. Nano Lett 2008;8:3498.
[139] Kumar S, Ostrikov K. Unidirectional arrays of vertically standing graphenes in reactive plasmas. Nanoscale 2011;3:4296.
[140] Novoselov K, Geim A, Morozov S, Jiang D, Zhang Y, Dubonos S, et al. Electric field effect in atomically thin carbon films. Science 2004;306:666–9.
[141] Zhang Y, Small J, Pontius W, Kim P. Fabrication and electric-field-dependent transport measurements of mesoscopic graphite devices. Appl Phys Lett 2005;86:073104.
[142] Sidorov A, Yazdanpanah M, Jalilian R, Ouseph P, Cohn R, Sumanasekara G. Electrostatic deposition of graphene. Nanotechnology 2007;18:4.
[143] Geim A, Novoselov K. The rise of Graphene. Nat Mater 2007;6:183–91.
[144] Hummers W, Offeman R. Preparation of graphitic oxide. J Am Chem Soc 1958;80:1339.
[145] Mohanty N, Nagaraja A, Armesto J, Berry V. High-throughput, ultrafast synthesis of solution-dispersed graphene via a facile hydride chemistry. Small 2009;6:226–31.
[146] Li Z, Yao Y, Lin Z, Moon K, Lin W, Wong C. Ultrafast, dry microwave synthesis of graphene sheets. J Mater Chem 2010;20:4781–3.
[147] Stankovich S, Dikin D, Piner R, Kohlhaas K, Kleinhammes A, Jia Y, et al. Synthesis of graphene-based nanosheets via chemical reduction of exfoliated graphite oxide. Carbon 2007;45:1558–65.
[148] Sutter P, Sadowski J, Sutter E. Graphene on Pt(111): Growth and substrate interaction. Phys Rev B 2009;80:245411.
[149] Sutter P, Fledge J, Sutter E. Epitaxial growth of graphene on ruthenium. Nat Mater 2008;7:406–11.
[150] Zhang Y, Gomez L, Ishikawa FN, Madaria A, Ryu K, Wang C, et al. Comparison of Graphene Growth on Single-Crystalline and Polycrystalline Ni by Chemical Vapor Deposition. J Phys Chem Lett 2010;1:3101–7.
[151] Coraux J, N'Diaye A, Engler M, Busse C, Wall D, Buckanie N, et al. Growth of graphene on Ir(111). New J Phys 2009;11:023006.
[152] Preobrajenski A, Ng M, Vinogradov A, Martensson N. Controlling graphene corrugation on lattice mismatched substrates. Phys Rev B 2008;78:073401.
[153] Ago H, Ito Y, Mizuta N, Yoshida K, Hu B, Orofeo CM, et al. Epitaxial Chemical Vapor Deposition Growth of Single-Layer Graphene over Cobalt Film Crystallized on Sapphire. ACS Nano 2010;4:7407–14.
[154] Kim KS, Zhao Y, Jang H, Lee SY, Kim JM, Kim KS, et al. Large-scale pattern growth of graphene films for strechable transparent electrodes. Nature 2009;457.
[155] Kang B, Mun J, Hwang C, Cho B. Monolayer graphene growth on sputtered thin film platinum. J Appl Phys 2009;106:104309.
[156] Wang X, You H, Liu F, Li M, Wan L, Li S, et al. Large-scale synthesis of few-layered graphene using CVD. Chem Vapor Deposit 2009;15:53–6.
[157] Li X, Cai W, An J, Kim S, Nah J, Yang D, et al. Large-area synthesis of high-quality and uniform graphene films on copper foils. Science 2009;324:1312–4.
[158] Shelton JC, Patil HR, Blakely JM. Equilibrium Segregation of Carbon to a Nickel (111) Surface-Surface PhaseTransition. Surf Sci 1974;43:493–520.
[159] Cao L, Cheng L, Zhang Z, Wang Y, Zhang X, Chen H, et al. Visual and high-throughput detection of cancer cells using a graphene oxide-based FRET aptasensing microfluidic chip. Lab Chip 2012;12:4864–9.

[160] Liu F, Piao Y, Choi KS, Seo TS. Fabrication of free-standing graphene composite films as electrochemical biosensors. Carbon 2012;50:123–33.

[161] Ohno Y, Maehashi K, Matsumoto K. Label-Free Biosensors Based on Aptamer-Modified Graphene Field-Effect Transistors. J Am Chem Soc 2010;132:18012.

[162] Wu L, Chu HS, Koh WS, Li EP. Highly sensitive graphene biosensors based on surface plasmon resonance. Opt Express 2010;18:14395.

[163] Rider AE, Kumar S, Furman SA, Ostrikov KK. Self-organized Au nanoarrays on vertical graphenes: an advanced three-dimensional sensing platform. Chem Commun 2012;48:2659–61.

[164] Seo DH, Kumar S, Rider AE, Han Z, Ostrikov K. Deterministic control of structural and optical properties of plasma-grown vertical graphene nanosheet networks via nitrogen gas variation. Opt Mater Express 2012;2:700.

[165] Xiang X, Luo M, Shi L, Ji X, He Z. Droplet-based microscale colorimetric biosensor for multiplexed DNA analysis via a graphene nanoprobe. Anal Chim Acta 2012;751:155–60.

[166] Shen J, Zhu Y, Yang X, Li C. Graphene quantum dots: emergent nanolights for bioimaging, sensors, catalysis and photovoltaic devices. Chem Commun 2012;48:3686–99.

[167] Gollavelli G, Ling YC. Multi-functional graphene as an in vitro and in vivo imaging probe. Biomaterials 2012;33:2532–45.

[168] Liu Q, Wei L, Wang J, Peng F, Luo D, Cui R, et al. Cell imaging by graphene oxide based on surface enhanced Raman scattering. Nanoscale 2012;4:7084–9.

[169] Wate PS, Banerjee SS, Jalota-Badhwar A, Mascarenhas RR, Zope KR, Khandare J, et al. Cellular imaging using biocompatible dendrimer-functionalized graphene oxide-based fluorescent probe anchored with magnetic nanoparticles. Nanotechnology 2012;23:415101.

[170] Qian J, Wang D, Cai FH, Xi W, Peng L, Zhu ZF, et al. Observation of multiphoton-induced fluorescence from graphene oxide nanoparticles and applications in in vivo functional bioimaging. Angew Chem Int 2012;51:10570–5.

[171] Sreejith S, Ma X, Zhao Y. Graphene oxide wrapping on squaraine-loaded mesoporous silica nanoparticles for bioimaging. J Am Chem Soc 2012;134:17346–9.

[172] Hu SH, Chen YW, Hung WT, Chen IW, Chen SY. In situ synthesis of graphene/single-walled carbon nanotube hybrid material by arc-discharge and its application in supercapacitors. Adv Mater 2012;24:1748–54.

[173] Zhang L, Lu Z, Zhao Q, Huang J, Shen H, Zhang Z. Enhanced chemotherapy efficacy by sequential delivery of siRNA and anticancer drugs using PEI-grafted graphene oxide. Small 2011;7:460–4.

[174] Kim H, Namgung R, Singha K, Oh IK, Kim WJ. Graphene oxide-polyethylenimine nanoconstruct as a gene delivery vector and bioimaging tool. Bioconjug Chem 2011;22:2558–67.

[175] Pandey H, Parashar V, Parashar R, Prakash R, Ramteke PW, Pandey AC. Controlled drug release characteristics and enhanced antibacterial effect of graphene nanosheets containing gentamicin sulfate. Nanoscale 2011;3:4104–8.

[176] Zhang Y, Nayak TR, Hong H, Cai W. Graphene: a versatile nanoplatform for biomedical applications. Nanoscale 2012;4:3833–42.

[177] Nayak TR, Andersen H, Makam VS, Khaw C, Bae S, Xu X, et al. Graphene for Controlled and Accelerated Osteogenic Differentiation of Human Mesenchymal Stem Cells. ACS Nano 2011;5:4670.

[178] Czarnecki JS, Lafdi K, Tsonis PA. A novel approach to control growth, orientation, and shape of human osteoblasts. Tissue Eng Part A 2008;14:255–65.

[179] Ryoo S-R, Kim Y-K, Kim M-H, Min D-H. Behaviors of NIH-3T3 Fibroblasts on Graphene/Carbon Nanotubes: Proliferation, Focal Adhesion, and Gene Transfection Studies. ACS Nano 2010;4:6587.

[180] Kalbacova M, Broz A, Kong J, Kalbac M. Graphene substrates promote adherence of human osteoblasts and mesenchymal stromal cells. Carbon 2010;48:4323–9.

[181] Park SY, Park J, Sim SH, Sung MG, Kim KS, Hong BH, et al. Enhanced differentiation of human neural stem cells into neurons on graphene. Adv Healthcare Mater 2011;23: H263–7.

Chapter 13

Characteristics and Applications of Carbon Nanotubes with Different Numbers of Walls

Chuan-Fu Sun, Brendan Meany, YuHuang Wang
Department of Chemistry and Biochemistry, University of Maryland, College Park, MD 20742, USA

13.1 INTRODUCTION

Carbon nanotubes (CNTs) are quasi-one-dimensional (1D) tubular forms of sp^2 carbon networks. They may contain one to several tens of concentric graphitic shells and are typically 1–50 nm in diameter and many micrometres in length. This fascinating class of new structures, first known as multi-walled carbon nanotubes (MWCNTs), was unambiguously elucidated using high-resolution transmission electron microscopy by Iijima in 1991 [1]. Two years later, single-walled carbon nanotubes (SWCNTs) were synthesized by Iijima and Ichihashi [2] and Bethune et al. [3]. Subsequently, crystalline ropes of closely packed SWCNTs containing tens to hundreds of tubes of similar diameters were synthesized by Smalley and co-workers and made widely available to the research community [4]. In between MWCNTs and SWCNTs lies a unique, but much less-explored member of this family, double-walled carbon nanotubes (DWCNTs), which consist of exactly two concentric SWCNTs, one nested within another [5].

Over the past 20 years, CNTs have attracted intense attention, and a significant amount of work has revealed the extraordinary and often unexpected structural, optical, electrical, mechanical and chemical properties [6–8]. The properties of CNTs depend strongly on structure, size and defect. Individual crystalline CNTs are among the strongest and most conductive materials. Defects doping, and chemical functionalization of the graphene lattice can drastically affect the properties of CNTs. As the production and characterization techniques for CNTs develop, a myriad of exciting applications have been proposed for CNTs, including nanocomposites, electronics and optoelectronics, energy storage and conversion devices, field emission displays, and sensors, among many others [6,8–11].

Carbon Nanotubes and Graphene. DOI: 10.1016/B978-0-08-098232-8.00013-9

With the massive volume of work in these areas, we will attempt to compare and contrast the structures and properties of SWCNTs, DWCNTs and MWCNTs, and their impacts on existing and potential applications. Given the limited space, we intend to emphasize current experiments and theoretical works that explore the limits of CNT properties. Readers who are interested in more comprehensive and in-depth accounts of specific topics are referred to excellent reviews already available in literature. This book chapter is organized in five sections including this brief introduction. We will first discuss the structure and properties of individual CNTs of different number of walls (Sec. 13.2). This is followed by a section (Sec. 13.3) to summarize current understanding of tube-to-tube and intra-tube wall-to-wall interactions. The collective properties of CNT networks and solids are ultimately determined by the interfacial electronic, mechanical and chemical couplings at these nanostructure interfaces, and these are areas that require significant amounts of research and where opportunities arise. We will discuss examples of current and potential applications of CNTs in Sec. 13.4. Great strides have been made to reveal the remarkable electrical, optical, mechanical, thermal and chemical properties of CNTs. Many emerging CNT applications still face enormous technical challenges and significant barriers to compete with existing markets. However, the outlook is bright.

13.2 STRUCTURE AND PROPERTIES OF INDIVIDUAL CNTs

Graphitic materials, such as graphene and CNTs, share common characters derived from carbon–carbon (C–C) sp^2-hybridized bonding. These bonds are one of the strongest in nature where a carbon atom is covalently connected to three neighbouring carbons by σ bonds in a massively extended network. π electrons in these systems are delocalized, giving rise to interesting electrical properties. CNTs are distinguished from other graphitic materials in part due to 1D quantum confinement of electrons and phonons [12,13].

A SWCNT is an atom-thick cylinder that may be conceptually rolled from a single sheet of graphene; hence, the structure of a SWCNT can be indexed by a pair of integers, (n, m), that describes the rolling vector [12,14,15]. Nanotubes are classified as armchair when $n = m$, zigzag when $m = 0$ or chiral for the remaining (n, m) species (see Figure 13.1). The electronic properties of SWCNTs are sensitive to their (n, m) chirality. SWCNTs can be either metallic or semiconducting with different bandgaps [16–18]. All armchair SWCNTs are metals; while those for which $n - m = 3j$, where j is non-zero, are semi-metals since tube curvature effects creates a small bandgap [14]. The remaining two-thirds of SWCNTs are semiconductors with bandgaps inversely proportional to their diameters, or approximately 0.9 eV/D, where D is the nanotube outer diameter in nanometres [14,19].

A CNT may contain multiple SWCNTs that are concentrically nested, with wall-to-wall distance approximating 0.34 nm, the separation between two adjacent layers in a graphite. Each wall within these structures has a unique (n, m) index; however, their individual properties are strongly influenced by

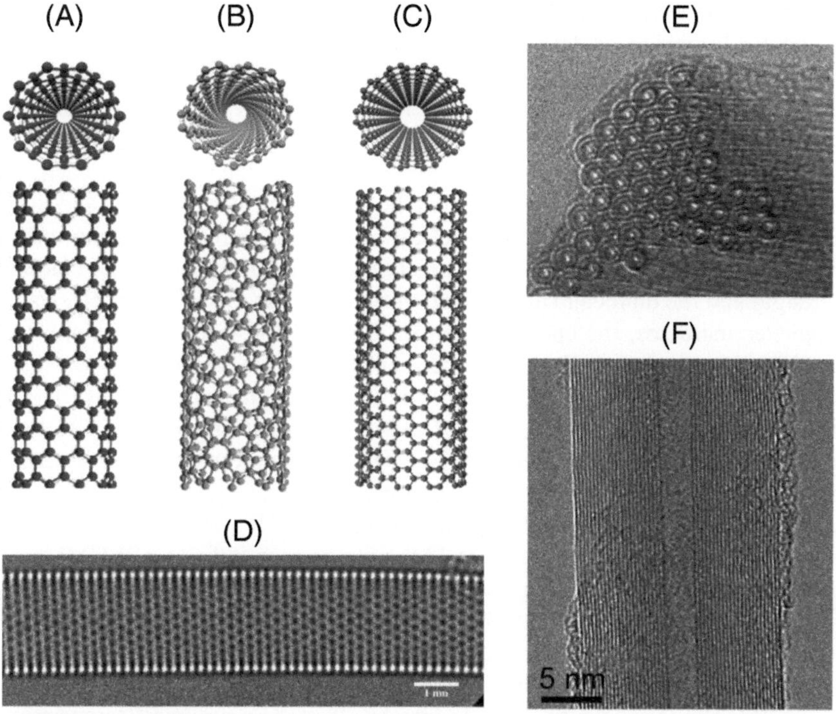

FIGURE 13.1 Schematic illustrations of the structures of (A) zigzag, (B) chiral and (C) armchair SWCNTs. (D) HRTEM image of a (28, 0) SWCNT. Adapted with permission [15]. Copyright 2011, Nature Publishing Group. (E) Cross-sectional view of a bundle of DWCNTs. Adapted with permission [20]. Copyright 2005, Nature Publishing Group. (F) HRTEM image of a MWCNT.

wall-to-wall interactions. DWCNTs are semiconducting or metallic, collectively depending on the chirality of both the inner and outer walls [21,20]. The majority of MWCNTs have overall metallic properties.

13.2.1 Electronic properties

The electronic transport in high-quality (defect-free) arc-discharge MWCNTs and catalytically produced metallic SWCNTs is found to occur ballistically. There is no electronic backscattering or energy dissipation in these CNTs on the microscale at room temperature [22–24]. In longer CNTs or at high bias, many scattering collisions can occur and the current flow could be described as diffusive transport, a characteristic of conventional conductors. The smallest possible resistance of SWCNTs is $h/4e^2$ or 6.5 kΩ, which is due to the quantum mechanical coupling between the two conducting sub-bands of a CNT and the device leads [25,26]. However, high-contact resistances usually prevent the observation of such a value. Electronic current in an individual SWCNT saturates at ~20 μA [27], while currents of up to 1 mA have been achieved for individual, arc-discharge MWCNTs with low defect density [23,28]. The current in arc-discharge MWCNTs is found to flow over the outer layer only [23,28,29]. Catalytically produced MWCNTs were found to be diffusive conductors with a resistance per micrometre of length on the order of 1 MΩ [30]. This increased resistance may derive from inter-layer electron transport caused by the disorder (defects) in nanotubes.

A DWCNT has one of four possible electronic configurations with each wall being either semiconducting (S) or metallic (M): S@S, M@S, M@M and S@M (inner-tube@outer-tube) [5]. Theoretical calculations predict that the energy band structure of a DWCNT could be affected by both the inter-wall coupling strength and the diameter (mean value of the inner and outer-tube) [21]. As the diameter increases, the energy gaps of S@S DWCNTs decrease quickly and vanish when the mean diameter is greater than 1.5 nm, as similarly occurs in SWCNTs. There are only a few reports on experimental measurements for electrical transport in DWCNTs [31–33]. Systematic measurements of 125 field-emission transistors fabricated using a catalytically produced DWCNT sample show that 42% are semiconducting, 35% are metallic, while the remaining 23% are neither purely semiconductors nor metals [32]. It was also confirmed by scanning tunnelling microscopy (STM) measurements that a DWCNT composed of two semiconducting walls can exhibit metallic properties overall [34].

The introduction of defects into the sp^2 carbon lattice is an effective way to tailor CNTs' intrinsic properties. Defects such as pentagons, heptagons, vacancies or dopant species can drastically modify a nanotube's electronic properties [8,35,36]. The conductance of CNTs may be affected by structural deformations involving bending and twisting. A smooth bending of the nanotube can lead to the formation of a local kink, inducing strong σ–π mixing and backscattering similar to that of a boron impurity [37]. Twisting has a much stronger effect. It was found that a metallic armchair nanotube develops a bandgap upon twisting [37]. For a realistic SWCNT, a finite doping is inescapable due to the charge transfer caused by the presence of adsorbates from the ambient. Semiconducting CNTs can be n-doped ($E_F > 0$, electron transfer to CNT) or p-doped ($E_F < 0$, electron transfer from CNT) [38–40]. Charge transfer characteristics in doped CNT bundles have been confirmed by Raman scattering through a vibrational mode shift [41]. When connected to external contacts, semiconducting SWCNTs are usually p-doped due to electron transfer from the CNT to the contacts [38]. Removing oxygen or doping the surface with alkali metals can cause the conversion of p-type to n-type [42], and interestingly, exposure to oxygen resulted in reversion to p-type characteristics. In addition, the resistance of SWCNTs is sensitive to various ambient gases [40].

13.2.2 Mechanical properties

Since the discovery of CNTs, it has been speculated that they have exceptional mechanical properties (high Young's modulus and high-tensile strength) because of a strong network of carbon–carbon bonds. Empirical force-constant model predicts that the Young's modulus of individual SWCNTs is largely insensitive to the nanotube diameter or chiral angle, except for very small nanotubes with diameters below ca. 0.6 nm, in which the strength of the C–C bonding is reduced due to the high-curvature effect [43]. The Young's modulus of individual, defect-free SWCNTs and MWCNTs was calculated to be ~1 and 1.11 TPa, respectively. Molecular dynamics simulations performed on SWCNTs and DWCNTs have also revealed that the breaking of nanotubes occurs at applied tensile strains of ~30% [44].

A large amount of experimental measurements have been conducted extensively to quantify the mechanical properties by stretching and bending CNTs [45–52]. The linear elastic behaviour of individual arc-discharge MWCNTs was studied by measuring the amplitude of the intrinsic vibrations of the nanotubes in the TEM, giving an average value for the axial Young's modulus of 1.8 TPa, much higher than that of carbon fibres (680 GPa) [45]. Later, the studies on individual SWCNTs gave an average Young's modulus of 1.25 TPa, which is close to the graphite basal plane elastic modulus [46]. AFM tips were used to apply tensile on individual MWCNTs and it was found that the Young's modulus of MWCNTs ranges from 270 to 950 GPa with a breaking strength between 11 and 63 GPa [50]. The fracture of embedded MWCNTs was observed with a tensile strength of 55 GPa [52]. MWCNTs with a mean fracture strength >100 GPa have also been reported. These results are in excellent agreement with quantum–mechanical calculations for nanotubes containing only an occasional vacancy defect, and are 80% of the values expected for defect-free nanotubes [53]. The effective bending modulus of MWCNTs was found to be diameter dependent. An effective bending modulus of 1 TPa was found for the nanotubes with diameter smaller than 12 nm, whereas a dramatically decreased value of 100 GPa was observed for those large-diameter nanotubes. A rippled or buckling distortion along the inner arc of the bent nanotubes was revealed by related high-resolution TEM [54]. Those results from bent nanotubes indicate that MWCNTs are difficult to stretch axially but easy to bend laterally and they can withstand large lateral distortions reversibly; i.e. nanotubes are resilient and will reversibly buckle rather than distort or break when they are bent. *In situ* Raman spectroscopy shows that CNTs lose their radial breathing modes (RBMs) at high pressures (1.5 GPa or higher) [55]. The pressure derivative of different phonon modes was also observed, suggesting that the strain components are different in the circumferential and axial directions of the nanotubes [56].

The mechanical response of individual DWCNTs and triple-walled carbon nanotubes (TWCNTs) under tensile load was measured using nanomanipulators in a scanning electron microscope. The Young's modulus and tensile strength were measured to be 0.73–1.25 TPa and 17–31 GPa for DWCNTs, and 0.83–1.33 TPa and 13–46 GPa for TWCNTs, respectively [57]. The low-tensile strength is most likely a result of defects present in the measured CNTs synthesized by chemical vapour deposition.

The mechanical properties of CNTs are sensitive to the density of defects, experimental parameters and synthesis techniques. Mechanical parameters

TABLE 13.1 Properties of individual CNTs with different numbers of walls

CNT	Electrical properties			Mechanical properties		Thermal properties	
	Current	Conductance	Resistance	Young's modulus	Tensile strength	Thermal stability	Thermal conductivity
SWCNT	20–25 mA [27]	~$2G_0$ [58]	~7 kΩ/μm^{-1} [59]	1.25 Tpa [46]	~1 TPa [51]	~1800 °C in Ar [60], 750 °C in air [61]	>3500 Wm^{-1}K^{-1} [62]
DWCNT		<$2G_0$ [63]	~4.2 KΩ [64]	0.73–1.25 Tpa [57]	17–31 Gpa [57]	~2000 °C in Ar [65], 800 °C in air [66]	~600 Wm^{-1}K^{-1} (produced by CVD) [62]
MWCNT	7.27 mA [63]	490G_0 [63]	34.4 Ω [63]	0.27–0.95 Tpa [57], 1.8 TPa (arc-discharge) [45]	11–63 Gpa [50]	Varies with diameter and length [50]	3000 Wm^{-1}K^{-1} (arc-discharge) [67], 42–343 Wm^{-1}K^{-1} (produced by CVD) [62]

from literatures are conflicting due to the use of two different ways (total occupied cross-sectional area or van der Waals area) to define Young's modulus and tensile strength. Experimental measurements as shown in Table 13.1 have not achieved the values predicted by theory, probably due to a combination of artefacts introduced during sample preparation and inadequate measurements. However, the experimentally measured values have already placed CNTs among the strongest materials known. As such, they are envisioned as ideal building blocks for high-performance engineering materials. The challenge is to translate the properties of individual CNTs to bulk solids, such as films and continuous fibres.

13.2.3 Thermal properties

So far, the thermal properties of CNTs are the least studied compared to their other properties. The thermal properties of CNTs include specific heat, thermal conductivity, thermal expansion and thermoelectric power. The specific heat and thermal conductivity of CNTs are generally dominated by phonon contribution. At high temperature, the specific heat should be similar to that of graphene. At low temperature, however, both the quantized phonon structure and the stiffening of the acoustic modes will cause the specific heat of CNTs to differ from that of graphene [68]. The measurements for MWCNTs at 10–300 K revealed a linear specific heat behaviour with the dependence on temperature [69]. This behaviour was attributed to the out-of-plane acoustic-phonon branches of the individual walls. The specific heat of SWCNT bundles with an average diameter of 1.25 nm has also been studied from 2 to 300 K. The experimental specific heat curve was close to the theoretical model for SWCNTs above 4 K and differed remarkably from that of graphene and graphite up to100 K [70].

Another important thermal property, thermal conductivity, has been predicted and measured to be superior to other materials. Experiments have determined ~3500 Wm^{-1} K^{-1} for a metallic SWCNT of diameter 1.7 nm and length 2.6 μm at room temperature [71]. This value is significantly greater than theoretical calculations for the thermal conductivity of diamond, graphite and few-layer graphene. For a (10, 10) armchair SWCNT, equilibrium and non-equilibrium molecular dynamics simulations give a thermal conductivity of ~6600 Wm^{-1} K^{-1} at room temperature [72]. However, this very high value has not been confirmed experimentally. Thermal resistance measurements have shown a substantial decrease in the thermal conductivity for DWCNTs and MWCNTs. The thermal conductivities of DWCNTs and MWCNTs produced by CVD method were measured to be ~600 and 42–343 Wm^{-1} K^{-1}, respectively [73]. However, arc-discharge MWCNTs showed a much higher thermal conductivity of over 3000 Wm^{-1} K^{-1} [67], indicating that the structural integrity of CNTs remarkably affects their thermal conductivities.

The thermal expansion of a SWCNT bundle has been measured using *in situ* X-ray diffraction techniques [74]. It was observed that there is almost no thermal expansion along the axial direction of tubes but a change of $(0.75 \pm 0.25) \times 10^{-5}$ K^{-1} along the radial direction in the range 300–950 K. The thermoelectric power of both SWCNT ropes and MWCNT bundles has been measured to be ~50 [75] and ~22 μV K^{-1} [76] at 300 K, respectively, and was found to increase with an increase in temperature.

13.2.4 Optical properties

CNTs have a strong tendency to form bundles due to van der Waals interactions. This bundling effect causes electronic coupling between nanotubes within the bundle. With the breakthrough of using surfactants to disperse SWCNTs individually in water, optical properties of SWCNTs are found to strongly depend on the (*n*, *m*) chirality [77,78]. Through absorption, photoluminescence (PL), Raman and ultra-fast optical spectroscopies the structure and 1D excitonic nature are revealed for the excitation states of CNTs [78,79].

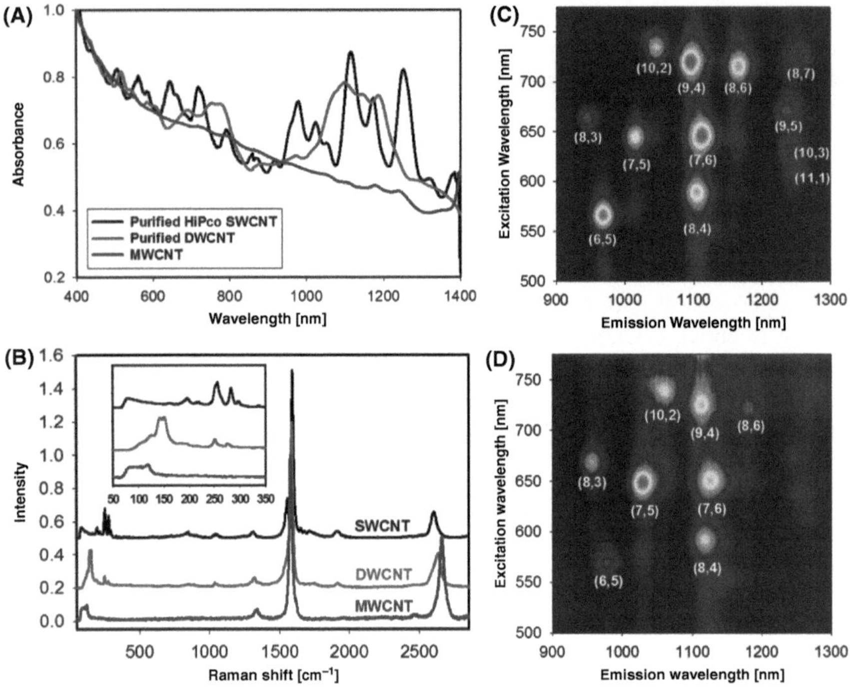

FIGURE 13.2 (A) UV-visible near-IR absorption spectra and (B) Raman spectra of HiPco SWCNTs, CVD-produced DWCNTs and arc-discharged MWCNTs. Inset shows the RBMs of CNTs. (C) PL map of HiPco SWCNTs in surfactant/D_2O solution. (D) PL map of CVD-produced DWCNTs in surfactant/D_2O solution.

Each SWCNT chirality has a unique set of absorption peaks. These peaks correlate with the van Hove singularities, or kinks, in the electronic density of states (DOS), a direct result of 1D confinement [12]. The first electronic transition, E_{11}, of a small-diameter SWCNT typically lies in the 800–1400 nm near-IR range, E_{22} in the visible range, and E_{33} and E_{44} in the UV range. Therefore, absorption spectroscopy provides useful information on the composition and distribution of SWCNTs individually dispersed in solution (Figure 13.2).

CNTs with more than one wall have increasingly less-defined absorption features due to wall-to-wall electronic coupling and spectral overlapping in a heterogeneous mixture. High-purity, small-diameter DWCNTs may show distinct characteristics [80]. FWCNTs and MWCNTs have smooth exponential decaylike absorption curves.

Semiconducting CNTs also show interesting fluorescence properties. Isolated semiconducting SWCNTs can have a quantum yield as large as 8%, with a strong dependence on chirality and environment, and a lifetime in the range of 10–100 ps [81–84]. However, in bulk solution the quantum yield is below 1%. The low-quantum yield is accounted for by several non-radiative decay processes including low-energy dark excitonic states, exciton–exciton annihilation and electron–phonon coupling [85–88]. In addition fluorescence quenching occurs due to chemical doping and defects in the crystal lattice [89]. DWCNTs have been shown to fluoresce from their inner walls but the majority of PL is quenched by the outer wall [90]. Only inner walls with a diameter of ~0.93 nm and a high chiral angle are able to produce measurable PL from a DWCNT [90]. MWCNTs do not emit PL because excitons are easily quenched by other concentric walls.

The origin of CNT excitations and optical properties was theoretically predicted to be excitonic in nature by Ando [91] long before experimental confirmation [79,92–94]. The excitons, or electron-hole pairs, are bound by Coulomb interactions. Unlike most other semiconducting materials, CNTs have large-exciton binding energies up to ~1 eV. The absorption and emission characteristics of CNTs are governed by excitons. However, only one bright (light-emitting) exciton state exists along with 15 dark states. Further experiments are needed for a more conclusive understanding of these species and how they might be manipulated. The excitonic nature of CNTs is an important characteristic for developing nanotube-based optoelectronic devices and applications.

Raman spectroscopy is a useful technique for studying CNTs. Similar to absorption and PL spectroscopy, initial use of Raman spectroscopy was hindered by CNT bundling. Individually dispersed surfactant-wrapped nanotubes in solution were used to assign (n, m) indices of metallic SWCNTs by Raman spectroscopy of the RBMs [95]. In conjunction with the Kataura plot, RBM Raman data may be used to determine (n, m) chirality of CNTs since $\omega_{RBM} \propto 1/d_t$. Experiments continue to utilize Raman scatter to characterize and understand the electronic structure and optical transitions of SWCNTs.

13.2.5 Superconductivity

When the curvature of CNTs leads to the creation of new electron–phonon scattering channels and consequent attractive electron–phonon interactions, superconductivity can be induced in CNTs [96–99]. A superconducting transition temperature (T_c) of ~0.55 K was observed in SWCNT ropes [100], and a T_c of ~5 K was reported for SWCNTs grown in zeolites [101]. Similarly, a bundle of ~60 DWCNTs show a significant drop in resistance below 6.8 K [64]. This array was a highly pure sample consisting mostly of metallic walls. Superconductivity has also been observed for MWCNTs [102]. However, superconductivity for CNTs was only observed at low temperatures. It was proposed that chemical doping of the nanotubes may raise T_c [103].

13.3 TUBE-TO-TUBE INTERACTIONS

Many applications will require the ability to translate the superlative electrical and mechanical properties observed in individual CNTs into macroscopic structures, such as thin films, composites and continuous fibres. CNT networks such as thin film and yarn have also attracted a great deal of attention due to their suitability for scalable integration into devices [104]. A random network film can be easily fabricated by either a post-production approach or direct growth. The first SWCNT film (buckypaper) was fabricated by a filtration process in 1998 [105]. Fan and co-workers and Baughman and co-workers reported the fabrication of super-aligned CNTs arrays, which were well-aligned bundles held together by van der Waals forces [106–108]. Continuous CNT yarns or films can be drawn from the super-aligned CNT arrays, similar to drawing silk from a cocoon. These yarns or films can be further condensed into shrunk yarns possessing high-tensile strengths and Young's moduli as well as good conductivity. Furthermore, techniques have been developed to fabricate CNT arrays on substrates with controlled shape, orientation and linkage [109–113]. The properties of these higher-order structures and architectures are dominated by the properties of the tube-to-tube interfaces where electrons scatter and the interaction are the weakest.

The normalized mechanical strengths of CNT bundles are typically a small fraction of those found with their individual building blocks. The Young's modulus of catalytically grown SWCNT ropes was reported with values in the range of 10–50 GPa [114], two orders of magnitude smaller than that of the laser evaporation grown individual SWCNTs (1.25 TPa) [46]. This large difference presumably derives from the influence of structural defects in catalytically grown nanotubes. The mechanical response of SWCNT ropes under tensile loads showed that the average Young's modulus was in the range of 320–1470 GPa with breaking strength values of 13–52 GPa, approximately 50 times larger than that of steel wires. The maximum breaking strain of 5.3% is close to the critical strain for defect nucleation (5%) for individual SWCNTs based on theoretical simulation [51]. The mechanical properties for DWCNT ropes consisting of only two DWCNTs have also been studied, providing a Young's modulus of 1.10–1.71 TPa and a tensile strength of 23–43 GPa [57].

The electrical conductivity of CNT networks combines the intrinsic conductivity of individual CNTs and that associated with the inter-tube junctions between CNTs. The electrical conductivity of random SWCNT networks ranges from 400 to ~6700 S cm^{-1} [115,116]. With respect to dense films of aligned SWCNTs, however, the anisotropic morphology leads to anisotropic electrical and thermal transport properties [117]. In the parallel direction, the electrical conductivity is on the same order as that of individual SWCNT ropes, and the thermal conductivity is over 200 $Wm^{-1}\,K^{-1}$, within an order of magnitude of highly crystalline graphite. Compared with SWCNT ropes, the conductivity of films or networks is usually 1 or 2 orders of magnitude lower. Purifying or compacting the films increases the conductivity, indicating that it is dominated by tube-to-tube interactions [115]. Compared with films made by a filtration process, the directly synthesized films have superior electrical and mechanical properties. With an electrical conductivity greater than 2000 S cm^{-1} and a strength of 360 MPa, these enhanced properties derive from strong, extended inter-tube interactions [118].

In general, the collective properties of CNT thin films and fibres are a relatively small fraction of those observed with their individual building blocks. A detailed understanding of the inter-tube interface, which is the weakest point, may open opportunities to translating the superlative properties of individual CNTs to macroscopic structures tailored for specific applications.

13.4 APPLICATIONS OF CNTS

The remarkable structural, mechanical, thermal and electronic properties of CNTs hold vast promise in fundamental studies and applications including nanocomposites, field emission devices, electronic and electrochemical devices as well as sensors and probes. A great number of prototype CNT devices have been fabricated. It is encouraging to note that some CNT-based devices, such as AFM tip and Hyperion's nanotube composite, have already entered the marketplace (Figure 13.3).

13.4.1 CNT composites

CNT composite fibres with enhanced mechanical and electrical properties were first produced in 1999 [120]. Methods have been developed to 'spin' nanotubes into macroscopic fibres or yarns, which is tougher than any known organic fibre [106,107]. The incorporation of CNTs into structural materials not only provides mechanical reinforcements but also enhances the electrical and thermal conductivity of the materials. By adding SWCNTs to industrial epoxy

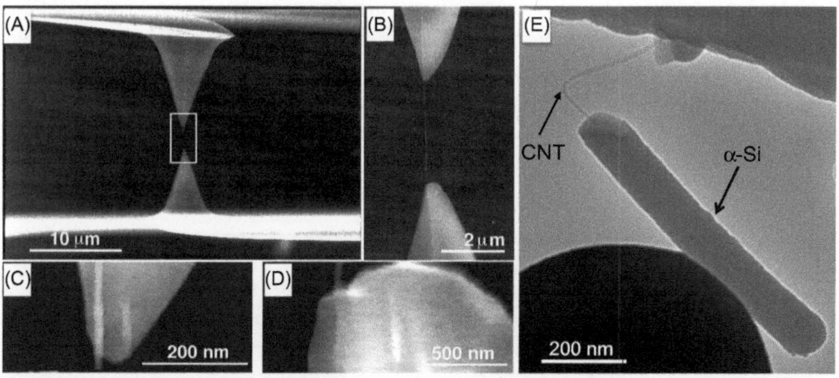

FIGURE 13.3 (A-D) An individual MWCNT mounted between two opposing AFM tips. Adapted with permission [50]. Copyright 2000, The American Association for the Advancement of Science. (A) SEM image of two AFM tips holding a MWCNT. (B) SEM image of the indicated region in (A), showing the MWCNT between the AFM tips. (C) SEM image showing the attachment of the MWCNT on the top AFM tip. There is an apparent thickening of the MWCNT section on the surface. (D) Close-up SEM image showing the attachment of the MWCNT on the lower AFM tip. (E) *In situ* mechanical study of CNT@α-Si core-shell structure. The broken α-Si shell is pulled out while the CNT core remains intact. Adapted with permission [119]. Copyright 2011, John Wiley and Sons.

the resulting composite can strengthen and enhance thermal conductivity by 125% [121]. Loading 1% MWCNT in polystyrene can increase the modulus and breaking stress by up to 42% and 25%, respectively [122]. The electrically conducting CNT–polymer composite is one type of commercially promising material. Conductivities of 0.01–0.1 S cm^{-1} can be achieved with 5% loading of MWCNTs into a polymer matrix [123]. Furthermore, moulded CNT composites also have potential applications in shielding electromagnetic radiation from cell phones and computers [124].

13.4.2 Mechanical applications

Given their extraordinary mechanical properties, CNTs have many potential applications in strain sensors, artificial muscles and even space elevators. Thin films of aligned SWCNTs have been shown to act as strain sensors [125]. Reacting to various modes of human motion these devices were fabricated into bandages, gloves and stockings with the ability to detect extension, flexion, breathing and jumping. Further research and development will lead to integration of CNTs into wearable batteries and lightweight electronic armour for soldiers. A second application utilizes individual SWCNTs as nanomechanical mass sensors [126]. At a length of ~150 nm, frequency of 2 GHz and a resolution of 1.7×10^{-24} g, a resonating nanotube is capable of sensing adsorption events and measuring binding energies.

Super-aligned CNT films can be used as new artificial muscles that are driven by an applied voltage and can provide giant elongations of 220% with elongation rates of (3.7×10^4)% per second at operating temperatures of 80–1900 K [127]. By spinning MWCNTs into yarns a rotating actuator has been created [128]. This device rotates at 590 revolutions/minute with up to 15,000° reversible rotations and can be used for mixing in fluidic chips. Instead of applying electricity, one could generate current by converting the mechanical motion of the torsional CNT muscle.

13.4.3 Thermal applications

CNTs are attractive materials for thermal applications. MWCNTs are able to convert chemically driven thermal energy into electrical energy [129]. Combustion of cyclotrimethylene trinitramine surrounding a MWCNT produces a thermopower wave propagating along the length of the nanotube. This produces thermally excited electrons that can be harvested for electrical energy in power pulses up to 7 kW kg^{-1}. Another application links nanoscale electronic devices with CNTs for thermal management. Joule heating in these devices is of strong concern and must be minimized. Through contacts between MWCNTs and a polar substrate, energy is dissipated remotely [130]. The coupling of vibrational modes between the nanotube and substrate has been shown to dissipate 84% of supplied electrical power from the MWCNT to a substrate.

13.4.4 Photonic and optoelectronic devices

The optical properties of CNTs may have important applications in nanophotonics, such as nanolasers, photodetectors, nanoantennas, solar energy harvesting and quantum optics. One such device uses a SWCNT as an infrared light-emitting diode (LED) by recombination of excitons [131]. The ambipolar field-effect transistor (FET) produced electroluminescence from a CNT by simultaneously injecting electrons and holes at either end of the tube. This has been improved upon with CNT p–n diodes, which allow efficient carrier-to-photon conversion, miniscule self-heating and reduced power dissipation [132,133]. The electroluminescence corresponds to the E_{11} energy transition of the nanotube with a power density of ~0.1 Wm^{-1}.

Due to coupling between electrons and phonons, energy transfer results in joule heating, which can be harnessed as thermal light emission [134]. Semimetallic SWCNTs emit in the infrared and visible wavelengths from their E_{11} and E_{22} transitions with exponential dependence on the devices power dissipation. Aligned MWCNTs on silicon have also shown optical emission from joule heating in the form of blackbody radiation [135]. The film emitted light in the 350–900 nm wavelength range while under resistive heating temperatures from 1550 to 2000 K.

In addition to light emission, CNTs can act as optical wires directing radiation between nanotubes or other materials within their environment. Parallel SWCNTs separated over large distances of 300 nm can be optically coupled by overlapping Rayleigh scatter [136]. In another example, SWCNTs have been incorporated into the scales of butterfly wings for mid-wave infrared imaging [137]. This bio-inspired nanoarchitecture converts infrared excitation into visible radiation through a complicated mechanism of changes in the optical cavity spacing in response to thermal expansion. Lastly, MWCNTs in a nanocoax cable transmit visible light through a sub-wavelength waveguide [138].

A semiconducting CNT can be used as a nanoscale photodetector that converts light into a current or voltage [139], and several implementations of individual CNT photodetectors have been demonstrated [139–141].

CNTs as transistors and conductors have been used to develop flat screen and flexible film devices [108,142]. CNT films are an improvement over other materials because of their flexibility and tuneable electronic properties. Arc-discharge SWCNTs have been used as a dopant transistor material in polymer matrixes of organic light-emitting diodes (OLEDs) [143]. With 87% transparency these OLEDs rival other materials such as indium-tin oxide (ITO). More recently a stretchable OLED display was created with SWCNTs as elastic conductors [144]. SWCNTs and a fluorinated copolymer were mixed into a rubber composite and printed on polydimethylsiloxane (PDMS) producing a monochromatic active matrix able to withstand bending, crumpling and stretching up to 50%. Another type of OLED active matrix uses SWCNTs as a network source electrode to achieve red, green, and blue emissions with very low power consumption and ~98% aperture emission [145] (Figure 13.4).

(A)
(B)
(C)
(D)
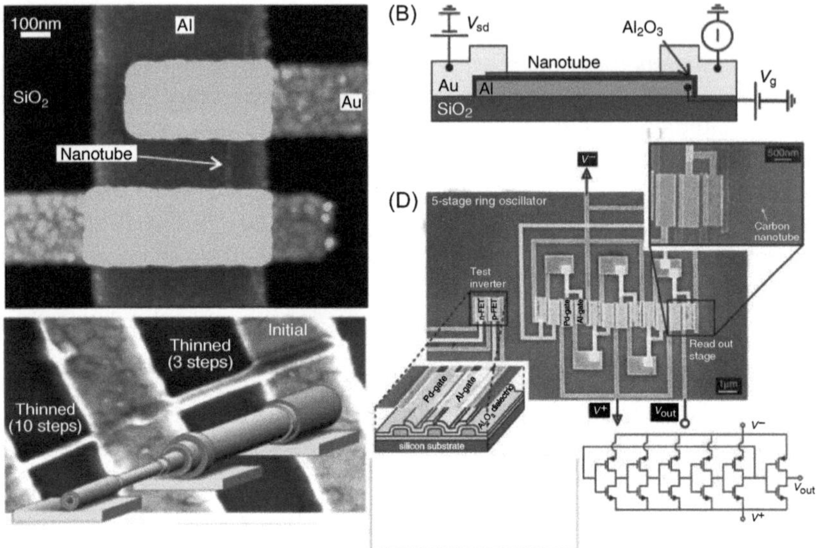

FIGURE 13.4 (A) AFM image of a SWCNT transistor. (B) Schematic side view of the logic circuit. Adapted with permission [146]. Copyright 2001, The American Association for the Advancement of Science. (C) Images and overlaying schematic representation for the effect of electrical pulses in removing successive layers of a MWCNT, so that layer having desired transport properties for devices can be revealed. Adapted with permission [147]. Copyright 2001, The American Association for the Advancement of Science. (D) A ring oscillator built on a long SWCNT. Adapted with permission [148]. Copyright 2006, The American Association for the Advancement of Science.

Super-aligned CNT films, which are transparent and conductive, have been suggested for flexible touch screens to replace the costly and brittle ITO [149]. CNT-based touch screens were 50 times better than ITO in scratch resistance and endurance tests. Furthermore, CNT-based touch screens integrated into mobile phones (Figure 13.5(E)) show superior performance in terms of endurance, flexibility and simple fabrication process.

13.4.5 Field-emission devices

Because of their attractive field-emitting properties, CNTs have been investigated for field-emission electron sources [150,151] in flat panel displays [152], lamps [153], gas discharge tubes [154] and X-ray generators [155]. Compared with tungsten and molybdenum tip arrays which require a vacuum of 10^{-10} torr, CNT field-emitting surfaces are relatively easier to manufacture and do not deteriorate under moderate vacuum (10^{-8} torr) [156]. Also, CNTs can provide stable emission, long lifetimes and low-emission threshold potentials [150,153] as well as high-current density of 4 A cm^{-2} (10 mA cm^{-2} required for flat-panel displays) [157].

Flat-panel displays are a promising application of CNTs. A potential applied between a CNT-coated surface and an anode produces high-local fields, which cause electrons to tunnel from the CNT tip into the vacuum. Electric fields

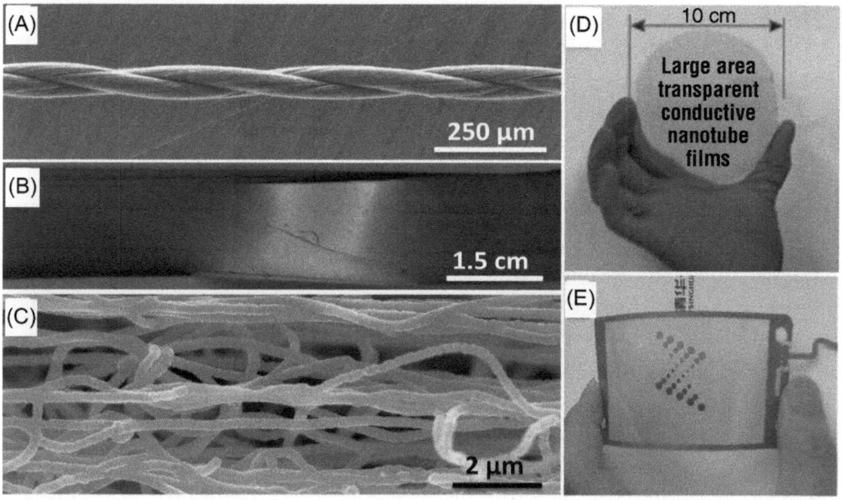

FIGURE 13.5 (A) MWCNT spun fibres by spin twisting. (B) Optical and (C) SEM images of Si-coated CNT electrodes for lithium ion batteries. (D) A large, 80-nm-thick film on a sapphire substrate 10 cm in diameter. Adapted with permission [142]. Copyright 2004, The American Association for the Advancement of Science. (E) Optical image of a flexible CNT touch screen. Adapted with permission [149]. Copyright 2010, John Wiley and Sons.

direct the field-emitted electrons towards the anode, where a phosphor produces light for the flat-panel application. The advantages of CNTs over liquid crystal displays are low-power consumption, high brightness, fast response rate, wide viewing angle and operating temperature range. Samsung has produced several generations of prototypes including a 9-inch red–blue–green colour display [152]. Samsung and Unidym have also fabricated the world's first CNT-based colour active matrix electrophoretic display (EPD) e-paper which was demonstrated on a 14.3 inch display [158]. Furthermore, prototypes of CNT-based fluorescent lamps with a high-performance lifetime of over 8000 hours [153] and CNT-based gas discharge tubes [154] have also been fabricated.

13.4.6 Electronic devices

In state-of-the-art copper interconnects used in microelectronics, the scattering lengths are a few nanometres, resulting in large resistances. Metallic CNTs offer the possibility of high-performance interconnects for large-scale integrated systems [159,160]. The advantages of CNT interconnects are a large-carrier mean-free path derived from ballistic transport, large carrier mobilities and high current density of 10^9 A cm^{-2} (10^5 A cm^{-2} for conventional metals) [22,27]. In addition, these interconnects are immune to electromigration due to the strong C–C bonds [103].

Electrical transport in semiconducting CNTs can be switched 'on' and 'off'; thus, semiconducting CNTs can be used as the conducting channel in transistors. The first room-temperature CNT FET was reported in 1998 [38]. It was demonstrated that a SWCNT draped across two gold electrodes could be gated. Since then, a large number of results for CNT–FETs have been reported [7,146,147,161–163]. One example is a CNT diode created by grafting a metallic to a semiconducting nanotube [164]. Other electrical devices include non-volatile random access memory [165] and three- and four-terminal electronic devices formed from crossed SWCNTs [162].

In a CNT–FET, semiconducting CNTs play the role of the channel [7]. The key advantages of CNT–FETs beyond Si-based metal oxide semiconductor FETs in logic applications are low scattering due to ballistic electrical transport, low-operating voltage and very low capacitance of ~10 aF as well as high mobility of the FET channel. Scattering by surface states and roughness, which plagues conventional FETs, especially at high-gate voltage, is absent in CNTs. In conventional metal oxide semiconductor FETs, the ability of the gate to control the potential in the channel is limited once the gate voltage exceeds the threshold voltage [166], whereas the optimum coupling between the gate and channel was allowed in CNT–FETs due to the small diameters of nanotubes, enabling the gate ability to control the channel potential. Furthermore, the chemically inert CNT surface is also advantageous in terms of reducing surface recombination, leakage current and sub-threshold conduction [167], and the lower resistance of CNTs can lead to faster device operation with low-power dissipation [103]. However, interfacial traps and inefficient contacts are often observed to significantly plague CNT–FETs devices. In addition, defects often induce loss of control in electrical properties and a source of unreliability. Particularly, ambient oxygen affects the performance of devices by inadvertently doping the contacts/CNTs. These issues represent significant technological challenges to the immediate applications of CNT–FETs in large-scale electronics.

The fabrication of CNT–FETs devices has developed beyond single devices to integrated logic circuits [146,148,168–170]. Patterned growth of CNTs on substrates marks a major step forward in integrated CNT electronics [109–111]. Logic circuits with SWCNT–FETs have been demonstrated [146]. The device layout features local gates that provide excellent capacitive coupling between the gate and nanotube, enabling strong electrostatic doping of the nanotube from p-doping to n-doping. The transistors show favourable device characteristics such as high gain (>10), a large on–off ratio (>10^5) and room-temperature operation. Importantly, the local-gate layout allows for integration of multiple devices on a single chip. In 2008, flexible medium-scale CNT sub-monolayer integrated circuits were fabricated [104], in which random SWCNT networks were used to fabricate medium-scale integrated digital circuits, composed of up to nearly 100 transistors on flexible plastic substrates. Transistors in these integrated circuits showed excellent properties: high mobility (80 $cm^2\,V^{-1}\,s^{-1}$), low sub-threshold slopes (140 $mV\,dec^{-1}$), low operating voltages (<5 V), high on–off ratios (10^5), good mechanical flexibility and switching speeds in the kilohertz range – all with levels of uniformity and reproducibility that enable high-yield fabrication of integrated circuits.

13.4.7 Sensors and probes

Beyond large-scale electronics, the high-carrier mobility and sensitivity of CNT–FETs over silicon-based FETs furnish CNTs very promising applications in chemical sensors [39,40,171–175]. Semiconducting SWCNTs have been used in gas sensors, where a three orders of magnitude change in the electrical resistance was observed upon exposure to oxidizing and reducing gases, such as NO_2 and NH_3 with a sensitivity of 2 ppm and 0.1%, respectively [40]. Strong oxidizers (NO_2) resulted in charge transfer from the SWCNTs, consequently inducing an increased hole density and enhancing electrical conductivity. An opposite effect was observed with reducing agents (NH_3), resulting in charge transfer of electrons to the nanotubes, decreasing the conductivity. DWCNTs have also been suggested to be novel sensors due to their unique double-walled structure. Compared with SWCNTs, this double-walled structure provides an outer wall that can be selectively functionalized with covalent chemistry to significantly improve the chemical selectivity of the sensor [5,176]. Because the outer wall is only one-atom thick, the inner tube remains efficient in sensing the chemical environment.

Several biosensors for detecting a whole host of pathogens and chemicals have been fabricated with CNTs as a base element [177]. CNTs were used to fabricate glucose sensors due to their fast electron transfer rate and the high-electrocatalytic effect [178]. Since the nanoscale detecting element has an enormous surface, CNT-based sensors are highly sensitive. Functionalization and changes in the electrical properties of CNTs can be used in the future for electronic sensors.

CNTs have been suggested to be novel scanning probes. Compared with conventional nanoprobes, the mechanical robustness of CNTs and the low buckling force enable minimizing sample damage during repeated hard crashes into substrates and thus dramatically increase the probe life. Imaging in narrow, deep crevices and high resolution can be achieved with CNT probes due to their tubular and nanoscale structures [179,180]. Furthermore, chemical modification of CNTs enables chemical and biological functions of nanotube tips for specific probes [181]. It is worth noting that CNT scanning probe tips for atomic probe microscopy are now commercially available.

13.4.8 Electrochemical devices

Owing to the ultrahigh surface area, mechanical properties, electronic conductivity and remarkable chemical stability as well as light weight, CNTs have shown promise as high-performance electrodes for electrochemical energy storage and conversion devices, such as lithium ion batteries [182,183], supercapacitors [184–186] and electromechanical actuators [6,187–190]. For lithium ion batteries, MWCNTs offer a high-theoretical capacity of 1000 mA hour g^{-1}, higher than that of graphite (372 mA hour g^{-1}) [182] and ball-milled graphite

(708 mA hour g^{-1}) [183]. Recently, CNT-based heterogeneous nanostructures, such as silicon-coated CNTs [119,191] and MnO_2-coated CNTs [192], have been proposed as promising electrodes for lithium ion batteries. These heterogeneous nanostructures combine the advantages of both CNTs and functional materials, and evidently improve the performance of lithium ion batteries.

Supercapacitors have high-power density and improved frequency response for better performance and more demanding applications. Their capacitances depend on the separation between the charge on the electrode and the countercharge in the electrolyte [6]. As electrodes for supercapacitors, the nanoscale size of CNTs provides a high surface area accessible to the electrolyte and consequently leads to very large capacitances. SWCNT and MWCNT electrodes have capacitances of 180 and 102 F g^{-1}, respectively [184,185]. A high specific power density over 20 kW kg^{-1} and a superior frequency response of about 7560 Hz were achieved by using MWCNT thin films as electrodes [186].

Under applied voltage, SWCNTs gain electronic charge and an array of these structures are capable of mechanical motion. Even with a few volts applied, there are large amounts of charge injection such that electrode expands and contracts to do mechanical work in electromechanical actuators [187]. CNT electromechanical actuators can operate at only a few volts and high temperatures up to 350 °C. It was also observed that strains in CNT actuators can exceed 1%, an order of magnitude over commercial actuators in work. The maximum observed isometric actuator stress of SWCNT actuators is 26 MPa, which is >10 times higher than previous reports for these actuators and ~100 times higher than that of natural muscle [6]. Recently, a SWCNT-based bimorph electromechanical actuator showed orders-of-magnitude improvements in many aspects, including superfast response (19 ms), wide available frequency range (dozens to hundreds of hertz), large stress-generating rate (1080 MPa s^{-1}), and high mechanical output power density (244 W kg^{-1}) [190]. Compared to ferroelectrics, the cycling rate of CNT actuators is much lower because of the ion diffusion–dependent actuation, which eliminates some applications.

13.5 CONCLUSION AND OUTLOOK

The structures, properties and applications of CNTs with different numbers of walls were introduced in this chapter. Because of their fascinating electrical, mechanical, thermal and optical properties, CNTs have captured and continue to attract a significant amount of interest. Many potential applications have been suggested, with several on the horizon. Nonetheless, there remains a gap between the promising potential and tentative reality of using nanotubes in diverse practical applications on a widespread scale. There are many issues in synthesis, purification and large-scale assembly of devices, which hamper further development of CNT applications. Numerous efforts are devoted to manufacturing monodispersed CNTs with defined chirality, diameter and length, to produce electrically homogeneous CNTs for effective, large-scale, processing techniques of CNT-based devices. This prerequisite bottlenecks further development and must be overcome to bring about a revolution in CNT material science. It is encouraging to note that chirality-controlled growth of SWCNTs have been demonstrated using SWCNTs as seeds for continued, epitaxial growth [193]. This cloning technique has been successfully applied in several experimental conditions and extended to single-chirality SWCNT seeds that have become available lately [194]. Continued progress at this frontier may ultimately allow specific SWCNT chirality to be directly synthesized in large scale.

After two decades of research, CNTs are in a transitive phase. While there remains a tremendous amount of work towards fundamental understanding of these structures, nanotubes are entering an infancy industrial stage for inclusion in various applications. Exciting research awaits the academic, the engineer and the entrepreneur.

ACKNOWLEDGEMENTS

Some of the work described in this book chapter was partially supported by the Office of Naval Research (N000141110465), the Department of Energy Office of Basic Energy Sciences (grant DESC0001160), the National Science Foundation (CAREER CHE 1055514), the ACS Petroleum Research Fund (Doctoral New Investigator Award) and the University of Maryland.

REFERENCES

[1] Iijima S. Helical microtubules of graphitic carbon. Nature 1991;354:56–8.
[2] Iijima S, Ichihashi T. Single-shell carbon nanotubes of 1-nm diameter. Nature 1993;363:603–5.
[3] Bethune DS, Kiang CH, Devries MS, Gorman G, Savoy R, Vazquez J, et al. Cobalt-catalyzed growth of carbon nanotubes with single-atomic-layerwalls. Nature 1993;363:605–7.
[4] Thess A, Lee R, Nikolaev P, Dai HJ, Petit P, Robert J, et al. Crystalline ropes of metallic carbon nanotubes. Science 1996;273:483–7.
[5] Shen C, Brozena AH, Wang YH. Double-walled carbon nanotubes: challenges and opportunities. Nanoscale 2011;3:503–18.
[6] Baughman RH, Zakhidov AA, de Heer WA. Carbon nanotubes - the route toward applications. Science 2002;297:787–92.
[7] Avouris P, Chen Z, Perebeinos V. Carbon-based electronics. Nat Nanotechnol 2007;2:605–15.
[8] Dresselhaus MS, Dresselhaus G, Avouris P. Carbon nanotubes: synthesis, structure, properties and applications. Berlin: Springer-Verlag; 2001.
[9] de Heer WA. Nanotubes and the pursuit of applications. Mrs Bull 2004;29:281–5.
[10] Dai HJ. Carbon nanotubes: opportunities and challenges. Surf Sci 2002;500:218–41.
[11] Popov VN. Carbon nanotubes: properties and application. Mater Sci Eng R 2004;43:61–102.
[12] Saito R, Dresselhaus G, Dresselhaus MS. Physical properties of carbon nanotubes. London: Imperial College Press; 1998.
[13] Samsonidze GG, Saito R, Jorio A, Filho AGS, Grüneis A, Pimenta MA, et al. Phonon trigonal warping effect in graphite and carbon nanotubes. Phys Rev Lett 2003;90:027403.

[14] Louie SG. Electronic properties, junctions, and defects of carbon nanotubes. Top Appl Phys 2001;80:113–45.
[15] Warner JH, Young NP, Kirkland AI, Briggs GAD. Resolving strain in carbon nanotubes at the atomic level. Nat Mater 2011;10:958–62.
[16] Saito R, Fujita M, Dresselhaus G, Dresselhaus MS. Electronic structure of chiral graphene tubules. Appl Phys Lett 1992;60:2204–6.
[17] Chico L, Crespi VH, Benedict LX, Louie SG, Cohen ML. Pure carbon nanoscale devices: nanotube heterojunctions. Phys Rev Lett 1996;76:971–4.
[18] Mintmire JW, Dunlap BI, White CT. Are fullerene tubules metallic. Phys Rev Lett 1992;68:631–4.
[19] Mintmire JW, White CT. Electronic and structural properties of carbon nanotubes. Carbon 1995;33:893–902.
[20] Endo M, Muramatsu H, Hayashi T, Kim YA, Terrones M, Dresselhaus MS. 'Buckypaper' from coaxial nanotubes. Nature 2005;433:476.
[21] Liang SD. Intrinsic properties of electronic structure in commensurate double-wall carbon nanotubes. Phys B Conden Mat 2004;352:305–11.
[22] Liang WJ, Bockrath M, Bozovic D, Hafner JH, Tinkham M, Park H. Fabry-Perot interference in a nanotube electron waveguide. Nature 2001;411:665–9.
[23] Frank S, Poncharal P, Wang ZL, de Heer WA. Carbon nanotube quantum resistors. Science 1998;280:1744–6.
[24] Bachtold A, Fuhrer MS, Plyasunov S, Forero M, Anderson EH, Zettl A, et al. Scanned probe microscopy of electronic transport in carbon nanotubes. Phys Rev Lett 2000;84:6082–5.
[25] Tans SJ, Devoret MH, Dai HJ, Thess A, Smalley RE, Geerligs LJ, et al. Individual single-wall carbon nanotubes as quantum wires. Nature 1997;386:474–7.
[26] Bockrath M, Cobden DH, McEuen PL, Chopra NG, Zettl A, Thess A, et al. Single-electron transport in ropes of carbon nanotubes. Science 1997;275:1922–5.
[27] Yao Z, Kane CL, Dekker C. High-field electrical transport in single-wall carbon nanotubes. Phys Rev Lett 2000;84:2941–4.
[28] Poncharal P, Berger C, Yi Y, Wang ZL, de Heer WA. Room temperature ballistic conduction in carbon nanotubes. J Phys Chem B 2002;106:12104–18.
[29] Schonenberger C, Bachtold A, Strunk C, Salvetat JP, Forro L. Interference and Interaction in multi-wall carbon nanotubes. Appl Phys A Mater Sci Process 1999;69:283–95.
[30] Dai HJ, Wong EW, Lieber CM. Probing electrical transport in nanomaterials: conductivity of individual carbon nanotubes. Science 1996;272:523–6.
[31] Kociak M, Suenaga K, Hirahara K, Saito Y, Nakahira T, Iijima S. Linking chiral indices and transport properties of double-walled carbon nanotubes. Phys Rev Lett 2002;89:155501.
[32] Wang S, Liang XL, Chen Q, Zhang ZY, Peng LM. Field-effect characteristics and screening in double-walled carbon nanotube field-effect transistors. J Phys Chem B 2005;109:17361–5.
[33] Shimada T, Sugai T, Ohno Y, Kishimoto S, Mizutani T, Yoshida H, et al. Double-wall carbon nanotube field-effect transistors: ambipolar transport characteristics. Appl Phys Lett 2004;84:2412–4.
[34] Tison Y, Giusca CE, Stolojan V, Hayashi Y, Silva SRP. The inner shell influence on the electronic structure of double-walled carbon nanotubes. Adv Mater 2008;20:189.
[35] Choi HJ, Ihm J, Louie SG, Cohen ML. Defects, quasibound states, and quantum conductance in metallic carbon nanotubes. Phys Rev Lett 2000;84:2917–20.
[36] Choi HJ, Ihm J. Ab initio pseudopotential method for the calculation of conductance in quantum wires. Phys Rev B 1999;59:2267–75.

[37] Rochefort A, Avouris P, Lesage F, Salahub DR. Electrical and mechanical properties of distorted carbon nanotubes. Phys Rev B 1999;60:13824–30.

[38] Tans SJ, Verschueren ARM, Dekker C. Room-temperature transistor based on a single carbon nanotube. Nature 1998;393:49–52.

[39] Collins PG, Bradley K, Ishigami M, Zettl A. Extreme oxygen sensitivity of electronic properties of carbon nanotubes. Science 2000;287:1801–4.

[40] Kong J, Franklin NR, Zhou CW, Chapline MG, Peng S, Cho KJ, et al. Nanotube molecular wires as chemical sensors. Science 2000;287:622–5.

[41] Rao AM, Eklund PC, Bandow S, Thess A, Smalley RE. Evidence for charge transfer in doped carbon nanotube bundles from Raman scattering. Nature 1997;388:257–9.

[42] Derycke V, Martel R, Appenzeller J, Avouris P. Controlling doping and carrier injection in carbon nanotube transistors. Appl Phys Lett 2002;80:2773–5.

[43] Lu JP. Elastic properties of carbon nanotubes and nanoropes. Phys Rev Lett 1997;79: 1297–300.

[44] Yakobson BI, Campbell MP, Brabec CJ, Bernholc J. High strain rate fracture and C-chain unraveling in carbon nanotubes. Comp Mater Sci 1997;8:341–8.

[45] Treacy MMJ, Ebbesen TW, Gibson JM. Exceptionally high Young's modulus observed for individual carbon nanotubes. Nature 1996;381:678–80.

[46] Krishnan A, Dujardin E, Ebbesen TW, Yianilos PN, Treacy MMJ. Young's modulus of single-walled nanotubes. Phys Rev B 1998;58:14013–9.

[47] Chopra NG, Zettl A. Measurement of the elastic modulus of a multi-wall boron nitride nanotube. Solid State Commun 1998;105:297–300.

[48] Wong EW, Sheehan PE, Lieber CM. Nanobeam mechanics: elasticity, strength, and toughness of nanorods and nanotubes. Science 1997;277:1971–5.

[49] Falvo MR, Clary GJ, Taylor RM, Chi V, Brooks FP, Washburn S, et al. Bending and buckling of carbon nanotubes under large strain. Nature 1997;389:582–4.

[50] Yu MF, Lourie O, Dyer MJ, Moloni K, Kelly TF, Ruoff RS. Strength and breaking mechanism of multiwalled carbon nanotubes under tensile load. Science 2000;287:637–40.

[51] Yu MF, Files BS, Arepalli S, Ruoff RS. Tensile loading of ropes of single wall carbon nanotubes and their mechanical properties. Phys Rev Lett 2000;84:5552–5.

[52] Wagner HD, Lourie O, Feldman Y, Tenne R. Stress-induced fragmentation of multiwall carbon nanotubes in a polymer matrix. Appl Phys Lett 1998;72:188–90.

[53] Peng B, Locascio M, Zapol P, Li S, Mielke SL, Schatz GC, et al. Measurements of near-ultimate strength for multiwalled carbon nanotubes and irradiation-induced crosslinking improvements. Nat Nanotechnol 2008;3:626–31.

[54] Poncharal P, Wang ZL, Ugarte D, de Heer WA. Electrostatic deflections and electromechanical resonances of carbon nanotubes. Science 1999;283:1513–6.

[55] Venkateswaran UD, Rao AM, Richter E, Menon M, Rinzler A, Smalley RE, et al. Probing the single-wall carbon nanotube bundle: Raman scattering under high pressure. Phys Rev B 1999;59:10928–34.

[56] Reich S, Jantoljak H, Thomsen C. Shear strain in carbon nanotubes under hydrostatic pressure. Phys Rev B 2000;61:13389–92.

[57] Wei X, Chen Q, Peng L-M, Cui R, Li Y. Tensile loading of double-walled and triple-walled carbon nanotubes and their mechanical properties. J Phys Chem C 2009;113:17002–5.

[58] Kong J, Yenilmez E, Tombler TW, Kim W, Dai H, Laughlin RB, et al. Quantum interference and ballistic transmission in nanotube electron waveguides. Phys Rev Lett 2001;87:106801.

[59] White CT, Todorov TN. Carbon nanotubes as long ballistic conductors. Nature 1998;393: 240–2.

[60] Méténier K, Bonnamy S, Béguin F, Journet C, Bernier P, Lamy de La Chapelle M, et al. Coalescence of single-walled carbon nanotubes and formation of multi-walled carbon nanotubes under high-temperature treatments. Carbon 2002;40:1765–73.

[61] Thostenson ET, Li C, Chou T-W. Nanocomposites in context. Compos Sci Technol 2005;65:491–516.

[62] Pettes MT, Shi L. Thermal and structural characterizations of individual single-, double-, and multi-walled carbon nanotubes. Adv Funct Mater 2009;19:3918–25.

[63] Li HJ, Lu WG, Li JJ, Bai XD, Gu CZ. Multichannel ballistic transport in multiwall carbon nanotubes. Phys Rev Lett 2005;95:086601.

[64] Shi W, Wang Z, Zhang Q, Zheng Y, Ieong C, He M, et al. Superconductivity in bundles of double-wall carbon nanotubes. Sci Rep 2012;2:625.

[65] Kim YA, Muramatsu H, Hayashi T, Endo M, Terrones M, Dresselhaus MS. Thermal stability and structural changes of double-walled carbon nanotubes by heat treatment. Chem Phys Lett 2004;398:87–92.

[66] Liu Q, Ren W, Li F, Cong H, Cheng H-M. Synthesis and high thermal stability of double-walled carbon nanotubes using nickel formate dihydrate as catalyst precursor. J Phys Chem C 2007;111:5006–13.

[67] Kim P, Shi L, Majumdar A, McEuen PL. Thermal transport measurements of individual multiwalled nanotubes. Phys Rev Lett 2001;87:215502.

[68] Benedict LX, Louie SG, Cohen ML. Heat capacity of carbon nanotubes. Solid State Commun 1996;100:177–80.

[69] Yi W, Lu L, Zhang DL, Pan ZW, Xie SS. Linear specific heat of carbon nanotubes. Phys Rev B 1999;59:R9015–8.

[70] Hone J, Batlogg B, Benes Z, Johnson AT, Fischer JE. Quantized phonon spectrum of single-wall carbon nanotubes. Science 2000;289:1730–3.

[71] Pop E, Mann D, Wang Q, Goodson KE, Dai HJ. Thermal conductance of an individual single-wall carbon nanotube above room temperature. Nano Lett 2006;6:96–100.

[72] Berber S, Kwon YK, Tomanek D. Unusually high thermal conductivity of carbon nanotubes. Phys Rev Lett 2000;84:4613–6.

[73] Pettes MT, Shi L. Thermal and structural characterizations of individual single-, double-, and multi-walled carbon nanotubes. Adv Funct Mater 2009;19:3918–25.

[74] Maniwa Y, Fujiwara R, Kira H, Tou H, Kataura H, Suzuki S, et al. Thermal expansion of single-walled carbon nanotube (SWNT) bundles: X-ray diffraction studies. Phys Rev B 2001;64:241402.

[75] Hone J, Ellwood I, Muno M, Mizel A, Cohen ML, Zettl A, et al. Thermoelectric power of single-walled carbon nanotubes. Phys Rev Lett 1998;80:1042–5.

[76] Tian ML, Li FQ, Chen L, Mao ZQ, Zhang YH. Thermoelectric power behavior in carbon nanotubule bundles from 4.2 to 300 K. Phys Rev B 1998;58:1166–8.

[77] O'Connell MJ, Bachilo SM, Huffman CB, Moore VC, Strano MS, Haroz EH, et al. Band gap fluorescence from individual single-walled carbon nanotubes. Science 2002;297:593–6.

[78] Bachilo SM, Strano MS, Kittrell C, Hauge RH, Smalley RE, Weisman RB. Structure-assigned optical spectra of single-walled carbon nanotubes. Science 2002;298:2361–6.

[79] Wang F, Dukovic G, Brus LE, Heinz TF. The optical resonances in carbon nanotubes arise from excitons. Science 2005;308:838–41.

[80] Green AA, Hersam MC. Processing and properties of highly enriched double-wall carbon nanotubes. Nat Nanotechnol 2009;4:64–70.

[81] Tsyboulski DA, Rocha J-DR, Bachilo SM, Cognet L, Weisman RB. Structure-dependent fluorescence efficiencies of individual single-walled carbon nanotubes. Nano Lett 2007;7:3080–5.

[82] Carlson LJ, Maccagnano SE, Zheng M, Silcox J, Krauss TD. Fluorescence efficiency of individual carbon nanotubes. Nano Lett 2007;7:3698–703.
[83] Wang F, Dukovic G, Brus LE, Heinz TF. Time-resolved fluorescence of carbon nanotubes and its implication for radiative lifetimes. Phys Rev Lett 2004;92:177401.
[84] Lefebvre J, Homma Y, Finnie P. Bright band gap photoluminescence from unprocessed single-walled carbon nanotubes. Phys Rev Lett 2003;90:217401.
[85] Piao YM, Meany B, Powell LR, Valley N, Kwon H, Schatz GC, Wang Y. Brightening of carbon nanotube photoluminescence through the incorporation of sp^3 defects. Nat Chem 2013;5:840–45.
[86] Wang F, Dukovic G, Knoesel E, Brus LE, Heinz TF. Observation of rapid Auger recombination in optically excited semiconducting carbon nanotubes. Phys Rev B 2004;70:241403.
[87] Ma YZ, Valkunas L, Dexheimer SL, Bachilo SM, Fleming GR. Femtosecond spectroscopy of optical excitations in single-walled carbon nanotubes: evidence for exciton-exciton annihilation. Phys Rev Lett 2005;94:157402.
[88] Perebeinos V, Avouris P. Phonon and electronic nonradiative decay mechanisms of excitons in carbon nanotubes. Phys Rev Lett 2008;101:057401.
[89] Crochet JJ, Duque JG, Werner JH, Doorn SK. Photoluminescence imaging of electronic-impurity-induced exciton quenching in single-walled carbon nanotubes. Nat Nanotechnol 2012;7:126–32.
[90] Yang S, Parks AN, Saba SA, Ferguson PL, Liu J. Photoluminescence from inner walls in double-walled carbon nanotubes: some do, some do not. Nano Lett 2011;11:4405–10.
[91] Ando T. Excitons in carbon nanotubes. J Phys Soc Jpn 1997;66:1066–73.
[92] Htoon H, O'Connell MJ, Cox PJ, Doorn SK, Klimov VI. Low temperature emission spectra of individual single-walled carbon nanotubes: multiplicity of subspecies within single-species nanotube ensembles. Phys Rev Lett 2004;93:027401.
[93] Ma YZ, Stenger J, Zimmermann J, Bachilo SM, Smalley RE, Weisman RB, et al. Ultrafast carrier dynamics in single-walled carbon nanotubes probed by femtosecond spectroscopy. J Chem Phys 2004;120:3368–73.
[94] Wang F, Cho DJ, Kessler B, Deslippe J, Schuck PJ, Louie SG, et al. Observation of excitons in one-dimensional metallic single-walled carbon nanotubes. Phys Rev Lett 2007;99:227401.
[95] Strano MS, Doorn SK, Haroz EH, Kittrell C, Hauge RH, Smalley RE. Assignment of (n, m) Raman and optical features of metallic single-walled carbon nanotubes. Nano Lett 2003;3:1091–6.
[96] Benedict LX, Crespi VH, Louie SG, Cohen ML. Static conductivity and superconductivity of carbon nanotubes: relations between tubes and sheets. Phys Rev B 1995;52:14935–40.
[97] Tinkham M. Introduction to superconductivity. Mineola, NY: Dover Publications Inc.; 2004.
[98] Morpurgo AF, Kong J, Marcus CM, Dai H. Gate-controlled superconducting proximity effect in carbon nanotubes. Science 1999;286:263–5.
[99] Sedeki A, Caron LG, Bourbonnais C. Superconductivity in armchair carbon nanotubes. Phys Rev B 2002;65:041406.
[100] Kociak M, Kasumov AY, Gueron S, Reulet B, Khodos II, Gorbatov YB, et al. Superconductivity in ropes of single-walled carbon nanotubes. Phys Rev Lett 2001;86:2416–9.
[101] Tang ZK, Zhang LY, Wang N, Zhang XX, Wen GH, Li GD, et al. Superconductivity in 4 angstrom single-walled carbon nanotubes. Science 2001;292:2462–5.
[102] Takesue I, Haruyama J, Kobayashi N, Chiashi S, Maruyama S, Sugai T, et al. Superconductivity in entirely end-bonded multiwalled carbon nanotubes. Phys Rev Lett 2006;96:057001.
[103] Bandaru PR. Electrical properties and applications of carbon nanotube structures. J Nanosci Nanotechnol 2007;7:1239–67.

[104] Cao Q, Kim HS, Pimparkar N, Kulkarni JP, Wang CJ, Shim M, et al. Medium-scale carbon nanotube thin-film integrated circuits on flexible plastic substrates. Nature 2008;454:495.

[105] Rinzler AG, Liu J, Dai H, Nikolaev P, Huffman CB, Rodriguez-Macias FJ, et al. Large-scale purification of single-wall carbon nanotubes: process, product, and characterization. Appl Phys A Mater Sci Process 1998;67:29–37.

[106] Jiang KL, Li QQ, Fan SS. Spinning continuous carbon nanotube yarns. Nature 2002;419:801.

[107] Zhang M, Atkinson KR, Baughman RH. Multifunctional carbon nanotube yarns by downsizing an ancient technology. Science 2004;306:1358–61.

[108] Zhang M, Fang SL, Zakhidov AA, Lee SB, Aliev AE, Williams CD, et al. Strong, transparent, multifunctional, carbon nanotube sheets. Science 2005;309:1215–9.

[109] Franklin NR, Li YM, Chen RJ, Javey A, Dai HJ. Patterned growth of single-walled carbon nanotubes on full 4-inch wafers. Appl Phys Lett 2001;79:4571–3.

[110] Rao SG, Huang L, Setyawan W, Hong SH. Large-scale assembly of carbon nanotubes. Nature 2003;425:36–7.

[111] Wang YH, Maspoch D, Zou SL, Schatz GC, Smalley RE, Mirkin CA. Controlling the shape, orientation, and linkage of carbon nanotube features with nano affinity templates. PNAS 2006;103:2026–31.

[112] Liu J, Casavant MJ, Cox M, Walters DA, Boul P, Lu W, et al. Controlled deposition of individual single-walled carbon nanotubes on chemically functionalized templates. Chem Phys Lett 1999;303:125–9.

[113] Meitl MA, Zhou YX, Gaur A, Jeon S, Usrey ML, Strano MS, et al. Solution casting and transfer printing single-walled carbon nanotube films. Nano Lett 2004;4:1643–7.

[114] Salvetat JP, Briggs GAD, Bonard JM, Bacsa RR, Kulik AJ, Stockli T, et al. Elastic and shear moduli of single-walled carbon nanotube ropes. Phys Rev Lett 1999;82:944–7.

[115] Wu ZC, Chen ZH, Du X, Logan JM, Sippel J, Nikolou M, et al. Transparent, conductive carbon nanotube films. Science 2004;305:1273–6.

[116] Bekyarova E, Itkis ME, Cabrera N, Zhao B, Yu AP, Gao JB, et al. Electronic properties of single-walled carbon nanotube networks. J Am Chem Soc 2005;127:5990–5.

[117] Hone J, Llaguno MC, Nemes NM, Johnson AT, Fischer JE, Walters DA, et al. Electrical and thermal transport properties of magnetically aligned single wall carbon nanotube films. Appl Phys Lett 2000;77:666–8.

[118] Ma WJ, Song L, Yang R, Zhang TH, Zhao YC, Sun LF, et al. Directly synthesized strong, highly conducting, transparent single-walled carbon nanotube films. Nano Lett 2007;7:2307–11.

[119] Liao H, Karki K, Zhang Y, Cumings J, Wang Y. Interfacial mechanics of carbon nanotube@amorphous-Si coaxial nanostructures. Adv Mater 2011;23:4318–22.

[120] Andrews R, Jacques D, Rao AM, Rantell T, Derbyshire F, Chen Y, et al. Nanotube composite carbon fibers. Appl Phys Lett 1999;75:1329–31.

[121] Biercuk MJ, Llaguno MC, Radosavljevic M, Hyun JK, Johnson AT, Fischer JE. Carbon nanotube composites for thermal management. Appl Phys Lett 2002;80:2767–9.

[122] Qian D, Dickey EC, Andrews R, Rantell T. Load transfer and deformation mechanisms in carbon nanotube-polystyrene composites. Appl Phys Lett 2000;76:2868–70.

[123] The website of Hyperion Catalysis International, Inc. is available at www.fibrils.com.

[124] Glatkowski, P. (24 July 2000). U.S. Patent 6265466.

[125] Yamada T, Hayamizu Y, Yamamoto Y, Yomogida Y, Izadi-Najafabadi A, Futaba DN, et al. A stretchable carbon nanotube strain sensor for human-motion detection. Nat Nanotechnol 2011;6:296–301.

[126] Chaste J, Eichler A, Moser J, Ceballos G, Rurali R, Bachtold A. A nanomechanical mass sensor with yoctogram resolution. Nat Nanotechnol 2012;7:300–3.

[127] Aliev AE, Oh J, Kozlov ME, Kuznetsov AA, Fang S, Fonseca AF, et al. Giant-stroke, superelastic carbon nanotube aerogel muscles. Science 2009;323:1575–8.

[128] Foroughi J, Spinks GM, Wallace GG, Oh J, Kozlov ME, Fang S, et al. Torsional carbon nanotube artificial muscles. Science 2011;334:494–7.

[129] Choi W, Hong S, Abrahamson JT, Han J-H, Song C, Nair N, et al. Chemically driven carbon-nanotube-guided thermopower waves. Nat Mater 2010;9:423–9.

[130] Baloch KH, Voskanian N, Bronsgeest M, Cumings J. Remote Joule heating by a carbon nanotube. Nat Nanotechnol 2012;7:315–8.

[131] Misewich JA, Martel R, Avouris P, Tsang JC, Heinze S, Tersoff J. Electrically induced optical emission from a carbon nanotube FET. Science 2003;300:783–6.

[132] Lee JU, Gipp PP, Heller CM. Carbon nanotube p-n junction diodes. Appl Phys Lett 2004;85:145–7.

[133] Mueller T, Kinoshita M, Steiner M, Perebeinos V, Bol AA, Farmer DB, et al. Efficient narrow-band light emission from a single carbon nanotube p-n diode. Nat Nanotechnol 2010;5:27–31.

[134] Mann D, Kato YK, Kinkhabwala A, Pop E, Cao J, Wang X, et al. Electrically driven thermal light emission from individual single-walled carbon nanotubes. Nat Nanotechnol 2007;2:33–8.

[135] Sveningsson M, Jonsson M, Nerushev OA, Rohmund F, Campbell EEB. Blackbody radiation from resistively heated multiwalled carbon nanotubes during field emission. Appl Phys Lett 2002;81:1095–7.

[136] Joh DY, Kinder J, Herman LH, Ju S-Y, Segal MA, Johnson JN, et al. Single-walled carbon nanotubes as excitonic optical wires. Nat Nanotechnol 2011;6:51–6.

[137] Pris AD, Utturkar Y, Surman C, Morris WG, Vert A, Zalyubovskiy S, et al. Towards high-speed imaging of infrared photons with bio-inspired nanoarchitectures. Nat Photonics 2012;6:195–200.

[138] Rybczynski J, Kempa K, Herczynski A, Wang Y, Naughton MJ, Ren ZF, et al. Subwavelength waveguide for visible light. Appl Phys Lett 2007;90:021104.

[139] Freitag M, Martin Y, Misewich JA, Martel R, Avouris PH. Photoconductivity of single carbon nanotubes. Nano Lett 2003;3:1067–71.

[140] Qiu XH, Freitag M, Perebeinos V, Avouris P. Photoconductivity spectra of single-carbon nanotubes: implications on the nature of their excited states. Nano Lett 2005;5:749–52.

[141] Lee JU. Photovoltaic effect in ideal carbon nanotube diodes. Appl Phys Lett 2005;87:073101.

[142] Wu Z, Chen Z, Du X, Logan JM, Sippel J, Nikolou M, et al. Transparent, conductive carbon nanotube films. Science 2004;305:1273–6.

[143] Zhang D, Ryu K, Liu X, Polikarpov E, Ly J, Tompson ME, et al. Transparent, conductive, and flexible carbon nanotube films and their application in organic light-emitting diodes. Nano Lett 2006;6:1880–6.

[144] Sekitani T, Nakajima H, Maeda H, Fukushima T, Aida T, Hata K, et al. Stretchable active-matrix organic light-emitting diode display using printable elastic conductors. Nat Mater 2009;8:494–9.

[145] McCarthy MA, Liu B, Donoghue EP, Kravchenko I, Kim DY, So F, et al. Low-voltage, low-power, organic light-emitting transistors for active matrix displays. Science 2011;332:570–3.

[146] Bachtold A, Hadley P, Nakanishi T, Dekker C. Logic circuits with carbon nanotube transistors. Science 2001;294:1317–20.

[147] Collins PG, Arnold MS, Avouris P. Engineering carbon nanotubes and nanotube circuits using electrical breakdown. Science 2001;292:706–9.

[148] Chen ZH, Appenzeller J, Lin YM, Sippel-Oakley J, Rinzler AG, Tang JY, et al. An integrated logic circuit assembled on a single carbon nanotube. Science 2006;311:1735–11735.

[149] Feng C, Liu K, Wu J-S, Liu L, Cheng J-S, Zhang Y, et al. Flexible, stretchable, transparent conducting films made from superaligned carbon nanotubes. Adv Funct Mater 2010;20:885–91.

[150] de Heer WA, Chatelain A, Ugarte D. A carbon nanotube field-emission electron source. Science 1995;270:1179–80.

[151] Rinzler AG, Hafner JH, Nikolaev P, Lou L, Kim SG, Tomanek D, et al. Unraveling nanotubes: field emission from an atomic wire. Science 1995;269:1550–3.

[152] Lee NS, Chung DS, Han IT, Kang JH, Choi YS, Kim HY, et al. Application of carbon nanotubes to field emission displays. Diam Relat Mater 2001;10:265–70.

[153] Saito Y, Uemura S. Field emission from carbon nanotubes and its application to electron sources. Carbon 2000;38:169–82.

[154] Rosen R, Simendinger W, Debbault C, Shimoda H, Fleming L, Stoner B, et al. Application of carbon nanotubes as electrodes in gas discharge tubes. Appl Phys Lett 2000;76:1668–70.

[155] Sugie H, Tanemura M, Filip V, Iwata K, Takahashi K, Okuyama F. Carbon nanotubes as electron source in an x-ray tube. Appl Phys Lett 2001;78:2578–80.

[156] Kwo JL, Tsou CC, Yokoyama M, Lin IN, Lee CC, Wang WC, et al. Field emission characteristics of carbon nanotube emitters synthesized by arc discharge. J Vacuum Sci Technol B 2001;19:23–6.

[157] Zhu W, Bower C, Zhou O, Kochanski G, Jin S. Large current density from carbon nanotube field emitters. Appl Phys Lett 1999;75:873–5.

[158] The website of Unidym. http://www.unidym.com/press/pr_081016.html.

[159] Javey A, Guo J, Wang Q, Lundstrom M, Dai HJ. Ballistic carbon nanotube field-effect transistors. Nature 2003;424:654–7.

[160] Naeemi A, Sarvati R, Meindl JD. Performance comparison between nanotubes and copper interconnects for GSI. IEDM Digest 2004;16:699–702.

[161] Martel R, Schmidt T, Shea HR, Hertel T, Avouris P. Single- and multi-wall carbon nanotube field-effect transistors. Appl Phys Lett 1998;73:2447–9.

[162] Fuhrer MS, Nygard J, Shih L, Forero M, Yoon YG, Mazzoni MSC, et al. Crossed nanotube junctions. Science 2000;288:494–7.

[163] Postma HWC, Teepen T, Yao Z, Grifoni M, Dekker C. Carbon nanotube single-electron transistors at room temperature. Science 2001;293:76–9.

[164] Yao Z, Postma HWC, Balents L, Dekker C. Carbon nanotube intramolecular junctions. Nature 1999;402:273–6.

[165] Rueckes T, Kim K, Joselevich E, Tseng GY, Cheung CL, Lieber CM. Carbon nanotube-based nonvolatile random access memory for molecular computing. Science 2000;289:94–7.

[166] Sze SM. Physics of semiconductor devices. New York: Wiley; 1981.

[167] Muller RS, Kamins TI. Device electronics for integrated circuits. New York: John Wiley; 1986.

[168] Derycke V, Martel R, Appenzeller J, Avouris P. Carbon nanotube inter- and intramolecular logic gates. Nano Lett 2001;1:453–6.

[169] Javey A, Kim H, Brink M, Wang Q, Ural A, Guo J, et al. High-kappa dielectrics for advanced carbon-nanotube transistors and logic gates. Nat Mater 2002;1:241–6.

[170] Javey A, Wang Q, Ural A, Li YM, Dai HJ. Carbon nanotube transistor arrays for multistage complementary logic and ring oscillators. Nano Lett 2002;2:929–32.

[171] Modi A, Koratkar N, Lass E, Wei BQ, Ajayan PM. Miniaturized gas ionization sensors using carbon nanotubes. Nature 2003;424:171–4.

[172] Kauffman DR, Star A. Carbon nanotube gas and vapor sensors. Angew Chem Int Ed 2008;47:6550–70.

[173] Cao Q, Rogers JA. Ultrathin films of single-walled carbon nanotubes for electronics and sensors: a review of fundamental and applied aspects. Adv Mater 2009;21:29–53.

[174] Lee CY, Choi W, Han J-H, Strano MS. Coherence resonance in a single-walled carbon nanotube ion channel. Science 2010;329:1320–4.

[175] Choi Y, Moody IS, Sims PC, Hunt SR, Corso BL, Perez I, et al. Single-molecule lysozyme dynamics monitored by an electronic circuit. Science 2012;335:319–24.

[176] Piao Y, Chen C-F, Green AA, Kwon H, Hersam MC, Lee CS, et al. Optical and electrical properties of inner tubes in outer wall-selectively functionalized double-wall carbon nanotubes. J Phys Chem Lett 2011;2:1577–82.

[177] Martin CR, Kohli P. The emerging field of nanotube biotechnology. Nat Rev Drug Discov 2003;2:29–37.

[178] Lin YH, Lu F, Tu Y, Ren ZF. Glucose biosensors based on carbon nanotube nanoelectrode ensembles. Nano Lett 2004;4:191–5.

[179] Dai HJ, Hafner JH, Rinzler AG, Colbert DT, Smalley RE. Nanotubes as nanoprobes in scanning probe microscopy. Nature 1996;384:147–50.

[180] Hafner JH, Cheung CL, Lieber CM. Growth of nanotubes for probe microscopy tips. Nature 1999;398:761–2.

[181] Wong SS, Joselevich E, Woolley AT, Cheung CL, Lieber CM. Covalently functionalized nanotubes as nanometre-sized probes in chemistry and biology. Nature 1998;394:52–5.

[182] Gao B, Kleinhammes A, Tang XP, Bower C, Fleming L, Wu Y, et al. Electrochemical intercalation of single-walled carbon nanotubes with lithium. Chem Phys Lett 1999;307:153–7.

[183] SalverDisma F, Lenain C, Beaudoin B, Aymard L, Tarascon JM. Unique effect of mechanical milling on the lithium intercalation properties of different carbons. Solid State Ion 1997;98:145–58.

[184] An KH, Kim WS, Park YS, Moon JM, Bae DJ, Lim SC, et al. Electrochemical properties of high-power supercapacitors using single-walled carbon nanotube electrodes. Adv Funct Mater 2001;11:387–92.

[185] Niu CM, Sichel EK, Hoch R, Moy D, Tennent H. High power electrochemical capacitors based on carbon nanotube electrodes. Appl Phys Lett 1997;70:1480–2.

[186] Du C, Pan N. Supercapacitors using carbon nanotubes films by electrophoretic deposition. J Power Sources 2006;160:1487–94.

[187] Baughman RH, Cui CX, Zakhidov AA, Iqbal Z, Barisci JN, Spinks GM, et al. Carbon nanotube actuators. Science 1999;284:1340–4.

[188] Fennimore AM, Yuzvinsky TD, Han WQ, Fuhrer MS, Cumings J, Zettl A. Rotational actuators based on carbon nanotubes. Nature 2003;424:408–10.

[189] Li C, Thostenson ET, Chou T-W. Sensors and actuators based on carbon nanotubes and their composites: a review. Compos Sci Technol 2008;68:1227–49.

[190] Li J, Ma W, Song L, Niu Z, Cai L, Zeng Q, et al. Superfast-response and ultrahigh-power-density electromechanical actuators based on hierarchal carbon nanotube electrodes and chitosan. Nano Lett 2011;11:4636–41.

[191] Hu L, Wu H, Gao Y, Cao A, Li H, McDough J, et al. Silicon-carbon nanotube coaxial sponge as Li-ion anodes with high areal capacity. Adv Energ Mater 2011;1:523–7.

[192] Reddy ALM, Shaijumon MM, Gowda SR, Ajayan PM. Coaxial MnO_2/carbon nanotube array electrodes for high-performance lithium batteries. Nano Lett 2009;9:1002–6.

[193] Wang YH, Kim MJ, Shan HW, Kittrell C, Fan H, Ericson LM, et al. Continued growth of single-walled carbon nanotubes. Nano Lett 2005;5:997–1002.

[194] Liu J, Wang C, Tu X, Liu B, Chen L, Zheng M, et al. Chirality-controlled synthesis of single-wall carbon nanotubes using vapour-phase epitaxy. Nat Commun 2012;3:1199.

Chapter 14

Graphene Oxide: Some New Insights into an Old Material

Kalyan Raidongia, Alvin T.L. Tan, Jiaxing Huang

Department of Materials Science and Engineering, Northwestern University, 2220 Campus Drive, Evanston, IL 60208, USA

14.1 INTRODUCTION

Graphene oxide (GO), sheets of what is formerly known as graphite oxide or graphitic acid, is the oxidation product of graphite [1–4]. The oxidation of graphite has been studied for over a century, with the first instance carried out by B.C. Brodie in 1859 [1]. Typically, graphite reacts with strong oxidizing agents, such as $KMnO_4$ as reported first by Hummers [3], and concentrated sulphuric acid, followed by purification and exfoliation in water, which results in a yellow colloidal dispersion (Figure 14.1(a)). Following the landmark discovery of graphene in 2004 [5,6], there has been resurgent interest in GO as a promising precursor for the large-scale chemical production of graphene [7–14]. GO, like graphene, comprises single-layer sheets of carbon. However, the chemical and physical properties of GO are markedly different from those of graphene. Unlike graphene (Figure 14.1(c), left), GO possesses a large number of oxygen-containing functional groups (Figure 14.1(c), right), which allows GO to be dispersed in solution [8, 15–18]. Furthermore, unlike graphene, which is an excellent electrical conductor, GO is insulating. Therefore, to restore some of the favourable electrical, mechanical and thermal properties of graphene, GO is commonly reduced by thermal annealing or chemical reducing agents [7–14]. The reduced product of GO is commonly termed as reduced graphene oxide (r-GO). Presently, the electrical conductivity of r-GO cannot match up to that of pristine graphene because the history of chemical treatment in r-GO renders its graphitic structure defective [8,19,20]. The disordered structure of GO can be seen with high-resolution transmission electron microscopy (TEM) (Figure 14.1(d)) [21].

The in-plane heterogeneity of GO and r-GO renders them very different chemical reactivity compared to graphene. For example, if graphene is mildly heated in air, combustion of carbon atoms would initiate from the edges given relatively smaller number of additional defect sites in the basal plane, then

Carbon Nanotubes and Graphene. DOI: 10.1016/B978-0-08-098232-8.00014-0

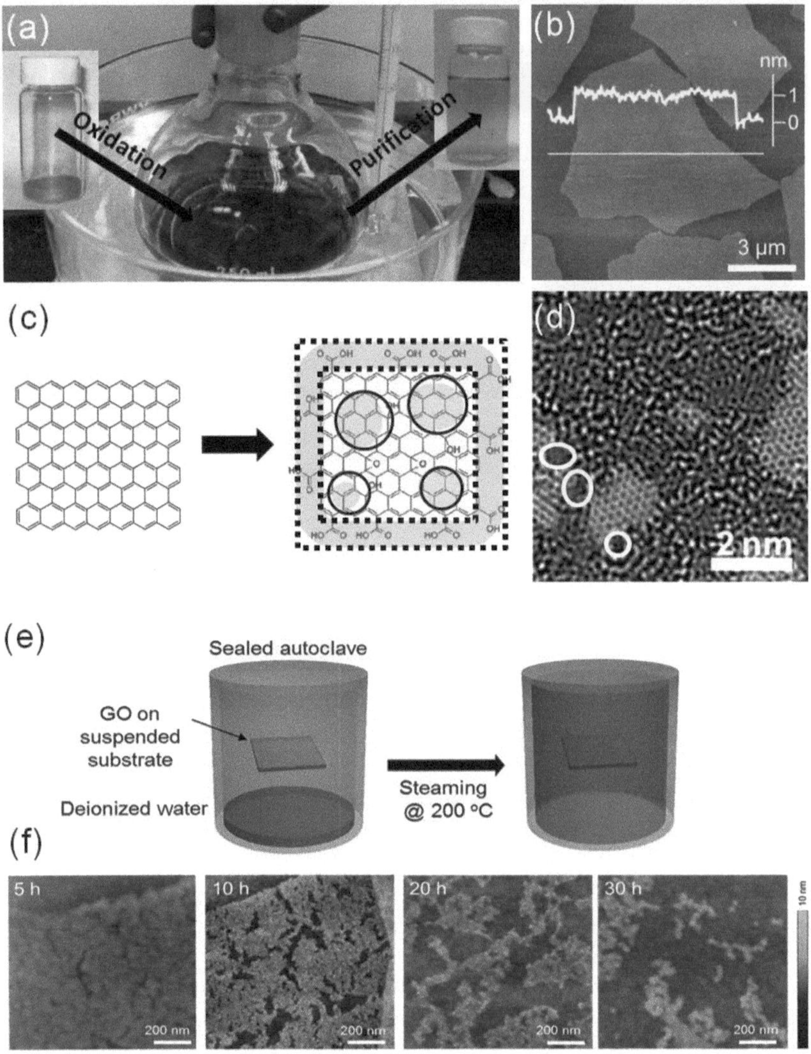

FIGURE 14.1 (a) Synthesis of GO: GO is typically synthesized by reacting graphite powder with strong oxidizing agents, such as $KMnO_4$ in concentrated H_2SO_4, followed by purification and exfoliation in water to yield a colloidal dispersion of single layers, as shown in the AFM image in (b). (c) Structural model of graphene (left) and GO (right). GO has hydrophilic (dotted) edges and a largely hydrophobic (solid circle) basal plane. (d) High-resolution TEM images showing the atomic structures of GO [21]. The bright, dark and circled areas depict ordered graphitic sp^2 domains, disordered highly oxidized sp^3 domains and holes on the sheet, respectively. (e) Steam etching turns GO into a nanoporous network, as shown in the (f) AFM images of GO after steaming for 5, 10, 20 and 30 hours, respectively. [(c)-(d) and (e)-(f) adapted and modified from [21] and [22], with permissions from Wiley-VCH and ACS Publications].

gradually shrinking the sheet before its complete disappearance [22,23]. However, for GO or r-GO, since they contain abundant defect sites distributed throughout the basal plane, gasification of carbon atoms would occur homogeneously across the full face of the sheet, leading to rapid fracturing into small pieces. For example, when GO sheets deposited on silicon wafer are steamed as shown in the schematic drawing in Figure 14.1(e), the initially continuous GO sheet would first develop pores and cracks, then turn into a network of interconnecting nano-sized particles, which finally broke into isolated dots upon further steaming (see atomic force microscopy (AFM) images in Figure 14.1(f)) [22]. While graphene consists of ordered sp^2-bonded carbon, the structure of GO consists of a majority of disordered sp^3-bonded carbon and a minority of ordered sp^2-bonded carbon. The reduction process can convert some of the sp^3-bonded carbon to sp^2-bonded carbon, but the structure of pristine graphene is seldom fully recovered in r-GO. Nevertheless, the conductivity of r-GO is sufficient for many applications.

Beyond being a precursor to graphene, GO is an interesting material in its own right [24–27]. The abundance of chemical functionalities in GO sheets makes these sheets a highly versatile chemical platform for creating interesting graphene-based materials. Furthermore, with two abruptly different length scales, GO is an intermediate between materials and molecules. As shown in the AFM height profile (Figure 14.1(b)), the thickness of GO sheets is in the order of 1 nm, which is the typical length scale of molecules. On the other hand, the lateral length of GO is in the order of micrometres, which is the typical length scale of colloidal particles. Therefore, depending on the length scale in question, GO can be treated as a molecule or a colloidal sheet. This molecule–particle duality [24,26] makes GO interesting for both chemists and materials scientists. For example, we first proposed to view GO from a soft materials perspective and observed many interesting phenomena associated with it [24,26,27]. In this chapter, we present some new insights into the synthesis, characterization, processing–structure–property relationship and some new applications of this century-old material.

14.2 SYNTHESIS OF GO: A FIRE HAZARD, OR A FLAME RETARDANT?

GO has classically been seen as an energetically unstable fire hazard. For example, in Brodie's original report published in 1859, he described that GO 'decomposed with ignition' [1]. Croft and Jimenez, among many others, also described in 1960 and 1987, respectively, the 'deflagration' of GO [4,28]. However, the combustion of GO has not been noted as much in recent literature. On the contrary, GO has been described as a flame-retardant adduct for polymeric materials [29–34]. In an attempt to reconcile these apparently contradictory observations, we have found that alkali salts introduced during the oxidation process causes GO to be energetically unstable [35,36]. This insight prompted us to develop a procedure for efficiently removing these salts, thus removing the fire hazard and improving the overall process for the synthesis of GO. On the other hand, energetically unstable GO can be exploited. Using pulses of light, GO can be reduced to r-GO, resulting in a technique for creating graphene–polymer composites in a rapid, clean and energy-efficient fashion [37].

14.2.1 Thermal instability of GO

GO is comprises both sp^2 and sp^3 hybridized carbon atoms. The partially oxidized sp^3 carbon atoms of GO are highly energetic and undergo disproportionation reactions, even at room temperature. During the disproportionation reaction, some of the partially oxidized carbon atoms are graphitized to sp^2 carbon atoms, while others are released as carbonaceous gases, such as CO_2. This is a highly exothermic process and can lead to explosion upon heating. The explosion of GO upon heating has been commonly observed from the beginning of its history [1,4,38]. Figure 14.2(a) illustrates one such explosion of GO, similar to what has been noted in various literature reports. A solid chunk of GO, covered with a glass Petri dish was placed on a preheated hotplate at 300 °C. The GO exploded within a few seconds by releasing lots of gaseous species, such as H_2O and CO_2, yielding the lightweight, fluffy black r-GO product. Before explosion, water molecules evaporated from the GO chunk and condensed on the top cover, as marked by the highlighted white broken circles in Figure 14.2(a). These photographs taken by a high-speed camera also reveal the red-hot transient state of the GO at the beginning of explosion. The thermal deoxygenation of GO was also studied in differential scanning calorimetry (DSC). The DSC study revealed that the energy released during the thermal reduction (6–8 kJ g^{-1}) of GO was around 10 times higher than that required to stimulate the deoxygenation reaction (Figure 14.2(b)). The heat input calculated from the DSC curve in Figure 14.2(b) also includes the energy required to evaporate the water absorbed in GO film, which was calculated to be around 10 wt% by thermogravimetric analysis (TGA) [35]. As the energy released from one GO layer deoxygenation stimulates few more surrounding GO layers, large amount of heat is generated in a short span of time, which leads to the violent explosions like the one shown in Figure 14.2(a). Figure 14.2(c) shows another experiment to demonstrate the self-propagating nature of the GO disproportionation reaction. When a heated (about 400 °C) soldering iron was gently tapped to one end of a thin strip of GO film, it immediately propagated the reaction to entire GO strip and the progress of the reaction was evident from the colour change of shiny dark brown GO to black r-GO. The self-propagating reaction is usually accompanied by huge volume expansion (~100–300 times) of the GO film, mainly due to the rapid release of various gaseous molecules [35]. The self-propagating reduction reaction can create r-GO powders with high-surface areas up to around 2000 m^2 g^{-1} measured by Brunauer–Emmett–Teller (BET) nitrogen gas adsorption method.

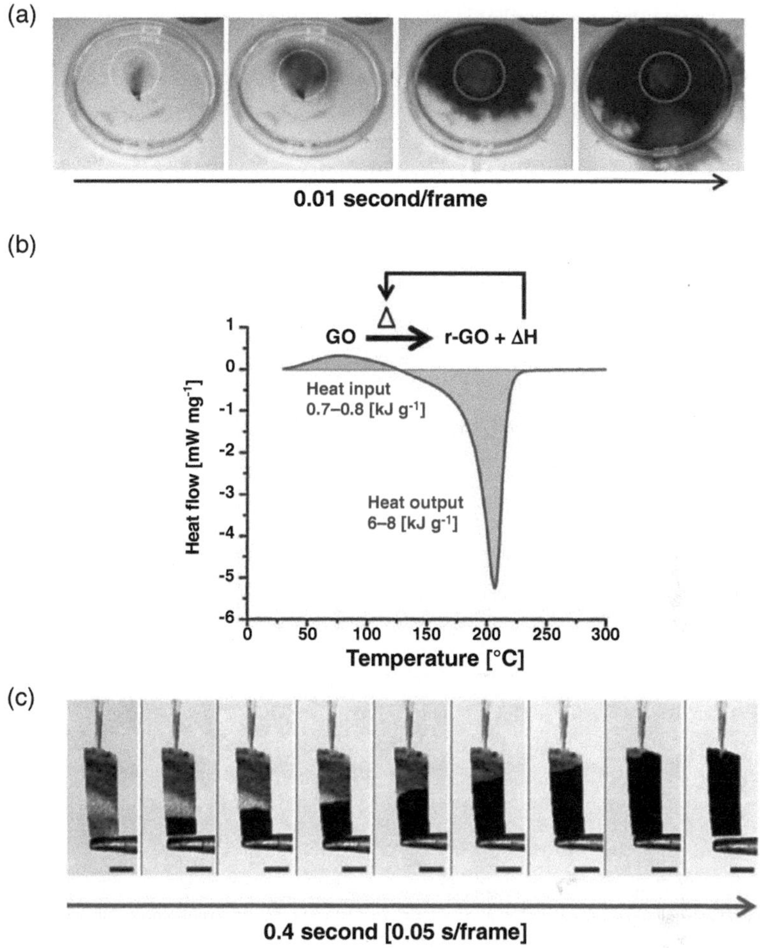

FIGURE 14.2 Thermal instability of GO. (a) snapshots showing the explosion of a solid piece of GO inside a petri dish, placed on a preheated hotplate at 300 °C. The white broken circle highlights the water vapour driven out of the GO paper. (b) DSC heating curve of GO in N_2 atmosphere. (c) Self-propagating reduction of a GO strip triggered by a gentle tap by a hot soldering iron of about 400 °C. Scale bars represent 5 mm. [Adapted and modified from [36], with permission from Elsevier].

14.2.2 The flame-retardant property of GO and r-GO composites

Although GO has been traditionally known as a flammable material, recent studies have shown that GO and its derivatives enhance the flame-retardant property of polymers. Again, careful observation of the GO explosion in Figure 14.2(a) shows that although GO was red heated, it did not combust. This contradiction to traditional belief needs more experimental studies. One way to study the flammability of GO and r-GO is to burn them with a natural gas flame. Natural gas flame triggers the self-propagating reduction to GO film, but the resulting r-GO product is quite stable against combustion. Figure 14.3(a) shows an r-GO film being repeatedly exposed to a natural gas flame, the exposed edge of the film turned red hot, but no combustion was observed upon removal of the flame. Even extended exposure to flames causes no significant material loss, which suggests the high thermal stability of the graphene sheets.

14.2.3 What makes GO flammable?

The above experiments conclude that GO and r-GO are not flammable, and heat treatment only reduces GO to r-GO. But, why was GO repeatedly reported as a flammable material in literature? One possible reason for the repeated observation of GO 'deflagration' could be the residual oxidants, such as $KMnO_4$ used in the synthesis of GO, which are highly flammable. Indeed, $KMnO_4$-contaminated

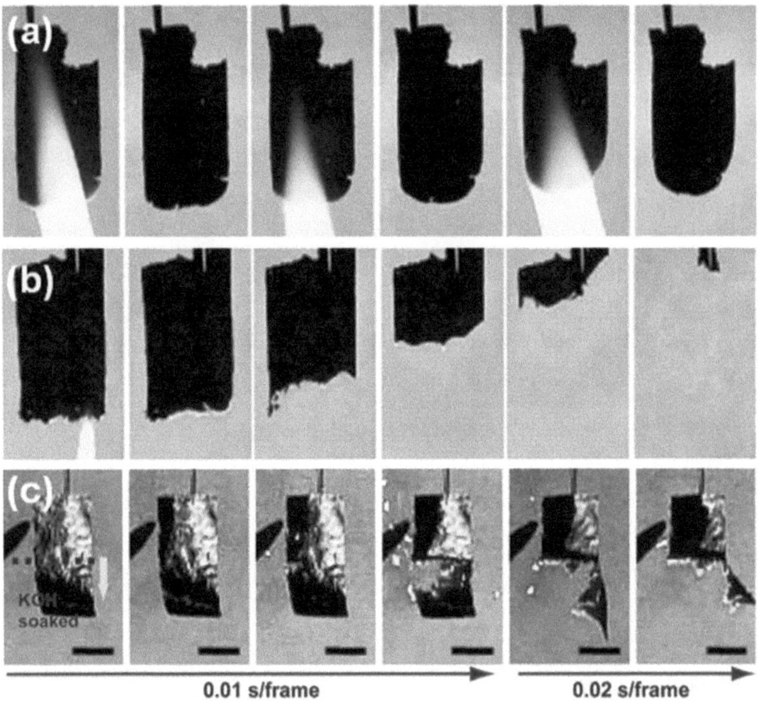

FIGURE 14.3 What makes GO inflammable? (a) Snapshots showing natural gas flame treatments on (a) r-GO film and (b) r-GO films contaminated with KOH. Pure r-GO is quite incombustible while K salt contaminated r-GO undergoes self-propagating combustion. (c) Snapshots showing the self-propagating reactions in a GO film, the bottom half of which had been dipped in KOH solution (0.01 M). Reduction reaction was triggered by a hot soldering iron and rapidly propagated to the boundary (dotted line), setting off violent reduction–combustion reactions in the KOH-treated area. [Adapted and modified from [35], with permission from Wiley-VCH].

GO can underwent a violent explosion and combustion upon ignition. However, it should be noted that in typical GO samples, $KMnO_4$ contamination is unlikely because after the oxidation step of the synthesis process, excessive hydrogen peroxide is added to ensure complete reduction of $KMnO_4$. The major inorganic by-products during the synthesis and purification of GO are alkaline metal salts, especially potassium salts such as KCl, K_2SO_4, KNO_3, etc. These water-soluble potassium salts are known to be good catalysts for 'carbon activation' and coal gasification reaction [39,40]. As GO tends to gel with water, removal of these salts by washing, filtration, centrifugation or dialysis is extremely tedious and can often be incomplete. Trace amounts of these impurities may significantly reduce the thermal stability of GO and r-GO. As shown in Figure 14.3(b), when an r-GO sample contaminated by KOH was subjected to a natural gas flame, ignition occurred immediately and the entire film was burned off very quickly. These experiments show that potassium-based impurities can significantly reduce the thermal stability of GO and r-GO. Figure 14.3(c) shows a neat GO film with the bottom half soaked with KOH. As soon as the hot soldering iron touched the upper half of the GO film, a self-propagating reduction reaction was triggered. When the reaction front reached the dividing line between the neat GO and KOH-soaked area, a violent, explosive combustion occurred, generating flying, red hot, fluffy debris. Finally, the bottom half completely burned away leaving only the upper half of the film. This reaction visualizes the GO deoxygenation as a self-propagating domino-like exothermic reaction [35,36].

14.2.4 Purification of GO

Neat GO and its derivatives are thermally stable, but metallic impurities generated as a by-product during the synthesis make them highly inflammable. Salt by-products were also found to interfere with dispersability and solution assembly of GO sheets. So, an improved purification process is vital for large-scale synthesis, storage and application of GO. Water-soluble salts and acids are two major by-products in GO syntheses. However, removing them by water-based washing is not a trivial task because GO tend to gel as the pH increases during washing (Figure 14.4(a)). This gelation makes commonly used purification techniques, such as centrifugation and filtration extremely tedious, time-consuming and non-effective. We have found that acetone and strong acid can greatly suppress gelation of GO and hence a two-step washing process has been developed (Figure 14.4(b)) [35,36]. To suppress the gelation, as prepared GO was rinsed with excess amounts of 3.4 wt% HCl, with pH of around 0. HCl solution is quite effective in removing inorganic ions from the products and does not leave any solid residue. After rinsing three to four times with HCl, the product was dried in air or under vacuum, which was followed by re-dispersion in acetone. The dispersion was then filtered and rinsed using excess acetone. Acetone can effectively remove the remaining HCl without causing gelation, and readily evaporates without leaving any residue. This two-step purification method is not only

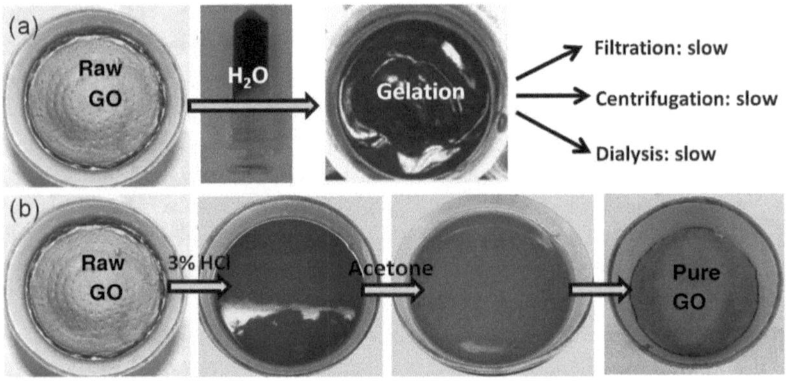

FIGURE 14.4 Improved purification of GO. (a) GO tends to gel when washed with water, and limits purification techniques such as filtration, centrifugation and dialysis. (b) GO gelation is suppressed in the two-step acid–acetone washing. [Adapted and modified from [14], with permission from ACS Publications].

very effective but also significantly speeds up the process. The potassium content of the two-step purified GO was undetectable by energy-dispersive X-ray (EDX) spectroscopy. Graphene-based materials produced from such purified GO samples are very stable against combustion, as those shown in Figure 14.3(a). Although we developed this two-step purification procedure independently, we shall point out that Nikolaev et al. also mentioned a similar purification method in his 1974 paper [41], which unfortunately has largely been overlooked.

14.2.5 Photothermal reduction of GO

GO has been widely studied in the last decade as a promising route to make graphene through reduction. Thermal annealing and chemical reductions are the most commonly used methods for GO reduction. However, the reduction of GO drastically reduces its solution processability, which in turn reduces its application possibilities in various fields, especially when GO has to blend with other materials. It is highly desirable to have a method to reduce GO inside a matrix without disturbing the surroundings much. In such cases, the excellent solution processability of GO can be exploited to blend it with the matrix. The previous study showed that reduction of GO is a self-propagating exothermic reaction and it requires only ~0.8 kJ g^{-1} of energy to initiate. This allows great flexibility in designing strategies for GO reduction, and camera flash reduction is one such strategy where GO can be photothermally reduced upon exposure to a pulsed Xenon flash at ambient conditions. Flash reduction of GO has many advantages over chemical and thermal reduction: it is rapid, clean and versatile. It is also much more energy efficient than conventional heat treatment in furnaces. The absorption spectrum of a 1-μm-thick GO film shows that it absorbs around 63% of the incident photon energy in the visible range (400–800 nm). A typical digital camera flash produces about 1000 mJ cm^{-2}

energy per pulse at a distance of around 2 mm, as measured by a pyroelectric energy meter. Therefore, the GO film can absorb ~630 mJ cm^{-2} of photon energy per pulse. There are no major alternative pathways to convert this photon energy, so most of the absorbed light energy must be converted to heat. The DSC experiment on GO suggests that it takes ~0.8 kJ g^{-1} heat to trigger the deoxygenating reaction (Figure 14.2(b)). Based on the density of the GO film (0.187 mg cm^{-2}) the mass-based heat input (0.8 kJ g^{-1}) can be converted to area-based heat input (150 J cm^{-2}). So, the energy produced by a typical camera flash is enough to trigger the self-propagating deoxygenation reaction of GO. This was indeed observed experimentally. The brown-coloured GO film shown in Figure 14.5(a) turned to a black and opaque r-GO film (Figure 14.5(b)) upon exposure to camera flash. The conductivity of flash-reduced r-GO (1000 S m^{-1}) is comparable to those made from common chemical or thermal treatments. Flash reduction also has additional advantages over common reduction methods. One such advantage is that the excellent solution processability of GO can be utilized to make graphene-based polymer composites, which is otherwise very tedious, due to the poor solvent compatibility of graphene. Here, after making the GO–polymers composite, flash reduction can

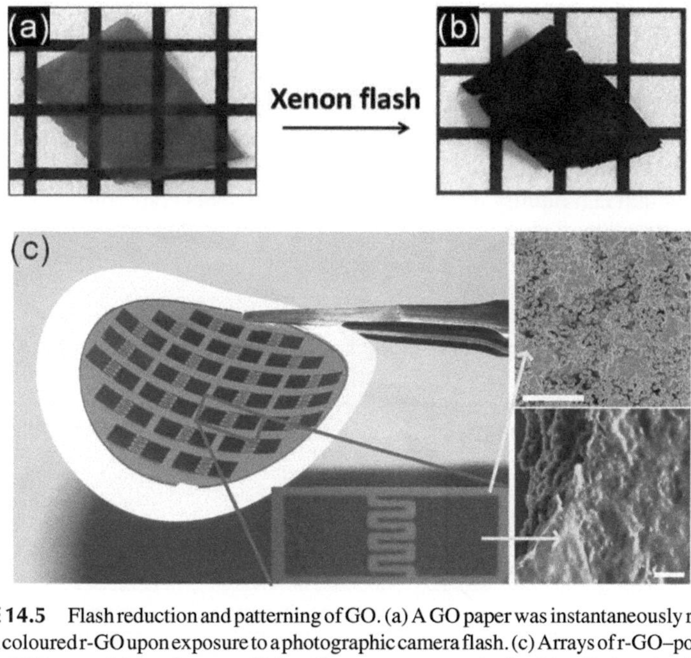

FIGURE 14.5 Flash reduction and patterning of GO. (a) A GO paper was instantaneously reduced to (b) black coloured r-GO upon exposure to a photographic camera flash. (c) Arrays of r-GO–polystyrene thin film IDE on a 1.5-in.-diameter Nylon filter paper were obtained by flash reduction of the GO–polystyrene composites through a photomask. The inset shows a single set of IDE with 5 mm × 5 mm contact pads and 100-μm-wide electrode fingers. The SEM images of flashed and unexposed areas confirm the welding of the polymer colloids with r-GO upon flash irradiation. [Adapted and modified from [37], with permission from ACS Publications].

selectively reduce GO to graphene, given that the polymer is not very strongly absorbing. With common reduction methods, such as chemical reduction and thermal annealing, this is not feasible. The polymeric matrix hinders the interaction of reducing agents to GO, and thermal heating may destroy the polymer itself. The excess heat from the flash or exothermic GO deoxygenation can sinter the r-GO sheets and polymer particles to form more homogeneous structures. Figure 14.5(c) shows an scanning electron microscopy (SEM) image of a piece of GO–polystyrene composite film supported on a piece of filter paper, made by filtration from their colloidal solutions. This film was exposed to Xenon flash irradiations through a photo-mask with transmitting patterns of interdigitated electrodes (IDE). Flash exposure clearly has converted GO to r-GO based on the colour change. SEM study (Figure 14.5(c), right panels) reveals that the polystyrene beads are welded together with the r-GO sheets, forming a smooth, continuous composite thin film.

14.3 CHARACTERIZATION OF GO: IMAGING GRAPHENE-BASED SHEETS

Microscopy is an invaluable tool for characterizing graphene-based single atomic sheets, such as GO. It enables a first-cut evaluation of sample quality, which is pertinent for further experiments on the sample. Here, we review some classical methods for imaging such thin sheets, including optical microscopy (OM), SEM, AFM and Raman imaging. As summarized in Table 14.1, these commonly used techniques all require the use of special type of substrates, and some other conditions such as low vibration and vacuum. We then present a new imaging technique discovered by us named fluorescence quenching microscopy (FQM), which can rapidly image graphene-based sheets on arbitrary substrates, including flexible plastic substrates.

TABLE 14.1 Common microscopy techniques for imaging GBS and their limitations

Technique	Throughput	Substrate	Other requirement
Optical microscopy	High	Dielectric-coated Si	Optimized dielectric thickness and wavelength
SEM	Medium	Conductive	Vacuum
AFM	Low	Smooth surface	Low vibration
Raman	Medium/high	Low fluorescence and efficient heat dissipation	Laser

14.3.1 Common imaging techniques for GBS

The study that discovered graphene in 2004 used the optical microscope to find graphene sheets mechanically exfoliated amidst a majority of graphite flakes [5]. Since then, the optical microscope has continued to be a vital bench top instrument for the microscopic examination of graphene. Commonly, sheets are deposited on a dielectric-coated silicon wafer and examined under the optical microscope in reflective illumination mode. Under visible light, even single-layer graphene sheets can be visible due to interference-based contrast [5,42,43]. OM allows very high throughput examination of graphene samples. However, the contrast of single-layer sheets, especially for the much-less absorbing GO, is often weak. Sheets also have to be deposited on substrates with a particular dielectric thickness that maximizes the contrast of the sheets. SEM is a moderate-throughput technique that is suitable for the examination of smaller graphene sheets that are beyond the resolution limit of OM [24]. However, conductive substrates and a vacuum chamber have to be used so as to minimize charging effects. These limitations preclude the examination of graphene materials on insulating substrates such as plastic and glass, or in solutions. Scanning probe techniques such as AFM are extensively used to characterize thin sheets of layered materials. However, the imaging throughput of AFM is limited by slow scan speeds and small scan areas. Furthermore, AFM requires good isolation against noise and samples need to be deposited on smooth substrates such silicon wafer, quartz, freshly cleaved mica or highly oriented pyrolytic graphite (HOPG). Raman imaging is especially useful for counting the number of layers in a graphene sample due to the layer-dependent properties of the more perfect graphene materials, such as those made from mechanical exfoliation or chemical vapour deposition. However, Raman imaging requires the use of focused laser beam, which could cause sample damage to graphene and GO. Another constraint in Raman imaging is that samples must be deposited on low-fluorescence substrates. Therefore, plastic substrates, which often have intrinsic fluorescence, could cause the Raman signal from graphene-based sheets to be overwhelmed.

All the above-mentioned imaging techniques require graphene-based sheets to be deposited on specific kinds of substrates. It would therefore be desirable for new imaging techniques to be developed that can image them on arbitrary substrates. For example, the flexible soft sheet nature of graphene-based sheets makes them attractive materials for use in flexible plastic electronics. However, it is challenging to see them on plastic substrates by OM, SEM, AFM or Raman imaging. A method to directly observe how thin-film processing strategies affect the morphology and microstructures of graphene can greatly advance progress in graphene-based devices on flexible plastic electronics. Another challenge in imaging is to observe their behaviour in dynamic environments such as in solution. For example, observing solvent-induced conformational change in GO has resulted in a technique for producing a crumpled paper ball-like structure of GO and r-GO (see Sec. 14.7) [44].

14.3.2 Fluorescence quenching microscopy

Fluorescence microscopy is a very common tool. Usually, fluorescent labels are used to brighten up the object of interest. However, the same strategy is not applicable for graphitic materials, such as graphite, graphene, GO or r-GO as they are strong quenchers of dye molecules [45]. Therefore, we developed a reverse strategy, FQM, where graphene-based sheets appear dark against a bright background [46–48]. The typical process for FQM is illustrated in Figure 14.6(a) and (b). A dye-doped polymer layer, such as fluorescein co-dissolved with poly(vinylpyrrolidone) (PVP), is spin-coated onto a substrate with pre-deposited graphene-based sheets. Illumination by the appropriate wavelength of light excites the dye molecules and reveals the sheets as dark sheets against a bright background. Figure 14.6(c) is an FQM image of GO deposited on a SiO_2/Si wafer, representing what is seen by a naked-eye observation through the eyepiece. The fluorescence quenching effect is strong enough to create high contrast of single-layer GO sheets against the background. Moreover, within the sheets, microstructural features such as wrinkles, folds and overlaps can be seen in striking detail. The higher degree of quenching by double layers suggests that the quenching mechanism proceeds via long-range mechanisms, such as energy transfer mediated by dipole–dipole interactions [46,49]. This is because only

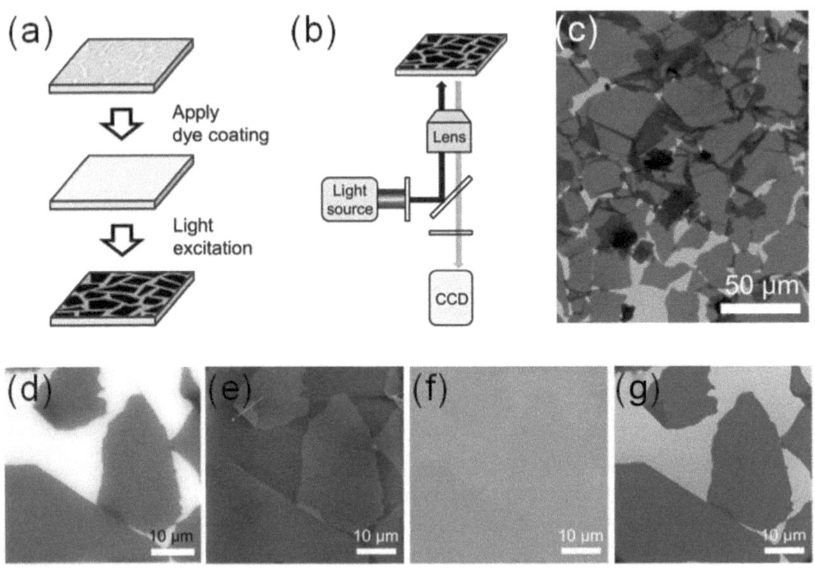

FIGURE 14.6 High-throughput visualization of graphene-based sheets by FQM. (a) In a typical FQM experiment, a fluorescent dye layer is applied onto a graphene sample covered substrate, which can reveal the underlying sheets upon excitation (b). (c) A FQM image of GO sheets deposited on SiO_2/Si substrate. (d–g) Images of the same GO sheets deposited on SiO_2/Si obtained by FQM, AFM, OM and SEM, respectively. [Adapted and modified from [46], with permission from ACS Publications].

the long quenching distance would allow the underlying sheet in a double layer to quench dye molecules above the top layer, resulting in an area that is even darker compared to an area quenched by a single layer. Short-range quenching mechanisms such as electron transfer rarely extend beyond van der Waals contact, which is typically within 1 nm, and thus are unlikely to allow single layers to be resolved from multi-layers. Indeed, when the dye polymer is too thin, all the sheets, regardless of number of layers, wrinkles and folds, appear with a uniform near-saturated contrast [46]. Figure 14.6(d)–(g) compares FQM to common imaging techniques such as (e) AFM, (f) OM and (g) SEM. The contrast of FQM is much improved compared to OM, and is even comparable to the contrast obtained with AFM or SEM. However, compared to AFM or SEM, FQM allows high-contrast imaging of graphene-based sheets with much more inexpensive set-up and much less time.

14.3.3 New opportunities created by FQM

FQM is able to address the challenge of imaging graphene on plastic substrates. Such a capability would provide valuable feedback for improving solution-processing strategies, which can be used to create thin films of graphene-based sheets for flexible and transparent plastic electronics. Imaging GBS on transparent plastics is particularly challenging. The low optical absorbance of graphene family sheets makes them practically invisible with OM, while the insulating nature of the plastic makes SEM an unsuitable technique. The plastic surface could also be too rough for AFM. However, using FQM, it is now possible to easily image graphene sheets on plastic substrates [46]. With FQM, it may even be possible to image graphene sheets in dynamic environments, such as in solution or in a polymer matrix. We have been able to observe the dewetting behaviour of suspended GO sheets near a contact line, which inspired a strategy for creating a highly crumpled morphology of graphene (Sec. 14.7) [46]. On the other hand, the behaviour of GBS in a polymer matrix has not been well studied. It would be interesting to be able to view, in real time, how these sheets disperse in the matrix and how they might respond when under external stress. This would aid the design of better graphene-based composites and boost our understanding of the dynamic behaviour of graphene-based sheets.

FQM involves only spin coating a dye-doped polymer layer and then examining the sample with a fluorescence microscope. These methods are well established and the instruments used are readily available. Thus, FQM can also be easily integrated into existing microfabrication techniques such as photolithography [47]. The benefit of FQM-based microfabrication is that there is no need for alignment marks and the process can be done on any substrate, including transparent substrates on which graphene is usually invisible. For FQM-based microfabrication, a dye, such as DCM, can be incorporated into a polymer photoresist, such as PMMA or SU-8. The dye and photoresist are chosen such that the excitation wavelength of the dye does not also degrade the photoresist. Once the desired sheets are located, a photo mask can be inserted into the optical path of the microscope and the illuminating wavelength can be switched to UV to expose the photoresist. The photoresist can then be developed to form the desired structures, such as electrode patterns, on the graphene-based sheets. Since FQM can be done on virtually any substrate, in principle, on-sheet device fabrication can be extended to a variety of situations, such as flexible and transparent substrates.

In summary, compared to current imaging techniques, FQM has the following advantages:

1. It is a low-cost, high-throughput optical technique that can rapidly image graphene-based sheets with very high contrast.
2. It is applicable to sheets deposited on arbitrary substrates and even in solution.
3. It gives strong layer contrast, allowing layer counting capability.
4. It works well for all graphene-based sheets, such as graphene made by mechanically exfoliation or chemical vapour deposition, and GO, r-GO. Recently we demonstrated that it is even applicable to other two-dimensional materials, such as MoS_2 [48].
5. FQM can produce sp^2/sp^3 contrast, and thus can tell the difference between sheets with different oxidation levels.
6. FQM can readily integrate with photolithography, using photoresist itself as part of the imaging sensitizer.
7. FQM can be non-destructive to the underlying graphene-based sheets, leaving very little or no contamination after the dye–polymer layer is washed off.

FQM was first reported by us at the end of 2009 [46], and by now it has been used and adopted by over 10 research groups worldwide to image graphene-based sheets.

14.4 NEW SOLUTION PROPERTIES OF GO: SURFACTANT SHEETS

14.4.1 GO as an unconventional soft material

GO is an intriguing material. As a functionalized single atomic layer of carbon, the thickness of GO is approximately 1 nm. On the other hand, the lateral length of GO sheets can span to micrometres (Figure 14.1(b)). Thus, GO straddles across the two abruptly different length scales typical of molecules and particulate materials. This molecule–particle duality makes GO an interesting material to study from the perspectives of both chemistry and materials science. Furthermore, the length scales associated with GO coincide with the length scales associated with soft materials, such as polymers, colloids, amphiphiles and liquid crystals, which are held together by weak interactions comparable to thermal fluctuation [27]. The softness of GO is also apparent from a structural mechanics point of view – many materials, even hard and brittle ones, can become soft and flexible when their aspect ratio increases. Most notably, the chemical structure of GO predicts that GO should exhibit many soft material properties.

GO can be viewed as a 2D polymer consisting of covalently tethered oxygen-derivatized carbon atoms (Figure 14.1(c), right). GO and graphite oxide have long been viewed to be hydrophilic due to its good dispersability in water. However, the prevalent structural model of GO shows that the edges of a GO sheet consist predominantly of ionizable carboxylic acid groups, which are hydrophilic. Meanwhile, the basal plane consists of both hydrophilic oxygen-containing functional groups and hydrophobic aromatic domains. Based on this structural model, a GO sheet has both hydrophilic and hydrophobic domains covalently tethered together, which should make it an amphiphile [26,50]. These amphiphilic properties allow GO sheets to be manipulated via interfacial assembly techniques such as Langmuir–Blodgett (LB) assembly [24,25]. Based on this approach, processing–structure–property relationships of GO and r-GO monolayers can be developed [51].

14.4.2 Interfacial activity of GO

The property-oriented definition of amphiphilies is that they can go to interfaces and lower the interfacial energies. To demonstrate that GO is amphiphilic, we followed this property definition to examine its interfacial activities. If GO is amphiphilic, we can expect that GO sheets dispersed in an aqueous solution should tend to migrate to the surface. Indeed, this has been observed with a Brewster angle microscope (BAM), which is a surface-sensitive imaging technique. A BAM image of clean water surface should be dark. Surface-active species such as islands of surfactant monolayers or floating nanoparticles will appear as bright features due to scattering of standing wave propagating along water surface [52]. BAM study reveals that for freshly prepared GO water, most of the GO sheets remain in the water sub-phase (Figure 14.7(a), left). This is because the high 'molecular' weight of GO results in very slow migration of the sheets to the air–water interface. To expedite the process, gas bubbles, such as CO_2 released from carbonated water, should be able to 'lift' GO sheets to the interface, if they are indeed amphiphilic and adhering to the bubble–water interface. The BAM image captured after flotation (Figure 14.7(a), right) shows that the GO water surface has indeed become densely populated with GO sheets trapped at the air–water interface. In oil–water systems, GO can also act as a colloidal surfactant to stabilize oil droplets, creating a Pickering emulsion. As shown in Figure 14.7(b), gentle shaking of toluene and GO water can produce sub-millimetre-sized toluene emulsion droplets, the size of which depends on the concentration of GO sheets in the original water phase [50]. This is characteristic of Pickering emulsions stabilized by colloidal particles. Breaking an

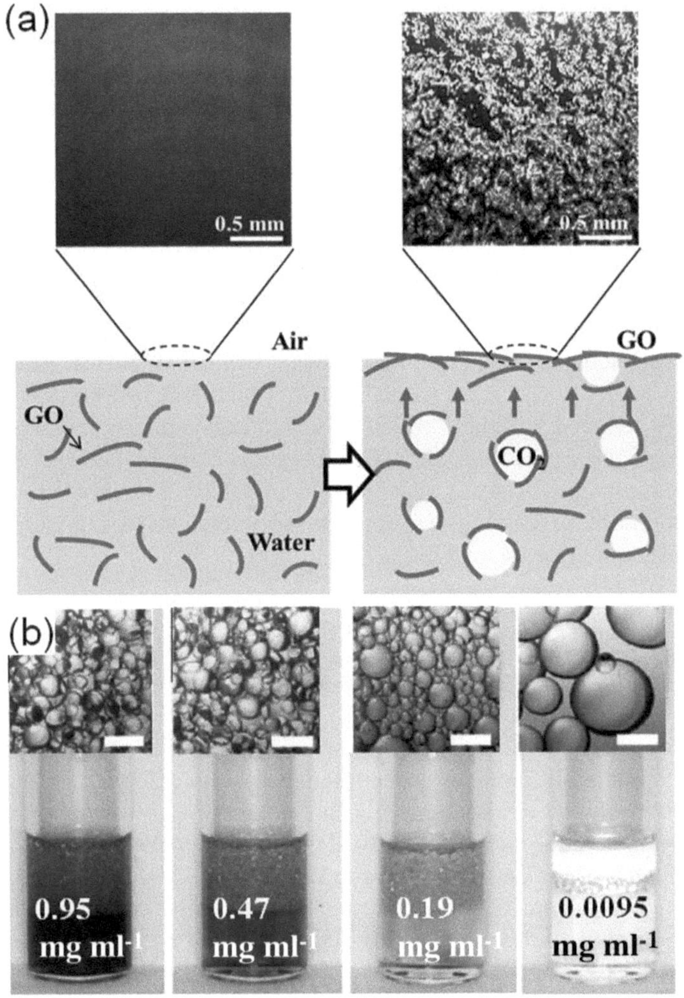

FIGURE 14.7 GO at interfaces. (a) BAM images of the water surface taken before (left) and after (right) bubbling the GO dispersion. The schematic illustration shows the flotation of GO by CO_2 bubbles in carbonated water. GO is first captured by the rising bubbles and then transported to the water surface. (b) Pickering emulsions of toluene droplets formed upon shaking a toluene–GO water mixture, with different GO concentrations. The insets show microscopic images of the toluene droplets. All scale bars are 1 mm. [Adapted and modified from [50], with permission from ACS Publications].

emulsion droplet with a glass substrate leads to the deposition of a GO thin film that is largely monolayer in thickness, suggesting that GO sheets are wrapping the oil droplets [50]. With the GO protecting layer, the oil droplets can be stable against coalescence for months to years. Using a drop shape analyser, we found that GO sheets can indeed lower the interfacial energy of toluene and water [50].

14.4.3 Novel surfactant properties of GO

The interfacial activity of GO sheets shown in Figure 14.7 demonstrates that they are indeed amphiphilic. As a novel 2D surfactant, GO has some unusual interfacial properties that are unattainable with classical molecular amphiphilies or particle emulsion stabilizers. For example, GO's amphiphilicity has been found to be pH and size dependent. Typically, the size of GO samples varies greatly from hundreds of nanometres to tens of micrometres. Since GO has a hydrophilic edge and a largely hydrophobic basal plane, its amphiphilicity should be affected by the lateral dimension of the sheet. Smaller sheets have a higher edge-to-area ratio and thus should be more hydrophilic. Indeed, when GO sheets are spread on a water surface, the water surface itself acts as an astonishingly simple size-separation filter (Figure 14.8(a)). The smaller, sub-micrometre sized, more hydrophilic sheets sink into the water sub-phase, while the larger, more amphiphilic sheets are found to stay on the water surface. This spontaneous size separation also occurs at oil–water interface (Figure 14.8(b)). The larger sheets tend to adhere to the oil–water interface, stabilizing the emulsion, while the smaller ones remain in the aqueous phase. Thus, the larger sheets can be extracted from water to the emulsion phase. The biphasic mixture can be re-agitated to re-form the emulsion phase and the larger sheets can be repeatedly extracted until they are depleted and Pickering emulsions can no longer form.

Since the ionization of –COOH groups is pH dependent, the charge density of GO sheets, and thus their amphiphilicity, should also be pH dependent. Figure 14.8(c) shows that when the aqueous phase is basic (pH = 10), the edge –COOH groups are dissociated and highly charged; thus, GO sheets become too hydrophilic to move to oil–water interface and remain in the aqueous medium. When the aqueous phase is acidic (pH = 5 and pH = 2), the –COOH are protonated and less charged. Therefore, the GO sheets can adhere to the oil–water interfaces and migrate to the emulsion phase. The emulsification process can be reversed by increasing the pH value back to 10, when GO sheets are returned to the aqueous phase. This suggests that GO can act as a pH-sensitive phase transfer agent that shuttles between the aqueous phase and organic phase.

14.4.4 Interfacial assembly of GO sheets

Since GO sheets are amphiphilic, like molecular surfactants, they can be spread directly on water surface for LB assembly. In typical LB assembly, surfactant molecules are deposited between two barriers onto a water surface. The packing density of the surfactant can then be controlled by moving the barriers toward or away from each other (Figure 14.9(a)) [24,25]. LB is not only a convenient technique for studying monolayer systems, but also a practical method for achieving continuous and dense film coverage. GO sheets, once assembled into a continuous monolayer, can be deposited onto a substrate to form uniform films.

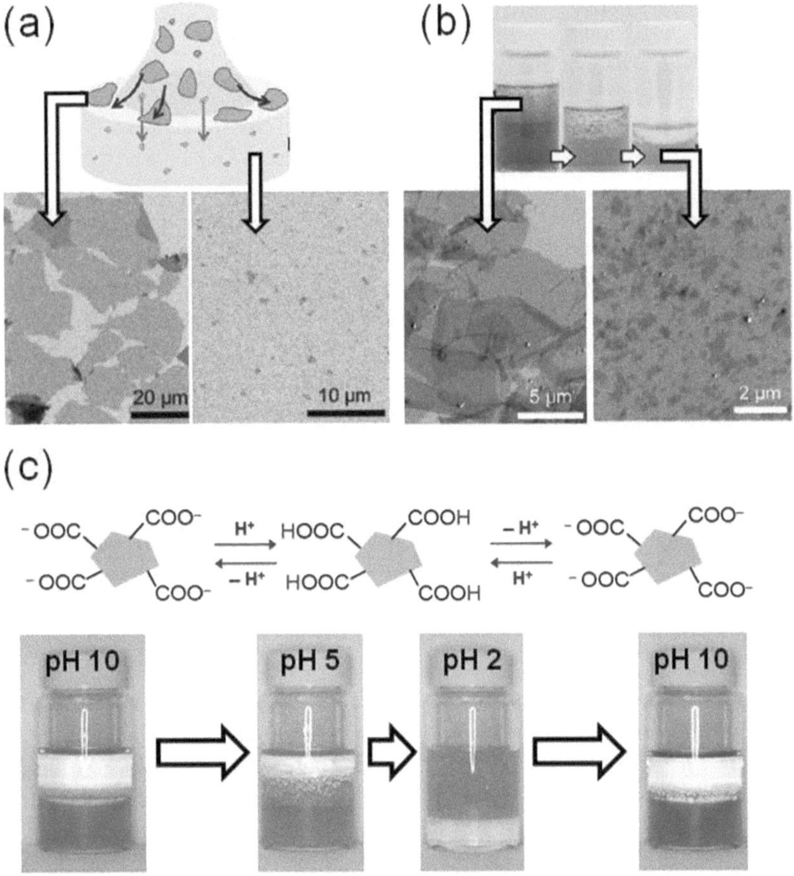

FIGURE 14.8 Size-dependent amphiphilicity of GO results in spontaneous size separation at (a) air–water and (b) oil–water interfaces, where smaller GO sheets tend to stay in water. (b) Large GO sheets can be selectively extracted from the Pickering emulsions. (c) pH-dependent amphiphilicity of GO as shown in the toluene–water biphasic system. [Adapted and modified from [50], with permission from ACS Publications].

The GO films can then be reduced to r-GO to form transparent conducting thin films. Such as-made GO and r-GO monolayers have been used to fabricate the windows of liquid sample cells for environmental spectroscopy measurements [53] and SEM imaging [54] on wet samples in high vacuum, and the hole transport layer in high-performance organic solar cells [55].

The interaction between the neighbouring GO sheets in the LB monolayer can be tuned by the pH of the sub-phase. This determines the microstructures of the tiled monolayer. When two GO sheets are pushed together on water surface, they can either slide on top of each other to form overlaps or buckle up to form wrinkles. Note that overlaps and wrinkles are the two fundamental morphology features of tiled monolayer. Tuning the sub-phase pH gives us a tool to quali-

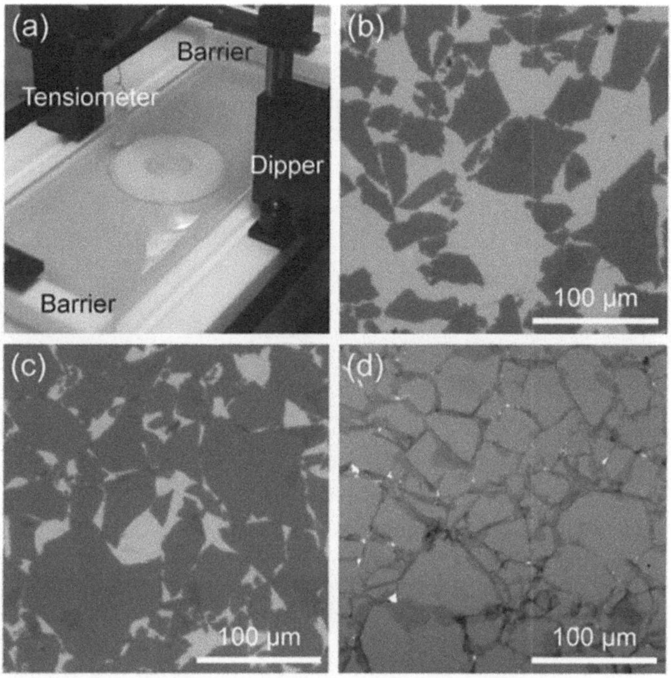

FIGURE 14.9 2D LB assembly of GO sheets. (a) Controllable tilling of the GO sheets can be achieved by directly spreading GO dispersion on the water surface of a LB tough. (b–d) FQM images LB monolayers of GO dip coated onto glass substrates with increasing surface coverage. [Adapted and modified from [24], with permission from ACS Publications].

tatively control the assembly of GO sheets and thus deconvolute the two types of microstructures. This helps to delineate the processing–structure–properties relationship of such assembled GO or r-GO thin films [56]. It was found that the sub-phase pH strongly influences the resulting morphology of the LB monolayers. Acidic conditions result in wrinkle structures, while basic conditions result in overlapping graphene layers. The overlapping layers result in better properties such as higher optical transparency, and after annealing, better conductivity. As a result, the overlapping layer morphology is more favourable for applications such as transparent conducting thin films.

14.4.5 GO as 2D surfactant and dispersing agent

The amphiphilic behaviour of GO suggests that it can act as a surfactant to disperse insoluble materials in water. Graphitic materials such as graphite and carbon nanotubes (CNTs) or other π-conjugated materials would be especially suitable to be dispersed by GO due to their strong π–π interactions. As shown in Figure 14.10, GO can disperse powders of graphite, unfunctionalized multi-walled carbon nanotubes (MWCNTs) and single-walled carbon nanotubes

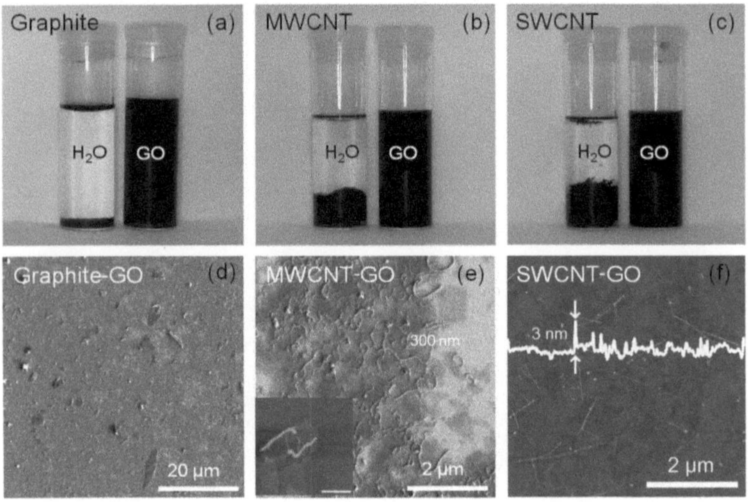

FIGURE 14.10 GO can act as a dispersing agent for insoluble materials such as (a) graphite, (b) MWCNTs and (c) SWCNTs. (d–e) SEM and (f) AFM images of samples casted from their respective GO-assisted dispersions shown in (a–c). Inset of (e) is an AFM image of MWCNT–GO sample. [Adapted and modified from [50], with permission from ACS Publications].

(SWCNTs) in water after sonication. SEM studies reveal that sonication with GO effectively disentangles the CNTs while yielding a stable GO–CNT dispersion in water. SEM and AFM examinations also show that the CNTs adhere well to GO, presumably through π–π interactions. Unlike conventional molecular or polymeric surfactants, GO can be reduced to conductive r-GO and contribute additional pathways for electrical conduction. In addition, GO sheets do not necessarily need to wrap the nanotubes, leaving much of their functional surface intact. Thus, using GO to disperse graphitic materials such as CNT is a useful strategy to obtain composite with enhanced electrical and mechanical properties.

14.5 WATER PROCESSABLE GO–SWCNTS INTERLAYERS FOR ORGANIC SOLAR CELLS

It has been reported that near-complete GO monolayers can be made by spin coating, which can replace currently used acidic conducting polymer-based hole transport layer in organic solar cells [57]. However, it is not trivial at all to create complete GO monolayer by spin coating. For example, we found that a small variation in the concentrations of GO dispersion for 0.1–0.2 wt% can result in large variations in thin film coverage and thickness [58]. Below 0.15 wt%, the GO dispersion yielded non-uniform, incomplete coverage of GO pieces. With 0.15 wt% GO dispersion, it started to yield nearly complete coverage of GO. A slight increase in GO concentration to 0.2 wt% yielded thicker GO films (ca. 3–4 nm) consisting of stacked multi-layers. This shows that the processing conditions of GO have to be limited to a very narrow window to ensure both near-complete coverage and near-monolayer thickness for achieving high device performance.

Figure 14.10(f) shows an interesting synergistic assembly effect between GO and SWCNTs during spin coating: While GO helps to disperse SWCNTs in water, the nanotubes also help to make more continuous GO films during spin coating by stitching the sheets through π–π stacking. We have found that adding SWCNTs, either functionalized or unfunctionalized ones, helps to create uniform GO thin films by spin coating [58]. Adding a non-percolating amount of SWCNTs into GO can significantly improve its thin-film processability. The SWCNTs also help to improve the vertical conductivity of GO thin films, which allows the use of thicker and thus easier-to-make GO films for the hole transport layer. This helps to reduce the strong dependence of the device performance on GO thickness and solution-processing conditions, leading to improved and more reproducible results. Figure 14.11(a) shows the device structure of an organic solar cell made with the GO–SWCNTs hole transport layer. The current–voltage curves shown in Figure 14.11(b) shows that GO–SWCNTs hole transport layer can offer comparable performance to the conventional PEDOT:PSS anode modifying layer.

We also demonstrated that water-processable GO–SWCNTs thin films can be used as an effective interfacial layer to construct tandem polymer solar cells in both regular and inverted geometry by all-solution processing routes [59]. In a tandem photovoltaic device, the sub-cells are stacked along the optical path to increase optical absorption. When the sub-cells are connected in series, the ideal open-circuit voltage (VOC) is equivalent to the sum of the VOCs of the

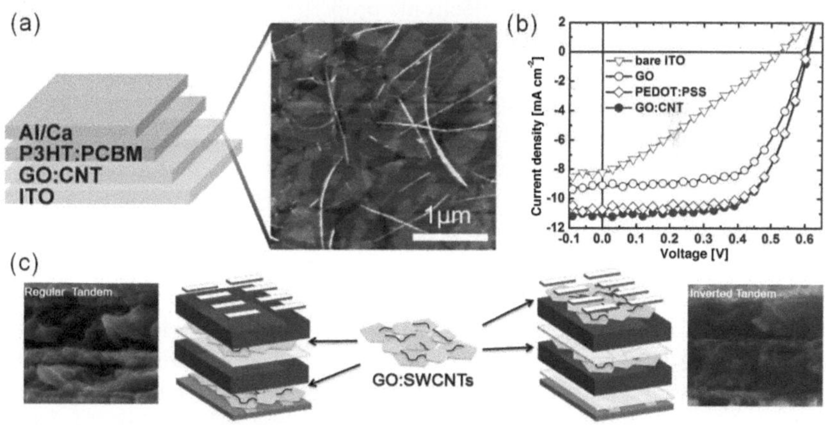

FIGURE 14.11 GO:SWCNT as the anode-modifying layer for polymer solar cells. (a) Schematic diagram illustrating the photovoltaic device with an AFM image of the GO:SWCNTs layer. (b) Current density–voltage characteristics of devices made on unmodified ITO, with GO layer, PEDOT:PSS layer and GO:SWCNT layer. (c) Water-processed GO:SWCNT thin films can also be used as interfacial layers to serially connect two P3HT:PCBM bulk heterojunction sub-cells in both regular and inverted tandem structures. [Adapted and modified from [58], with permission from Wiley-VCH].

individual sub-cells, which can potentially increase the power conversion efficiency (PCE). Central to the construction of tandem devices is the interconnect layer that spatially separates but electrically connects the two sub-cells. The interconnect layers need to have (a) high-optical transparency over the entire solar spectrum; (b) sufficient vertical conductivity; (c) complete coverage to avoid intermixing of the two sub-cells; (d) low surface roughness to avoid interrupting the deposition of neighbouring sub-cells and (e) appropriate energetics to facilitate efficient charge recombination. For solution-processable interconnect materials, there are additional challenges such as orthogonal solubility, chemical and mechanical stability. As shown in Figure 14.11(c), the water-processable GO–SWCNTs thin films can be used to construct such interconnect layers for tandem solar cells in both regular and inverted configurations. In both cases, nearly doubled VOCs were obtained using two identical sub-cells made of poly(3-hexylthiophene):[6,6]-phenyl C61 butyric acid methyl ester (P3HT:PCBM). This suggests that GO:SWCNTs thin films can serve as good mechanical separator and electrical interconnect of the two sub-cells. Because of their high transparency in the NIR region, GO:SWCNTs-based interconnect should also work well with complementary tandem cells using lower bandgap polymers.

14.6 TOWARDS SOLUTION PROCESSED ALL-CARBON SOLAR CELLS

The use of GO as a dispersing and solution-processing agent has led to the discovery of an all-carbon composite that can act as the light-absorbing layer of solar cells, which is the most important milestone for creating truly all-carbon solar cells [60,61]. The unique electronic properties and high thermal stability of carbon nanostructures make them attractive to use in photovoltaic devices. Different carbon nanostructures have different electronic band structure. For example, C_{60} is an n-type ultrafast electron donor, while semiconducting SWCNTs are typically p-type under ambient conditions. Thus, if C_{60} and SWCNTs can be co-assembled, numerous nanoscopic p–n junctions can be formed to create a photovoltaic material. Organic chemists have long been attempting to prepare such carbon–carbon heterojunctions between fullerenes and nanotubes or graphene [62–66]. However, if covalent chemistry is used to create such junctions, the functional conjugated carbon network will be damaged. A non-covalent method that can bring these carbon nanostructures intimately together will not hinder charge transfer between pristine surfaces of these carbon nanomaterials. Since we now know that GO is amphiphilic, using GO to process these carbon materials provides a clean and green route for co-assembling graphitic nanostructures in solution. As shown in Figure 14.12(a), SWCNT and C_{60} can be sonicated together in GO solution to form a C_{60}–SWCNT–GO colloidal dispersion. The ternary hybrid is solution processable and can be spin-coated into thin films (Figure 14.12(b)), which can then be annealed to reduce GO to r-GO, resulting in a conductive C_{60}–SWCNT–r-GO all-carbon composite. Here GO

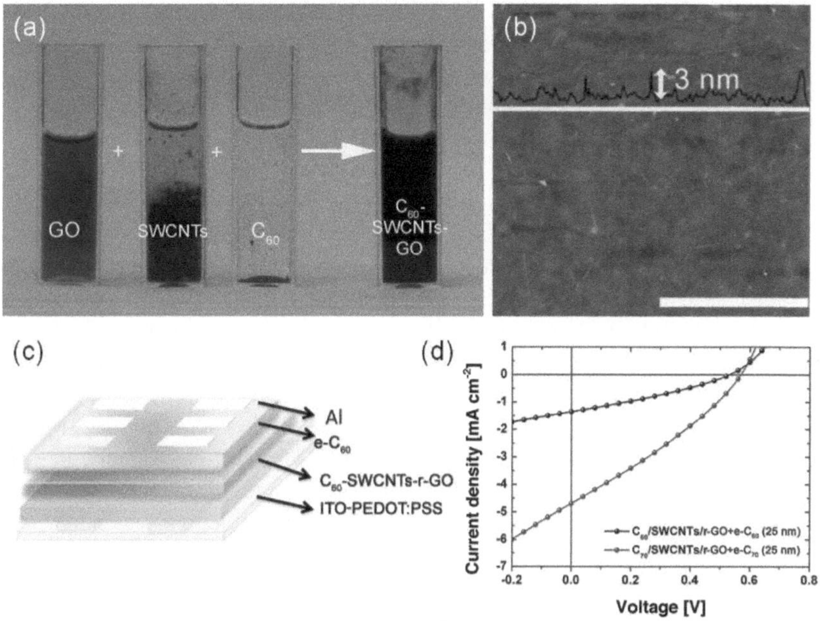

FIGURE 14.12 Photovoltaic all-carbon composite. (a) Picture of cuvettes containing aqueous dispersions of GO, SWCNTs, C_{60} and their ternary blend C_{60}–SWCNTs–GO, respectively. (b) AFM images of spin-coated thin films of C_{60}–SWCNTs–GO, with height profile along the white line. (c) Schematic illustration of the photovoltaic device structure with C_{60}–SWCNTs–r-GO as the active layer with an additional layer of thermally evaporated C_{60} as the protection/hole blocking layer. (d) J–V characteristics of devices made with C_{60} and C_{70}, respectively. [(a)-(c) and (d) adapted and modified from [61] and [60], with permissions from ACS Publications and Royal Society of Chemistry].

acts as both the dispersing and planarization agent to ensure thin films with low surface roughness can be obtained.

The all-carbon composite can then be used as the active layer of a solar cell. A prototype device architecture is shown in Figure 14.12(c). A layer of evaporated C_{60} (e-C_{60}) is deposited as the electron transport layer on top of the 5–8 nm layer of the all-carbon composite C_{60}–SWCNT–r-GO. The device with C_{60} produced a short-circuit current (J_{SC}) of 1.23 mA cm^{-2}, VOC of 0.59 V, fill factor (FF) of 0.29 and an overall PCE of 0.21% [61]. This was already the highest PCE among solar cells made with all known all-carbon active layer. Later, by changing C_{60} to a better absorber C_{70}, the PCE was increased to 0.85% (Figure 14.12(d)) [60]. It is now entirely possible to make a solution processed solar cell with all the layers – from transparent electrodes, interfacial layers, to the light-absorbing active layer with graphitic materials [60]. Since the initial two reports on all-carbon-based active layer that absorbs in the visible [61] and IR [67] region of the solar spectrum, there have been rapid progresses towards making all-carbon solar cells [56,68–71]. The highest to date PCE of 1.3% was achieved on devices with the active layer made of r-GO, SWCNTs and PCBM [71]. All-carbon solar cells can potentially be made with low cost materials and processing conditions. They are also expected to have higher chemical and thermal stability, and thus longer lifetime than similarly solution-processed organic solar cells.

14.7 AGGREGATION-RESISTANT CRUMPLED GRAPHENE BALLS

The unprecedentedly high surface to volume ratio is the key to most of the outstanding properties of graphene and its derivatives. Preservation of this extremely large surface area uncovered is vital for its success in many potential applications. But, it is very difficult to save this extra-large surface area, as these sheets tend to aggregate through strong van der Waals interaction. In the typical synthesis and post-synthesis processing steps of graphene, the aggregation of the sheets through van der Waals interaction is uncontrollable, random and irreversible. Therefore, the surface-area-based properties of graphene-based materials are highly inconsistent. The situation becomes worst when processing steps involved drying and palletizing. To solve this problem, the van der Waals attraction between these sheets needs to be reduced. The attraction potential is related to the interacting geometry of the sheets by the following:

$$E \sim S/d^4 \tag{14.1}$$

E is the interacting potential, S the overlapping area between two sheets and d the separating distance [72]. There have been a few strategies to make graphene aggregation resistant in the solution phase, which include both reduction of 'S' (i.e. reducing the size of the sheets) [73], and increasing 'd' (i.e. employing dispersing agents, or tailoring the solvent–graphene interactions) [11,13]. Another strategy is to use colloidal chemistry [74] techniques to introduce repulsive interactions between the sheets to overcome van der Waals attraction. In order to standardize the properties and performance for large-scale applications of graphene, it is highly essential to make them compression and aggregation-resistant in both liquid and solid state.

An alternative way to solve the aggregation problem is inspired by our daily life experience with crumpled paper balls [75]. Mechanically deforming a piece of paper turns it into a highly stable fractal dimensional structure that cannot be easily unfolded and never aggregate, even under compression [76]. So, if we consider graphene-like layers as thin sheets of paper, crumpling them up to paper ball-like structures should also significantly reduce re-stacking possibilities and render them excellent compression and aggregation-resistant properties. Preciously, using FQM, we have directly observed that capillary force can squeeze suspended GO sheets into heavily folded and wrinkled structure during evaporation [46,47]. Based on this, we have developed an aerosol-assisted capillary compression method to synthesize crumpled GO or r-GO balls [44]. In this method, a nebulizer is used to make free-floating aerosol droplets of aqueous dispersion of GO sheets. Drying of the droplets isotropically compress the sheets inside into crumpled balls. Figure 14.13 (top panel) shows the experimental set-up and procedure for producing crumpled balls from GO sheets. An aqueous dispersion of micrometre-sized GO sheets was nebulized to generate aerosol droplets, which were carried by N_2 gas flow through a tube furnace pre-heated at a given temperature to either dry off the water droplets or further thermally reduce GO. Rapid evaporation causes shrinkage of the droplets, which in turn concentrates and finally compresses GO sheets into crumpled balls (Figure 14.13(a) and (b)). The size and the extent of crumpling of these particles can be tuned by changing the initial concentration of the GO solution used for producing aerosols. With increasing GO concentration, the degree of crumpling reduces and particles became bigger. At higher concentration, many GO sheets undergo overlapping before crumpling, so the higher bending modulus of thicker GO flakes makes them stiffer against deformation and thus leads to the less-crumpled, larger particles.

Crumpling of thin GO sheets into ball-like structures significantly minimizes their van der Waals attraction but maintains the high surface area. The

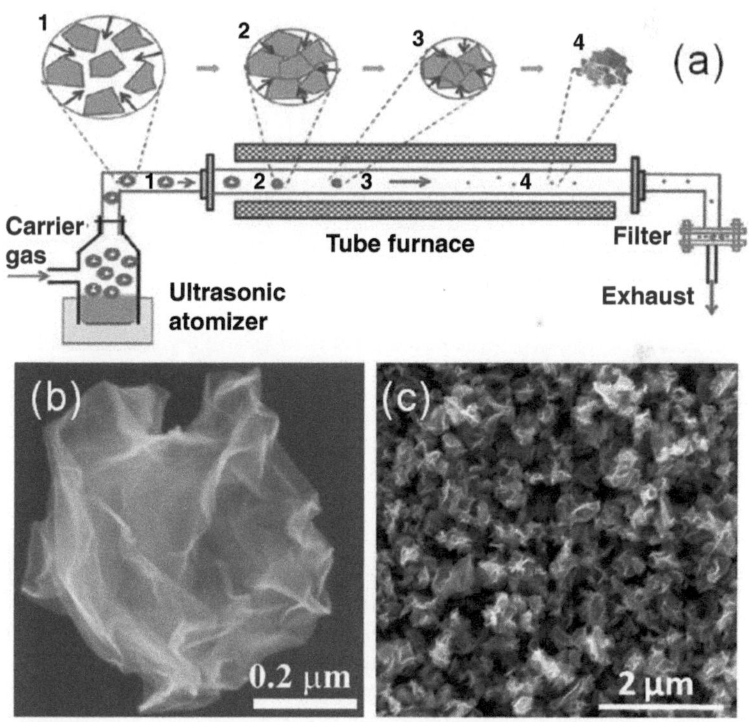

FIGURE 14.13 (a) Schematic illustration of evaporation-induced crumpling process and the experimental setup. (b) A representative high-magnification SEM image of a crumpled particle. (c) Low-magnification SEM overview of crumpled, ball-like particles. [Adapted and modified from [44], with permission from ACS Publications].

large density of ridges and vertices present on the surface of crumples diminishes probable overlapping area '*S*' and the 3D spherical shape increases the inter-sheet separation '*d*' enormously [14,44]. The reduced van der Waals attraction makes them aggregation-resistant and highly processable. Unlike flat sheets, crumpled particles do not stick to each other. The crumpled particles are very open and robust. They do not unfold or collapse under common material-processing conditions, such as solution processing, thermal shock, microwave heating and hydrazine treatment. With this morphology, chemically modified graphene sheets can be dispersed in a wide range of solvents by gentle 'hand shaking,' which is otherwise very difficult [14].

The crumpled graphene particles possess excellent aggregation resistance not only in solution state but also after being compressed in solid state. Figure 14.14 compares the compression resistance of crumpled graphene to sheetlike graphene. Both the samples were subjected to high pressure (55 MPa) to form a pellet, followed by SEM examination on the microstructures. The crumpled graphene balls exhibit similar microstructure at the surface and the cross-section (Figure 14.14(a) and (b)). On the other hand, wrinkled graphene sheets yields an anisotropic microstructure upon pelletization, having very smooth, featureless surface but laminated cross-section (Figure 14.14(c) and (d)). Even after this high-pressure compression, the crumpled graphene particles can still be readily re-dispersed in many solvents. The inset of Figure 14.14(e) is a photo of compressed crumpled particles re-dispersed in water by just 'hand shaking'.

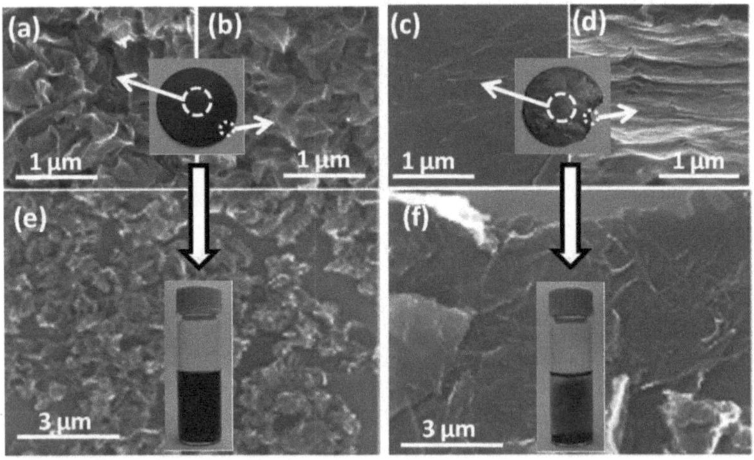

FIGURE 14.14 Aggregation-resistant properties of crumpled graphene balls. SEM images of the (a) surface and (b) cross-section of a pellet made from crumpled graphene particles compressed at 55 MPa. (c) Surface and (d) cross-sectional view of a pellet made from wrinkled sheets under similar conditions. (e) The pellet of crumpled graphene balls can be readily re-dispersed in water as individual particles by hand-shaking, while (f) the pellet of wrinkled sheets cannot be re-dispersed even after sonication due to extensive aggregation. [Adapted and modified from [44], with permission from ACS Publications].

SEM image of such re-dispersed crumpled particles clearly shows individual particles, slightly flattened but largely unaffected by the compression. In contrast, pellets of the wrinkled sheets prepared under the same conditions cannot be re-dispersed even after prolonged sonication (inset Figure 14.14(f)). The SEM examination shows that the sheets have re-stacked into thick graphite-like flakes.

In addition to the outstanding processability, the crumpled graphene balls have a few additional advantages. The ease of aggregation of graphene sheets makes their properties, especially the surface area, highly dependent on processing history. For example, the surface area of even wrinkled graphene sheets will rapidly decrease when they are subject to common material processing steps such as solvent washing and pelletization. In contrast, crumpled graphene balls are much more robust, showing consistently higher surface area [44]. It suggests crumpled graphene balls can help to standardize its bulk properties by making them more independent of processing history. This is especially important for making scalable graphene-based energy storage devices, where high electrode surface area at large mass loading level is crucial. In such applications crumpled graphene offer two major advantages over flat sheets. First, their higher surface area provides higher device performances. Second, the aggregation-resistant property of these particles provide more proportional enhancement of the performance with increasing mass loading. In the case of sheetlike graphene, the device performance significantly decreases with increasing mass loading because the higher loading level usually leads to higher degree of aggregation. In contrast, the specific capacitance of the crumpled graphene balls is more consistent with increasing loading level [77]. This clearly demonstrates that making graphene aggregation resistant is crucial for applications where bulk quantities of materials are used. Crumpled graphene balls have already stirred significant interests. We have demonstrated that they have superior performance as an expandable coating for nanoparticles used in lithium ion battery electrodes [78], biosensors [79, 80, 84, 85], as modifier for microbial fuel cells [81] and as a support for electrocatalysts [82]. Using a removable template, the aerosol process can readily be modified to produce hollow graphene capsulates [83] for oil absorption.

14.8 GO-BASED NANOFLUIDIC IONIC CONDUCTORS

In the earlier sections of this chapter, we described how revisiting of GO, the centuries-old materials, leads to discovery of many new facets of the materials. Here, we discuss another novel property and application of GO. Recently, we discovered that GO sheets can act as building blocks to construct massive arrays of few-nanometre-wide nanochannels. In spite of being highly polydisperse in lateral dimensions and negatively charged, GO sheets form an interconnecting network of 2D nanochannels upon re-stacking (Figure 14.15(a)). These 2D nanochannels can be easily explored for nanofluidics ion transport and other related applications [84]. In our experiment, an aqueous dispersion of GO was vacuum filtered to obtain a paperlike flexible film of stacked sheets. An SEM

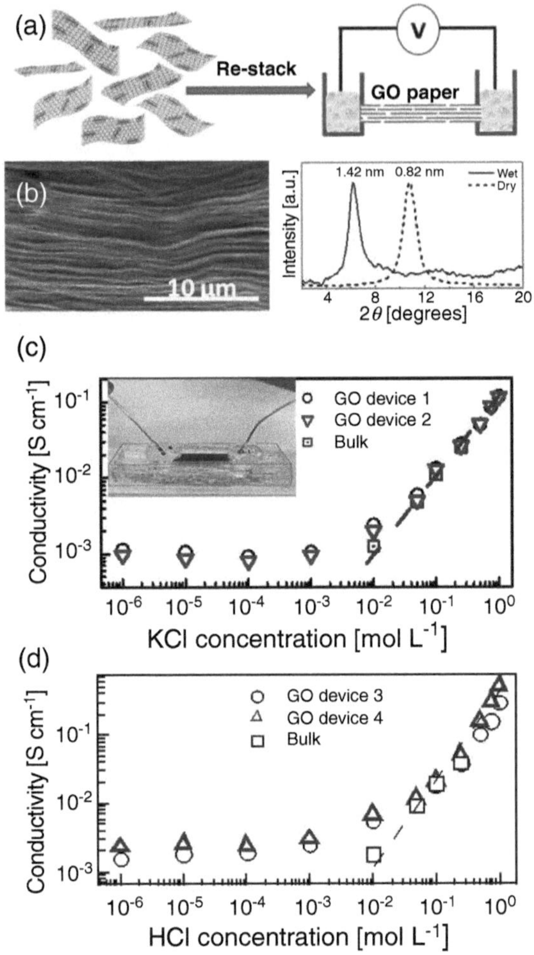

FIGURE 14.15 GO-based nanofluidic ion channels. (a) Schematic illustration of the nanofluidics device constructed from re-stacking exfoliated GO sheets. (b) Cross-sectional SEM image of a GO strip with the X-ray diffraction (XRD) patterns of air-dried and fully hydrated GO papers, respectively. Ionic conductivities of GO nanochannels as a function of electrolyte concentration for (c) KCl and (d) HCl solutions, respectively. Inset of (c) is a photograph of a GO paper-based nanofluidics device. [Adapted and modified from [84], with permission from ACS Publications].

image of the well-stacked GO layers at the cross-section of the film is shown in Figure 14.15(b). The average interlayer spacing of vacuum-filtered GO film was found to be ~0.82 nm by X-ray diffraction, which was increased to ~1.42 nm upon hydration by soaking in water for overnight (Figure 14.15(b)). The paper of GO was cut by scissors into rectangular pieces and embedded into freshly prepared PDMS pre-polymer and curing agent followed by gentle warming at 50–60 °C. Once the strip was completely embedded in the PDMS elastomer, two reservoirs of about 0.3 ml in volume were carved out using regular blades such a way that only both ends of the GO strip get exposed to electrolyte. Two Ag/AgCl electrodes were inserted into the source and drain reservoirs to measure ionic current through the GO paper (see schematic of Figure 14.15(a)). A photograph of the GO paper-based nanofluidics device is shown at the inset of Figure 14.15(c).

Ionic currents were measured through GO paper as function of electrolyte concentrations and compared with bulk ionic currents in Figure 14.15(c) and (d) for KCl and HCl, respectively. For each electrolyte, data is shown for two devices. The data in Figure 14.15(c) and (d) clearly shows that GO nanochannels exhibits characteristics of surface-charge-governed ionic transport, which is significantly different from that of the bulk. In bulk liquid, conductivity is directly proportional to concentration of the solution, but in GO nanofluidic devices, it show two distinct regimes of different behaviours. At high concentration just like bulk, conductivity varies linearly with changing concentrations. However, as the concentration drops below 0.05 M, it diverges from bulk and gradually plateaus at lower concentrations. At this concentration regime conductivity becomes independent of the ionic concentration and is always higher than corresponding bulk concentrations. The negative charges of GO sheets draw positively charged ions, but repel the anions. As a result, concentration of cations dramatically enhances inside the tiny volume of nanochannels and it is now determined by the electrostatic attraction of the GO sheets. Therefore, the conductivity remains unaffected by the changing bath concentrations and always higher than bulk, in that concentration regime.

Nanofluidic devices based on re-stacked GO sheets have distinct advantages such as low cost, facile fabrication, ease of scaling up to support high ionic currents and high flexibility. GO-based nanofluidic devices are very robust, and can be handled under high-bending-strain conditions [84]. These nanochannels exhibiting orders of magnitude higher ionic current than bulk liquids promise applications in batteries and fuel cells and new systems for harvesting energy, water purification and DNA sorting. With the rich selection of layered materials, their many possible ways for chemical functionalization and their intrinsic optoelectrical properties, layered materials or their hybrids should offer many exciting new opportunities for studying and manufacturing nanofluidic devices.

14.9 CONCLUSIONS

Other than being a graphene precursor, GO itself is a water-dispersable, single atomic, layered soft material with large lateral dimension that can be easily converted to conduct electricity. As a relatively new comer to carbon materials research since 2008, we benefit tremendously from the knowledge foundation established by numerous carbon nanomaterials researchers in the last few decades. We were also fortunate to see some new angles of GO, and even more fortunate to be able to bring some new insights to this field. Due to space limit, we are not able to discuss in detail some other new ideas derived from the amphiphilicity of GO, which uses GO sheets as a soft, amphiphilic template to alter the molecular configuration of partially aromatic polymers in water, such as conducting polymers and the carbonization product of hydrothermally treated carbon hydrates [87, 88, 89]. The bottom line about GO is that other than being a graphene precursor, GO itself is a water-dispersable, single atomic, layered soft material with large lateral dimensions that can be easily converted to conduct electricity [27]. Therefore, in addition to the new properties, tools, concepts and applications discussed in this chapter, we fully expect more interesting science and potentially useful 'beyond graphene' applications of GO to be uncovered in the near future.

ACKNOWLEDGEMENTS

The work described here has been mainly supported by a start-up fund from the Northwestern University, the Initiative for Sustainability and Energy at Northwestern (ISEN) through the Inaugural ISEN Early Career Investigator Award and a Booster Award, the National Science Foundation (CAREER DMR 0955612 for the discovery of GO-based surfactant and FQM; CMMI 1130407 for all-carbon photovoltaics) and the Office of Naval Research (N00014-13-1-0556 for crumpled graphene balls). Additional support is provided from Alfred P. Sloan Research Foundation through a Sloan Fellowship and the Sony Corporation through gift donation. J.H. thanks his current and former research group members and collaborators for the works described here.

REFERENCES

[1] Brodie BC. On the atomic weight of graphite. Philos Trans R Soc London 1859;149:249–59.
[2] Staudenmaier L. . Berichte der Deutschen Chemischen Gesellschaft 1898.
[3] Hummers WS, Offeman RE. Preparation of graphitic oxide. J Am Chem Soc 1958;80: 1339–1339.
[4] Croft RC. Lamellar compounds of graphite. Q Rev Chem Soc 1960;14:1–45.
[5] Novoselov KS, Geim AK, Morozov SV, Jiang D, Zhang Y, Dubonos SV, et al. Electric field effect in atomically thin carbon films. Science 2004;306:666–9.
[6] Geim AK, Novoselov KS. The rise of graphene. Nat Mater 2007;6:183–91.
[7] Li D, Kaner RB. Graphene-based materials. Science 2008;320:1170–1.
[8] Gao W, Alemany LB, Ci L, Ajayan PM. New insights into the structure and reduction of graphite oxide. Nat Chem 2009;1:403–8.
[9] Park S, Ruoff RS. Chemical methods for the production of graphenes. Nat Nanotechnol 2009;4:217–24.
[10] Rao CNR, Sood AK, Subrahmanyam KS, Govindaraj A. Graphene: the new two-dimensional nanomaterial. Angew Chem Int Ed 2009;48:7752–77.
[11] Allen MJ, Tung VC, Kaner RB. Honeycomb carbon: a review of graphene. Chem Rev 2010;110:132–45.
[12] Compton OC, Nguyen ST. Graphene oxide, highly reduced graphene oxide, and graphene: versatile building blocks for carbon-based materials. Small 2010;6:711–23.

[13] Eda G, Chhowalla M. Chemically derived graphene oxide: towards large-area thin-film electronics and optoelectronics. Adv Mater 2010;22:2392–415.

[14] Luo JY, Jang HD, Huang JX. Material processing of chemically modified graphene: some challenges and solutions. Acc Chem Res 2013;46(10):2225–34.

[15] Lerf A, He H, Forster M, Klinowski J. Structure of graphite oxide revisited. J Phys Chem B 1998;102:4477–82.

[16] Stankovich S, Dikin DA, Piner RD, Kohlhaas KA, Kleinhammes A, Jia Y, et al. Synthesis of graphene-based nanosheets via chemical reduction of exfoliated graphite oxide. Carbon 2007;45:1558–65.

[17] Wang X, Zhi L, Mullen K. Transparent, conductive graphene electrodes for dye-sensitized solar cells. Nano Lett 2007;8:323–7.

[18] Li D, Muller MB, Gilje S, Kaner RB, Wallace GG. Processable aqueous dispersions of graphene nanosheets. Nat Nanotechnol 2008;3:101–5.

[19] Becerril HcA, Mao J, Liu Z, Stoltenberg RM, Bao Z, Chen Y. Evaluation of solution-processed reduced graphene oxide films as transparent conductors. ACS Nano 2008;2:463–70.

[20] Mattevi C, Eda G, Agnoli S, Miller S, Mkhoyan KA, Celik O, et al. Evolution of electrical, chemical, and structural properties of transparent and conducting chemically derived graphene thin films. Adv Funct Mater 2009;19:2577–83.

[21] Erickson K, Erni R, Lee Z, Alem N, Gannett W, Zettl A. Determination of the local chemical structure of graphene oxide and reduced graphene oxide. Adv Mater 2010;22:4467–72.

[22] Han TH, Huang YK, Tan ATL, Dravid VP, Huang JX. Steam etched porous graphene oxide network for chemical sensing. J Am Chem Soc 2011;133:15264–7.

[23] Liu L, Ryu S, Tomasik MR, Stolyarova E, Jung N, Hybertsen MS, et al. Graphene oxidation: thickness-dependent etching and strong chemical doping. Nano Lett 2008;8:1965–70.

[24] Cote LJ, Kim F, Huang JX. Langmuir-Blodgett assembly of graphite oxide single layers. J Am Chem Soc 2009;131:1043–9.

[25] Kim F, Cote LJ, Huang JX. Graphene oxide: surface activity and two-dimensional assembly. Adv Mater 2010;22:1954–8.

[26] Cote LJ, Kim J, Tung VC, Luo JY, Kim F, Huang JX. Graphene oxide as surfactant sheets. Pure Appl Chem 2011;83:95–110.

[27] Kim J, Cote LJ, Huang JX. Two dimensional soft material: new faces of graphene oxide. Acc Chem Res 2012;45:1356–64.

[28] Jimenez PSV. Thermal-decomposition of graphite oxidation-products DSC studies of internal-combustion of graphite oxide. Mater Res Bull 1987;22:601–8.

[29] Uhl FM, Wilkie CA. Preparation of nanocomposites from styrene and modified graphite oxides. Polym Degrad Stab 2004;84:215–26.

[30] Zhang R, Hu Y, Xu JY, Fan WC, Chen ZY. Flammability and thermal stability studies of styrene-butyl acrylate copolymer/graphite oxide nanocomposite. Polym Degrad Stab 2004;85: 583–8.

[31] Zhang R, Hu Y, Xu JY, Fan WC, Chen ZY, Wang QN. Preparation and combustion properties of flame retardant styrene-butyl acrylate copolymer/graphite oxide nanocomposites. Macromol Mater Eng 2004;289:355–9.

[32] Cui WG, Guo F, Chen JF. Flame retardancy and toughening of high impact polystyrene. Polym Compos 2007;28:551–9.

[33] Dasari A, Yu ZZ, Mai YW, Cai GP, Song HH. Roles of graphite oxide, clay and POSS during the combustion of polyamide 6. Polymer 2009;50:1577–87.

[34] Higginbotham AL, Lomeda JR, Morgan AB, Tour JM. Graphite oxide flame-retardant polymer nanocomposites. ACS Appl Mater Inter 2009;1:2256–61.

[35] Kim F, Luo JY, Cruz-Silva R, Cote LJ, Sohn K, Huang JX. Self-propagating domino-like reactions in oxidized graphite. Adv Funct Mater 2010;20:2867–73.

[36] Krishnan D, Kim F, Luo JY, Cruz-Silva R, Cote LJ, Jang HD, et al. Energetic graphene oxide: challenges and opportunities. Nano Today 2012;7:137–52.

[37] Cote LJ, Cruz-Silva R, Huang J. Flash reduction and patterning of graphite oxide and its polymer composite. J Am Chem Soc 2009;131:11027–32.

[38] Boehm HP, Scholz W. Deflagration point" of graphite oxide. Z Anorg Allg Chem 1965;335:74–9.

[39] Veraa MJ, Bell AT. Effect of alkali metal catalysts on gasification of coal char. Fuel 1978;57:194–200.

[40] Marsh H, Rodríguez-Reinoso F. Activated carbon. Amsterdam; London: Elsevier; 2006. p. 536.

[41] Nikolaev AV, Nazarov AS, Lisitsa VV. Preparation of graphite oxide. Russ J Inorg Chem 1974;19:3396–7.

[42] Nair RR, Blake P, Grigorenko AN, Novoselov KS, Booth TJ, Stauber T, et al. Fine structure constant defines visual transparency of graphene. Science 2008;320:1308–1308.

[43] Jung I, Pelton M, Piner R, Dikin DA, Stankovich S, Watcharotone S, et al. Simple approach for high-contrast optical imaging and characterization of graphene-based sheets. Nano Lett 2007;7:3569–75.

[44] Luo J, Jang HD, Sun T, Xiao L, He Z, Katsoulidis AP, et al. Compression and aggregation-resistant particles of crumpled soft sheets. ACS Nano 2011;5:8943–9.

[45] Kagan MR, McCreery RL. Reduction of fluorescence interference in Raman spectroscopy via analyte adsorption on graphitic carbon. Anal Chem 1994;66:4159–65.

[46] Kim J, Cote LJ, Kim F, Huang J. Visualizing graphene based sheets by fluorescence quenching microscopy. J Am Chem Soc 2010;132:260–7.

[47] Kim J, Kim F, Huang J. Seeing graphene-based sheets. Mater Today 2010;13:28–38.

[48] Tan ATL, Kim J, Huang J-K, Li L-J, Huang J. Seeing two-dimensional sheets on arbitrary substrates by fluorescence quenching microscopy. Small 2013;9(19):3253–8.

[49] Swathi RS, Sebastian KL. Long range resonance energy transfer from a dye molecule to graphene has (distance)$^{-4}$ dependence. J Chem Phys 2009;130:086101.

[50] Kim J, Cote LJ, Kim F, Yuan W, Shull KR, Huang JX. Graphene oxide sheets at interfaces. J Am Chem Soc 2010;132:8180–6.

[51] Cote LJ, Kim J, Zhang Z, Sun C, Huang JX. Tunable assembly of graphene oxide surfactant sheets: wrinkles, overlaps and impacts on thin film properties. Soft Matter 2010;6:6096–101.

[52] Lipp MM, Lee KYC, Zasadzinski JA, Waring AJ. Design and performance of an integrated fluorescence, polarized fluorescence, and Brewster angle microscope Langmuir trough assembly for the study of lung surfactant monolayers. Rev Sci Instrum 1997;68:2574–82.

[53] Kolmakov A, Dikin DA, Cote LJ, Huang JX, Abyaneh MK, Amati M, et al. Graphene oxide windows for in situ environmental cell photoelectron spectroscopy. Nat Nanotechnol 2011;6:651–7.

[54] Krueger M, Berg S, Stone DA, Strelcov E, Dikin DA, Kim J, et al. Drop-casted self-assembling graphene oxide membranes for scanning electron microscopy on wet and dense gaseous samples. ACS Nano 2011;5:10047–54.

[55] Murray IP, Lou SJ, Cote LJ, Loser S, Kadleck CJ, Xu T, et al. Graphene oxide interlayers for robust, high-efficiency organic photovoltaics. J Phys Chem Lett 2011;2:3006–12.

[56] Dan B, Behabtu N, Martinez A, Evans JS, Kosynkin DV, Tour JM, et al. Liquid crystals of aqueous, giant graphene oxide flakes. Soft Matter 2011;7:11154–9.

[57] Li S-S, Tu K-H, Lin C-C, Chen C-W, Chhowalla M. Solution-processable graphene oxide as an efficient hole transport layer in polymer solar cells. ACS Nano 2010;4:3169–74.

[58] Kim J, Tung VC, Huang JX. Water processable graphene oxide: single walled carbon nanotube composite as anode modifier for polymer solar cells. Adv Energy Mater 2011;1:1052–7.
[59] Tung VC, Kim J, Huang JX. Graphene Oxide: Single Walled Carbon Nanotube Based Interfacial Layer for All-solution-processed Multijunction Solar Cells in Both Regular and Inverted Geometries. Adv Energy Mater 2012;2:299–303.
[60] Tung VC, Huang JH, Kim J, Smith AJ, Chu CW, Huang JX. Towards solution processed all-carbon solar cells: a perspective. Energy Environ Sci 2012;5:7810–8.
[61] Tung VC, Huang J-H, Tevis I, Kim F, Kim J, Chu C-W, et al. Surfactant-free water-processable photoconductive all-carbon composite. J Am Chem Soc 2011;133:4940–7.
[62] Umeyama T, Tezuka N, Seki S, Matano Y, Nishi M, Hirao K, et al. Selective formation and efficient photocurrent generation of [70]fullerene-single-walled carbon nanotube composites. Adv Mater 2010;22:1767–70.
[63] Kalita G, Adhikari S, Aryal HR, Umeno M, Afre R, Soga T, et al. Fullerene (C(60)) decoration in oxygen plasma treated multiwalled carbon nanotubes for photovoltaic application. Appl Phys Lett 2008;92:063508.
[64] Umeyama T, Tezuka N, Fujita M, Hayashi S, Kadota N, Matano Y, et al. Clusterization, electrophoretic deposition, and photoelectrochemical properties of fullerene-functionalized carbon nanotube composites. Chem Eur J 2008;14:4875–85.
[65] Yamamoto Y, Zhang GX, Jin WS, Fukushima T, Ishii N, Saeki A, et al. Ambipolar-transporting coaxial nanotubes with a tailored molecular graphene-fullerene heterojunction. Proc Natl Acad Sci 2009;106:21051–6.
[66] Yamamoto Y, Fukushima T, Suna Y, Ishii N, Saeki A, Seki S, et al. Photoconductive coaxial nanotubes of molecularly connected electron donor and acceptor layers. Science 2006;314:1761–4.
[67] Bindl DJ, Wu M-Y, Prehn FC, Arnold MS. Efficiently harvesting excitons from electronic type-controlled semiconducting carbon nanotube films. Nano Lett 2010;11:455–60.
[68] Shea MJ, Arnold MS. 1% Solar cells derived from ultrathin carbon nanotube photoabsorbing films. Appl Phys Lett 2013;102:243101–5.
[69] Ramuz MP, Vosgueritchian M, Wei P, Wang C, Gao Y, Wu Y, et al. Evaluation of solution-processable carbon-based electrodes for all-carbon solar cells. ACS Nano 2012;6:10384–95.
[70] Jain RM, Howden R, Tvrdy K, Shimizu S, Hilmer AJ, McNicholas TP, et al. Polymer-free near-infrared photovoltaics with single chirality (6,5) semiconducting carbon nanotube active layers. Adv Mater 2012;24:4436–9.
[71] Bernardi M, Lohrman J, Kumar PV, Kirkeminde A, Ferralis N, Grossman JC, et al. Nanocarbon-based photovoltaics. ACS Nano 2012;6:8896–903.
[72] Brennan RO. The interlayer binding in graphite. J Chem Phys 1952;20:40–8.
[73] Luo JY, Cote LJ, Tung VC, Tan ATL, Goins PE, Wu JS, et al. Graphene oxide nanocolloids. J Am Chem Soc 2010;132:17667–9.
[74] Israelachvili JN. Intermolecular and surface forces. 3rd ed. Academic Press; 2010. p. 674.
[75] Evans J. Curious minds. Chem Industry 2012;76:32–5.
[76] Matan K, Williams RB, Witten TA, Nagel SR. Crumpling a thin sheet. Phys Rev Lett 2002;88:076101.
[77] Luo JY, Jang HD, Huang JX. Effect of sheet morphology on the scalability of graphene-based ultracapacitors. ACS Nano 2013;7:1464–71.
[78] Luo JY, Zhao X, Wu JS, Jang HD, Kung HH, Huang JX. Crumpled graphene-encapsulated Si nanoparticles for lithium ion battery anodes. J Phys Chem Lett 2012;3:1824–9.
[79] Chen YT, Guo F, Jachak A, Kim SP, Datta D, Liu JY, et al. Aerosol synthesis of cargo-filled graphene nanosacks. Nano Lett 2012;12:1996–2002.

[80] Jang HD, Kim SK, Chang H, Roh KM, Choi JW, Huang JX. A glucose biosensor based on Tio_2-graphene composite. Biosens Bioelectron 2012;38:184–8.

[81] Xiao L, Damien J, Luo JY, Jang HD, Huang JX, He Z. Crumpled graphene particles for microbial fuel cell electrodes. J Power Sources 2012;208:187–92.

[82] Jang HD, Kim SK, Chang H, Choi JW, Luo JY, Huang JX. One-step synthesis of Pt-nanoparticles-laden graphene crumples by aerosol spray pyrolysis and evaluation of their electrocatalytic activity. Aerosol Sci Technol 2013;47:93–8.

[83] Sohn K, Na YJ, Chang H, Roh KM, Jang HD, Huang JX. Oil absorbing graphene capsules by capillary molding. Chem Commun 2012;48:5968–70.

[84] Kim SK, Chang H, Choi J-W, Huang J, Jang HD. Aerosol Processing of Graphene and Its Application to Oil Absorbent and Glucose Biosensor. KONA 2014;31:111–25.

[85] Jang HD, Kim SK, Chang H, Choi J-W, Huang J. Synthesis of graphene based noble metal composites for glucose biosensor. Mater Lett 2013;106:277–80.

[86] Raidongia K, Huang JX. Nanofluidic ion transport through reconstructed layered materials. J Am Chem Soc 2012;134:16528–31.

[87] Tung VC, Kim J, Cote LJ, Huang J. Sticky Interconnect for Solution-Processed Tandem Solar. Cells J Am Chem Soc 2011;133:9262–5.

[88] Luo J, Tung VC, Koltonow AR, Jang HD, Huang J. Graphene Oxide based Conductive Glue as Binder for Ultracapacitor Electrodes. J Mater Chem 2012;22:12993–6.

[89] Krishnan D, Raidongia K, Shao J, Huang J. Graphene Oxide Assisted Hydrothermal Carbonization of Carbon Hydrates. ACS Nano. 2014;8:449–57.

Chapter 15

Graphene Nanoribbon and Nanographene

Kazuyoshi Tanaka
Department of Molecular Engineering, Kyoto University, Nishikyo-ku, Kyoto 615-8510, Japan

15.1 INTRODUCTION

Not only a complete graphene but also graphene fragments belong to intriguing carbon materials group. The graphene fragments can be classified into two categories as shown in Figure 15.1, i.e. graphene nanoribbon and nanographene. The degree of hybridization of these materials is of sp^2 and they are on the vertical line in Figure 1.1 as shown in Chapter 1. Although many of these materials belonging to this group have not been actually synthesized yet, they are of considerable importance and hence described in this chapter.

Graphene nanoribbon system in Figure 15.1(a) generally has the structure developed toward one-dimensional direction, whereas nanographene system in Figure 15.1(b) develops rather two dimensionally, both starting from a benzene molecule. Unless specially noted, we consider that edge carbon is connected with hydrogen atoms for termination.

Infinite growth of both the graphene nanoribbons and nanographenes eventually leads to the formation of an infinite graphene sheet. In this sense, the larger size they become, the more resemblance of their characteristics to that of graphene should occur.

15.2 GRAPHENE NANORIBBON

This material is characterized by the shapes with lengthened along one direction and was once referred to as one-dimensional graphite or graphene. The most basic member is polyacene (PA), illustrated in Figure 15.2(a), having infinite length along the long-axis direction, i.e. PA is a polymer. A structural isomer of PA can be considered to be polyphenanthrene (PPh) in Figure 15.2(b). Both the PA and PPh have the [H]/[C] atomic ratio of 0.5 lying just on the middle of the line connecting polyacetylene ([H]/[C] = 1) and graphene ([H]/[C] = 0) (see Figure 1.1). Both of these polymers have not yet been synthesized; we will mainly mention the conjectured electronic properties of these and the relatives in Sections 15.2.1 and 15.2.2, respectively.

Carbon Nanotubes and Graphene. DOI: 10.1016/B978-0-08-098232-8.00015-2

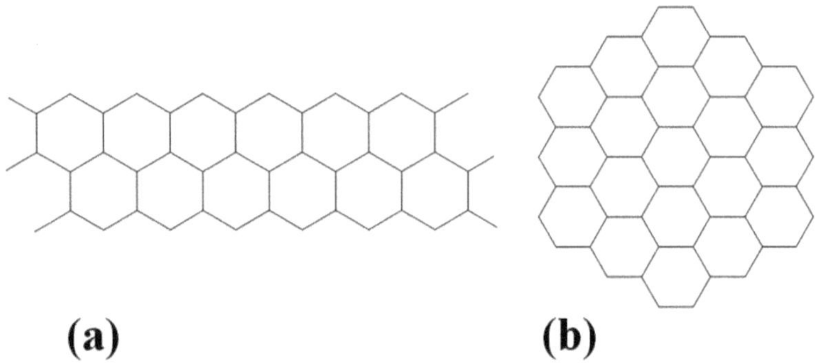

FIGURE 15.1 Modelled graphene fragments: (a) graphene nanoribbon and (b) nanographene. Note that the bond alternation at the skeletons and the terminal hydrogen atoms are abbreviated.

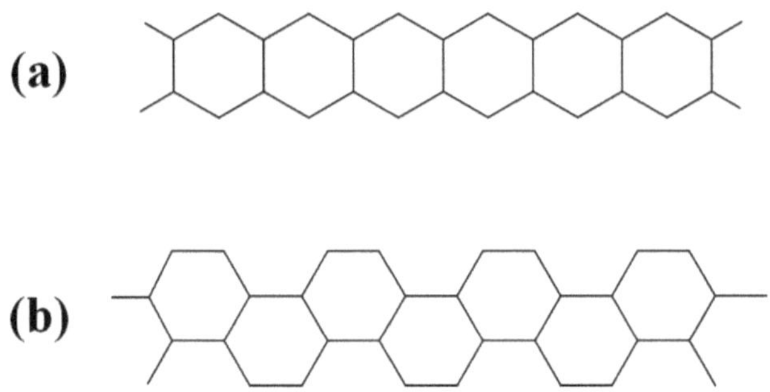

FIGURE 15.2 Typical graphene nanoribbons: (a) PA and (b) PPh. See the legend of Figure 15.1.

15.2.1 Polyacene (PA)

This is a fictitious polymer having not synthesized yet, but its electronic structure has theoretically been examined from the 1960s using the Hückel crystal-orbital (CO) method [1]. The conclusion at those days was that, when PA has no bond alternations, the energy difference of the top of the highest occupied (HO) band and the bottom of the lowest unoccupied (LU) band, i.e. the bandgap value, becomes zero. But later it became clear [2] that this is just an apparent feature when the interactions between carbon atoms is cut off within the nearest neighbouring couplings used in the Hückel method.

After appearance of polyacetylene as organic conductive polymer in the late 1970s, PA has again become a focus of interest as its relative material and the electronic properties have been examined [2–4]. What have become clear from the analyses for PA with the use of CO method within the one-electron approximation [2,3,5] are the following:

1. There likely exists carbon–carbon bond alternation, although it is in a small scale.

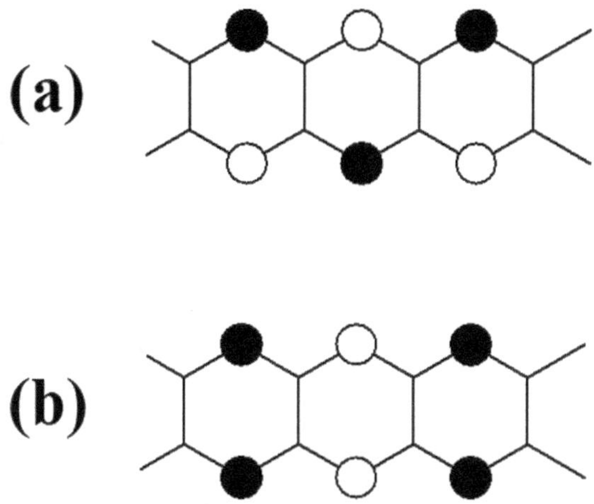

FIGURE 15.3 (a) The HOCO and (b) the LUCO patterns of PA. Note that these are views of the 2pπ orbitals from the above. Also see the legend of Figure 15.1.

2. The HOCO and the LUCO are both of π-nature, with the substantial bandgap being less than 1 eV irrespective of the calculation method.
3. The orbital patterns of both the HOCO and the LUCO only exist on the carbon atoms at the edge of the PA structure as shown in Figure 15.3.

The above (2) implies the pseudo-degeneracy of the HOCO and the LUCO signifying that PA will show strong electronic correlation and, furthermore, possibility of the anti-ferromagnetic ground state with the open-shell singlet configuration. This situation would require more elaborate discussion with detailed check of the electronic structure.

A simple Hückel CO calculation for wider nanoribbon PA(N), with N being the width of the ribbon as shown in Figure 15.4, gives rapid decrease in the bandgap [5]. In other words, the electronic structure of PA(N) rather rapidly approaches that of graphene and degeneracy of the HOCO and the LUCO is enhanced. Figure 15.5(a) and (b) illustrates the patterns of the HOCO and the LUCO of PA(4) within the framework of the one-electron approximation. These clearly show the tendency of the patterns to exist only on the edge carbons and, as a result, spin polarization clearly tends to occur on those atoms. Note that the edges of all PA(N) family are zigzag shapes, being somewhat similar to the edge of carbon nanotube (CNT) classified by its chiral indices. The image of the spin alignment is illustrated in Figure 15.5(c). The anti-ferromagnetic spin correlation should occur between both the edges of PA(N) on account of the quite small bandgap and the CO patterns as stated above.

15.2.2 Polyphenanthrene (PPh)

PPh in Figure 15.2(b) is a structural isomer of PA with the same [H]/[C] atomic ratio. However, their electronic properties are pretty different. For instance, the bandgap of PPh is 3.59 eV reminiscent of its aromaticity and stability [6]. Hence,

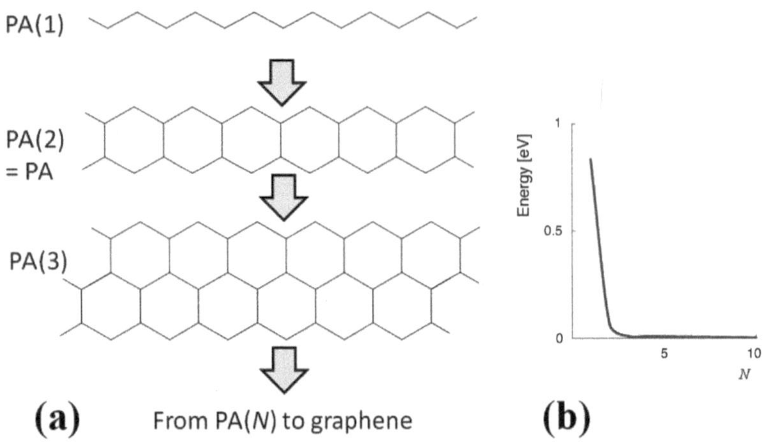

FIGURE 15.4 (a) The mode of increase in N (ribbon width) of PA(N) series and (b) change in the bandgap values depending on N. Note that this value rapidly decreases at PA(2), i.e. at PA.

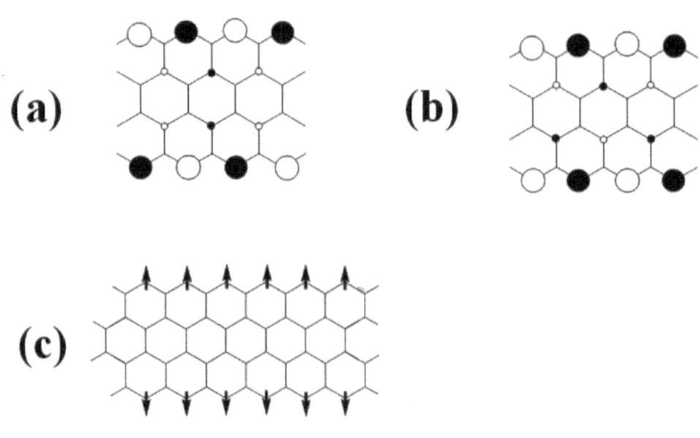

FIGURE 15.5 (a) The HOCO and (b) the LUCO patterns of PA(4). Schematic spin alignments with the anti-ferromagnetic correlation are shown in (c). See also the legend of Figure 15.3.

PPh will show the semiconductive property providing with another possibility to the electronic material. Moreover, the HOCO and the LUCO patterns tend to delocalize on the whole carbon atoms of the polymer skeleton compared with PA [2,7].

The bandgap values show characteristic oscillatory mode with an increase in the width of ribbon, i.e. increase in N in PPh(N) as in Figure 15.6, and eventually become close to zero to turn into graphene [5,7]. Both the HOCO and the LUCO patterns of PPh(4) shown in Figure 15.7, for instance, delocalize onto the carbon atoms inside the ribbon compared with those of PA(4) in Figure 15.5. Also note that the edges of all PPh(N) family are armchair shapes, again somewhat similar to the edge of CNT classified by its chiral indices. PPh, like PA, has not been synthesized and currently syntheses of oligophenanthrenes are being attempted as described in the next section.

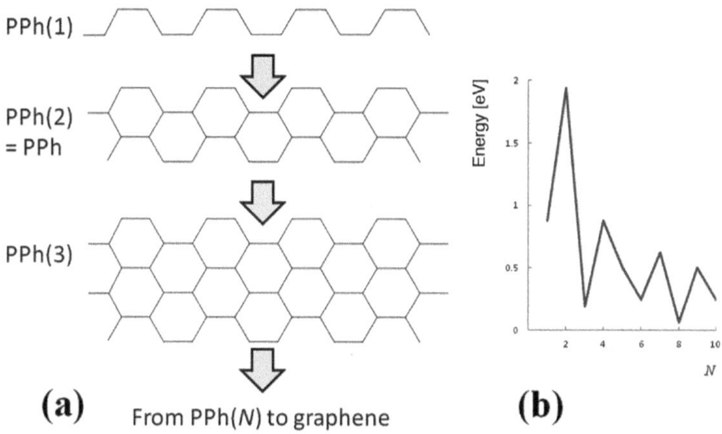

FIGURE 15.6 (a) The mode of increase in *N* (ribbon width) of PPh(*N*) series and (b) change in the bandgap values depending on *N*.

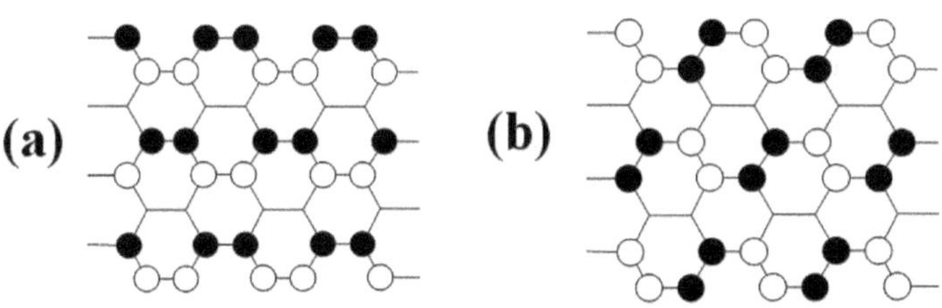

FIGURE 15.7 (a) The HOCO and (b) the LUCO patterns of PPh(4). See the legend of Figure 15.3.

15.2.3 Oligoacene and oligophenanthrene

Although the complete PA has not yet been synthesized, oligoacenes with finite length are being prepared hitherto step by step. Nonacene with nine condensed aromatic rings in Figure 15.8(a), which is the longest oligoacene up to now, has been synthesized [8]. Chemical reactivity of nonacene or octacene is quite high, and therefore they have to be prepared under inert atmosphere, such as in argon matrix of 30 K, and then transformed into the precursor (bridged α-diketone) by the photoradiation. Pentacene and rubrene in Figure 15.8(b) and (c), respectively, have large electron mobility (3 and 15 $cm^2\ V^{-1}\ s^{-1}$, respectively) [9,10] and are promising as the semiconductive material to build up organic transistor.

Theoretical investigations on oligoacenes have also been continued with respect to carbon–carbon bond alternation pattern and the electronic ground state [3,4]. These oligoacenes have been predicted to have anti-ferromagnetic ground state with the singlet open shell [11]. This is somewhat similar to that of PA described in Sec. 15.2.1.

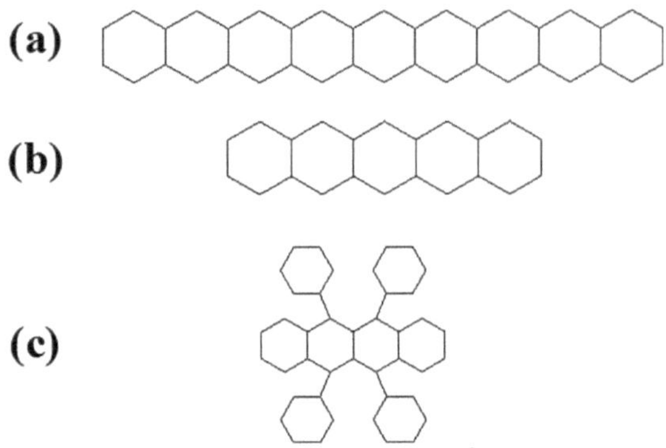

FIGURE 15.8 Oligoacenes: (a) nonacene, (b) pentacene and (c) rubrene. See the legend of Figure 15.1.

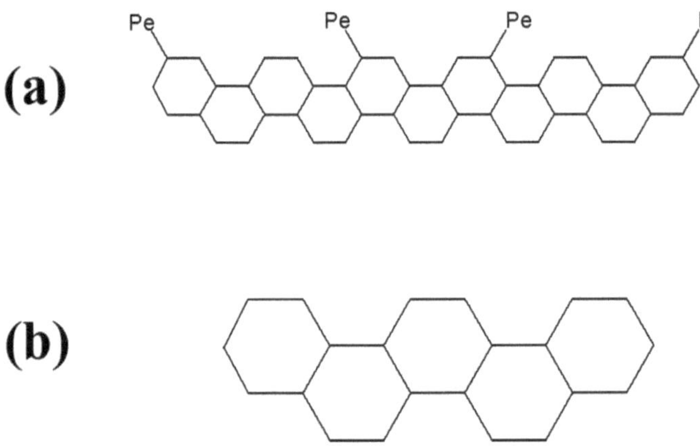

FIGURE 15.9 Oligophenanthrenes: (a) 2, 17, 21, 24-tetrapentyl[11]phenacene and (b) picene ([5] phenacene). See the legend of Figure 15.1.

Also, only some oligophenanthrenes (oligophenacenes) have been synthesized. The current longest oligophenanthrene is 2,17,21,24-tetrapentyl[11] phenacene having the structure with condensed 11 benzene rings and it is chemically stable (Figure 15.9(a)) [12]. The electronic transition spectrum of this [11]phenacene with the longest wavelength is at 400 nm, the energy of which is larger than the highest occupied molecular orbital (HOMO)–lowest unoccupied (LU) MO gap values of oligoacenes. The value of the unsubstituted [11]phenacene has been estimated to be 382 nm (3.25 eV) [11]. Such larger HOMO–LUMO gap suggests larger aromaticity of oligophenanthrene than oligoacene, which is a characteristic difference between these two [13].

From the viewpoint of organic chemistry, the more numbers of the 'limit structures' of benzene molecule within the condensed aromatic rings signify higher aromaticity [14]. This situation is shown in Figure 15.10(a) and (b) with respect to, for instance, anthracene and phenanthrene. Phenanthrene is more

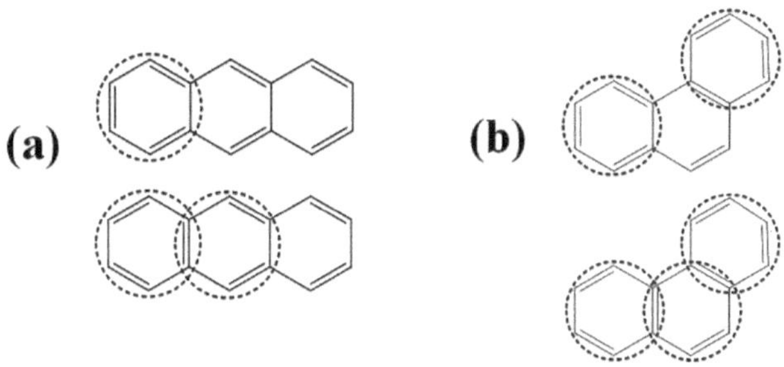

FIGURE 15.10 Comparison of aromaticity between (a) anthracene and (b) phenanthrene. There are more numbers of benzene-type 'limit structures' (surrounded with broken circle) in phenanthrene indicating more aromaticity of this molecule.

aromatic than anthracene as depicted in Figure 15.10. Similar discussion has been given that PPh is more aromatic than PA [2]. What is additionally interesting is that superconductivity with the transition temperature of 18 K appears in potassium- or rubidium-doped [5]phenacene (= picene) in Figure 15.9(b) which is a rather short oligomer with five condensed benzene rings [15].

15.2.4 Other graphene nanoribbons

Polyfluoranthene belongs to a kind of graphene nanoribbon also theoretically designed [16], which contains five-membered rings in the structure and alignment of oligoacenes along both the polymer chain direction and its orthogonal direction as in Figure 15.11. It has been reported that adjustment of the p and m values enables tuning of the bandgap values. There have also been attempts at actual synthesis with respect to both of these oligomers and polymers [17,18].

There are other graphene nanoribbon members called polyarylene series PAr(N) which can be regarded as those starting from poly(p-phenylene) as in Figure 15.12. The bandgap value of polyperylene, PAr(3), has theoretically been estimated as 0.44 eV [19] and there has also been attempt to synthesize the precursor toward polyperylene [20].

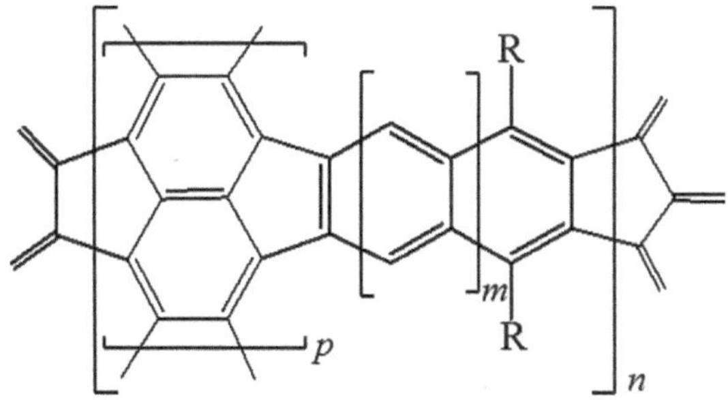

FIGURE 15.11 Polyfluoranthene.

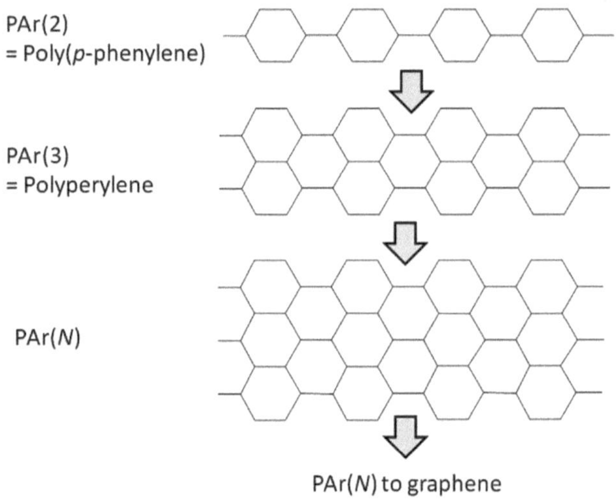

FIGURE 15.12 Polyarylene series PAr(*N*).

15.3 NANOGRAPHENE

Compared with graphene nanoribbon described in Sec. 15.2, nanographene is a name of graphene fragments rather developed to the two-dimensional direction. Members of nanographene can be broadly considered to consist of large condensed (fused) aromatic rings starting from benzene as shown in Figure 15.13.

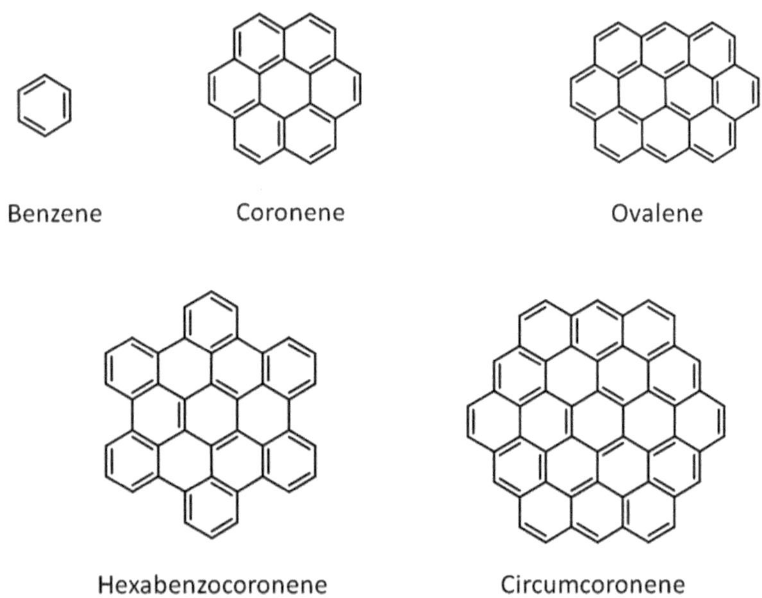

FIGURE 15.13 Aromatic hydrocarbons toward nanographene.

For these members they can be classified as those with zigzag-type edge and armchair-type edge, as shown in Figure 15.14, in a similar sense to graphene nanoribbon families. Some nanographenes with zigzag and armchair edges terminated with hydrogen atoms have actually been observed with scanning tunnelling microscope (STM) method [21].

The HOMO and the LUMO patterns are shown in Figure 15.15 [22]. It is clearly seen that these patterns concentrate on the edge of the nanographene with the zigzag type and that they uniformly delocalize inside the fragment of those with the armchair type. These features seem to be common to those in graphene nanoribbon, which is considered to represent the universal characteristics of the partial graphenes with two kinds of edges. Larger nanographenes with zigzag edge have smaller HOMO–LUMO gap and show bundle-like assembly of orbital levels near the frontier orbital levels, but those with armchair edge

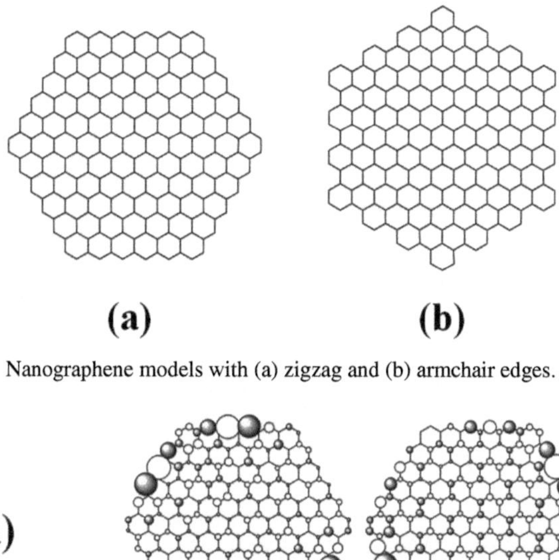

FIGURE 15.14 Nanographene models with (a) zigzag and (b) armchair edges.

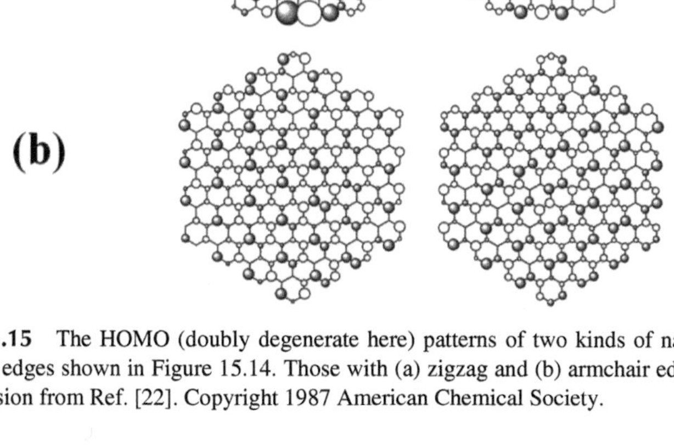

FIGURE 15.15 The HOMO (doubly degenerate here) patterns of two kinds of nanographenes with typical edges shown in Figure 15.14. Those with (a) zigzag and (b) armchair edges. Adapted with permission from Ref. [22]. Copyright 1987 American Chemical Society.

possess relatively wider HOMO–LUMO gap [22,23] or, in other words, more aromatic characteristics. The electronic structures of larger nanographenes will eventually approach those of graphene itself.

It can be supposed that there are some similarities between graphene nanoribbon and nanographene, when they are classified with their edge shape. This will provide with fundamental importance to consider these carbon materials and is desirable to further pursuit this aspect as well [24].

REFERENCES

[1] Salem L, Longuet-Higgins HC. The alternation of bond lengths in long conjugated molecules. II. The polyacenes. Proc R Soc (London) 1960;A255:435.
[2] Tanaka K, Ohzeki K, Nankai S, Yamabe T, Shirakawa H. The electronic structures of polyacene and polyphenanthrene. J Phys Chem Solids 1983;44:1069.
[3] Kertesz M, Choi CH, Yang S. Conjugated polymers and aromaticity. Chem Rev 2005;105:3448.
[4] Bettinger HF. Electronic structure of higher acenes and polyacene: the perspective developed by theoretical analyses. Pure Appl Chem 2010;82:905.
[5] Tanaka K, Yamashita S, Yamabe H, Yamabe T. Electronic properties of one-dimensional graphite family. Synth Met 1987;17:143.
[6] Tian Y-H, Park G, Kertesz M. Electronic structure of helicenes, C_2S helicenes, and thiaheterohelicenes. Chem Mater 2008;20:3266.
[7] Yoshizawa K, Yahara K, Tanaka K, Yamabe T. Bandgap oscillation in polyphenanthrenes. J Phys Chem B 1998;102:498.
[8] Tönshoff C, Bettinger HF. Photogeneration of octacene and nonacene. Angew Chem Int Ed 2010;49:4125.
[9] Klauk H, Halik M, Zschieschang U, Schmid G, Radlik W. High-mobility polymer gate dielectric pentacene thin film transistors. J Appl Phys 2002;92:5259.
[10] Sundar VC, Zaumseil J, Podzorov V, Menard E, Willett RL, Someya T, et al. Elastomeric transistor stamps: reversible probing of charge transport in organic crystals. Science 2004;303:1644.
[11] Jiang D, Dai S. Electronic ground state of higher acenes. J Phys Chem A 2008;112:332.
[12] Mallory FB, Butler KF, Evans AC, Brondyke EJ, Mallory CW, Yang C, et al. Phenacenes: a family of graphite ribbons. 2. Syntheses of some [7]phenacenes and an [11]phenacene by stilbene-like photocyclizations. J Am Chem Soc 1997;119:2119.
[13] Wiberg KB. Properties of some condensed aromatic systems. J Org Chem 1997;62:5720.
[14] Clar E. The aromatic sextet. New York: Wiley; 1972.
[15] Mitsuhashi R, Suzuki Y, Yamanari Y, Mitamura H, Kambe T, Ikeda N, et al. Superconductivity in alkali-metal-doped picene. Nature 2010;464:76.
[16] Kertesz M, Ashertehrani A. Electronic structure of polyfluoranthene ladder polymers. Macromolecules 1996;29:940.
[17] Schlüter A-D, Löffler M, Enkelmann V. Synthesis of a fully unsaturated all-carbon ladder polymer. Nature 1994;368:831.
[18] Löffler M, Schlüter A-D, Gessler K, Saenger W, Toussaint J-M, Brédas JL. Synthesis of a fully unsaturated "molecular board". Angew Chem Int Ed 1994;33:2209.

[19] Brédas J-L, Baughman RH. Theoretical study of the electronic properties and crystal structure of poly(perynaphthalene): on the origin of high observed conductivities. J Chem Phys 1985;83:1316.

[20] Former C, Becker S, Grimsdale AC, Müllen K. Cyclodehydrogenation of poly(perylene) to poly(quaterrylene): toward poly(*peri*-naphthalene). Macromolecules 2002;35:1576.

[21] Kobayashi Y, Fukui K, Enoki T, Kusakabe K, Kaburagi Y. Observation of zigzag and armchair edges of graphite using scanning tunnelling microscopy and spectroscopy. Phys. Rev. B 2005;71:193406.

[22] Stein SE, Brown RL. π-Electron properties of large condensed polyaromatic hydrocarbons. J Am Chem Soc 1987;109:3721.

[23] Yamabe T, Yamashita S, Yamabe H, Fukui K, Tanaka K. Electronic structures of electrically conductive fragment models in pyrolyzed polymers. Collection Czechoslovak Chem Commun 1988;53:1881.

[24] Tanaka K, Kobashi M, Yamabe T. Electronic structures and ring currents in planar microcluster models. Intern J Quantum Chem Quantum Chem Symp 1989;23:641.

Chapter 16

Application of Functional Hybrids Incorporating Carbon Nanotubes or Graphene

Cameron J. Shearer, Alexey Cherevan, Dominik Eder
Institut für Physikalische Chemie, Westfälische Wilhelms-Universität Münster, Corrensstrasse 28-30, 48149 Münster, Germany

16.1 INTRODUCTION

The intrinsic properties of carbon nanotubes (CNTs) and graphene (G), such as excellent electric and thermal conductivity, mechanical strength and toughness, thermal and chemical stability, high aspect ratio and a large surface area, have triggered a tremendous interest for both fundamental and applied research. Their performance in applications as well the number of potential applications can be further increased by combining nanocarbons with a second (or third etc.) functional material. This can be achieved in two ways: (1) the random mixing of CNTs or graphene with a second material to form a composite or (2) the designed interaction of CNTs or graphene with a second material to form a hybrid (Table 16.1). Since nomenclature used within current literature is often inconsistent between nanocomposites and hybrids, the next section shall offer suitable definitions that will be subsequently used in this chapter [1,2].

1. Nanocarbon nanocomposites are usually produced by mechanically mixing the filler nanocarbon into a matrix material (e.g. polymer, ceramic), which often results in a heterogeneous distribution of the filler and consequently in non-uniform properties. The resulting nanocomposite typically displays the intrinsic properties of the individual components, such as the high electrical conductivity of CNTs to create a conducting polymer, or the mechanical properties to reinforce metals and ceramics. Nanocomposites have considerable potential impact in industrial applications, such as lightweight structural materials in aerospace [3–5] and conductive plastics [6–8].
2. Nanocarbon hybrids are produced via the designed coating of the hybrid component onto the nanocarbon or the encapsulation of the functional component with a nanocarbon. In contrast to nanocomposites, in which the volume fraction of the nanocarbon is typically less than a few per cent, hybrids

Carbon Nanotubes and Graphene. DOI: 10.1016/B978-0-08-098232-8.00016-4

TABLE 16.1 **Comparison of key characteristics of nanocarbon composites and nanocarbon hybrids**

	Composite	Hybrid
Structure	CNT or graphene filler dispersed in second matrix	CNT or graphene coated by functional material (or vice versa)
Synthesis	Mechanical mixing and dispersion	(1) Ex situ linking of CNT/G with hybrid component (2) In situ growth of one component upon the second
Properties	Individual properties of the components	Individual property of components plus increased interfacial charge/energy transfer resulting in new properties
Application	Increasing strength or conductivity of material	Catalysis, supercapacitors, batteries, sensors, PVs

are formed by both components with similar volume fractions. Typically, the hybrid component is deposited from molecular precursors directly onto the surface of the nanocarbon (in situ synthesis), although the attachment of pre-synthesized building blocks via anchor molecules is also possible (ex situ). The close proximity and similar size domains/volume fraction of the two phases introduce the interface as a powerful new parameter. Interfacial processes such as charge and energy transfer create synergistic effects that improve the properties of the individual components and even create new properties. As will be discussed in this chapter, these new hybrid materials have already shown great potential as next-generation functional materials in catalysis, batteries, supercapacitors, photovoltaics (PVs) and sensors.

Nanocarbon hybrids offer the most exciting, wide-ranging applications and will be the focus of this chapter with special emphasis being placed upon the most intriguing results to date. Prior to this, however, this chapter will first summarize the various hybrid synthesis methods along with their advantages and challenges, since the chosen technique and process conditions often play a pivotal role in the performance of the resulting hybrid.

16.2 SYNTHESIS OF NANOCARBON HYBRIDS

The various synthesis strategies for nanocarbon hybrids can be categorized into two areas, namely, ex-situ and in-situ techniques. In the ex-situ (building block) approach, the two hybrid components are first produced separately with their desired dimensions and morphology, then modified with a suitable functional group or linker molecule, and finally combined using covalent, non-covalent or electrostatic interactions. In contrast, in the in-situ approach one of the hybrid components is synthesized in the presence of the other; sometimes both components are produced simultaneously. As a consequence, one component affects the synthesis of the other in terms of size, crystal structure and especially morphology (i.e. as particles, nanowires, nanotubes or thin amorphous or crystalline films). As this chapter offers only a brief description of both approaches, various review articles are recommended for more detail, such as the comprehensive review on CNT–inorganic [1], the specialized reviews on CNT–G-polymer [9,10] and CNT–metals [11–13] or the comparative review on nanocarbon nanocomposites and hybrids [2].

Of paramount importance for both synthesis approaches is the type of nanocarbon and the degree of functionalization prior to hybridization. The chemical inertness and low solubility of pristine CNTs (especially multi-walled CNTs (MWCNTs)) and graphene (especially those with two or more layers) render many common hybridization approaches (particularly wet-chemical) very challenging. A particularly popular route to overcome these problems is to oxidize the nanocarbon in order to introduce oxygen-containing functional groups on the surface, which increases the solubility of the nanocarbon and also provides active sites for subsequent hybridization processes. Oxidation is most commonly achieved via sonication or refluxing in strong oxidizing acids (i.e. 3:1 H_2SO_4:HNO_3). This method is also used to produce graphene oxide (GO) via oxidative exfoliation of graphite, often called the 'Hummer's method', which results in a variety of single- to few-layer graphene sheets decorated with many oxygen-containing functional groups [14]. Although acid treatment of nanocarbons renders them hydrophilic, it is not possible to control the type, number and location of the functional groups, which often creates non-homogeneous coatings. Other covalent functionalization reactions, i.e. via organic radicals, improve the distribution and uniformity of functional groups, but are limited to single-layer graphene and single-walled CNTs (SWCNTs). In any case, covalent functionalization disrupts the nanocarbon's π-network, thus reducing its mechanical, thermal and electronic properties. In contrast, non-covalent functionalization, i.e. via $\pi - \pi$ or van der Waals interactions, is suitable for MWCNTs and multi-layer graphene and does not alter the properties significantly, but provides a weaker driving force for hybridization. Therefore, the choice of hybridization technique depends on the type of nanocarbon, applied functionalization method, physical properties and desired application.

16.3 EX-SITU APPROACH

In this 'building-block' approach, the components are synthesized separately and then hybridized via linking agents that utilize covalent, non-covalent (van der Waals, π–π interactions, hydrogen bonding) or electrostatic interactions.

The attachment of these building blocks often requires the chemical modification of at least one component to overcome the differences in surface chemistry. As a consequence, deposition is often limited to the first layer. Excess nanoparticles (NPs) can be removed by filtration or centrifugation. The greatest advantage of the ex-situ approach is that the pre-formed hybridizing component is of pre-defined shape/size/function. Therefore, hybrids consisting of complex structures, such as nanocubes, polymers of known molecular weight and biomolecules, can be prepared.

16.3.1 Ex-situ hybridization of pristine nanocarbons

The most direct hybridization method is covalent linkage. A comparatively simple and popular general approach to functionalize pristine nanocarbons is based upon diazonium/nitrene chemistry [15,16]. This reaction is thought to occur by injection of an electron from the nanocarbon into an aryl diazonium salt that then induces the formation of a reactive aryl radical, thereby releasing N_2 and resulting in the formation of a new nanocarbon–arene bond (Figure 16.1(a-i)) [17]. Polymers or NPs can be terminated or capped with diazonium groups for direct hybridization. Alternatively the diazonium-linking molecule can express a second functional group (e.g. amine, alkyne etc.) for further hybridization.

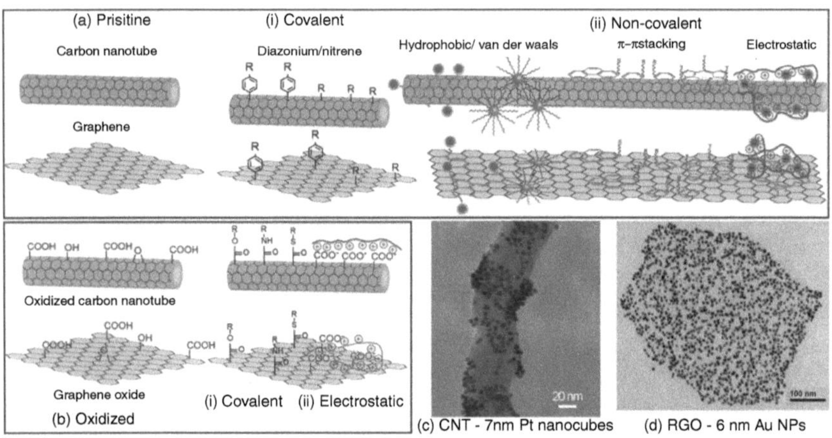

FIGURE 16.1 Summary of ex-situ nanocarbon hybridization schemes. (a) Pristine nanocarbons can be hybridized via (i) covalent interactions through chemical modification with diazonium or nitrene functional groups and (ii) non-covalent methods including hydrophobic, van der Walls, π–π stacking or electrostatic interactions. (b) Oxidized nanocarbons can also be hybridized directly via (i) covalent interactions such as ester, amide or thioester bonds and (ii) electrostatic interactions. In all cases, the *R* group can represent a number of possible hybrid components including biomolecules, polymers, NPs and functional groups for further hybridization. (c) and (d) TEM of examples of pre-defined NPs hybridized via ex-situ techniques showing (c) 7-nm Pt nanocubes on CNT and (d) 6-nm Au NPs on RGO. (c) Adapted with permission from Ref. [44], copyright (2008) John Wiley and Sons. (d) Reprinted with permission from Ref. [45], copyright (2010) American Chemical Society.

Non-covalent interactions offer a non-destructive route towards hybridization. The surface chemistry of nanocarbons supports van der Waals, π–π and hydrophobic interactions. Surfactant polymers which consist of both a hydrophobic tail and a hydrophilic head group are often used to wrap around CNTs and graphene to facilitate dispersion. Many studies have used surfactants that wrap around nanocarbons via van der Waals interactions [18]. For instance, surfactants such as sodium dodecylsulphate (SDS) are commonly used to disperse CNTs in aqueous solutions [19,20] while other surfactants, such as Pluronic-123, are used to mechanically exfoliate graphene from graphite flakes (Figure 16.1(a-ii)) [21,22]. The head group of the surfactant can be used for hybridization via electrostatic or in-situ means.

Another approach is based on π–π stacking, which utilizes the moderately strong interactions between delocalized π-electrons of nanocarbons and those in aromatic organic compounds, such as derivatives of pyrene [23,24], porphyrins [25,26], benzyl alcohol [27] or triphenylphosphine [28]. These molecules are often modified with long alkyl chains that are terminated with thiol, amine or acid groups, which can then be used for further hybridization (Figure 16.1(a-ii)). The π electrons of nanocarbons can lead to interesting CNT–graphene hybrids [29–32]. MWCNT–graphene films for use as actuators have been produced by the group of Lu et al. [33]. GO and MWCNTs were dispersed via sonication prior to GO reduction by the addition of hydrazine during which π–π interactions led to the non-covalent formation of the hybrid [33].

16.3.2 Ex-situ hybridization of oxidized nanocarbons

The oxygen-containing groups, particularly carboxylic acid functionalities of oxidized CNTs and GO, are potential binding sites for a number of covalent interactions particularly amide and ester bonds (Figure 16.1(b-i)). This coupling is commonly assisted by activating the carboxyl group with carbodiimide coupling agents, such as dicyclohexylcarbodiimide (DCC) [34]. This has been used to successfully link, e.g. NPs [35,36], quantum dots (QDs) [37], polymers [38], proteins [39,40] and porphyrins [41,42], to oxidized CNTs and GO. This is particularly useful in the hybridization of nanocarbons with biomaterials. For example, Liu et al. utilized carbodiimide coupling to link glucose oxidase (GOx) with GO via an amide linkage between amine functionalities on the enzyme and carboxylic acid functionalities on GO [43].

Positively charged hybridizing components can directly interact with oxidized CNTs or GO. For example, poly(sodium 4-styrenesulphonate) (PSS) coating of CNTs has been shown to facilitate the electrostatic adsorption of cetyltrimethylammonium bromide (CTAB) capped Pt nanocubes (Figure 16.1(c)) [44]. This work demonstrates one of the biggest advantages of the ex-situ hybridization route, which is the formation and attachment of metal NPs with uniform size (i.e. 6–7 nm) and unique morphology (i.e. nanocubes).

An example for a multi-step hybridization procedure is the work by Liu et al., who non-covalently hybridized GO with bovine serum albumin (BSA) via a range of π–π and van der Waals interactions [45]. The BSA was then used as a 'universal adhesive' to further hybridize the GO with a range of NPs including 6 nm Au NPs (Figure 16.1(d)).

In summary, these few examples demonstrate the simplicity and feasibility of the ex-situ approach, which remains the only method suitable for the hybridization of biomolecules. The main advantage is the possibility of using prepared NPs with controlled morphology, structure, shape and size, and therefore a good structure–property relationship. The downside of this route, however, is the need to chemically modify either the nanocarbon or the hybridizing component.

16.4 IN-SITU APPROACH

In this approach, the hybridizing component is formed directly on the surface of a pristine or modified nanocarbon using molecular precursors, such as organic monomers, metal salts or metal organic complexes. The resulting hybrid has a large interfacial area creating robust hybrids capable of intimate charge and energy transfer. In the case of inorganic hybrids amorphous, polycrystalline or single-crystalline films with uniform thickness can be deposited. As with ex-situ hybridization, the surface chemistry of the nanocarbon needs to be suitable for the chosen molecular precursor. While, in contrast to the ex-situ approach, the functionalization of the nanocarbon is not explicitly required, the presence of functional groups and the use of linking agents provide attractive reaction centres that can facilitate the initial stages in hybrid synthesis.

16.4.1 In-situ polymerization

The in-situ growth via covalent binding of a hybridizing component to a nanocarbon can be achieved in the case of polymers, dendrons and various other macromolecules, which are synthesized in a step-wise manner. The in-situ synthesis of such macromolecules potentially increases binding site density while steric effects of the nanocarbon can lead to increased variation in polymer chain length (polydispersity) [46–48].

There are many polymerization techniques that have been successfully completed from nanocarbons. These have been covered in detail elsewhere for CNTs and graphene [9,10]. The simplest method is the chemical grafting of a monomer followed by polymerization, such as the polymerization of polystyrene (PS) in the presence of GO (Figure 16.2) [49]. In addition, 'living' polymerization reactions have also been completed including atom transfer radical polymerization (ATRP) [50] and reversible addition-fragmentation chain transfer (RAFT) [51]. In addition to polymers, the step-wise growth of dendrons [52] and various hyperbranched macromolecules [53] originating from nanocarbons has also been demonstrated.

FIGURE 16.2 Summary of in-situ nanocarbon hybridization techniques showing TEM and SEM images of unhybridized carbon nanotubes and graphene. (a) In-situ/grafting from polymerization showing an RGO–PS hybrid. (b) The microwave-assisted co-reduction of Zn^{2+} onto GO demonstrates chemical processes. (c) Sol–gel processes are demonstrated by the growth of thin V_2O_5 platelets onto benzyl alcohol-functionalized CNTs via a solvothermal sol–gel route. (d) The stepwise deposition of thin films of ZnO and Al_2O_3 achieved by ALD onto pyrene carboxylic acid-functionalized CNTs is used as an example of gas-phase processes.

16.4.2 Chemical oxidation/reduction processes

Inorganic–nanocarbon hybrids can be produced from the simplest metallic building blocks, metal salts. These are readily available, and are generally highly soluble and best suited to form metal NPs. Designing nanocarbon–metal complex interactions enables effective control over nucleation and growth [54,55]. In general, metal NPs are obtained via reduction of metal complexes, such as metal chlorides, by chemical agents (chemical reduction/oxidation) [56,57], heat (thermal reduction) [58], light (photoreduction) [59] or by electrons (electrodeposition) [60]. Hybrids of metal oxides (MOs) are obtained by oxidation, network formation or precipitation of precursors such as metal nitrates and acetates. Common reducing agents include $NaBH_4$ [61], ethylene glycol [56], thiourea [62], sodium citrate [63] and formic acid [64].

The wet-chemical reduction process offers a special opportunity for graphene hybrids. The majority of applications require graphene or the graphenelike reduced GO (RGO), which is produced via reduction of GO. Therefore, a popular strategy is the concurrent reduction of both, GO and the hybrid precursor, with reducing agents, such as $NaBH_4$. As an example, Fe–Ni alloy NPs have been produced on poly(ethyleneimine) (PEI)-wrapped GO by Zhou et al.; first an aqueous solution of PEI-wrapped GO was prepared, $FeSO_4$ and $NiCl_2$ were then slowly added and finally reduction was initiated via addition of $NaBH_4$ [65].

Semiconducting NPs can also be deposited on nanocarbons via reduction, but typically require additional control of pH or the addition of specific chemical reagents. For example, hybrids of CNT with $Cd_{0.8}Zn_{0.2}S$ QDs have been obtained by first adsorbing Cd^{2+} and Zn^{2+} ions onto oxidized MWCNTs followed by reduction via addition of Na_2S to precipitate the QDs onto the MWCNT surface [66]. Figure 16.2 highlights the preparation of RGO–ZnS hybrids via the microwave-assisted co-reduction of Zn^{2+} onto GO using thioacetamide as reduction agent [67].

Selective deposition of NPs inside wide-diameter CNTs is also possible [68], as demonstrated initially by Tessonnier et al. [69]. A concentrated aqueous solution of the metal salt is dropped slowly into a suspension of large-diameter, pre-opened MWCNTs under vigorous stirring. Upon drying at room temperature, capillary forces draw the metal ions into the MWCNTs where they can be reduced to metal NPs [69].

16.4.3 Sol–gel processes

The sol − gel process is a versatile, solution-based process for producing various NP, thin-film coatings, fibres or aerogels and involves the transition of a liquid, colloidal 'sol' into a solid 'gel' phase. Typical starting materials for the preparation of the sol include metal salts or metal organic compounds, such as metal alkoxides, which undergo a series of hydrolysis and condensation reactions to form a colloidal or polymeric sol. Upon ageing, the sol forms a wet inorganic continuous network with oxo ($M - O - M$) or hydroxo ($M - OH - M$) bridges, creating a gel. The growth of NPs and thin films onto nanocarbons requires a hydrophilic carbon surface, such as that of oxygen-containing groups in oxidized CNTs and GO [70,71].

Alternatively, the addition of aromatic linking agents offers an interesting non-covalent route [27]. Benzyl alcohol is an excellent candidate as it densely adsorbs on the CNT surface in a cagelike morphology with a large number of hydroxyl groups for the hydrolysis of metal-organic precursors [72]. This route has since been applied for the deposition of a whole variety of MOs in the form of NPs [73], nanowires and platelets, such as for V_2O_5 in Figure 16.2. Also complex inorganic materials, such as metal-oxide frameworks and zeolites, have been deposited onto nanocarbons via this route, as recently demonstrated for the photoactive titanium silicate-1 (TS-1) onto CNTs and graphene [74,75]. In all these cases the benzyl alcohol-assisted sol–gel method produced a homogenous, thin coating with uniform thickness. One of the most intriguing results of this route is that the nanocarbon can act as a template, which facilitates the formation of seed crystals, and as a heat sink, which removes excess heat during crystallization and phase-transformation processes, thus preventing crystal growth [73]. For example, the particle size of TS-1 on CNTs and graphene was approximately 10 and 20 times, respectively, smaller than when formed in absence of the nanocarbon. This has led to uniquely small, thermally stable zeolite crystals of about 8–10 nm in diameter [74]. Interestingly, graphene affected not only the size but also the morphology of the zeolite particles, which changed with increasing graphene concentration from large hexagonal prisms (0 wt%), to ellipsoidal particles (1 wt%), to rectangular plates (~5 wt%) and finally small spherical NPs (~10–20 wt%) [75].

16.4.4 Gas-phase deposition

Increased control of film composition, structure and size can be achieved by limiting the rate of reaction. This is possible using gas-phase deposition where the amount of reactant is relatively low. Gas-phase deposition loosely covers any hybridization strategy where at least one of the hybrid components is in the gas phase. This includes chemical vapour deposition (CVD), physical vapour deposition (PVD) and atomic layer deposition (ALD) as well as various plasma, sputtering and evaporation processes.

Gas phase approaches have the advantage that the nanocarbons do not need to be filtered or washed after hybridization, making them ideal for nanocarbons produced on substrates, such as CVD-grown graphene films, and for porous architectures, such as CNT forests and fibres [76]. These techniques are generally independent of the surface chemistry of the nanocarbon, and thus do not require functionalizations prior to hybridization.

ALD is a very promising technique, which adds the reactants sequentially with intermitting evacuation steps and is, thus, in contrast to CVD, a strictly surface-sensitive process. Metals, MOs and other inorganic compounds can be deposited in a layer-by-layer mode with very uniform and atomically controlled thickness. In contrast to CVD, the deposition via ALD proceeds at very low temperatures (<200 °C) and thus usually requires hydrophilic surface groups. Only a few examples for hybrid synthesis by ALD have been reported, such as hybrids of acid-treated CNTs and GO with TiO_x [77,78]. Others used non-covalently functionalized nanocarbons [79], including Dai et al., who have monitored the influence of a monolayer of perylene tetracarboxylic acid (PTCA) on the ALD deposition of Al_2O_3 on CVD-grown graphene [80]. For comparison, without PTCA, Al_2O_3 deposition occurred only on the edge of the graphene and near-defect sites where dangling bonds interacted with the ALD precursor.

16.5 COMPARISON OF TECHNIQUES

The main advantage of the ex-situ approach is that structure, size and morphology of the hybridizing component is pre-defined. This is a vital aspect when considering the hybridization of uniquely shaped moieties such as nanocubes or biomolecules. However, ex-situ linking is a multi-step process that generally requires modifications of both the nanocarbon and the hybridizing compound. Consequently, the coating is restricted to the first layer and the nanocarbon properties (mechanical, electronic) are often negatively affected. The in-situ wet-chemical approach requires less nanocarbon modification, especially for gas-phase processes, and can produce thin, uniform, multi-layer films.

Furthermore, the role of the nanocarbon as templates and heat sinks can result in smaller particle sizes, new morphologies and stabilized high-temperature phases and so can create interesting hybrids with new properties. The in-situ approach is also the method of choice for nanocarbon–polymer hybrids as the increased interfacial area reduces problems of nanocarbon insolubility and subsequent aggregation. Wet-chemical processes are cheap, simple and very versatile, yet gas-phase deposition offers the greatest control of film thickness and allows the fabrication of 3D hybrid architectures.

16.6 APPLICATION OF NANOCARBON HYBRIDS

The main focus of this chapter is to summarize the most important applications for nanocarbon hybrids. Due to the extensive amount of work that has been reported in this field, especially in the last 2 years, only some of the (to these authors) most intriguing examples will be presented. Furthermore, challenges and future directions will be discussed.

16.6.1 Catalysis

Catalysis is a process that decreases the activation energy of a specific reaction by providing an alternative reaction path. Since it reduces both, time and energy, for a desired process, catalysis, especially heterogeneous catalysis, has become one of the industrially most important concepts. Nanocarbon hybrids have found application in heterogeneous catalysis, electrocatalysis and photocatalysis, in all of which the nanocarbon provides a mechanically stable, high surface area platform for active catalyst NPs. In electrocatalysis, they additionally provide excellent electron conductivity, while additional synergistic charge transfer effects lead to the generation of superior photocatalysts.

16.6.1.1 Heterogeneous Catalysis

Industrial chemical synthesis often utilizes NPs deposited on activated carbon (AC) and carbon black as low-cost, high-surface-area, robust and chemically inert supports [81]. As the cost of nanocarbon production continues to decrease, experiments in utilizing CNTs and graphene as catalytic supports are gaining interest. Additionally, reports have indicated that defect sites as well as O and N surface groups on nanocarbons are capable of a range of catalytic oxidative dehydrogenation reactions [82–84].

For example, Siamaki et al. have grown Pd NPs on GO via microwave-assisted chemical reduction and investigated the catalytic properties of the hybrid for C–C cross-coupling Suzuki and Heck reactions [85]. The many surface oxygen groups of the GO and fast reduction assisted by microwave energy produced well-dispersed, small Pd NPs (7–9 nm) on the surface. The Pd–G hybrids achieved a complete Suzuki coupling conversion after 25 min compared to 30% conversion after 30 min for Pd-supported on AC. In addition to improved catalytic rate, the recyclability of the hybrid catalyst was also shown to be enhanced. The improved catalytic properties of the hybrid were attributed to the higher number of small, well-dispersed Pd(0) NPs on the graphene sheets [85].

Chemically oxidized MWCNTs covalently hybridized with immidazolium-based ionic liquids have been used to catalyse the formation of various cyclic carbonates via reactions of epoxides and CO_2. The turnover rate of synthesis using the CNT hybrid was 2–10 times greater than commercial silica or polymer bead supports. The authors hypothesized that this is an effect of the high surface area and synergistic enhancement from remaining surface oxygen functionalities towards cycloaddition reactions [86].

The well-defined nanosized channels of CNTs provide an intriguing confined environment for catalysis. NPs formed inside CNTs are of a restricted size, and limited reactant diffusion towards the active NPs can potentially alter reaction products and pathways [68]. For example, Chen et al. have formed Fe_2O_3 NPs both inside (~70%) and outside (~100%) of MWCNTs for the catalytic Fischer–Tropsch synthesis of hydrocarbons from syngas (CO) [87]. The confined hybrid converted 1.5 times more CO with double the preference for longer carbon chains (C_{5+}) compared to the outer-decorated hybrid [87]. A similar conversion improvement for NH_3 decomposition was reported for Fe–Co alloy NPs confined within MWCNTs [88]. Confinement has also shown interesting results for liquid-phase catalysis. Tessonnier et al. have deposited Pd NPs inside MWCNTs for the hydrogenation of cinnamaldehyde to hydrocinnamaldehyde (Figure 16.3(a)) [69]. Figure 16.3(b) and (c) shows how the confined hybrid exhibited a faster hydrogenation rate and selectivity compared to AC-supported Pd catalysts. The authors concluded that the selectivity of the reaction was related to the lack of oxygen functionality on the CNT support and charge transfer processes between CNT and NP [69]. The hydrogenation of cinnamaldehyde has also been studied using confined Pt–Ru alloy catalysts where improved rate and selectivity were observed upon NP confinement [89]. Interestingly, in this report confinement led to the selective formation of cinnamyl alcohol, a different product to that of Tessonnier et al. using confined Pd NPs for the same reaction.

16.6.1.2 Electrocatalysis

Direct alcohol fuel cells (DAFCs) have attracted a great deal of attention due to their high energy density, low fossil fuel consumption and low operating temperature [91]. The oxygen reduction reaction (ORR) occurs at the cathode in DAFCs, which is catalysed by supported Pt catalysts. The support material must have high surface area, high conductivity and be corrosion resistant. The major obstacle for the successful commercialization of DAFCs is the slow kinetics of the electro-oxidation of alcohols (e.g. methanol or ethanol) and the poisoning of the catalyst electrode (Pt) by CO [92]. CNTs or graphene have significantly increased the reaction kinetics, and nanocarbon hybrids with a second active catalyst for CO oxidation (e.g. Ru) can offer a promising solution [93–95].

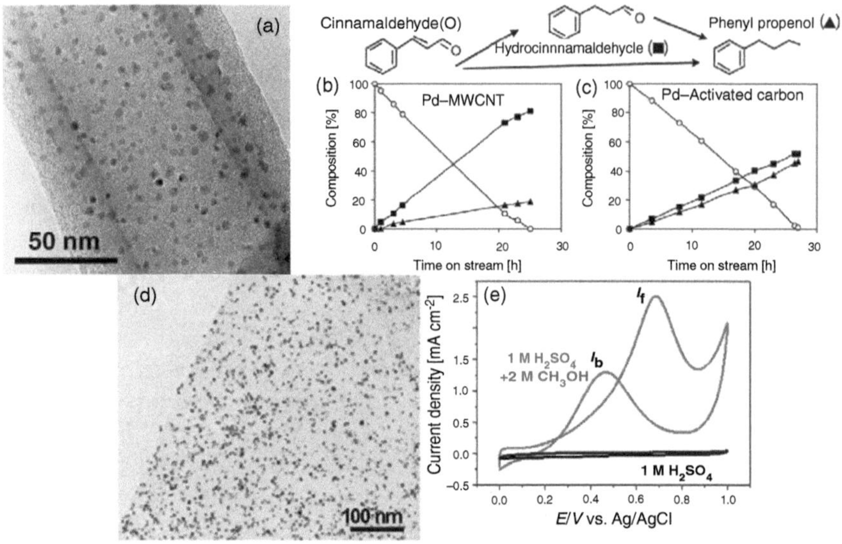

FIGURE 16.3 (a)–(c) Pd NPs selectively deposited inside MWCNT for selective hydrogenation of cinnamaldehyde [69]. (a) TEM of Pd NPs inside MWCNTs with possible hydrogenation reaction pathways of cinnamaldehyde. Comparison between (b) Pd–MWCNT and (c) Pd-activated carbon catalysts for the selective hydrogenation of cinnamaldehyde (open circles) to hydrocinnamaldehyde (filled squares) with the undesired by-product phenyl propanol (filled triangles) as determined by gas chromatography. (d) and (e) Pt NPs dispersed on PDDA-functionalized GO for electrocatalytic methanol oxidation reaction [90]. (d) TEM of low Pt loading hybrid. (e) Cyclic voltammetry measurement of Pd–GO hybrid in 1 M H_2SO_4 with (grey series) and without (black series) 2 M methanol. (a)–(c) Reprinted with permission from Ref. [69], copyright (2005) Elsevier. (d) and (e) Reprinted and modified with permission from Ref. [90], copyright (2011) American Chemical Society.

When supported on nanocarbons, noble metal particles have shown significantly higher power densities at higher backpressures, lower onset potentials and with higher anodic currents compared to AC supports [96]. In the case of hydrogen polymer electrolyte fuel cell, CNT-supported Pt catalyst with 12 wt% Pt loading gave a 10% higher fuel cell voltage and twice the power density than 29 wt% Pt loaded on a carbon black support [97]. Qiu et al. have prepared a Pt–GO hybrid via in-situ reduction of H_2PtCl_6 onto poly(diallyldimethylammonium chloride) (PDDA)-modified GO (Figure 16.3(c)) [90]. The resulting hybrid consisted of NPs (~4.6 nm) evenly dispersed on the GO and exhibited superior methanol oxidation reaction properties than Pt NPs deposited on RGO and chemically converted graphene (Figure 16.3(d)) [90]. Improvements in cell recyclability have also been demonstrated for hybrids with metal alloys, such as Pt–Sn–graphene hybrid [98].

An alternative to co-deposition of a second metal is the hybridization of CNTs with polyaniline (PANI) which has been shown to (1) act as a binding site for the electrochemical deposition of Au NPs and (2) oxidize CO to CO_2 to reduce CO-related catalyst poisoning in methanol fuel cells [99]. Additionally, many reports indicate a lower degree of CO catalyst poisoning upon forming nanocarbon hybrids with smaller NPs [55,58,100,101]. A recent theoretical study has shown that a vacancy defect in graphene strongly stabilizes a single Pt adatom and increases its positive charge, which weakens CO adsorption and facilitates O_2 adsorption, thus enhancing activity for O_2 oxidation and alleviating the CO poisoning of the platinum catalyst [102].

Fuel cells that do not require precious metal catalysts clearly have economic benefit. Experimental results have shown that N-doped CNTs and graphene are efficient ORR catalysts even without metal NPs [103,104]. This has inspired further work towards investigation of N-rich polymer–nanocarbon hybrids as electrocatalysts. Hybrids of PDDA with CNTs and RGO have been shown to exhibit comparable ORR catalytic performance to standard Pt NP cathodes with higher fuel selectivity, lifetime and tolerance to CO [105,106].

Hybrids with MOs offer a low-cost alternative to Pt and Pd NPs for DAFCs [107]. Liang et al. prepared a robust hybrid of CoO with nitrogen-doped CNTs [108]. The CNTs were originally mildly chemically oxidized but became N-doped during annealing of the hybrid in NH_3 atmosphere. The methanol oxidation reaction was shown to be superior for the N-doped CNTs platform than for graphene-supported CoO. This was attributed to greater charge transfer and increased number of active (N-doped) sites [108].

16.6.1.3 *Photocatalysis*

The discipline of photocatalysis originates from the studies of Stone on photoadsorption/desorption of oxygen on ZnO and TiO_2 [109] as well as photocatalytic oxidation of organic compounds such as CO [110] and acetone on these MOs [111]. But the most known and the most-cited publication in the field refers to the discovery of Honda and Fujishima in 1972 in their work on the photoelectrolysis of water using a TiO_2 anode under UV irradiation [112]. Stemming from the original developments in Japan, photocatalysis has become a major field of study with application in solar energy conversion and environmental protection.

Photocatalysis is a light-assisted catalysis where reactive intermediates are formed on the surface upon photoillumination. Some semiconductors (e.g. TiO_2, ZnO, WO_3, CdS, $NaTaO_3$, etc.) can act as a photocatalyst due to their intrinsic electronic structure composed of a filled valence band (VB) and an empty conduction band (CB). The bands are separated by a bandgap (E_g) so that electrons cannot freely move from the VB to the CB (Figure 16.4(a)). However, if the semiconductor is illuminated by light with energy higher than its bandgap ($h\nu \geq E_g$), the VB electrons can be excited to the CB, leaving a positively charged hole in the VB, thereby creating free-moving charges in both energy bands. The photo-generated holes and electrons play a key role in solar energy conversion applications and are also able to drive reduction/oxidation processes on the particle surface through either direct charge transfer to reactants or formation of reactive intermediates (e.g. $OOH^{\cdot}$, $OH^{\cdot}$, $O_2^{\cdot}$), which then catalyse a reaction.

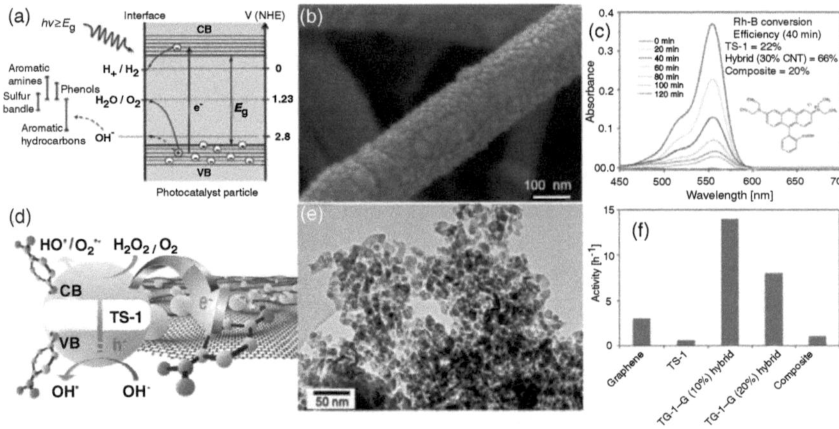

FIGURE 16.4 (a) Band diagram of semiconductor showing excitation of electron from VB to CB to produce an electron–hole pair (exciton). Depending upon the potential of the band positions, the electrons could be used to reduce hydrogen or the holes can be used to create reactive oxygen radicals for water purification or oxidize oxygen in water-splitting experiments. (b) SEM of hybrid of CNTs with TS-1 used for the (c) photocatalytic degradation of rhodamine B under visible light [74]. (d) Proposed mechanism of synergistic charge transfer between semiconductor and nanocarbon. (e) TEM image of TS-1–graphene (10%) hybrid and (f) comparison of photocatalytic decomposition of 4-NP under UV illumination between graphene, TS-1 as well as hybrids and composites of the two. (b) and (c) Modified and reprinted with permission from Ref. [74], copyright (2010) Elsevier. (d)–(f) Modified and reprinted with permission from Ref. [75], copyright (2012) Royal Society of Chemistry.

There are several crucial factors that determine the photocatalytic activity of a given material, which are potentially improved when CNTs or graphene are incorporated with the photocatalysts. (a) Graphene and CNTs can act as a photosensitizer for an extended absorption range of a photocatalyst providing additional electrons to the VB of a semiconductor [113,114]. (b) The high conductivity of the nanocarbon allows them to extract and store photo-generated electrons reducing electron–hole pair recombination process, one of the major drawbacks of semiconductor photocatalysis, and subsequently improve quantum yield and lifetime of the charge carriers [115,116].

Nanocarbon–inorganic hybrids have recently demonstrated superior performance in both environmental (e.g. degradation of industrial pollutants) and energy (e.g. hydrogen production from water splitting) applications in comparison to common semiconductors. The photocatalytic enhancement depends on several factors including quality of a carbon material (oxidized or pristine), quality of hybrid coating (thin film or NP), test reaction (band position of hybrid) and the experimental procedure (possible use of sacrificial agents). A thorough comparison of the huge amount of results from literature to date revealed that the photocatalytic activity for the oxidation reactions of volatile organic compounds (VOCs) was on average 1.5–4 times higher for nanocarbon hybrids than for the bare MO photocatalysts (Table 16.2). A similar trend has been observed for other reactions, such as photocatalytic water splitting.

TABLE 16.2 Summary of nanocarbon hybrids for photocatalytic processes including chemical conversion and water splitting

Nanocarbon	Photo-catalyst	Process	Light conditions	Activity enhancement*	Ref.
GO	P25	OX MB	UV	1.4	[117]
G	P25	OX MB	UV-Vis	3.4	[118]
CNT				2.8	
G	TiO_2	OX MB	UV	1.6	[119]
G	TiO_2	OX MB OX MO	UV-Vis	2.9 1.96	[120]
G	TiO_2	OX MO	UV	1.89 (compared to composite)	[121]
G	TiO_2	OX RhB	Visible	1.20	[122]
G	TiO_2	OX RhB	Visible	4	[123]
CNT	TiO_2	OX acetone	UV	1.6	[124]
CNT	TiO_2	OX acetone	UV	1.81	[125]
		OX phenol		2.5	
CNT	TiO_2	OX phenol	Visible	4.1	[113]
CNT	TiO_2	OX propene	UV	0.8 (TiO_2 tubes from CNT template)	[126]
CNT	TiO_2	OX dyes	-	2.50	[127]
CNT	TS-1	OX 4-NP	UV	2.43	[74]
		OX RhB	Visible	3.09	
CNTs	TS-1	OX 4-NP	UV	25	[75]
G	ZnO	OX MB	UV	4.45	[128]
G CNTs	ZnO	OX MB	UV	1.26 1.19	[129]
CNTs	ZnO	OX MB	UV-Vis	2	[130]
G	Bi_2WO_6	OX RhB	Visible	3.00	[131]
CNTs	WO_3	OX MB	Visible	4	[132]
G	$InNbO_4$	OX MB	Visible	1.87	[133]
RGO	P25	WS sacrificial H_2	UV	10.8	[134]

(continued)

TABLE 16.2 Summary of nanocarbon hybrids for photocatalytic processes including chemical conversion and water splitting *(cont.)*

Nanocarbon	Photo-catalyst	Process	Light conditions	Activity enhancement*	Ref.
G	TiO_2	WS sacrificial H_2	UV-Vis	1.6	[135]
G	TiO_2	WS sacrificial H_2 WS overall	-	2.3 5	[136]
G	TiO_2	WS sacrificial H_2	UV-Vis	1.68	[137]
G	TiO_2	WS sacrificial H_2	UV	1.9	[138]
G	CdS	WS sacrificial H_2	Visible	4.87	[139]
CNT	CdS	WS sacrificial H_2	Visible	3	[140]
G	C_3N_4	WS sacrificial H_2	Visible	4.87	[141]
CNTs	C_3N_4	WS sacrificial H_2	Visible	2	[142]
CNTs	C_3N_4	WS sacrificial H_2	Visible	3.7	[143]
G	N-doped $Sr_2Ta_2O_7$	WS sacrificial H_2	Visible	1.6	[144]

**Activity enhancement calculated by activity of hybrid/activity of active material (generally MO), WS, water splitting; OX, oxidation; P25, Degussa TiO_2; MB, methylene blue; MO, methylene orange; RhB, Rhodamine B; TS-1, titanium silicate-1; 4-NP, 4-nitrophenol.*

Titanium dioxide is the most-studied semiconductor photocatalyst material in general as well as in nanocarbon hybrids. G–TiO_2 hybrids have been tested for the photocatalytic degradation of methylene blue [117–120], methylene orange [120,121] and rhodamine B [122,123], and in selective oxidation of alcohols to aldehydes [145], whereas CNTs–TiO_2 hybrids have been investigated in photocatalytic oxidation of acetone [124,125] phenol [113,125], propene [126] and various dyes [118,127]. However, TiO_2 has several limitations as a photocatalyst, i.e. the relatively large bandgap (~3.2 eV for anatase) that allows the absorption of only a small portion of visible light, and recent research efforts have been directed towards developing new photocatalysts, including hybrids, which are able to utilize the wider range of incident sunlight energy and also have better charge transfer properties. Apart from TiO_2, nanocarbon hybrids with ZnO [128,129] SnO_2 [122], Bi_2WO_3 [131], $InNbO_4$ [133] and $ZnFe_2O_4$ [146] based on graphene and ZnO [129,130] and WO_3 [132] based on CNTs have been produced and tested for water treatment applications.

One of the most pronounced improvements of photocatalytic activity has been observed for nanocarbon hybrids with the photoactive zeolite, TS-1, as produced via in-situ microwave-assisted growth [74,75]. Figure 16.4(b) shows an SEM image of a hybrid with 20 wt% CNTs and about 15-nm small TS-1 NPs. The activity of TS-1 for the photocatalytic degradation of 4-NP under UV irradiation was enhanced six times for CNT hybrids and an unprecedented 25 times for hybrids with graphene with respect to the pure TS-1, presumably due to interfacial charge transfer processes and the resulting increased lifetimes (Figure 16.4(f)). It is important to note that the activity of TS-1 itself was about 70% of that of the benchmark photocatalyst TiO_2–P25; thus, the improvement though hybridization is remarkable. Furthermore, the degradation of rhodamine B under visible light was enhanced 5.5 times for the CNT–TS-1 hybrid (Figure 16.4(c) and (d)) [74,75], which confirmed the additional role of the nanocarbon as a photosensitizer.

Another important photocatalytic application that remains a challenge for hybrid materials is the photocatalytic splitting of water to convert H_2O into H_2 and O_2. Among the many potential MO, TiO_2 has again been used most commonly in hybrids (Table 16.2), followed by CdS [139,140,147]. In most cases, the hybrids demonstrated excellent performance in sacrificial or non-sacrificial, UV or visible light, photocatalytic water splitting. As can be seen in Table 16.2, the activity for hydrogen production over nanocarbon hybrids has on average improved by a factor of about 3.5 with respect to the bare oxide. Furthermore, graphitic carbon nitride (g-C_3N_4) has been hybridized with both MWCNTs [142,143] and graphene [141], which has also resulted in superior (factor 3.5) photocatalytic performance for hydrogen production via water splitting under visible irradiation.

16.6.2 Energy storage

Reliable and affordable electricity storage is a prerequisite for optimizing the integration of renewable energy systems. Electrochemical energy storage devices such as supercapacitors and batteries have been extensively developed for many such applications. The electrical properties, high surface area, surface chemistry and mechanical stability allow nanocarbons to improve energy storage capability, power delivery and cycle life.

16.6.2.1 Supercapacitors

Supercapacitors combine the high energy storage capability of conventional batteries with the high power delivery capability of conventional capacitors and can have lifetimes of over a million cycles [148,149]. According to the mechanism of energy storage, supercapacitors can be categorized into two classes: (a) electrochemical double-layer capacitors (EDLC), based on double-layer capacitance due to charge separation at the electrode/electrolyte interface, which thereby need materials with high specific surface area (e.g. AC, CNTs, graphene); and (b) pseudocapacitors, based on the pseudocapacitance of faradaic processes in active electrode materials such as transition MOs and conducting polymers [150,151]. Nanocarbon hybrids have demonstrated superior performance in both classes of supercapacitors. CNT–G hybrids are especially good candidates for EDLC-type supercapacitors due to the very large interfacial area of the two-carbon components, while nanocarbon hybrids with MOs or conjugated polymers are the materials of choice for pseudocapacitors.

ACs are the major electrode materials for commercial supercapacitors. Owing to their conductivity and surface area CNTs and graphene can be used in the place of or along with AC [152–154]. However, due to strong π–π and van der Waals interactions, nanocarbons tend to agglomerate, thus reducing effective surface area. This can be prevented by creating hierarchical 3D structures of CNTs and graphene [155]. For example, Niu et al. electrophoretically deposited layers of RGO and Au NPs in a step-wise manner, such that the Au NPs increased the spacing between the RGO layers, thus increasing the active surface area without compromising charge transport [156]. An interesting idea is the use of CNTs to separate graphene layers in order to increase the active surface area of graphene without reducing electrical conductivity and mechanical stability (Figure 16.5(a)) [157].

The simplest method is the step-wise deposition of alternating CNT and graphene films [158]. Yu et al. first reduced GO in the presence of the cationic polymer PEI to form a water-soluble, well-dispersed, positively charged hybrid [159], which was then dip coated onto an electrode followed by the deposition of negatively charged acid-oxidized MWCNTs. The process was repeated several times to form a multi-layered structure before it was heat treated to create amide bonds between the PEI and MWCNTs. Analysis by cyclic voltammetry on a hybrid with nine layers showed a relatively high capacitance of 120 F g^{-1} which maintained a square shape at high scan rates (1 V s^{-1}), indicating fast charging and discharging [159].

The growth of CNTs on graphene is another method to prepare CNT–G hybrids for EDLCs (Figure 16.5(a)) [160,161]. Kim et al. first deposited Ni NPs on single-layer graphene grown by CVD. These Ni NPs then acted as the catalyst for the growth of CNTs, again via CVD. The resulting hybrid had a specific capacitance of 653.7 μF cm^{-2}, which amounts to an increase by a factor of ~6 over the bare graphene [162]. Further improvement was made possible with multi-layered structures such as those prepared by Du et al. The authors grew vertically aligned CNTs within thermally expanded highly oriented pyrolytic graphite (HOPG) (Figure 16.5(b) and (c)) [163]. The resulting ordered hybrid structure had a large surface area (~210 m^2 g^{-1}) and a specific capacitance of 110 F g^{-1} (at 10 mV s^{-1}). Upon electrochemical deposition of $Ni(OH)_2$ the

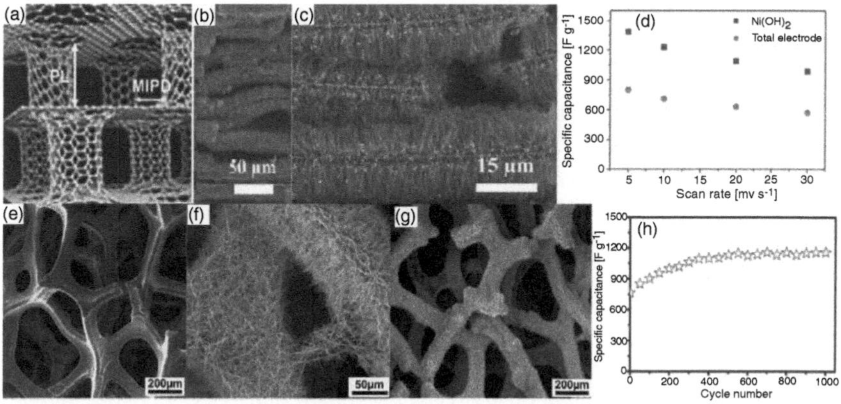

FIGURE 16.5 Summary of nanocarbon hybrid based supercapacitors. (a)–(d) Supercapacitor based upon layered CNT and graphene structure showing (a) model of ideal structure; (b) and (c) SEM images of CNTs grown vertically from expanded HOPG layers and (d) specific capacitance of G–CNT–$Ni(OH)_2$ electrode as calculated using total mass (filled circle series) and mass of $Ni(OH)_2$ (filled square series) [163]. (e)–(h) Hybrid graphene–Co_3O_4 supercapacitor based upon the hybridization of (e) graphene foam (SEM) with (f) and (g) Co_3O_4 needles via solvothermal deposition. (h) Specific capacitance cycling ability recorded at 10 A g^{-1} for the hybrid [164]. (a)–(d) Reprinted with permission from Ref. [163], copyright (2011) American Chemical Society. (e)–(h) Reprinted with permission from Ref. [164], copyright (2012) American Chemical Society.

specific capacitance increased to 1384 F g^{-1}, effectively making a mixed EDLC/pseudocapacitor device (Figure 16.5(d)).

An alternative is to prepare nanocarbon hybrids with conducting polymers, whose low cost and relative low toxicity make them a cheap and 'green' option. Hybrids of CNTs and graphene with polymers such as PANI, polypyrrole and polythiothene derivatives have been prepared with specific capacity values in the range of 100–300 F g^{-1} common [49,165–168]. The supercapacitive properties strongly depend on surface area and coating uniformity on the nanocarbon. Dong et al. have grown few-layer graphene via CVD on commercial Ni foam prior to the deposition of PANI to form a porous, surface-bound G–PANI electrode with improved surface area [169]. The measured specific capacitance of the hybrid was calculated to be 346 F g^{-1} (10 times higher than the bare graphene), which decreased to 245 F g^{-1} after 120 cycles [169]. The capacitance reduction is related to degradation of the organic polymers and is more apparent at higher charge and discharge rates.

Electrochemically stable MOs suffer less from capacitance degradation but are generally semiconducting and thus require a conductive support unless supported on nanocarbons. A wide range of oxides has been investigated for use in nanocarbon hybrids, including NiO_x [170,171], CoO_x [172,173], ZnO [174,175], V_2O_5 [176,177], MnO_2 [178–180] and RuO_2 [181,182]. Among these transition MOs, ruthenium oxides are regarded as the most promising electrode material for supercapacitor applications due to their high-specific capacitance, highly reversible redox reactions, wide potential window and very long cycle life [183]. Kim et al. electrochemically deposited 3 nm small RuO_2 NPs on three substrates (1) a Pt electrode, (2) carbon paper and (3) a porous 2.5-μm-thick CNT film [184]. The recorded specific capacitances for the three electrodes were (1) 420, (2) 840 and (3) 1170 F g^{-1}. In view of the rather high costs and environmental issues concerning RuO_2, and despite the lower specific capacitances, there has been a clear trend towards the use of manganese or cobalt oxides in the last few years [185].

Yu et al. synthesized an interesting ternary supercapacitor electrode by first electrochemically depositing flowerlike particles of MnO_2 onto a porous graphene substrate, which was followed by the deposition of a thin film of either CNTs or a conducting polymer via dip coating [186]. The third layer acted to both increase conductance, as demonstrated by the square shape of CV curves, and provide extra charge storage (either via EDLC or pseudocapacitance). The specific capacitance increased from ~270 F g^{-1} for the G–MnO_2 system to ~320 F g^{-1} for the G–MnO_2–CNT system to ~380 F g^{-1} for the G–MnO_2–polymer system. However, the polymer deposition significantly reduced the cycling stability and operating voltage range [186].

The specific capacitance of nanocarbon–MO electrodes depends on the active surface area. Choi et al. showed that the specific capacitance of GO increased by 300% when produced using PS spheres as sacrificial templates and thus exhibited an increased surface area [187]. Another route towards high surface area pseudocapacitive materials is the use of Co_3S_4 or MnO_2 nanostructures, such as nanospheres [188,189]. Dong et al. have combined these approaches by solvothermally growing Co_3O_4 nanowires on a high surface area (~850 m^2 g^{-1}) graphene foam (grown via CVD on a commercial Ni foam template, Figure 16.5(e) and (g)) [164,190]. The initial specific capacitance of the hybrid electrode was ~760 F g^{-1} at 10 A g^{-1}, which interestingly increased to ~1100 F g^{-1} after 500 cycles and did not degrade after up to 1000 cycles (Figure 16.5(h)).

In summary, high surface area nanocarbons can themselves be used as supercapacitors with greater capacitance and stability than ACs. The hybridization of CNTs with graphene is an attractive route to increase surface area without decreasing electrode conductivity. Nanocarbons can also be used as platforms for the deposition of pseudocapacitive materials,such as conjugated polymers and MOs. The high surface area and conductivity of nanocarbons allow for a greater loading of active material to produce superior energy storage devices.

16.6.2.2 Li-Ion Batteries

Lithium-ion (Li-ion) batteries have attracted a tremendous interest in the last few years as the premier rechargeable battery [191]. A Li-ion battery consists of three components: the Li^+ intercalation anode (typically graphite), the Li^+ intercalation cathode (typically MO) and the electrolyte (Figure 16.6(a) and (b)) [192]. Li^+ ions move from the cathode to the anode during charging and back when discharging. Graphite is currently the preferred anode material with a low

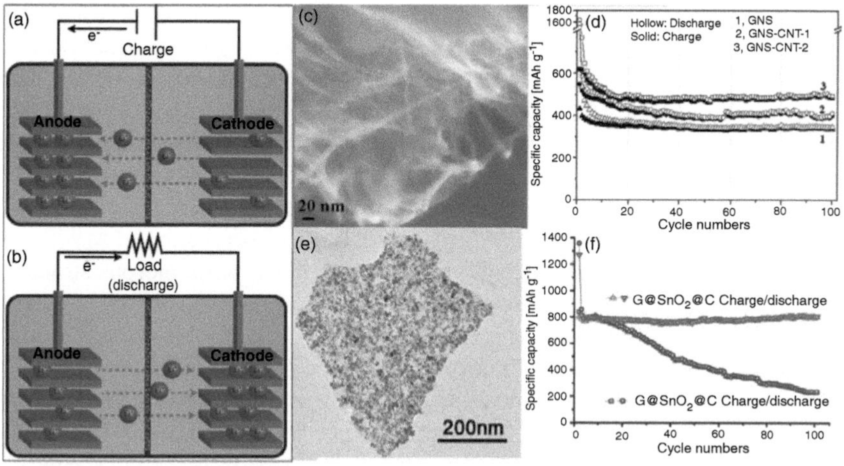

FIGURE 16.6 (a) and (b) Schematic of Li-ion battery working principle. (a) During charge, Li^+ ions are stored in the anode material, during (b) discharge the ions transport to the cathode upon application of a load. (c) and (d) All nanocarbon Li-ion battery anode showing (c) SEM of 2:1 RGO:CNT hybrid and (d) charge/discharge recyclability of RGO only (GNS), 1:1 RGO:CNT (GNS–CNT-1) and 2:1 RGO:CNT (GNS–CNT-2) [31]. (e) and (f) RGO–SnO_2–carbon hybrid as anode material in Li-ion battery showing (e) TEM of RGO–SnO_2 hybrid and (f) comparison of recyclability of hybrid with (G@SnO2@C) and without (G@SnO2) carbon layer [201]. (c) and (d) Reprinted with permission from Ref. [31], copyright (2012) Elsevier. (e) and (f) Modified and reprinted with permission from Ref. [201], copyright (2012) American Chemical Society.

redox potential (~0.2 vs. Li/Li^+), good capacity (theoretical max. 372 mAh g^{-1}) and good cyclability [193]. Under standard conditions, Li^+ intercalation occurs until equilibrium at LiC_6; however, at high recharge rates metallic lithium can be formed onto the anode creating a safety hazard. For high-power applications, higher capacity and higher charge/discharge rates are required [194]. Transition MOs (Co, Mn, V) are commonly used cathode materials for Li-ion batteries [191]. MOs offer a high redox potential (~4 V vs. Li/Li^+) and high theoretical intercalation capacity (i.e. 590 mAh g^{-1} for V_2O_5) [195]. However, these materials suffer from low conductivity, which limits charge/discharge rates. Additionally, Li intercalation results in an increase in MO volume, which causes cracking in pristine MO films.

Nanocarbons have shown high ion capacities (e.g. 1000 mAh g^{-1} for ball-milled SWCNTs [196]); however, a large portion of the intercalated ions form a passivating layer, which reduces subsequent charging and decreases the capacity. Additionally, ions intercalate and release at different rates, causing a hysteresis between charge and discharge cycles. Li^+ intercalation is thought to occur in the voids within the nanocarbon networks, at defect sites and within the tube interior after tip opening (CNTs). One way to minimize these problems and increase the capacity is to use hybrids containing both CNTs and graphene, which can be used as low-potential Li^+ intercalation anodes. Similar to EDLC devices, the deposition of CNTs between graphene sheets provides large voids as well as a large, accessible active surface area and has shown to increase the performance of Li-ion anodes [30,197]. Chen et al. compared RGO (surface area ~436 $m^2 g^{-1}$) with a RGO–CNT hybrid (2:1 weight ratio, surface area ~399.5 $m^2 g^{-1}$) and observed a fully reversible capacity after 100 cycles that was higher in the case of the hybrid (618 mAh g^{-1}) than for pure RGO (432 mAh g^{-1}) (Figure 16.6(c) and (d)) [31]. Since the initial capacity of RGO and RGO–MWCNT was both about 1550 mAh g^{-1}, the incorporation of MWCNTs had significantly reduced the irreversible capacity due to the porous structure that allowed easy release of intercalated ions (Figure 16.6(c) and (d)) [31]. The incorporation of NPs of SiO_2 [198,199] and Cu_2O [200] between graphene sheets to create voids has also improved the anode performance.

Nanocarbons also provide conductive, mechanically stable and high surface area supports for MOs [202,203] or conducting polymers [168,204] in high-performance Li-ion batteries. Depending on their redox potential versus Li/Li^+, hybrids with MOs can be used as both anodes and cathodes. The most popular MOs are SnO_2 [205,206], Fe_3O_4 [207,208], Co_3O_4 [209,210], MnO_2 [211,212] and V_2O_5 [213–215]. Tang et al. demonstrated that the active surface area is a crucial parameter and that creating a hybrid with porous 3D graphene foams is an attractive route to develop high-performance electrodes [216,217]. Another interesting example is the work by Su et al., who have prepared a ternary hybrid consisting of SnO_2 NPs, which were sandwiched between a PDDA-modified RGO sheet and a graphitized polymer layer (Figure 16.6(e)) [201]. Figure 16.6(f) compares the specific capacity recyclability of both C-capped and uncapped hybrids. For both materials, the specific capacity after the first and second cycles was ~1300 and ~800 mAh g^{-1}, respectively. However, after about 20 cycles the specific capacity of the uncapped hybrid decreased gradually to about 235 mAh g^{-1} after 100 cycles, while the hybrid with the carbon layer maintained a level of about 800 mAh g^{-1}. The authors believe that the carbon layer reduced the diffusion length of charge carriers, enhancing electron transport rate, and improved the structural integrity of the electrode by suppressing MO aggregation (as supported by impedance measurements) [201].

16.6.3 Photovoltaics

In simple words, a PV cell directly converts light into electricity and so provides a promising solution to sustainable energy production. Over the past decades, intensive research has created five generations of PV cells; however, current applied PV technology is dominated by inorganic solar cells which consist of a transparent conducting electrode, a photoactive p–n junction and a reflective back electrode (Figure 16.7(a)). Although the power conversion efficiency of these first- and second-generation cells is quite high (35%), the cost associated in production is currently too high to compete with conventional grid electricity [218,219]. Lower cost of production and materials is possible by using organic

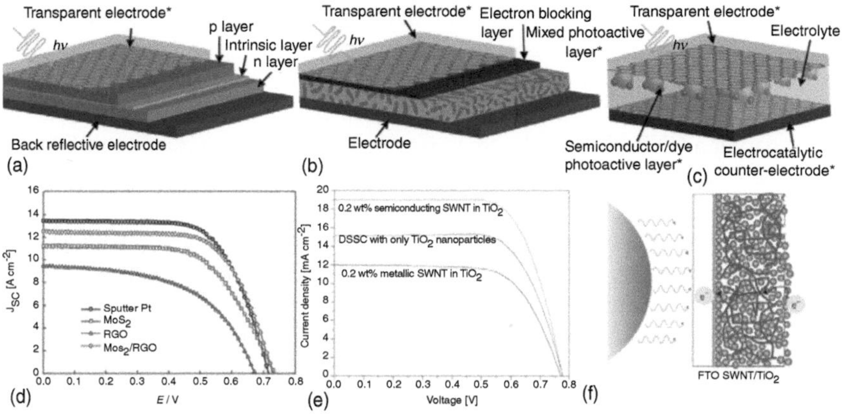

FIGURE 16.7 Nanocarbon hybrid based PVs. Schematics of (a) inorganic, (b) organic and (c) DSSCs showing major components of each system. Components for which nanocarbon hybrids can be incorporated are marked with an asterisk. (d) Example of MoS_2–GO hybrid as counter electrode in DSSC showing similar *I–V* characteristics to sputtered Pt [222]. (e) *I–V* curve comparing DSSCs based upon addition of TiO_2-coated SWCNTs showing improved current when using semiconducting SWCNTs [223]. (f) Schematic of improved charge transfer in DSSCs upon incorporation of CNTs [223]. (d) Reprinted with permission from Ref. [222], copyright (2012) Royal Society of Chemistry. (e) and (f) Reprinted with permission from Ref. [223], copyright (2011) Nature Publishing Group.

PV cells. Organic PV cells typically consist of a transparent conducting electrode, an electron-blocking layer, a photoactive layer consisting of an electron-donating and electron-withdrawing polymer and a conducting back electrode (Figure 16.7(b)). It is possible for organic solar cells to be based purely on carbon [220]. Another alternative is the dye-sensitized solar cell (DSSC) in which a highly porous semiconducting photoanode (generally dye-modified TiO_2) is deposited on a conductive transparent electrode. A liquid electrolyte (typically I^-/I_3^-) completes the circuit and is restored at an electrocatalytic counter-electrode (Figure 16.7(c)) [221]. Organic PV cells and DSSCs can be produced at a lower cost but currently exhibit much lower conversion efficiencies (~10%) [218]. Nanocarbon hybrids can be utilized in all three PV cell configurations to fulfil multiple functions including as transparent-conducting electrodes, electrocatalytic counter-electrodes, channels for charge transport and as a part of the photoactive material.

16.6.3.1 *Transparent Conducting and Counter Electrodes*

Transparent-conductive electrodes are currently based upon a thin film of indium (or fluorine)-doped tin oxide (ITO or FTO), which has a low resistivity and high transparency but is inflexible and prone to cracking at high temperatures. Films of pristine CNTs and graphene have been prepared as transparent electrodes via techniques such as drop casting, spin coating or direct transfer (of bucky papers or CVD-grown graphene sheets) [224–226]. However, the sheet resistance recorded of such films is relatively high due to charge percolation between individual CNTs and graphene flakes [227]. Increased sheet conductance of CNT films have been reported for hybrids with metal NPs and conducting polymers [228,229]. Tung et al. created a transparent conducting electrode by spin coating a CNT–G hybrid, which was prepared via co-reduction of GO and oxidized CNTs in hydrazine, onto a substrate, and demonstrated that it outperformed ITO both in electric conductivity and as a flexible electrode [230].

The counter-electrode in DSSCs is used to electrocatalytically reduce the electrolyte and often consists of a Pt-coated ITO (or FTO) electrode. Due to the increasing cost of platinum, alternative counter-electrodes for DSSCs are required. The high surface area and electronic properties of CNTs and graphene make them ideal candidates. Interestingly, in contrast to the theoretical prediction, defect-rich nanocarbons have been shown to outperform pristine nanocarbons in electrocatalysis [231]. An even lower film resistance was measured with a hybrid consisting of GO covered with a conducting polymer [232] and hybrids with active electrocatalysts such as MoN [233]. Liu et al. prepared a MoS_2–RGO hybrid via mixing of GO with Mo precursor prior to co-reduction via heat treatment in H_2 atmosphere [222]. The hybrid was drop-cast onto FTO and used as a counter electrode in a DSSC with an I^-/I_3^- electrolyte. The solar cell performance of the hybrid (efficiency = 6.04%) was superior to the individual MoS_2 (5.09%) and RGO (3.43%) components and comparable to a platinum coated (6.38%) counter electrode, proving that it is a viable, low-cost, flexible alternative (Figure 16.7(d)) [222].

16.6.3.2 Photoactive and Charge Separation Material

The efficiency of organic PV cells can be further increased by reducing the electron–hole recombination rate, such as through interfacial charge transfer in nanocarbon hybrids. In organic/polymeric solar cells the fullerene C_{60} and its derivatives are common additives as electron acceptors. Efficiencies of such cells were very high with values up to ~8% [234], yet they can be improved further by increasing the thickness of the photoactive layer. However, the low-electron mobility of C_{60} reduces the effective diffusion length of the photogenerated charge. CNTs and graphene can act not only as electron acceptors but also as charge transporters allowing the fabrication of thicker devices that absorb more photons while the charge recombination loss is minimized through fast charge diffusion/collection [219].

Hybrids of CNTs or GO with various conducting polymers has been implemented in organic solar cells [235–238]. For example, Kuila et al. covalently linked poly(3-hexylthiophene) (P3HT) to acid-oxidized CNTs to form a photoactive acceptor layer, which improved the cell efficiency by 38% over pure P3HT and by 45% over a mechanically mixed composite of P3HT and CNTs [239]. GO hybrids with P3HT prepared in a similar fashion showed an even greater improvement (300%) compared to both, the pure polymer and a mechanical mixture (composite) [240]. The improvement was attributed to the strong electronic interaction and good bandgap matching between P3HT and GO. In a further study, C_{60}–graphene hybrid nanosheets were used as electron acceptors in a P3HT-based solar cell to further improve efficiency [241].

In DSSCs electron–hole pair recombination is not the major issue, as energy bands of the components typically match well. Instead, the efficiency can be significantly improved by reducing the series resistance related to charge transport from the semiconducting TiO_2 to the anode. The incorporation of CNTs and graphene into DSSCs as a charge transfer bridge is a promising but challenging approach to improve device performance [242]. There have been many such reports of producing hybrids of CNT–TiO_2 [243,244] and G–TiO_2 [245,246] for DSSCs. Dang et al. coated pristine semiconducting SWCNTs with TiO_2 via a sol–gel process and mixed the hybrids with TiO_2 NPs to form a DSSC with low (~0.2 wt%) CNT content [223]. Increased short circuit current density was observed when the semiconducting SWCNT–TiO_2 hybrid was incorporated (Figure 16.7(e)). The improvement was attributed to increased charge transfer rate to the electrode (Figure 16.7(f)). Interestingly, a reduction in short circuit current density and overall efficiency was observed when metallic SWCNTs were incorporated, most likely due to recombination processes between the metallic SWCNTs and the dye and electrolyte [223].

In summary, nanocarbon hybrids have the potential to significantly impact the future design of inorganic, organic and dye-sensitized solar cells. The design of flexible solar cells is attractive for incorporation into handheld electronic devices. The long-term future of solar cells requires a shift away from non-renewable precious metals such as those currently used in DSSC-counter electrodes and nanocarbon hybrids offer an effective alternative. For nanocarbon hybrids to be successfully implemented as photoactive materials in organic and DSSCs, a greater research effort is required with focus on the associated charge transfer processes and on optimizing interfacial processes.

16.6.4 Sensors

The quantitative and qualitative detection of chemical and biological molecules with high sensitivity and selectivity is crucial for a range of practical applications including determining and monitoring air/water/soil quality, medical diagnosis and chemical synthesis [247]. There is an ever increasing need for the development of miniaturized portable sensors. CNTs and graphene hold great promise for the development of such systems because of their high surface area (e.g. greater interaction zone), electrical properties (e.g. faster electron transfer), mechanical properties (e.g. robustness, flexibility) and their greater modulation of properties upon interaction with analytes (e.g. capacitance, resistance) [248,249].

Sensors have been made with pristine and functionalized CNTs and graphene [250,251]. However, the selectivity of such devices tends to be low because there is no specific interaction between the nanocarbon and the analyte. Selectivity and sensitivity can be greatly enhanced by hybridization with a sensing element. The sensing element in a CNT–graphene hybrid sensor can range from MOs [252], metal NPs [253], small molecules [254] or peptides [255] and other biomolecules [256].

The advantageous nature of utilizing a nanocarbon hybrid is dependent on the sensing mechanism. Although various optical sensors (e.g. Förster resonances energy transfer, FRET [249,257]) based on nanocarbon hybrids have been reported, the most common are electronic and electrochemical sensors.

16.6.4.1 Electronic Sensors

The most common configuration of nanocarbon hybrid electronic sensors is based upon field-effect transistors (FETs), the so-called 'chemresistor'. In this setup two metal electrodes (the source and drain) are connected via a semiconducting channel (CNT or graphene) on an insulating substrate (Figure 16.8(a)). In the case of a nanocarbon-based FET sensor, adsorption on the surface of the nanocarbon results in a change in the recorded FET conductance. The exact mechanism is currently being debated between (1) charge transfer from the adsorbate to the nanocarbon or (2) adsorbate-induced Schottky barrier modulation at the nanocarbon/metal contacts or (3) a combination of both [258–260].

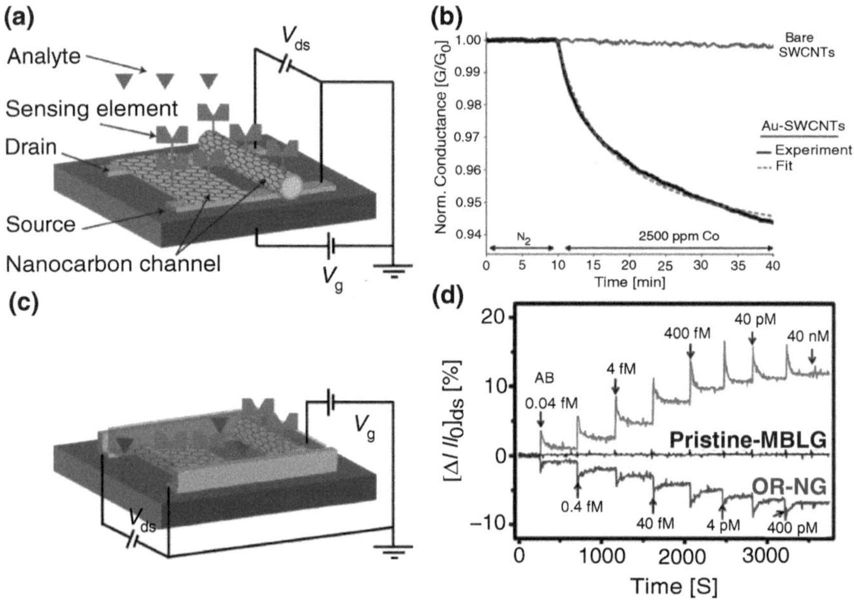

FIGURE 16.8 FET-based nanocarbon hybrid sensors showing (a) schematic of gas sensor and (b) example of CO gas sensor based on Au–SWCNT hybrid [261]. (c) Schematic of aqueous sensor for chemical or biological sensors with (d) sensor response of amyl butyrate sensor from an OR-modified graphene FET sensor [272]. (b) Reprinted with permission from Ref. [261], copyright (2010) American Chemical Society. (d) Reprinted with permission from Ref. [272], copyright (2012) American Chemical Society.

Kauffman et al. prepared SWCNT–Au, SWCNT–Pt and SWCNT–Rh hybrids via electrochemical deposition and tested their response as FET CO sensors (Figure 16.8(b)) [261]. Each showed a decrease in conductance upon exposure to CO vapour as a consequence of electron transfer from CO to the NP and further to the SWCNTs. No change in conductance was observed without the presence of NPs [261]. Similar experiments have been conducted for various gases using MWCNT–SnO_2 for NO_2 [252], SWCNT–Pd and graphene–Pd for H_2 [253,262], MWCNT–WO_3 for NH_3 [263], RGO–Pd for NO [264], RGO–ZnO for H_2S [265] and graphene–DNA [266].

FET-based nanocarbon hybrid chemical sensors can also operate in aqueous conditions for the detection of chemical and biological species. Selectivity is often improved when nanocarbons are hybridized with a functional compound due to its specific interactions with the target analyte. For example, Zhang et al. functionalized mechanically exfoliated graphene with 1-octadecanethiol, whose thiol groups have a high binding affinity for heavy metals such as Hg^{2+} [267]. Greater specificity in detection can be obtained by hybridizing CNTs and graphene with specific peptides. Sudibya et al. detected Ca^{2+} ions in the micromolar range using an RGO–calmodulin (Ca^{2+}-binding protein) hybrid FET sensor [268].

Complex natural proteins such as enzymes often exhibit specificities greater than synthetic materials. Besteman et al. have prepared a working glucose sensor by non-covalently hybridizing a SWCNT–FET sensor with GOx [269]. As glucose is oxidized by the enzyme (not irreversibly immobilized), the conductance change is believed to correlate with local changes in the chemical environment such as the formation of hydrogen peroxide [270,271]. Park et al. have immobilized the human olfactory receptors (ORs) 2AG1 on a graphene FET sensor to create a 'bioelectric nose' [272]. The OR binds specifically to amyl butyrate and a detection limit as low as 0.04 fM was recorded (Figure 16.8(d)) [272].

The detection of peptides and proteins is of particular interest to medical diagnosis. Biotin immobilized upon CNT and RGO–FET sensors have been shown to detect the protein streptavidin [273,274]. Unspecific adsorption of proteins on FET sensors is an issue which can be overcome by hybridizing the sensor with polyethylene oxide chains [275]. Mao et al. prepared an RGO–FET sensor for specific proteins via first hybridizing the RGO with Au NPs before modifying the NPs with thiol functionalized anti-immunoglobulin graphene (anti-IgG). The obtained detection limit for complementary IgG was in the range of nanogram per millilitres with little interaction observed for other proteins [276]. Another example is aptamers, which are artificial oligonucleotides that bind to a wide variety of entities with high selectivity and are an alternative to natural biomolecules, which are comparatively much larger. The smaller size of the aptamers means that the interaction with the target occurs closer to the nanocarbon and within the electrical double layer, such that a greater effect on conductance can occur [277]. So et al. have covalently immobilized an aptamer to a CNT FET sensor which was modified with carbodiimidazole-activated Tween 20. The Tween 20 reduced unspecific adsorption and the aptamer, which was designed to bind to thrombin (a blood-clotting factor), produced a sensor with a limit of detection of 10 nM [278]. Finally, single-stranded DNA has been hybridized to a CNT or graphene FET sensor for the detection of complementary DNA [279–281].

Nanocarbon hybrid FET-based sensors are capable of detecting gases, chemicals and biomolecules with very high sensitivity; however, further improvements are required for commercialization. Moreover, the reproducibility of the sensors needs to be improved.

16.6.4.2 Electrochemical Sensors

Nanocarbon hybrid electrochemical sensors consist of an electrode-bound nanocarbon hybridized with a sensing element which acts as the working electrode in a three-electrode electrochemical cell. A potential is applied and the oxidation or reduction of an analyte is measured as an increase or a decrease in current. The nanocarbon provides a mechanically stable, conducting and high surface area platform while the hybrid component increases selectivity of the produced sensor.

Electrochemical sensors are efficient in the detection of heavy metals using nanocarbon–peptide hybrids. Flavel et al. prepared a SWCNT–Gly–Gly–His (Cu^{2+} capture tripeptide) hybrid-modified electrode prior to detection of Cu^{2+} in aqueous solution via electrochemically monitoring the oxidation and reduction of Cu^{2+} captured by the peptide [282]. In this configuration, the nanocarbons act as both an electron bridge and a high surface area platform for the sensing element. Similar electrode bound hybrids of CNTs and graphene have been prepared to detect various other chemicals including cysteine [283], trinitrotoluene [284], tyrosine [285], methyl jasmonate [286], nitromethane [287], organophosphate [288] and carbohydrates [289].

The most popular nanocarbon hybrid electrochemical sensor is the biosensor [290]. Enzymes such as GOx [104,291], horseradish peroxidase [256,292], cholesterol oxidase [291,293], fructose dehydrogenase [294,295] and nitrite reductase [296] have been used in nanocarbon electrochemical biosensors. There are two basic sensing mechanisms for the enzyme, i.e. direct (mediated) electron transfer and oxidation of redox active products (e.g. H_2O_2). In the case of direct electron transfer, the nano-sized nanotube or graphene can potentially extend into the otherwise inaccessible redox-active centre of the enzyme for efficient charge transfer without the use of electron mediators [297,298]. Claussen et al. prepared a sophisticated graphene hybrid glucose biosensor by first growing graphene (multi-layer) plates vertically from a Ti-coated Si wafer via plasma-enhanced CVD [299]. The graphene plates had a height of 500 nm, length of 0.1–5 μm and a thickness of 1–10 nm. After oxygen plasma treatment, Pt NPs were deposited electrochemically followed by electrochemical deposition of poly(3,4-ethylenedioxthiophene) (PEDOT)-encapsulated GOx. The graphene plates provided a high surface area and conducting bridge while the Pt NPs were found to efficiently catalyse the oxidation of H_2O_2 (formed in the enzymatic conversion of glucose) with a limit of detection of 0.3 μM and linear range from 0.01 to 50 mM [299].

Electrochemical sensors from nanocarbon hybrid electrodes have been demonstrated for many target analytes. The commercial market already has examples of biosensors (e.g. glucose) that are based on the monitoring of H_2O_2 formation. Instead, focus is required for the development of biosensors utilizing enzymatic processes that do not produce H_2O_2 such as for the sulphite oxidase-mediated conversion of sulphite [300].

16.6.5 Other applications

The potential applications of nanocarbon-hybrid materials are broad and cover a variety of topics such as electrochromism [301], actuators [33,302], thermoelectrics [303,304] and oil absorption materials [305].

16.6.5.1 Memory-Switching Devices

The drive for miniaturization of electronics has stimulated interest in carbon-based memory devices. The change in nanocarbon conductance upon charge doping by a coating material is a possible writing mechanism. For example, Zheng et al. coated the ferroelectric material poly(vinylidene fluoride trifluorethylene) onto a graphene channel between two electrodes. By altering the electric field the effect of charge doping by the ferroelectric material could alter the conductivity of the graphene between a high-resistivity and low-resistivity state, which can be used as a '1' or '0' in a memory device [306]. Similar reports have been released using Au NPs [307] and MoS_2 [308] as switching hybrid material. Alternatively, Kumar et al. have utilized conductive atomic force microscopy to read and write using a CNT–ferroelectric hybrid, creating a memory device [309].

16.6.5.2 Field Emission

Field emission is the extraction of electrons from a material upon the application of an external electric field. One-dimensional conducting tips are required as they exhibit the highest electric field enhancement factor, β, which reduces the external electric field required for emission. CNTs exhibit a very low threshold for emission but suffer from poor reliability as the emitting CNTs degrade via a range of mechanisms [310]. The improvement of CNT emission stability without reducing β is a popular field of study [311]. One method to improve field emission stability is to coat CNTs with a stable MO film such as ZnO [312,313] or SiO_x [314]. Graphene is a potential conducting, flexible substrate for a field emission device. Examples include the vertical growth of ZnO nanowires [315,316] or CNTs [317] from a graphene substrate.

16.7 CONCLUSIONS AND FUTURE OUTLOOK

Nanocarbon hybrids have the potential to become next-generation functional materials for a variety of applications with considerable impact on society, such as catalysis, energy storage, PVs and sensors. As the push towards a 'greener' future continues, the commercial use of nanocarbon hybrids will escalate. However, there are a number of improvements that are vital for continued development. The reproducibility of hybrid devices is of paramount importance, as their properties often change dramatically between experiments, yet it is rarely discussed in literature. Therefore, hybrids should be synthesized following standard methods that are more reproducible, or more focus should be placed on improving the reproducibility of systems that have known issues (such as preparing nanocarbon-FET devices).

The type and quality of the nanocarbon have a clear influence on the final performance of the nanocarbon hybrid. The performance broadly relates to two factors: (1) the properties of the nanocarbon, which are related to synthesis procedure, defects, functionalizations etc.; and (2) the quality of the coating, e.g. dispersion, size, concentration of NPs. Currently most publications focus on synthesizing hybrids with little thought towards the quality of the nanocarbon. This is evident in the percentage of publications related to hybrids consisting of GO, which is both easy to prepare and simple to hybridize. However, the mechanical and electrical properties of GO (and also RGO) are inferior to their pristine counterparts and become more analogous to activated carbon. Therefore, pristine/partially functionalized nanocarbons should be preferentially used in applications that require improvement in mechanical stability and electrical conductivity.

One of the most important challenges is to understand and ultimately control the interfacial charge and energy transfer processes between the nanocarbon and the hybrid component, i.e. by designing hybrids with interlayers and unique structures. Furthermore, the functionality of the hybrid needs to be improved, i.e. by increasing the complexity of the inorganic coating or the addition of two or more components. Such systems will require greater control of hybrid synthesis to ensure that the multi-component system does not interfere with itself.

ABBREVIATIONS/SYMBOLS

AC	Activated carbon
ALD	Atomic layer deposition
ATRP	Atom transfer radical polymerization
BSA	Bovine serum albumin
CB	Conduction band
CNT	Carbon nanotube
CTAB	Cetyltrimethylammonium bromide
CVD	Chemical vapour deposition
DAFC	Direct alcohol fuel cell
DCC	Dicyclohexylcarbodiimide
DSSC	Dye-sensitized solar cell
EDLC	Electrochemical double-layer capacitors
FET	Field-effect transistor
FTO	Fluorine-doped tin oxide

G	Graphene
GO	Graphene oxide
GOx	Glucose oxidase
HOPG	Highly oriented pyrolytic graphite
IgG	Immunoglobulin graphene
ITO	Indium-doped tin oxide
MO	Metal oxide
NP	Nanoparticle
ORR	Oxygen reduction reaction
OX	Oxidation
PANI	Polyaniline
P3HT	Poly(3-hexylthiophene)
PDDA	Poly(diallyldimethylammonium chloride)
PEDOT	Poly(3,4-ethylenedioxthiophene)
PEI	Poly(ethyleneimine)
PS	Polystyrene
PSS	Poly(sodium 4-styrenesulphonate)
PTCA	Perylene tetracarboxylic acid
PV	Photovoltaic
PVD	Physical vapour deposition
QDs	Quantum dots
RAFT	Reversible addition-fragmentation chain transfer
RGO	Reduced graphene oxide
SDS	Sodium dodecylsulphate
TS-1	Titanium silicate-1
VB	Valence band
VOC	Volatile organic compound
WS	Water splitting
β	Electric field enhancement factor

REFERENCES

[1] Eder D. Carbon nanotube – inorganic hybrids. Chem Rev 2010;110:1348–85.

[2] Vilatela JJ, Eder D. Nanocarbon composites and hybrids in sustainability: a review. ChemSusChem 2012;5:456–78.

[3] Ci L, Suhr J, Pushparaj V, Zhang X, Ajayan PM. Continuous carbon nanotube reinforced composites. Nano Lett 2008;8:2762–6.

[4] Bekyarova E, Thostenson ET, Yu A, Kim H, Gao J, Tang J, et al. Multiscale carbon nanotube – carbon fiber reinforcement for advanced epoxy composites. Langmuir 2007;23:3970–4.

[5] Wajid AS, Ahmed HST, Das S, Irin F, Jankowski AF, Green MJ. High-performance pristine graphene/epoxy composites with enhanced mechanical and electrical properties. Macromolec Mater Eng 2013;298:339–47.

[6] Stankovich S, Dikin DA, Dommett GHB, Kohlhaas KM, Zimney EJ, Stach EA, et al. Graphene-based composite materials. Nature 2006;442:282–6.

[7] Watcharotone S, Dikin DA, Stankovich S, Piner R, Jung I, Dommett GHB, et al. Graphene – silica composite thin films as transparent conductors. Nano Lett 2007;7:1888–92.

[8] Huang YY, Terentjev EM. Tailoring the electrical properties of carbon nanotube–polymer composites. Adv Funct Mater 2010;20:4062–8.

[9] Kuilla T, Bhadra S, Yao D, Kim NH, Bose S, Lee JH. Recent advances in graphene based polymer composites. Progr Polym Sci 2010;35:1350–75.
[10] Spitalsky Z, Tasis D, Papagelis K, Galiotis C. Carbon nanotube–polymer composites: chemistry, processing, mechanical and electrical properties. Progr Polym Sci 2010;35:357–401.
[11] Georgakilas V, Gournis D, Tzitzios V, Pasquato L, Guldi DM, Prato M. Decorating carbon nanotubes with metal or semiconductor nanoparticles. J Mater Chem 2007;17:2679–94.
[12] Wildgoose GG, Banks CE, Compton RG. Metal nanopartictes and related materials supported on carbon nanotubes: methods and applications. Small 2006;2:182–93.
[13] Li X, Qin Y, Picraux ST, Guo Z-X. Noncovalent assembly of carbon nanotube-inorganic hybrids. J Mater Chem 2011;21:7527–47.
[14] Hummers WS, Offeman RE. Preparation of graphitic oxide. J Am Chem Soc 1958;80:1339–11339.
[15] Peng X, Wong SS. Functional covalent chemistry of carbon nanotube surfaces. Adv Mater 2009;21:625–42.
[16] Wang QH, Jin Z, Kim KK, Hilmer AJ, Paulus GLC, Shih CJ, et al. Understanding and controlling the substrate effect on graphene electron-transfer chemistry via reactivity imprint lithography. Nat Chem 2012;4:724–32.
[17] Dyke CA, Stewart MP, Maya F, Tour JM. Diazonium-based functionalization of carbon nanotubes: Xps and Gc-Ms analysis and mechanistic implications. Synlett 2004;2004:155–60.
[18] Moore VC, Strano MS, Haroz EH, Hauge RH, Smalley RE, Schmidt J, et al. Individually suspended single-walled carbon nanotubes in various surfactants. Nano Lett 2003;3:1379–82.
[19] Islam MF, Rojas E, Bergey DM, Johnson AT, Yodh AG. High weight fraction surfactant solubilization of single-wall carbon nanotubes in water. Nano Lett 2003;3:269–73.
[20] Blanch AJ, Lenehan CE, Quinton JS. Optimizing surfactant concentrations for dispersion of single-walled carbon nanotubes in aqueous solution. J Phys Chem B 2010;114:9805–11.
[21] Fernández-Merino MJ, Paredes JI, Villar-Rodil S, Guardia L, Solís-Fernández P, Salinas-Torres D, et al. Investigating the influence of surfactants on the stabilization of aqueous reduced graphene oxide dispersions and the characteristics of their composite films. Carbon 2012;50:3184–94.
[22] Guardia L, Fernández-Merino MJ, Paredes JI, Solís-Fernández P, Villar-Rodil S, Martínez-Alonso A, et al. High-throughput production of pristine graphene in an aqueous dispersion assisted by non-ionic surfactants. Carbon 2011;49:1653–62.
[23] Guo CX, Ng SR, Khoo SY, Zheng X, Chen P, Li CM. Rgd-peptide functionalized graphene biomimetic live-cell sensor for real-time detection of nitric oxide molecules. ACS Nano 2012;6:6944–51.
[24] Yang DQ, Hennequin B, Sacher E. Xps demonstration of $\pi-\pi$ interaction between benzyl mercaptan and multiwalled carbon nanotubes and their use in the adhesion of Pt nanoparticles. Chem Mater 2006;18:5033–8.
[25] Zhu M, Dong Y, Xiao B, Yukou D, Yang P, Wang X. Enhanced photocatalytic hydrogen evolution performance based on Ru-trisdicarboxybipyridine-reduced graphene oxide hybrid. J Mater Chem 2012;22:23773–9.
[26] Roquelet C, Lauret J-S, Alain-Rizzo V, Voisin C, Fleurier R, Delarue M, et al. π-Stacking functionalization of carbon nanotubes through micelle swelling. Chem Phys Chem 2010;11:1667–72.
[27] Eder D, Windle AH. Carbon–inorganic hybrid materials: the carbon-nanotube/TiO_2 interface. Adv Mater 2008;20:1787–93.
[28] Mu Y, Liang H, Hu J, Jiang L, Wan L. Controllable Pt nanoparticle deposition on carbon nanotubes as an anode catalyst for direct methanol fuel cells. J Phys Chem B 2005;109:22212–6.
[29] Zhang D, Yan T, Shi L-Y, Peng Z, Wen X, Zhang J. Enhanced capacitive deionization performance of graphene/carbon nanotube composites. J Mater Chem 2012;22:14696–704.

[30] Vinayan BP, Nagar R, Raman V, Rajalakshmi N, Dhathathreyan KS, Ramaprabhu S. Synthesis of graphene-multiwalled carbon nanotubes hybrid nanostructure by strengthened electrostatic interaction and its lithium ion battery application. J Mater Chem 2012;22:9949–56.
[31] Chen S, Yeoh W, Liu Q, Wang G. Chemical-free synthesis of graphene–carbon nanotube hybrid materials for reversible lithium storage in lithium-ion batteries. Carbon 2012;50: 4557–65.
[32] Mani V, Devadas B, Chen S-M. Direct electrochemistry of glucose oxidase at electrochemically reduced graphene oxide-multiwalled carbon nanotubes hybrid material modified electrode for glucose biosensor. Biosens Bioelectron 2012;41:309–15.
[33] Lu L, Liu J, Hu Y, Zhang Y, Randriamahazaka H, Chen W. Highly stable air working bimorph actuator based on a graphene nanosheet/carbon nanotube hybrid electrode. Adv Mater 2012;24:4317–21.
[34] Valeur E, Bradley M. Amide bond formation: beyond the myth of coupling reagents. Chem Soc Rev 2009;38:606–31.
[35] He F, Fan J, Ma D, Zhang L, Leung C, Chan HL. The attachment of Fe_3O_4 nanoparticles to graphene oxide by covalent bonding. Carbon 2010;48:3139–44.
[36] Yang C, Chai Y, Yuan R, Xu W, Zhang T, Jia F. Conjugates of graphene oxide covalently linked ligands and gold nanoparticles to construct silver ion graphene paste electrode. Talanta 2012;97:406–13.
[37] Banerjee S, Wong SS. Synthesis and characterization carbon nanotube-nanocrystal heterostructures. Nano Lett 2002;2:195–200.
[38] Hu H, Wang X, Wang J, Liu F, Zhang M, Xu C. Microwave-assisted covalent modification of graphene nanosheets with chitosan and its electrorheological characteristics. Appl Surf Sci 2011;257:2637–42.
[39] Sharma P, Tuteja SK, Bhalla V, Shekhawat G, Dravid VP, Suri CR. Bio-functionalized graphene–graphene oxide nanocomposite based electrochemical immunosensing. Biosens Bioelectron 2013;39:99–105.
[40] Gooding JJ, Wibowo R, Liu, Yang W, Losic D, Orbons S, et al. Protein electrochemistry using aligned carbon nanotube arrays. J Am Chem Soc 2003;125:9006–7.
[41] Xu Y, Liu Z, Zhang X, Wang Y, Tian J, Huang Y, et al. A graphene hybrid material covalently functionalized with porphyrin: synthesis and optical limiting property. Adv Mater 2009;21:1275–9.
[42] Zhao H, Zhu Y, Chen C, He L, Zheng J. Synthesis, characterization, and photophysical properties of covalent-linked ferrocene–porphyrin–single-walled carbon nanotube triad hybrid. Carbon 2012;50:4894–902.
[43] Liu Y, Yu D, Zeng C, Miao Z, Dai L. Biocompatible graphene oxide-based glucose biosensors. Langmuir 2010;26:6158–60.
[44] Yang W, Wang X, Yang F, Yang C, Yang X. Carbon nanotubes decorated with Pt nanocubes by a noncovalent functionalization method and their role in oxygen reduction. Adv Mater 2008;20:2579–87.
[45] Liu J, Fu S, Yuan B, Li Y, Deng Z. Toward a universal "adhesive nanosheet" for the assembly of multiple nanoparticles based on a protein-induced reduction/decoration of graphene oxide. J Am Chem Soc 2010;132:7279–81.
[46] Turgman-Cohen S, Genzer J. Computer simulation of controlled radical polymerization: effect of chain confinement due to initiator grafting density and solvent quality in "grafting from" method. Macromolecules 2010;43:9567–77.
[47] Minko S. Grafting on solid surfaces: "grafting to" and "grafting from" methods polymer surfaces and interfaces. Berlin Heidelberg: Springer; 2008.

[48] Yao Z, Braidy N, Botton GA, Adronov A. Polymerization from the surface of single-walled carbon nanotubes – preparation and characterization of nanocomposites. J Am Chem Soc 2003;125:16015–24.

[49] Kumar NA, Choi HJ, Bund A, Baek J-B, Jeong YT. Electrochemical supercapacitors based on a novel graphene/conjugated polymer composite system. J Mater Chem 2012;22:12268–74.

[50] Qin S, Qin D, Ford WT, Resasco DE, Herrera JE. Polymer Brushes on Single-Walled Carbon Nanotubes by Atom Transfer Radical Polymerization of *N*-butyl methacrylate. J Am Chem Soc 2003;126:170–6.

[51] Zhang B, Chen Y, Xu L, Zeng L, He Y, Kang E-T, et al. Growing poly(*N*-vinylcarbazole) from the surface of graphene oxide via raft polymerization. J Polym Sci A Polym Chem 2011;49:2043–50.

[52] Bifeng P, Daxiang C, Feng G, Rong H. Growth of multi-amine terminated poly(amidoamine) dendrimers on the surface of carbon nanotubes. Nanotechnology 2006;17:2483.

[53] Hijazi I, Jousselme B, Jegou P, Filoramo A, Campidelli S. Formation of linear and hyperbranched porphyrin polymers onto carbon nanotubes via Cuaac "grafting from" approach. J Mater Chem 2012;22:20936.

[54] Walsh FC, Herron ME. Electrocrystallization and Electrochemical Control of Crystal Growth: Fundamental Considerations and Electrodeposition of Metals. J Phys D Appl Phys 1991;24:217.

[55] Guo D-j, Li H-l. Electrochemical synthesis of Pd nanoparticles on functional MWNT surfaces. Electrochem Commun 2004;6:999–1003.

[56] Zhang S, Shao Y, Yin G, Lin Y. Carbon nanotubes decorated with Pt nanoparticles via electrostatic self-assembly: a highly active oxygen reduction electrocatalyst. J Mater Chem 2010;20:2826–30.

[57] Veličković Z, Vukovi GD, Marinković AD, Moldovan M-S, Perić-Grujić AA, Uskokovi PS, et al. Adsorption of arsenate on iron(III) oxide coated ethylenediamine functionalized multiwall carbon nanotubes. Chem Eng J 2012;181-182:174–81.

[58] Sheng W, Woo Lee S, Crumlin EJ, Chen S, Shao-Horn Y. Synthesis, activity and durability of Pt nanoparticles supported on multi-walled carbon nanotubes for oxygen reduction. J Electrochem Soc 2011;158:B1398–404.

[59] Tjoa V, Chua J, Pramana SS, Wei J, Mhaisalkar SG, Mathews N. Facile photochemical synthesis of graphene-Pt nanoparticle composite for counter electrode in dye sensitized solar cell. ACS Appl Mater Interfaces 2012;4(7):3447–52.

[60] Wang X, Dong X, Wen Y, Li C, Xiong Q, Chen P. A graphene-cobalt oxide based needle electrode for non-enzymatic glucose detection in micro-droplets. Chem Commun 2012;48:6490–2.

[61] Jiao J, Zhang H, Yu L, Wang X, Wang R. Decorating multi-walled carbon nanotubes with Au nanoparticles by amphiphilic ionic liquid self-assembly. Colloids Surf A Physicochem Eng Aspects 2012;408:1–7.

[62] Wu H, Liu D, Zhang H, Wei C, Zeng B, Shi J, et al. Solvothermal synthesis and optical limiting properties of carbon nanotube-based hybrids containing ternary chalcogenides. Carbon 2012;50:4847–55.

[63] Liu Y, Brown CM, Neumann DA, Geohegan DB, Puretzky AA, Rouleau CM, et al. Metal-assisted hydrogen storage on Pt-decorated single-walled carbon nanohorns. Carbon 2012;50:4953–64.

[64] Qiu J-D, Shi L, Liang R-P, Wang G-C, Xia X-H. Controllable deposition of a platinum nanoparticle ensemble on a polyaniline/graphene hybrid as a novel electrode material for electrochemical sensing. Chem Eur J 2012;18:7950–9.

[65] Zhou X, Chen Z, Yan D, Lu H. Deposition of Fe-Ni nanoparticles on polyethyleneimine-decorated graphene oxide and application in catalytic dehydrogenation of ammonia borane. J Mater Chem 2012;22:13506–16.
[66] Liu X, Zeng P, Peng T, Zhang X, Deng K. Preparation of multiwalled carbon nanotubes/Cd0.8zn0. 2s nanocomposite and its photocatalytic hydrogen production under visible-light. Int J Hydrogen Energy 2012;37:1375–84.
[67] Hu H, Wang X, Liu F, Wang J, Xu C. Rapid microwave-assisted synthesis of graphene nanosheets–zinc sulfide nanocomposites: optical and photocatalytic properties. Synth Metals 2011;161:404–10.
[68] Pan X, Bao X. The effects of confinement inside carbon nanotubes on catalysis. Acc Chem Res 2011;44:553–62.
[69] Tessonnier J-P, Pesant L, Ehret G, Ledoux MJ, Pham-Huu C. Pd nanoparticles introduced inside multi-walled carbon nanotubes for selective hydrogenation of cinnamaldehyde into hydrocinnamaldehyde. Appl Catal A Gen 2005;288:203–10.
[70] Shakir I, Muhammad S, Ali Z, Kang DJ. MoO3-MWCNTs nanocomposites photocatalyst with control of light-harvesting under visible and natural sunlight irradiation. J Mater Chem 2012;22:20549–53.
[71] Nethravathi C, Viswanath B, Michael J, Rajamath M. Hydrothermal synthesis of a monoclinic VO_2 nanotube–graphene hybrid for use as cathode material in lithium ion batteries. Carbon 2012;50:4839–46.
[72] Cooke DJ, Eder D, Elliott JA. Role of benzyl alcohol in controlling the growth of TiO_2 on carbon nanotubes. J Phys Chem C 2010;114:2462–70.
[73] Aksel S, Eder D. Catalytic effect of metal oxides on the oxidation resistance in carbon nanotube-inorganic hybrids. J Mater Chem 2010;20:9149–54.
[74] Krissanasaeranee M, Wongkasemjit S, Cheetham AK, Eder D. Complex carbon nanotube-inorganic hybrid materials as next-generation photocatalysts. Chem Phys Lett 2010;496:133–8.
[75] Ren Z, Kim E, Pattinson SW, Subrahmanyam KS, Rao CNR, Cheetham AK, et al. Hybridizing photoactive zeolites with graphene: a powerful strategy towards superior photocatalytic properties. Chem Sci 2012;3:209–16.
[76] Neocleus S, Pattinson SW, Moisala Motta AM, Windle AH, Eder D. Hierarchical carbon nanotube-inorganic hybrid structures involving Cnt arrays and Cnt fibers. Funct Mater Lett 2011;4:83–9.
[77] Jin SH, Jun GH, Hong SH, Jeon S. Conformal coating of titanium suboxide on carbon nanotube networks by atomic layer deposition for inverted organic photovoltaic cells. Carbon 2012;50:4483–8.
[78] Hsu C-Y, Lien D-H, Lu S-Y, Chen C-Y, Kang C-F, Chueh Y-L, et al. Supersensitive, ultrafast, and broad-band light-harvesting scheme employing carbon nanotube/TiO_2 core–shell nanowire geometry. ACS Nano 2012;6:6687–92.
[79] Alaboson JMP, Wang QH, Emery JD, Lipson AL, Bedzyk MJ, Elam JW, et al. Seeding atomic layer deposition of high-K dielectrics on epitaxial graphene with organic self-assembled monolayers. ACS Nano 2011;5:5223–32.
[80] Wang X, Tabakman SM, Dai H. Atomic layer deposition of metal oxides on pristine and functionalized graphene. J Am Chem Soc 2008;130:8152–3.
[81] Rodríguez-reinoso F. The role of carbon materials in heterogeneous catalysis. Carbon 1998;36:159–75.
[82] Zhang J, Su D, Zhang A, Wang D, Schlögl R, Hébert C. Nanocarbon as robust catalyst: mechanistic insight into carbon-mediated catalysis. Angew Chem Int Ed 2007;46:7319–23.

[83] Zhang J, Liu X, Blume R, Zhang A, Schlögl R, Su DS. Surface-modified carbon nanotubes catalyze oxidative dehydrogenation of n-butane. Science 2008;322:73–7.

[84] Machado BF, Serp P. Graphene-based materials for catalysis. Catal Sci Technol 2012;2:54–75.

[85] Siamaki AR, Khder AERS, Abdelsayed V, El-Shall MS, Gupton BF. Microwave-assisted synthesis of palladium nanoparticles supported on graphene: a highly active and recyclable catalyst for carbon–carbon cross-coupling reactions. J Catal 2011;279:1–11.

[86] Han L, Li H, Choi S-J, Park M-S, Lee S-M, Kim Y-J, et al. Ionic liquids grafted on carbon nanotubes as highly efficient heterogeneous catalysts for the synthesis of cyclic carbonates. Appl Catal A: Gen 2012;429-430:67–72.

[87] Chen W, Fan Z, Pan X, Bao X. Effect of confinement in carbon nanotubes on the activity of Fischer − Tropsch iron catalyst. J Am Chem Soc 2008;130:9414–9.

[88] Zhang J, Müller J-O, Zheng W, Wang D, Su D, Schlögl R. Individual Fe − Co alloy nanoparticles on carbon nanotubes: structural and catalytic properties. Nano Lett 2008;8:2738–43.

[89] Castillejos E, Debouttière P-J, Roiban L, Solhy A, Martinez V, Kihn Y, et al. An efficient strategy to drive nanoparticles into carbon nanotubes and the remarkable effect of confinement on their catalytic performance. Angew Chem Int Ed 2009;48:2529–33.

[90] Qiu J-D, Wang G-C, Liang R-P, Xia X-H, Yu H-W. Controllable deposition of platinum nanoparticles on graphene as an electrocatalyst for direct methanol fuel cells. J Phys Chem C 2011;115:15639–45.

[91] Shao Z, Haile SM. A high-performance cathode for the next generation of solid-oxide fuel cells. Nature 2004;431:170–3.

[92] Jusys Z, Behm RJ. Methanol oxidation on a carbon-supported Pt fuel cell catalyst a kinetic and mechanistic study by differential electrochemical mass spectrometry. J Phys Chem B 2001;105:10874–83.

[93] Cao L, Scheiba F, Roth C, Schweiger F, Cremers C, Stimming U, et al. Novel nanocomposite Pt/RuO_2 × H_2O/carbon nanotube catalysts for direct methanol fuel cells. Angew Chem Int Ed 2006;45:5315–9.

[94] Nethravathi C, Anumol EA, Rajamathi M, Ravishankar N. Highly dispersed ultrafine Pt and PtRu nanoparticles on graphene: formation mechanism and electrocatalytic activity. Nanoscale 2011;3:569–71.

[95] Awasthi R, Singh RN. Graphene-supported Pd–Ru nanoparticles with superior methanol electrooxidation activity. Carbon 2013;51:282–9.

[96] Yu H, Zeng K, Fu X, Zhang Y, Peng F, Wang H, et al. RuO_2·× H_2O supported on carbon nanotubes as a highly active catalyst for methanol oxidation. J Phys Chem C 2008;112:11875–80.

[97] Matsumoto T, Komatsu T, Arai K, Yamazaki T, Kijima M, Shimizu H, et al. Reduction of Pt usage in fuel cell electrocatalysts with carbon nanotube electrodes. Chem Commun 2004;840–1.

[98] Han F, Wang X, Lian J, Wang Y. The effect of Sn content on the electrocatalytic properties of Pt–Sn nanoparticles dispersed on graphene nanosheets for the methanol oxidation reaction. Carbon 2012;50:5498–504.

[99] Santhosh P, Gopalan A, Lee K-P. Gold nanoparticles dispersed polyaniline grafted multiwall carbon nanotubes as newer electrocatalysts: preparation and performances for methanol oxidation. J Catal 2006;238:177–85.

[100] Guo D-J, Li H-L. High dispersion and electrocatalytic properties of Pt nanoparticles on Swnt bundles. J Electroanal Chem 2004;573:197–202.

[101] Yoo E, Okada T, Akita T, Kohyama M, Honma I, Nakamura J. Sub-nano-Pt cluster supported on graphene nanosheets for Co tolerant catalysts in polymer electrolyte fuel cells. J Power Sources 2011;196:110–5.

[102] Tang Y, Yang Z, Dai X. A theoretical simulation on the catalytic oxidation of Co on Pt/graphene. Phys Chem Chem Phys 2012;14:16566–72.

[103] Maldonado S, Stevenson KJ. Influence of nitrogen doping on oxygen reduction electrocatalysis at carbon nanofiber electrodes. J Phys Chem B 2005;109:4707–16.

[104] Alwarappan S, Liu C, Kumar A, Li C-Z. Enzyme-doped graphene nanosheets for enhanced glucose biosensing. J Phys Chem C 2010;114:12920–4.

[105] Wang S, Yu D, Dai L. Polyelectrolyte functionalized carbon nanotubes as efficient metal-free electrocatalysts for oxygen reduction. J Am Chem Soc 2011;133:5182–5.

[106] Wang S, Yu D, Dai L, Chang DW, Baek J-B. Polyelectrolyte-functionalized graphene as metal-free electrocatalysts for oxygen reduction. ACS Nano 2011;5:6202–9.

[107] Tong X, Qin Y, Guo X, Moutanabbir O, Ao X, Pippel E, et al. Enhanced catalytic activity for methanol electro-oxidation of uniformly dispersed nickel oxide nanoparticles—carbon nanotube hybrid materials. Small 2012;8(22):3390–5.

[108] Liang Y, Wang H, Diao P, Chang W, Hong G, Li Y, et al. Oxygen reduction electrocatalyst based on strongly coupled cobalt oxide nanocrystals and carbon nanotubes. J Am Chem Soc 2012;134:15849–57.

[109] Barry TI, Stone FS. The reactions of oxygen at dark and irradiated zinc oxide surface. Proc R Soc London 1960;255:124–44.

[110] Doerffler W, Hauffe K. Heterogeneous photocatalysis II. The Mechanism of the Carbon Monoxide Oxidation at Dark and Illuminated Zinc Oxide Surfaces. Journal of Catalysis 1964;3:171–8.

[111] Doerffler W, Hauffe K. Heterogeneous Photocatalysis I. The Influence of Oxidizing and Reducing Gases on the Electrical Conductivity of Dark and Illuminated Zinc Oxide Surfaces. Journal of Catalysis 1964;3:156–70.

[112] Fujishima A, Honda K. Electrochemical Photolysis of Water at a Semiconductor Electrode. Nature 1972;238:37–8.

[113] Wang W, Serp P, Kalck P, Faria JL. Visible Light Photodegradation of Phenol on Mwnt-Tio2 Composite Catalysts Prepared by a Modified Sol–Gel Method. Journal of Molecular Catalysis A: Chemical 2005;235:194–9.

[114] Zhang Y, Zhang N, Tang Z-R, Xu Y-J. Graphene transforms wide band gap ZnS to a visible light photocatalyst. the new role of graphene as a macromolecular photosensitizer. ACS Nano 2012;6:9777–89.

[115] Leary R, Westwood A. Carbonaceous nanomaterials for the enhancement of TiO_2 photocatalysis. Carbon 2011;49:741–72.

[116] Roquelet C, Vialla F, Diederichs C, Roussignol P, Delalande C, Deleporte E, et al. Local field effects in the energy transfer between a chromophore and a carbon nanotube: a single-nanocompound investigation. ACS Nano 2012;6:8796–802.

[117] Liu J, Bai H, Wang Y, Liu Z, Zhang X, Sun DD. Self-assembling TiO_2 nanorods on large graphene oxide sheets at a two-phase interface and their anti-recombination in photocatalytic applications. Adv Funct Mater 2010;20:4175–81.

[118] Zhang H, Lv X, Li Y, Wang Y, Li J. P25-Graphene composite as a high performance photocatalyst. ACS Nano 2009;4:380–6.

[119] Du J, Lai X, Yang N, Zhai J, Kisailus D, Su F, et al. Hierarchically ordered macro − mesoporous TiO_2 − graphene composite films: improved mass transfer, reduced charge recombination, and their enhanced photocatalytic activities. ACS Nano 2010;5:590–6.

[120] Liu B, Huang Y, Wen Y, Du L, Zeng W, Shi Y, et al. Highly dispersive {001} facets-exposed nanocrystalline TiO_2 on high quality graphene as a high performance photocatalyst. J Mater Chem 2012;22:7484–91.

[121] Zhang H, Xu P, Du G, Chen Z, Oh K, Pan D, et al. A facile one-step synthesis of TiO_2/graphene composites for photodegradation of methyl orange. Nano Res 2011;4:274–83.

[122] Zhang J, Xiong Z, Zhao XS. Graphene-metal-oxide composites for the degradation of dyes under visible light irradiation. J Mater Chem 2011;21:3634–40.

[123] Liang Y, Wang H, Sanchez Casalongue H, Chen Z, Dai H. TiO_2 nanocrystals grown on graphene as advanced photocatalytic hybrid materials. Nano Res 2010;3:701–5.

[124] Yu Y, Yu JC, Yu J-G, Kwok Y-C, Che Y-K, Zhao J-C, et al. Enhancement of photocatalytic activity of mesoporous TiO_2 by using carbon nanotubes. Appl Catal A Gen 2005;289:186–96.

[125] Yao Y, Li G, Ciston S, Lueptow RM, Gray KA. Photoreactive TiO_2/carbon nanotube composites: synthesis and reactivity. Environ Sci Technol 2008;42:4952–7.

[126] Bouazza N, Ouzzine M, Lillo-Ródenas MA, Eder D, Linares-Solano A. TiO_2 nanotubes and Cnt-TiO_2 hybrid materials for the photocatalytic oxidation of propene at low concentration. Appl Catal B Environ 2009;92:377–83.

[127] Woan K, Pyrgiotakis G, Sigmund W. Photocatalytic carbon-nanotube–TiO_2 composites. Adv Mater 2009;21:2233–9.

[128] Fan H, Zhao X, Yang J, Shan X, Yang L, Zhang Y, et al. ZnO - graphene composite for photocatalytic degradation of methylene blue dye. Catal Commun 2012;29:29–34.

[129] Lv T, Pan L, Liu X, Sun Z. Enhanced photocatalytic degradation of methylene blue by ZnO-reduced graphene oxide-carbon nanotube composites synthesized via microwave-assisted reaction. Catal Sci Technol 2012;2:2297–301.

[130] Jiang L, Gao L. Fabrication and characterization of ZnO-coated multi-walled carbon nanotubes with enhanced photocatalytic activity. Mater Chem Phys 2005;91:313–6.

[131] Gao E, Wang W, Shang M, Xu J. Synthesis and enhanced photocatalytic performance of graphene-Bi_2WO_6 composite. Phys Chem Chem Phys 2011;13:2887–93.

[132] Tian L, Ye L, Liu J, Zan L. Solvothermal synthesis of Cnts–WO_3 hybrid nanostructures with high photocatalytic activity under visible light. Catal Commun 2012;17:99–103.

[133] Zhang X, Quan X, Chen S, Yu H. Constructing graphene/$InNbO_4$ composite with excellent adsorptivity and charge separation performance for enhanced visible-light-driven photocatalytic ability. App Catal B Environ 2011;105:237–42.

[134] Fan W, Lai Q, Zhang Q, Wang Y. Nanocomposites of TiO_2 and reduced graphene oxide as efficient photocatalysts for hydrogen evolution. J Phys Chem C 2011;115:10694–701.

[135] Zhang X, Sun Y, Cui X, Jiang Z. A green and facile synthesis of TiO_2/graphene nanocomposites and their photocatalytic activity for hydrogen evolution. Int J Hydrogen Energy 2012;37:811–5.

[136] Li N, Liu G, Zhen C, Li F, Zhang L, Cheng H-M. Battery performance and photocatalytic activity of mesoporous anatase TiO_2 nanospheres/graphene composites by template-free self-assembly. Adv Funct Mater 2011;21:1717–22.

[137] Cheng P, Yang Z, Wang H, Cheng W, Chen M, Shangguan W, et al. TiO_2–graphene nanocomposites for photocatalytic hydrogen production from splitting water. Int J Hydrogen Energy 2012;37:2224–30.

[138] Zhang X-Y, Li H-P, Cui X-L, Lin Y. Graphene/TiO_2 nanocomposites: synthesis, characterization and application in hydrogen evolution from water photocatalytic splitting. J Mater Chem 2010;20:2801–6.

[139] Li Q, Guo B, Yu J, Ran J, Zhang B, Yan H, et al. Highly efficient visible-light-driven photocatalytic hydrogen production of CdS-cluster-decorated graphene nanosheets. J Am Chem Soc 2011;133:10878–84.

[140] Kim YK, Park H. Light-harvesting multi-walled carbon nanotubes and CdS hybrids: application to photocatalytic hydrogen production from water. Energy Environ Sci 2011;4:685–94.

[141] Xiang Q, Yu J, Jaroniec M. Preparation and enhanced visible-light photocatalytic H_2-production activity of graphene/C_3N_4 composites. J Phys Chem C 2011;115:7355–63.
[142] Suryawanshi A, Dhanasekaran P, Mhamane D, Kelkar S, Patil S, Gupta N, et al. Doubling of photocatalytic H_2 evolution from G-C_3N_4 via its nanocomposite formation with multiwall carbon nanotubes: electronic and morphological effects. Int J Hydrogen Energy 2012;37:9584–9.
[143] Ge L, Han C. Synthesis of MWNTs/G-C_3N_4 composite photocatalysts with efficient visible light photocatalytic hydrogen evolution activity. Appl Catal B Environ 2012;117-118:268–74.
[144] Mukherji A, Seger B, Lu GQ, Wang L. Nitrogen doped $Sr_2Ta_2O_7$ coupled with graphene sheets as photocatalysts for increased photocatalytic hydrogen production. ACS Nano 2011;5:3483–92.
[145] Zhang Y, Zhang N, Tang Z-R, Xu Y-J. Improving the photocatalytic performance of graphene-TiO_2 nanocomposites via a combined strategy of decreasing defects of graphene and increasing interfacial contact. Phys Chem Chem Phys 2012;14:9167–75.
[146] Fu Y, Wang X. Magnetically separable $ZnFe_2O_4$–graphene catalyst and its high photocatalytic performance under visible light irradiation. Indus Eng Chem Res 2011;50:7210–8.
[147] Peng T, Li K, Zeng P, Zhang Q, Zhang X. Enhanced photocatalytic hydrogen production over graphene oxide–cadmium sulfide nanocomposite under visible light irradiation. J Phys Chem C 2012;116:22720–6.
[148] Miller JR, Simon P. Electrochemical capacitors for energy management. Science 2008; 321:651–2.
[149] Kötz R, Carlen M. Principles and applications of electrochemical capacitors. Electrochim Acta 2000;45:2483–98.
[150] Conway BE. Transition from "supercapacitor" to "battery" behavior in electrochemical energy storage. J Electrochem Soc 1991;138:1539–48.
[151] Yu G, Xie X, Pan L, Bao Z, Cui Y. Hybrid nanostructured materials for high-performance electrochemical capacitors. Nano Energy 2012;2(2):213–34.
[152] An KH, Kim WS, Park YS, Moon JM, Bae DJ, Lim SC, et al. Electrochemical properties of high-power supercapacitors using single-walled carbon nanotube electrodes. Adv Funct Mater 2001;11:387–92.
[153] Zhu Y, Murali S, Stoller MD, Ganesh KJ, Cai W, Ferreira PJ, et al. Carbon-based supercapacitors produced by activation of graphene. Science 2011;332:1537–41.
[154] Portet C, Taberna PL, Simon P, Flahaut E. Influence of carbon nanotubes addition on carbon–carbon supercapacitor performances in organic electrolyte. J Power Sources 2005;139:371–8.
[155] Chen C-M, Zhang Q, Huang C-H, Zhao X-C, Zhang B-S, Kong Q-Q, et al. Macroporous 'bubble' graphene film via template-directed ordered-assembly for high rate supercapacitors. Chem Commun 2012;48:7149–51.
[156] Niu Z, Du J, Cao X, Sun Y, Zhou W, Hng HH, et al. Electrophoretic build-up of alternately multilayered films and micropatterns based on graphene sheets and nanoparticles and their applications in flexible supercapacitors. Small 2012;8:3201–8.
[157] Fan Z, Yan J, Zhi L, Zhang Q, Wei T, Feng J, et al. A three-dimensional carbon nanotube/graphene sandwich and its application as electrode in supercapacitors. Adv Mater 2010;22:3723–8.
[158] Huang Z-D, Zhang B, Oh S-W, Zheng Q-B, Lin X-Y, Yousefi N, et al. Self-assembled reduced graphene oxide/carbon nanotube thin films as electrodes for supercapacitors. J Mater Chem 2012;22:3591–9.
[159] Yu D, Dai L. Self-assembled graphene/carbon nanotube hybrid films for supercapacitors. J Phys Chem Lett 2009;1:467–70.

[160] Sridhar V, Kim H-J, Jung J-H, Lee C, Park S, Oh I-K. Defect-engineered three-dimensional graphene–nanotube–palladium nanostructures with ultrahigh capacitance. ACS Nano 2012;6(12):10562–70.

[161] Wang W, Guo S, Penchev M, Ruiz I, Bozhilov KN, Yan D, et al. Three dimensional few layer graphene and carbon nanotube foam architectures for high fidelity supercapacitors. Nano Energy 2012;2(2):294–303.

[162] Youn-Su K, Kitu K, Frank TF, Eui-Hyeok Y. Out-of-plane growth of Cnts on graphene for supercapacitor applications. Nanotechnology 2012;23:015301.

[163] Du F, Yu D, Dai L, Ganguli S, Varshney V, Roy AK. Preparation of tunable 3D pillared carbon nanotube–graphene networks for high-performance capacitance. Chem Mater 2011;23:4810–6.

[164] Dong X-C, Xu H, Wang X-W, Huang Y-X, Chan-Park MB, Zhang H, et al. 3D graphene–cobalt oxide electrode for high-performance supercapacitor and enzymeless glucose detection. ACS Nano 2012;6:3206–13.

[165] Jurewicz K, Delpeux S, Bertagna V, Béguin F, Frackowiak E. Supercapacitors from nanotubes/polypyrrole composites. Chem Phys Lett 2001;347:36–40.

[166] Wang D-W, Li F, Zhao J, Ren W, Chen Z-G, Tan J, et al. Fabrication of graphene/polyaniline composite paper via in situ anodic electropolymerization for high-performance flexible electrode. ACS Nano 2009;3:1745–52.

[167] Lee T, Yun T, Park B, Sharma B, Song H-K, Kim B-S. Hybrid multilayer thin film supercapacitor of graphene nanosheets with polyaniline: importance of establishing intimate electronic contact through nanoscale blending. J Mater Chem 2012;22:21092–9.

[168] Hughes M, Chen GZ, Shaffer MSP, Fray DJ, Windle AH. Electrochemical capacitance of a nanoporous composite of carbon nanotubes and polypyrrole. Chem Mater 2002;14:1610–3.

[169] Dong X, Wang J, Wang J, Chan-Park MB, Li X, Wang L, et al. Supercapacitor electrode based on three-dimensional graphene–polyaniline hybrid. Mater Chem Phys 2012;134:576–80.

[170] Zhang H, Zhang X, Zhang D, Sun X, Lin H, Wang C, et al. One-step electrophoretic deposition of reduced graphene oxide and $Ni(OH)_2$ composite films for controlled syntheses supercapacitor electrodes. J Phys Chem B 2012;117(6):1616–27.

[171] Nam K-W, Kim K-H, Lee E-S, Yoon W-S, Yang X-Q, Kim K-B. Pseudocapacitive properties of electrochemically prepared nickel oxides on 3-dimensional carbon nanotube film substrates. J Power Sources 2008;182:642–52.

[172] Zhao C, Wang X, Wang S, Wang Y, Zhao Y, Zheng W. Synthesis of $Co(OH)_2$/graphene/Ni foam nano-electrodes with excellent pseudocapacitive behavior and high cycling stability for supercapacitors. Int J Hydrogen Energy 2012;37:11846–52.

[173] Lang J, Yan X, Xue Q. Facile preparation and electrochemical characterization of cobalt oxide/multi-walled carbon nanotube composites for supercapacitors. J Power Sources 2011;196:7841–6.

[174] Dong X, Cao Y, Wang J, Chan-Park MB, Wang L, Huang W, et al. Hybrid structure of zinc oxide nanorods and three dimensional graphene foam for supercapacitor and electrochemical sensor applications. RSC Adv 2012;2:4364–9.

[175] Zhang Y, Sun X, Pan L, Li H, Sun Z, Sun C, et al. Carbon nanotube–ZnO nanocomposite electrodes for supercapacitors. Solid State Ionics 2009;180:1525–8.

[176] Fang W-C. Synthesis and electrochemical characterization of vanadium oxide/carbon nanotube composites for supercapacitors. J Phys Chem C 2008;112:11552–5.

[177] Bonso JS, Rahy A, Perera SD, Nour N, Seitz O, Chabal YJ, et al. Exfoliated graphite nanoplatelets–V_2O_5 nanotube composite electrodes for supercapacitors. J Power Sources 2012;203:227–32.

[178] Zhang H, Cao G, Wang Z, Yang Y, Shi Z, Gu Z. Growth of manganese oxide nanoflowers on vertically-aligned carbon nanotube arrays for high-rate electrochemical capacitive energy storage. Nano Lett 2008;8:2664–8.

[179] Lei Z, Shi F, Lu L. Incorporation of MnO_2-coated carbon nanotubes between graphene sheets as supercapacitor electrode. ACS Appl Mater Interfaces 2012;4:1058–64.

[180] Lee CY, Tsai HM, Chuang HJ, Li SY, Lin P, Tseng TY. Characteristics and electrochemical performance of supercapacitors with manganese oxide-carbon nanotube nanocomposite electrodes. J Electrochem Soc 2005;152:A716–20.

[181] Arabale G, Wagh D, Kulkarni M, Mulla IS, Vernekar SP, Vijayamohanan K, et al. Enhanced supercapacitance of multiwalled carbon nanotubes functionalized with ruthenium oxide. Chem Phys Lett 2003;376:207–13.

[182] Kim Y-T, Tadai K, Mitani T. Highly dispersed ruthenium oxide nanoparticles on carboxylated carbon nanotubes for supercapacitor electrode materials. J Mater Chem 2005;15:4914–21.

[183] Sun Z, Liu Z, Han B, Miao S, Du J, Miao Z. Microstructural and electrochemical characterization of RuO_2/Cnt composites synthesized in supercritical diethyl amine. Carbon 2006;44:888–93.

[184] Kim I-H, Kim J-H, Lee Y-H, Kim K-B. Synthesis and characterization of electrochemically prepared ruthenium oxide on carbon nanotube film substrate for supercapacitor applications. J Electrochem Soc 2005;152:A2170–8.

[185] Li Z, Wang J, Liu X, Liu S, Ou J, Yang S. Electrostatic layer-by-layer self-assembly multilayer films based on graphene and manganese dioxide sheets as novel electrode materials for supercapacitors. J Mater Chem 2011;21:3397–403.

[186] Yu G, Hu L, Liu N, Wang H, Vosgueritchian M, Yang Y, et al. Enhancing the supercapacitor performance of graphene/MnO_2 nanostructured electrodes by conductive wrapping. Nano Lett 2011;11:4438–42.

[187] Choi BG, Yang M, Hong WH, Choi JW, Huh YS. 3D macroporous graphene frameworks for supercapacitors with high energy and power densities. ACS Nano 2012;6:4020–8.

[188] Wang Q, Jiao L, Du H, Si Y, Wang Y, Yuan H. Co_3S_4 hollow nanospheres grown on graphene as advanced electrode materials for supercapacitors. J Mater Chem 2012;22:21387–91.

[189] Chen H, Zhou S, Chen M, Wu L. Reduced graphene oxide-MnO_2 hollow sphere hybrid nanostructures as high-performance electrochemical capacitors. J Mater Chem 2012;22: 25207–16.

[190] Chen Z, Ren W, Gao L, Liu B, Pei S, Cheng H-M. Three-dimensional flexible and conductive interconnected graphene networks grown by chemical vapour deposition. Nat Mater 2011;10:424–8.

[191] Brodd RJ, Bullock KR, Leising RA, Middaugh RL, Miller JR, Takeuchi E. Batteries, 1977 to 2002. J Electrochem Soc 2004;151:K1–K11.

[192] Marom R, Amalraj SF, Leifer N, Jacob D, Aurbach D. A review of advanced and practical lithium battery materials. J Mater Chem 2011;21:9938–54.

[193] Wakihara M. Recent developments in lithium ion batteries. Mater Sci Eng 2001;33:109–34.

[194] Frackowiak E, Béguin F. Electrochemical storage of energy in carbon nanotubes and nanostructured carbons. Carbon 2002;40:1775–87.

[195] Le DB, Passerini S, Guo J, Ressler J, Owens BB, Smyrl WH. High surface area V_2O_5 aerogel intercalation electrodes. J Electrochem Soc 1996;143:2099–104.

[196] Gao B, Kleinhammes A, Tang XP, Bower C, Fleming L, Wu Y, et al. Electrochemical intercalation of single-walled carbon nanotubes with lithium. Chem Phys Lett 1999;307:153–7.

[197] Yoo E, Kim J, Hosono E, Zhou H-s, Kudo T, Honma I. Large reversible Li storage of graphene nanosheet families for use in rechargeable lithium ion batteries. Nano Lett 2008;8: 2277–82.

[198] Zhou XS, Yin YX, Wan LJ, Guo YG. Self-assembled nanocomposite of silicon nanoparticles encapsulated in graphene through electrostatic attraction for lithium-ion batteries. Adv Energy Mater 2012;2:1086–90.

[199] Xin X, Zhou X, Wang F, Yao X, Xu X, Zhu Y, et al. A 3D porous architecture of Si/graphene nanocomposite as high-performance anode materials for Li-ion batteries. J Mater Chem 2012;22:7724–30.

[200] Zhang Y, Wang X, Zeng L, Song S, Liu D. Green and controlled synthesis of Cu_2O-graphene hierarchical nanohybrids as high-performance anode materials for lithium-ion batteries via an ultrasound assisted approach. Dalton Trans 2012;41:4316–9.

[201] Su Y, Li S, Wu D, Zhang F, Liang H, Gao P, et al. Two-dimensional carbon-coated graphene/metal oxide hybrids for enhanced lithium storage. ACS Nano 2012;6:8349–56.

[202] Chen W, Li S, Chen C, Yan L. Self-assembly and embedding of nanoparticles by in situ reduced graphene for preparation of a 3D graphene/nanoparticle aerogel. Adv Mater 2011;23:5679–83.

[203] Wu ZS, Zhou G, Yin LC, Ren W, Li F, Cheng HM. Graphene/metal oxide composite electrode materials for energy storage. Nano Energy 2012;1:107–31.

[204] Song Z, Xu T, Gordin ML, Jiang Y-B, Bae I-T, Xiao Q, et al. Polymer–graphene nanocomposites as ultrafast-charge and -discharge cathodes for rechargeable lithium batteries. Nano Lett 2012;12:2205–11.

[205] Li X, Meng X, Liu J, Geng D, Zhang Y, Banis MN, et al. Tin oxide with controlled morphology and crystallinity by atomic layer deposition onto graphene nanosheets for enhanced lithium storage. Adv Funct Mater 2012;22:1647–54.

[206] Wu P, Du N, Zhang H, Yu J, Yang D. Cnts@Sno2@C coaxial nanocables with highly reversible lithium storage. J Phys Chem C 2010;114:22535–8.

[207] Zhang M, Lei D, Yin X, Chen L, Li Q, Wang Y, et al. Magnetite/graphene composites: microwave irradiation synthesis and enhanced cycling and rate performances for lithium ion batteries. J Mater Chem 2010;20:5538–43.

[208] Zhao X, Johnston C, Grant PS. A novel hybrid supercapacitor with a carbon nanotube cathode and an iron oxide/carbon nanotube composite anode. J Mater Chem 2009;19:8755–60.

[209] Wu Z-S, Ren W, Wen L, Gao L, Zhao J, Chen Z, et al. Graphene anchored with Co_3O_4 nanoparticles as anode of lithium ion batteries with enhanced reversible capacity and cyclic performance. ACS Nano 2010;4:3187–94.

[210] Wang G, Shen X, Yao J, Wexler D, Ahn J-h. Hydrothermal synthesis of carbon nanotube/cobalt oxide core-shell one-dimensional nanocomposite and application as an anode material for lithium-ion batteries. Electrochem Commun 2009;11:546–9.

[211] Yu A, Park HW, Davies A, Higgins DC, Chen Z, Xiao X. Free-standing layer-by-layer hybrid thin film of graphene-MnO_2 nanotube as anode for lithium ion batteries. J Phys Chem Lett 2011;2:1855–60.

[212] Reddy ALM, Shaijumon MM, Gowda SR, Ajayan PM. Coaxial MnO_2/carbon nanotube array electrodes for high-performance lithium batteries. Nano Lett 2009;9:1002–6.

[213] Liu H, Yang W. Ultralong single crystalline V_2O_5 nanowire/graphene composite fabricated by a facile green approach and its lithium storage behavior. Energy Environ Sci 2011;4:4000–8.

[214] Sakamoto JS, Dunn B. Vanadium oxide-carbon nanotube composite electrodes for use in secondary lithium batteries. J Electrochem Soc 2002;149:A26–30.

[215] Chen X, Zhu H, Chen Y-C, Shang Y, Cao A, Hu L, et al. MWCNT/V_2O_5 core/shell sponge for high areal capacity and power density Li-ion cathodes. ACS Nano 2012;6:7948–55.

[216] Tang Y, Huang F, Bi H, Liu Z, Wan D. Highly conductive three-dimensional graphene for enhancing the rate performance of $LiFePO_4$ cathode. J Power Sources 2012;203:130–4.

[217] Ji H, Zhang L, Pettes MT, Li H, Chen S, Shi L, et al. Ultrathin graphite foam: a three-dimensional conductive network for battery electrodes. Nano Lett 2012;12:2446–51.

[218] Green MA, Emery K, Hishikawa Y, Warta W, Dunlop ED. Solar cell efficiency tables (version 40). Progr Photovoltaics Res Appl 2012;20:606–14.

[219] Dai L, Chang DW, Baek J-B, Lu W. Carbon nanomaterials for advanced energy conversion and storage. Small 2012;8:1130–66.

[220] Ramuz MP, Vosgueritchian M, Wei P, Wang C, Gao Y, Wu Y, et al. Evaluation of solution-processable carbon-based electrodes for all-carbon solar cells. ACS Nano 2012;6:10384–95.

[221] O'Regan B, Gratzel M. A low-cost, high-efficiency solar cell based on dye-sensitized colloidal TiO_2 films. Nature 1991;353:737–40.

[222] Liu C-J, Tai S-Y, Chou S-W, Yu Y-C, Chang K-D, Wang S, et al. Facile synthesis of MoS_2/graphene nanocomposite with high catalytic activity toward triiodide reduction in dye-sensitized solar cells. J Mater Chem 2012;22:21057–64.

[223] Dang X, Yi H, Ham M-H, Qi J, Yun DS, Ladewski R, et al. Virus-templated self-assembled single-walled carbon nanotubes for highly efficient electron collection in photovoltaic devices. Nat Nano 2011;6:377–84.

[224] Zhao Z, Fite JD, Haldar P, Lee JU. Enhanced ultraviolet response using graphene electrodes in organic solar cells. Appl Phys Lett 2012;101:063305.

[225] Gomez De Arco L, Zhang Y, Schlenker CW, Ryu K, Thompson ME, Zhou C. Continuous, highly flexible, and transparent graphene films by chemical vapor deposition for organic photovoltaics. ACS Nano 2010;4:2865–73.

[226] Rowell MW, Topinka MA, McGehee MD, Prall H-J, Dennler G, Sariciftci NS, et al. Organic solar cells with carbon nanotube network electrodes. Appl Phys Lett 2006;88:233506.

[227] Brennan LJ, Byrne MT, Bari M, Gun'ko YK. Carbon nanomaterials for dye-sensitized solar cell applications: a bright future. Adv Energy Mater 2011;1:472–85.

[228] Byrne MT, Hanley CA, Gun'ko YK. Preparation and properties of buckypaper-gold nanoparticle composites. J Mater Chem 2010;20:2949–51.

[229] Ferrer-Anglada N, Kaempgen M, Skákalová V, Dettlaf-Weglikowska U, Roth S. Synthesis and characterization of carbon nanotube-conducting polymer thin films. Diamond Relat Mater 2004;13:256–60.

[230] Tung VC, Chen L-M, Allen MJ, Wassei JK, Nelson K, Kaner RB, et al. Low-temperature solution processing of graphene − carbon nanotube hybrid materials for high-performance transparent conductors. Nano Lett 2009;9:1949–55.

[231] Roy-Mayhew JD, Bozym DJ, Punckt C, Aksay IA. Functionalized graphene as a catalytic counter electrode in dye-sensitized solar cells. ACS Nano 2010;4:6203–11.

[232] Hong W, Xu Y, Lu G, Li C, Shi G. Transparent graphene/PEDOT–PSS composite films as counter electrodes of dye-sensitized solar cells. Electrochem Commun 2008;10:1555–8.

[233] Song J, Li GR, Xiong FY, Gao XP. Synergistic effect of molybdenum nitride and carbon nanotubes on electrocatalysis for dye-sensitized solar cells. J Mater Chem 2012;22:20580–5.

[234] He Y, Li Y. Fullerene derivative acceptors for high performance polymer solar cells. Phys Chem Chem Phys 2011;13:1970–83.

[235] Liu Z, Liu Q, Huang Y, Ma Y, Yin S, Zhang X, et al. Organic photovoltaic devices based on a novel acceptor material: graphene. Adv Mater 2008;20:3924–30.

[236] Stylianakis MM, Stratakis E, Koudoumas E, Kymakis E, Anastasiadis SH. Organic bulk heterojunction photovoltaic devices based on polythiophene–graphene composites. ACS Appl Mater Interfaces 2012;4:4864–70.

[237] Stylianakis MM, Spyropoulos GD, Stratakis E, Kymakis E. Solution-processable graphene linked to 3,5-dinitrobenzoyl as an electron acceptor in organic bulk heterojunction photovoltaic devices. Carbon 2012;50:5554–61.

[238] Ren S, Bernardi M, Lunt RR, Bulovic V, Grossman JC, Gradečak S. Toward efficient carbon nanotube/P_3Ht solar cells: active layer morphology, electrical, and optical properties. Nano Lett 2011;11:5316–21.

[239] Kuila BK, Park K, Dai L. Soluble P_3Ht-grafted carbon nanotubes: synthesis and photovoltaic application. Macromolecules 2010;43:6699–705.

[240] Yu D, Yang Y, Durstock M, Baek J-B, Dai L. Soluble P_3Ht-grafted graphene for efficient bilayer − heterojunction photovoltaic devices. ACS Nano 2010;4:5633–40.

[241] Yu D, Park K, Durstock M, Dai L. Fullerene-grafted graphene for efficient bulk heterojunction polymer photovoltaic devices. J Phys Chem Lett 2011;2:1113–8.

[242] Chen J, Li B, Zheng J, Zhao J, Zhu Z. Role of carbon nanotubes in dye-sensitized TiO_2-based solar cells. J Phys Chem C 2012;116:14848–56.

[243] Kongkanand A, Martínez Domínguez R, Kamat PV. Single wall carbon nanotube scaffolds for photoelectrochemical solar cells. Capture and transport of photogenerated electrons. Nano Lett 2007;7:676–80.

[244] Brown P, Takechi K, Kamat PV. Single-walled carbon nanotube scaffolds for dye-sensitized solar cells. J Phys Chem C 2008;112:4776–82.

[245] Fan J, Liu S, Yu J. Enhanced photovoltaic performance of dye-sensitized solar cells based on TiO_2 nanosheets/graphene composite films. J Mater Chem 2012;22:17027–36.

[246] Yang N, Zhai J, Wang D, Chen Y, Jiang L. Two-dimensional graphene bridges enhanced photoinduced charge transport in dye-sensitized solar cells. ACS Nano 2010;4:887–94.

[247] He Q, Wu S, Yin Z, Zhang H. Graphene-based electronic sensors. Chem Sci 2012;3:1764–72.

[248] Ting Z, Syed M, Nosang VM, Marc AD. Recent progress in carbon nanotube-based gas sensors. Nanotechnology 2008;19:332001.

[249] Liu Y, Dong X, Chen P. Biological and chemical sensors based on graphene materials. Chem Soc Rev 2012;41:2283–307.

[250] Someya T, Small J, Kim P, Nuckolls C, Yardley JT. Alcohol vapor sensors based on single-walled carbon nanotube field effect transistors. Nano Lett 2003;3:877–81.

[251] Yavari F, Chen ZP, Thomas AV, Ren WC, Cheng HM, Koratkar N. High sensitivity gas detection using a macroscopic three-dimensional graphene foam network. Scienti Rep 2011;1:166.

[252] Sharma A, Tomar M, Gupta V. Room temperature trace level detection of NO_2 gas using SnO_2 modified carbon nanotubes based sensor. J Mater Chem 2012;22:23608–16.

[253] Chung MG, Kim D-H, Seo DK, Kim T, Im HU, Lee HM, et al. Flexible hydrogen sensors using graphene with palladium nanoparticle decoration. Sensors Actuat B Chem 2012;169:387–92.

[254] Fan L, Zhang Q, Wang K, Li F, Niu L. Ferrocene functionalized graphene: preparation, characterization and efficient electron transfer toward sensors of H_2O_2. J Mater Chem 2012;22:6165–70.

[255] Yu J, Zvarec O, Huang DM, Bissett MA, Scanlon DB, Shapter JG, et al. Electron transfer through [small alpha]-peptides attached to vertically aligned carbon nanotube arrays: a mechanistic transition. Chem Commun 2012;48:1132–4.

[256] Li M, Xu S, Tang M, Liu L, Gao F, Wang Y. Direct electrochemistry of horseradish peroxidase on graphene-modified electrode for electrocatalytic reduction towards H_2O_2. Electrochim Acta 2011;56:1144–9.

[257] Zhao X-H, Kong R-M, Zhang X-B, Meng H-M, Liu W-N, Tan W, et al. Graphene–dnazyme based biosensor for amplified fluorescence "turn-on" detection of Pb^{2+} with a high selectivity. Analy Chem 2011;83:5062–6.

[258] Peng N, Zhang Q, Chow CL, Tan OK, Marzari N. Sensing mechanisms for carbon nanotube based NH_3 gas detection. Nano Lett 2009;9:1626–30.

[259] Salehi-Khojin A, Khalili-Araghi F, Kuroda MA, Lin KY, Leburton J-P, Masel RI. On the sensing mechanism in carbon nanotube chemiresistors. ACS Nano 2010;5:153–8.

[260] Chang H, Lee JD, Lee SM, Lee YH. Adsorption of NH_3 and NO_2 molecules on carbon nanotubes. Appl Phys Lett 2001;79:3863–5.

[261] Kauffman DR, Sorescu DC, Schofield DP, Allen BL, Jordan KD, Star A. Understanding the sensor response of metal-decorated carbon nanotubes. Nano Lett 2010;10:958–63.

[262] Kong J, Chapline MG, Dai H. Functionalized carbon nanotubes for molecular hydrogen sensors. Adv Mater 2001;13:1384–6.

[263] Bittencourt C, Felten A, Espinosa EH, Ionescu R, Llobet E, Correig X, et al. WO_3 films modified with functionalised multi-wall carbon nanotubes: morphological, compositional and gas response studies. Sensors Actuat B Chem 2006;115:33–41.

[264] Li W, Geng X, Guo Y, Rong J, Gong Y, Wu L, et al. Reduced graphene oxide electrically contacted graphene sensor for highly sensitive nitric oxide detection. ACS Nano 2011;5:6955–61.

[265] Cuong TV, Pham VH, Chung JS, Shin EW, Yoo DH, Hahn SH, et al. Solution-processed ZnO-chemically converted graphene gas sensor. Mater Lett 2010;64:2479–82.

[266] Lu Y, Goldsmith BR, Kybert NJ, Johnson ATC. DNA-decorated graphene chemical sensors. Appl Phys Lett 2010;97. 083107-083103.

[267] Zhang T, Cheng Z, Wang Y, Li Z, Wang C, Li Y, et al. Self-assembled 1-octadecanethiol monolayers on graphene for mercury detection. Nano Lett 2010;10:4738–41.

[268] Sudibya HG, He Q, Zhang H, Chen P. Electrical detection of metal ions using field-effect transistors based on micropatterned reduced graphene oxide films. ACS Nano 2011;5:1990–4.

[269] Besteman K, Lee J-O, Wiertz FGM, Heering HA, Dekker C. Enzyme-coated carbon nanotubes as single-molecule biosensors. Nano Lett 2003;3:727–30.

[270] Huang Y, Dong X, Shi Y, Li CM, Li L-J, Chen P. Nanoelectronic biosensors based on CVD grown graphene. Nanoscale 2010;2:1485–8.

[271] Boussaad S, Diner BA, Fan J. Influence of redox molecules on the electronic conductance of single-walled carbon nanotube field-effect transistors: application to chemical and biological sensing. J Am Chem Soc 2008;130:3780–7.

[272] Park SJ, Kwon OS, Lee SH, Song HS, Park TH, Jang J. Ultrasensitive flexible graphene based field-effect transistor (Fet)-type bioelectronic nose. Nano Lett 2012;12:5082–90.

[273] Star A, Gabriel J-CP, Bradley K, Grüner G. Electronic detection of specific protein binding using nanotube FET devices. Nano Lett 2003;3:459–63.

[274] He Q, Wu S, Gao S, Cao X, Yin Z, Li H, et al. Transparent, flexible, all-reduced graphene oxide thin film transistors. ACS Nano 2011;5:5038–44.

[275] Chen RJ, Bangsaruntip S, Drouvalakis KA, Wong Shi Kam N, Shim M, Li Y, et al. Noncovalent functionalization of carbon nanotubes for highly specific electronic biosensors. Proc Natl Acad Sci 2003;100:4984–9.

[276] Mao S, Lu G, Yu K, Bo Z, Chen J. Specific protein detection using thermally reduced graphene oxide sheet decorated with gold nanoparticle-antibody conjugates. Adv Mater 2010;22:3521–6.

[277] Ohno Y, Maehashi K, Matsumoto K. Label-free biosensors based on aptamer-modified graphene field-effect transistors. J Am Chem Soc 2010;132:18012–3.

[278] So H-M, Won K, Kim YH, Kim B-K, Ryu BH, Na PS, et al. Single-walled carbon nanotube biosensors using aptamers as molecular recognition elements. J Am Chem Soc 2005;127:11906–7.

[279] Stine R, Robinson JT, Sheehan PE, Tamanaha CR. Real-time DNA detection using reduced graphene oxide field effect transistors. Adv Mater 2010;22:5297–300.

[280] Yin Z, He Q, Huang X, Zhang J, Wu S, Chen P, et al. Real-time DNA detection using Pt nanoparticle-decorated reduced graphene oxide field-effect transistors. Nanoscale 2012;4:293–7.

[281] Dong X, Shi Y, Huang W, Chen P, Li L-J. Electrical detection of DNA hybridization with single-base specificity using transistors based on CVD-grown graphene sheets. Adv Mater 2010;22:1649–53.

[282] Flavel B, Nambiar M, Shapter J. Electrochemical detection of copper using a Gly-Gly-His modified carbon nanotube biosensor. Silicon 2011;3:163–71.

[283] Xian Y, Zhou Y, Wang H, Zhou L, Liu F, Jin L. Nanostructured electrode based on multi-wall carbon nanotubes/Pt microparticles nanocomposite for electrochemical determination of thiols in rat striatum by high performance liquid chromatography separation. J Chromatogr B 2005;817:239–46.

[284] Guo S, Wen D, Zhai Y, Dong S, Wang E. Platinum nanoparticle ensemble-on-graphene hybrid nanosheet: one-pot, rapid synthesis, and used as new electrode material for electrochemical sensing. ACS Nano 2010;4:3959–68.

[285] Junhua W, Jingjing Q, Li L, Liqiang R, Xianwen Z, Jharna C, et al. A reduced graphene oxide based electrochemical biosensor for tyrosine detection. Nanotechnology 2012;23:335707.

[286] Gan T, Hu C, Chen Z, Hu S. Fabrication and application of a novel plant hormone sensor for the determination of methyl jasmonate based on self-assembling of phosphotungstic acid–graphene oxide nanohybrid on graphite electrode. Sensors Actuat B 2010;151:8–14.

[287] Wang L, Zhang X, Xiong H, Wang S. A novel nitromethane biosensor based on biocompatible conductive redox graphene-chitosan/hemoglobin/graphene/room temperature ionic liquid matrix. Biosensors Bioelectron 2010;26:991–5.

[288] Choi BG, Park H, Park TJ, Yang MH, Kim JS, Jang S-Y, et al. Solution chemistry of self-assembled graphene nanohybrids for high-performance flexible biosensors. ACS Nano 2010;4:2910–8.

[289] Male KB, Hrapovic S, Liu Y, Wang D, Luong JHT. Electrochemical detection of carbohydrates using copper nanoparticles and carbon nanotubes. Anal Chim Acta 2004;516:35–41.

[290] Rivas GA, Rubianes MD, Rodríguez MC, Ferreyra NF, Luque GL, Pedano ML, et al. Carbon nanotubes for electrochemical biosensing. Talanta 2007;74:291–307.

[291] Lin Y, Lu F, Tu Y, Ren Z. Glucose biosensors based on carbon nanotube nanoelectrode ensembles. Nano Lett 2003;4:191–5.

[292] Yamamoto K, Shi G, Zhou T, Xu F, Xu J, Kato T, et al. Study of carbon nanotubes-Hrp modified electrode and its application for novel on-line biosensors. Analyst 2003;128:249–54.

[293] Dey RS, Raj CR. Development of an amperometric cholesterol biosensor based on graphene − Pt nanoparticle hybrid material. J Phys Chem C 2010;114:21427–33.

[294] Antiochia R, Lavagnini I, Magno F. Amperometric mediated carbon nanotube paste biosensor for fructose determination. Anal Lett 2004;37:1657–69.

[295] Tominaga M, Nomura S, Taniguchi I. D-Fructose detection based on the direct heterogeneous electron transfer reaction of fructose dehydrogenase adsorbed onto multi-walled carbon nanotubes synthesized on platinum electrode. Biosensors Bioelectron 2009;24:1184–8.

[296] Silveira CM, Baur J, Holzinger M, Moura JJG, Cosnier S, Almeida MG. Enhanced direct electron transfer of a multihemic nitrite reductase on single-walled carbon nanotube modified electrodes. Electroanalysis 2010;22:2973–8.

[297] Sotiropoulou S, Chaniotakis N. Carbon nanotube array-based biosensor. Anal Bioanal Chem 2003;375:103–5.

[298] Shan C, Yang H, Song J, Han D, Ivaska A, Niu L. Direct electrochemistry of glucose oxidase and biosensing for glucose based on graphene. Anal Chem 2009;81:2378–82.

[299] Claussen JC, Kumar A, Jaroch DB, Khawaja MH, Hibbard AB, Porterfield DM, et al. Nanostructuring platinum nanoparticles on multilayered graphene petal nanosheets for electrochemical biosensing. Adv Funct Mater 2012;22:3399–405.

[300] Adeloju SB, Shaw SJ, Wallace GG. Polypyrrole-based amperometric biosensor for sulfite determination. Electroanalysis 1994;6:865–70.

[301] Yao Z, Di J, Yong Z, Zhao Z, Li Q. Aligned coaxial tungsten oxide-carbon nanotube sheet: a flexible and gradient electrochromic film. Chem Commun 2012;48:8252–4.

[302] Liu J, Wang Z, Zhao Y, Cheng H, Hu C, Jiang L, et al. Three-dimensional graphene-polypyrrole hybrid electrochemical actuator. Nanoscale 2012;4:7563–8.

[303] Hewitt CA, Kaiser AB, Roth S, Craps M, Czerw R, Carroll DL. Multilayered carbon nanotube/polymer composite based thermoelectric fabrics. Nano Lett 2012;12:1307–10.

[304] Hong S, Kim ES, Kim W, Jeon S-J, Lim SC, Kim KH, et al. A hybridized graphene carrier highway for enhanced thermoelectric power generation. Phys Chem Chem Phys 2012;14:13527–31.

[305] Dong X, Chen J, Ma Y, Wang J, Chan-Park MB, Liu X, et al. Superhydrophobic and superoleophilic hybrid foam of graphene and carbon nanotube for selective removal of oils or organic solvents from the surface of water. Chem Commun 2012;48:10660–2.

[306] Zheng Y, Ni G-X, Toh C-T, Zeng M-G, Chen S-T, Yao K, et al. Gate-controlled nonvolatile graphene-ferroelectric memory. Appl Phys Lett 2009;94:163505.

[307] Myung S, Park J, Lee H, Kim KS, Hong S. Ambipolar memory devices based on reduced graphene oxide and nanoparticles. Adv Mater 2010;22:2045–9.

[308] Lee HS, Min S-W, Park MK, Lee YT, Jeon PJ, Kim JH, et al. MoS_2 nanosheets for top-gate nonvolatile memory transistor channel. Small 2012;8:3111–5.

[309] Ashok K, Sai GS, Margarita C, Oscar R, Youngjin C, Matthew TC, et al. Ferroelectric–carbon nanotube memory devices. Nanotechnology 2012;23:165702.

[310] Shearer CJ, Fahy A, Barr MG, Moore KE, Dastoor PC, Shapter JG. Field emission from single-, double-, and multi-walled carbon nanotubes chemically attached to silicon. J Appl Phys 2012;111.

[311] Shearer C, Fahy A, Barr M, Dastoor P, Shapter J. Improved field emission stability from single-walled carbon nanotubes chemically attached to silicon. Nanoscale Res Lett 2012;7:432.

[312] Yan X, Tay B-K, Miele P. Field emission from ordered carbon nanotube-ZnO heterojunction arrays. Carbon 2008;46:753–8.

[313] Yu K, Zhang YS, Xu F, Li Q, Zhu ZQ, Wan Q. Significant improvement of field emission by depositing zinc oxide nanostructures on screen-printed carbon nanotube films. Appl Phys Lett 2006;88:153123–1153123.

[314] Moon JS, Alegaonkar PS, Han JH, Lee TY, Yoo JB, Kim JM. Enhanced field emission properties of thin-multiwalled carbon nanotubes: role of SiO_X coating. J Appl Phys 2006;100:104303–7.

[315] Hwang JO, Lee DH, Kim JY, Han TH, Kim BH, Park M, et al. Vertical ZnO nanowires/graphene hybrids for transparent and flexible field emission. J Mater Chem 2011;21:3432–7.

[316] Wu C, Li F, Zhang Y, Guo T. Improving the field emission of graphene by depositing zinc oxide nanorods on its surface. Carbon 2012;50:3622–6.

[317] Nguyen DD, Tai N-H, Chen S-Y, Chueh Y-L. Controlled growth of carbon nanotube-graphene hybrid materials for flexible and transparent conductors and electron field emitters. Nanoscale 2012;4:632–8.

Index

D

E

F

G

H

O

P

Q

R

S

FSC
www.fsc.org
MIX
Papier | Fördert
gute Waldnutzung
FSC® C083411

Kolibri 360 GmbH
Ferdinand-Jühlke-Straße 7
99095 Erfurt, Deutschland
produktsicherheit@kolibri360.de